STUDENT'S SOLUTIONS MANUAL

PART TWO

ARDIS • BORZELLINO • BUCHANAN • MOGILL • NELSON

to accompany

THOMAS' CALCULUS

ELEVENTH EDITION

AND

THOMAS' CALCULUS: EARLY TRANSCENDENTALS

ELEVENTH EDITION

BASED ON THE ORIGINAL WORK BY

George B. Thomas, Jr.

Massachusetts Institute of Technology

AS REVISED BY

Maurice D. Weir

Naval Postgraduate School

Joel Hass

University of California, Davis

Frank R. Giordano

Naval Postgraduate School

PEARSON

Addison
Wesley

Boston San Francisco New York
London Toronto Sydney Tokyo Singapore Madrid
Mexico City Munich Paris Cape Town Hong Kong Montreal

Copyright © 2005 Pearson Education, Inc.
Publishing as Pearson Addison-Wesley, 75 Arlington Street, Boston, MA 02116.

ISBN   0-321-22647-X

10 11 12 13 14 15 16 BB 11 10 09 08

PREFACE TO THE STUDENT

The Student's Solutions Manual contains the solutions to all of the odd-numbered exercise in the 11th Edition of THOMAS' CALCULUS by Maurice Weir, Joel Hass and Frank Giordano, excluding the Computer Algebra System (CAS) exercises. We have worked each solution to ensure that it

- conforms exactly to the methods, procedures and steps presented in the text

- is mathematically correct

- includes all of the steps necessary so you can follow the logical argument and algebra

- includes a graph or figure whenever called for by the exercise, or if needed to help with the explanation

- is formatted in an appropriate style to aid in its understanding

How to use a solution's manual

- solve the assigned problem yourself

- if you get stuck along the way, refer to the solution in the manual as an aid but continue to solve the problem on your own

- if you cannot continue, reread the textbook section, or work through that section in the Student Study Guide, or consult your instructor

- if your answer is correct by your solution procedure seems to differ from the one in the manual, and you are unsure your method is correct, consult your instructor

- if your answer is incorrect and you cannot find your error, consult your instructor

Acknowledgments

Solutions Writers
      William Ardis, Collin County Community College-Preston Ridge Campus
      Joseph Borzellino, California Polytechnic State University
      Linda Buchanan, Howard College
      Tim Mogill
      Patricia Nelson, University of Wisconsin-La Crosse

Accuracy Checkers
      Karl Kattchee, University of Wisconsin-La Crosse
      Marie Vanisko, California State University, Stanislaus
      Tom Weigleitner, VISTA Information Technologies

Thanks to Rachel Reeve, Christine O'Brien, Sheila Spinney, Elka Block, and Joe Vetere for all their guidance and help at every step.

TABLE OF CONTENTS

15 Multiple Integrals 463

16 Integration in Vector Fields 489

CHAPTER 11 INFINITE SEQUENCES AND SERIES

11.1 SEQUENCES

1. $a_1 = \frac{1-1}{1^2} = 0$, $a_2 = \frac{1-2}{2^2} = -\frac{1}{4}$, $a_3 = \frac{1-3}{3^2} = -\frac{2}{9}$, $a_4 = \frac{1-4}{4^2} = -\frac{3}{16}$

3. $a_1 = \frac{(-1)^2}{2-1} = 1$, $a_2 = \frac{(-1)^3}{4-1} = -\frac{1}{3}$, $a_3 = \frac{(-1)^4}{6-1} = \frac{1}{5}$, $a_4 = \frac{(-1)^5}{8-1} = -\frac{1}{7}$

5. $a_1 = \frac{2}{2^2} = \frac{1}{2}$, $a_2 = \frac{2^2}{2^3} = \frac{1}{2}$, $a_3 = \frac{2^3}{2^4} = \frac{1}{2}$, $a_4 = \frac{2^4}{2^5} = \frac{1}{2}$

7. $a_1 = 1$, $a_2 = 1 + \frac{1}{2} = \frac{3}{2}$, $a_3 = \frac{3}{2} + \frac{1}{2^2} = \frac{7}{4}$, $a_4 = \frac{7}{4} + \frac{1}{2^3} = \frac{15}{8}$, $a_5 = \frac{15}{8} + \frac{1}{2^4} = \frac{31}{16}$, $a_6 = \frac{63}{32}$,

   $a_7 = \frac{127}{64}$, $a_8 = \frac{255}{128}$, $a_9 = \frac{511}{256}$, $a_{10} = \frac{1023}{512}$

9. $a_1 = 2$, $a_2 = \frac{(-1)^2(2)}{2} = 1$, $a_3 = \frac{(-1)^3(1)}{2} = -\frac{1}{2}$, $a_4 = \frac{(-1)^4\left(-\frac{1}{2}\right)}{2} = -\frac{1}{4}$, $a_5 = \frac{(-1)^5\left(-\frac{1}{4}\right)}{2} = \frac{1}{8}$,

   $a_6 = \frac{1}{16}$, $a_7 = -\frac{1}{32}$, $a_8 = -\frac{1}{64}$, $a_9 = \frac{1}{128}$, $a_{10} = \frac{1}{256}$

11. $a_1 = 1$, $a_2 = 1$, $a_3 = 1 + 1 = 2$, $a_4 = 2 + 1 = 3$, $a_5 = 3 + 2 = 5$, $a_6 = 8$, $a_7 = 13$, $a_8 = 21$, $a_9 = 34$, $a_{10} = 55$

13. $a_n = (-1)^{n+1}$, $n = 1, 2, \ldots$

15. $a_n = (-1)^{n+1}n^2$, $n = 1, 2, \ldots$

17. $a_n = n^2 - 1$, $n = 1, 2, \ldots$

19. $a_n = 4n - 3$, $n = 1, 2, \ldots$

21. $a_n = \frac{1 + (-1)^{n+1}}{2}$, $n = 1, 2, \ldots$

23. $\lim\limits_{n \to \infty} 2 + (0.1)^n = 2 \Rightarrow$ converges   (Theorem 5, #4)

25. $\lim\limits_{n \to \infty} \frac{1 - 2n}{1 + 2n} = \lim\limits_{n \to \infty} \frac{\left(\frac{1}{n}\right) - 2}{\left(\frac{1}{n}\right) + 2} = \lim\limits_{n \to \infty} \frac{-2}{2} = -1 \Rightarrow$ converges

27. $\lim\limits_{n \to \infty} \frac{1 - 5n^4}{n^4 + 8n^3} = \lim\limits_{n \to \infty} \frac{\left(\frac{1}{n^4}\right) - 5}{1 + \left(\frac{8}{n}\right)} = -5 \Rightarrow$ converges

29. $\lim\limits_{n \to \infty} \frac{n^2 - 2n + 1}{n - 1} = \lim\limits_{n \to \infty} \frac{(n-1)(n-1)}{n-1} = \lim\limits_{n \to \infty} (n - 1) = \infty \Rightarrow$ diverges

31. $\lim\limits_{n \to \infty} (1 + (-1)^n)$ does not exist $\Rightarrow$ diverges

33. $\lim\limits_{n \to \infty} \left(\frac{n+1}{2n}\right)\left(1 - \frac{1}{n}\right) = \lim\limits_{n \to \infty} \left(\frac{1}{2} + \frac{1}{2n}\right)\left(1 - \frac{1}{n}\right) = \frac{1}{2} \Rightarrow$ converges

35. $\lim\limits_{n \to \infty} \frac{(-1)^{n+1}}{2n - 1} = 0 \Rightarrow$ converges

37. $\lim\limits_{n \to \infty} \sqrt{\frac{2n}{n+1}} = \sqrt{\lim\limits_{n \to \infty} \frac{2n}{n+1}} = \sqrt{\lim\limits_{n \to \infty} \left(\frac{2}{1+\frac{1}{n}}\right)} = \sqrt{2} \Rightarrow$ converges

39. $\lim\limits_{n \to \infty} \sin\left(\frac{\pi}{2} + \frac{1}{n}\right) = \sin\left(\lim\limits_{n \to \infty} \left(\frac{\pi}{2} + \frac{1}{n}\right)\right) = \sin\frac{\pi}{2} = 1 \Rightarrow$ converges

41. $\lim\limits_{n \to \infty} \frac{\sin n}{n} = 0$ because $-\frac{1}{n} \le \frac{\sin n}{n} \le \frac{1}{n} \Rightarrow$ converges by the Sandwich Theorem for sequences

43. $\lim\limits_{n \to \infty} \frac{n}{2^n} = \lim\limits_{n \to \infty} \frac{1}{2^n \ln 2} = 0 \Rightarrow$ converges (using l'Hôpital's rule)

45. $\lim\limits_{n \to \infty} \frac{\ln(n+1)}{\sqrt{n}} = \lim\limits_{n \to \infty} \frac{\left(\frac{1}{n+1}\right)}{\left(\frac{1}{2\sqrt{n}}\right)} = \lim\limits_{n \to \infty} \frac{2\sqrt{n}}{n+1} = \lim\limits_{n \to \infty} \frac{\left(\frac{2}{\sqrt{n}}\right)}{1+\left(\frac{1}{n}\right)} = 0 \Rightarrow$ converges

47. $\lim\limits_{n \to \infty} 8^{1/n} = 1 \Rightarrow$ converges   (Theorem 5, #3)

49. $\lim\limits_{n \to \infty} \left(1 + \frac{7}{n}\right)^n = e^7 \Rightarrow$ converges   (Theorem 5, #5)

51. $\lim\limits_{n \to \infty} \sqrt[n]{10n} = \lim\limits_{n \to \infty} 10^{1/n} \cdot n^{1/n} = 1 \cdot 1 = 1 \Rightarrow$ converges   (Theorem 5, #3 and #2)

53. $\lim\limits_{n \to \infty} \left(\frac{3}{n}\right)^{1/n} = \frac{\lim\limits_{n \to \infty} 3^{1/n}}{\lim\limits_{n \to \infty} n^{1/n}} = \frac{1}{1} = 1 \Rightarrow$ converges   (Theorem 5, #3 and #2)

55. $\lim\limits_{n \to \infty} \frac{\ln n}{n^{1/n}} = \frac{\lim\limits_{n \to \infty} \ln n}{\lim\limits_{n \to \infty} n^{1/n}} = \frac{\infty}{1} = \infty \Rightarrow$ diverges   (Theorem 5, #2)

57. $\lim\limits_{n \to \infty} \sqrt[n]{4^n\, n} = \lim\limits_{n \to \infty} 4 \sqrt[n]{n} = 4 \cdot 1 = 4 \Rightarrow$ converges   (Theorem 5, #2)

59. $\lim\limits_{n \to \infty} \frac{n!}{n^n} = \lim\limits_{n \to \infty} \frac{1 \cdot 2 \cdot 3 \cdots (n-1)(n)}{n \cdot n \cdot n \cdots n \cdot n} \le \lim\limits_{n \to \infty} \left(\frac{1}{n}\right) = 0$ and $\frac{n!}{n^n} \ge 0 \Rightarrow \lim\limits_{n \to \infty} \frac{n!}{n^n} = 0 \Rightarrow$ converges

61. $\lim\limits_{n \to \infty} \frac{n!}{10^{6n}} = \lim\limits_{n \to \infty} \frac{1}{\left(\frac{10^{6n}}{n!}\right)} = \infty \Rightarrow$ diverges   (Theorem 5, #6)

63. $\lim\limits_{n \to \infty} \left(\frac{1}{n}\right)^{1/(\ln n)} = \lim\limits_{n \to \infty} \exp\left(\frac{1}{\ln n} \ln\left(\frac{1}{n}\right)\right) = \lim\limits_{n \to \infty} \exp\left(\frac{\ln 1 - \ln n}{\ln n}\right) = e^{-1} \Rightarrow$ converges

65. $\lim\limits_{n \to \infty} \left(\frac{3n+1}{3n-1}\right)^n = \lim\limits_{n \to \infty} \exp\left(n \ln\left(\frac{3n+1}{3n-1}\right)\right) = \lim\limits_{n \to \infty} \exp\left(\frac{\ln(3n+1) - \ln(3n-1)}{\frac{1}{n}}\right)$

$= \lim\limits_{n \to \infty} \exp\left(\frac{\frac{3}{3n+1} - \frac{3}{3n-1}}{\left(-\frac{1}{n^2}\right)}\right) = \lim\limits_{n \to \infty} \exp\left(\frac{6n^2}{(3n+1)(3n-1)}\right) = \exp\left(\frac{6}{9}\right) = e^{2/3} \Rightarrow$ converges

67. $\lim\limits_{n \to \infty} \left(\frac{x^n}{2n+1}\right)^{1/n} = \lim\limits_{n \to \infty} x \left(\frac{1}{2n+1}\right)^{1/n} = x \lim\limits_{n \to \infty} \exp\left(\frac{1}{n} \ln\left(\frac{1}{2n+1}\right)\right) = x \lim\limits_{n \to \infty} \exp\left(\frac{-\ln(2n+1)}{n}\right)$

$= x \lim\limits_{n \to \infty} \exp\left(\frac{-2}{2n+1}\right) = xe^0 = x, x > 0 \Rightarrow$ converges

69. $\lim\limits_{n \to \infty} \frac{3^n \cdot 6^n}{2^{-n} \cdot n!} = \lim\limits_{n \to \infty} \frac{36^n}{n!} = 0 \Rightarrow$ converges   (Theorem 5, #6)

71. $\lim\limits_{n \to \infty} \tanh n = \lim\limits_{n \to \infty} \frac{e^n - e^{-n}}{e^n + e^{-n}} = \lim\limits_{n \to \infty} \frac{e^{2n} - 1}{e^{2n} + 1} = \lim\limits_{n \to \infty} \frac{2e^{2n}}{2e^{2n}} = \lim\limits_{n \to \infty} 1 = 1 \Rightarrow$ converges

73. $\lim\limits_{n \to \infty} \frac{n^2 \sin\left(\frac{1}{n}\right)}{2n-1} = \lim\limits_{n \to \infty} \frac{\sin\left(\frac{1}{n}\right)}{\left(\frac{2}{n} - \frac{1}{n^2}\right)} = \lim\limits_{n \to \infty} \frac{-\left(\cos\left(\frac{1}{n}\right)\right)\left(\frac{1}{n^2}\right)}{\left(-\frac{2}{n^2} + \frac{2}{n^3}\right)} = \lim\limits_{n \to \infty} \frac{-\cos\left(\frac{1}{n}\right)}{-2 + \left(\frac{2}{n}\right)} = \frac{1}{2} \Rightarrow$ converges

75. $\lim\limits_{n \to \infty} \tan^{-1} n = \frac{\pi}{2} \Rightarrow$ converges

77. $\lim\limits_{n \to \infty} \left(\frac{1}{3}\right)^n + \frac{1}{\sqrt{2^n}} = \lim\limits_{n \to \infty} \left(\left(\frac{1}{3}\right)^n + \left(\frac{1}{\sqrt{2}}\right)^n\right) = 0 \Rightarrow$ converges   (Theorem 5, #4)

79. $\lim\limits_{n \to \infty} \frac{(\ln n)^{200}}{n} = \lim\limits_{n \to \infty} \frac{200(\ln n)^{199}}{n} = \lim\limits_{n \to \infty} \frac{200 \cdot 199 (\ln n)^{198}}{n} = \ldots = \lim\limits_{n \to \infty} \frac{200!}{n} = 0 \Rightarrow$ converges

81. $\lim\limits_{n \to \infty} \left(n - \sqrt{n^2 - n}\right) = \lim\limits_{n \to \infty} \left(n - \sqrt{n^2 - n}\right)\left(\frac{n + \sqrt{n^2 - n}}{n + \sqrt{n^2 - n}}\right) = \lim\limits_{n \to \infty} \frac{n}{n + \sqrt{n^2 - n}} = \lim\limits_{n \to \infty} \frac{1}{1 + \sqrt{1 - \frac{1}{n}}}$
$= \frac{1}{2} \Rightarrow$ converges

83. $\lim\limits_{n \to \infty} \frac{1}{n} \int_1^n \frac{1}{x} \, dx = \lim\limits_{n \to \infty} \frac{\ln n}{n} = \lim\limits_{n \to \infty} \frac{1}{n} = 0 \Rightarrow$ converges   (Theorem 5, #1)

85. $1, 1, 2, 4, 8, 16, 32, \ldots = 1, 2^0, 2^1, 2^2, 2^3, 2^4, 2^5, \ldots \Rightarrow x_1 = 1$ and $x_n = 2^{n-2}$ for $n \geq 2$

87. (a) $f(x) = x^2 - 2$; the sequence converges to $1.414213562 \approx \sqrt{2}$
    (b) $f(x) = \tan(x) - 1$; the sequence converges to $0.7853981635 \approx \frac{\pi}{4}$
    (c) $f(x) = e^x$; the sequence $1, 0, -1, -2, -3, -4, -5, \ldots$ diverges

89. (a) If $a = 2n + 1$, then $b = \left\lfloor \frac{a^2}{2} \right\rfloor = \left\lfloor \frac{4n^2 + 4n + 1}{2} \right\rfloor = \left\lfloor 2n^2 + 2n + \frac{1}{2} \right\rfloor = 2n^2 + 2n, c = \left\lceil \frac{a^2}{2} \right\rceil = \left\lceil 2n^2 + 2n + \frac{1}{2} \right\rceil$
    $= 2n^2 + 2n + 1$ and $a^2 + b^2 = (2n + 1)^2 + (2n^2 + 2n)^2 = 4n^2 + 4n + 1 + 4n^4 + 8n^3 + 4n^2$
    $= 4n^4 + 8n^3 + 8n^2 + 4n + 1 = (2n^2 + 2n + 1)^2 = c^2.$

    (b) $\lim\limits_{a \to \infty} \frac{\left\lfloor \frac{a^2}{2} \right\rfloor}{\left\lceil \frac{a^2}{2} \right\rceil} = \lim\limits_{a \to \infty} \frac{2n^2 + 2n}{2n^2 + 2n + 1} = 1$ or $\lim\limits_{a \to \infty} \frac{\left\lfloor \frac{a^2}{2} \right\rfloor}{\left\lceil \frac{a^2}{2} \right\rceil} = \lim\limits_{a \to \infty} \sin\theta = \lim\limits_{\theta \to \pi/2} \sin\theta = 1$

91. (a) $\lim\limits_{n \to \infty} \frac{\ln n}{n^c} = \lim\limits_{n \to \infty} \frac{\left(\frac{1}{n}\right)}{cn^{c-1}} = \lim\limits_{n \to \infty} \frac{1}{cn^c} = 0$
    (b) For all $\epsilon > 0$, there exists an $N = e^{-(\ln \epsilon)/c}$ such that $n > e^{-(\ln \epsilon)/c} \Rightarrow \ln n > -\frac{\ln \epsilon}{c} \Rightarrow \ln n^c > \ln\left(\frac{1}{\epsilon}\right)$
    $\Rightarrow n^c > \frac{1}{\epsilon} \Rightarrow \frac{1}{n^c} < \epsilon \Rightarrow \left|\frac{1}{n^c} - 0\right| < \epsilon \Rightarrow \lim\limits_{n \to \infty} \frac{1}{n^c} = 0$

93. $\lim\limits_{n \to \infty} n^{1/n} = \lim\limits_{n \to \infty} \exp\left(\frac{1}{n} \ln n\right) = \lim\limits_{n \to \infty} \exp\left(\frac{1}{n}\right) = e^0 = 1$

95. Assume the hypotheses of the theorem and let ϵ be a positive number. For all ϵ there exists a N_1 such that when $n > N_1$ then $|a_n - L| < \epsilon \Rightarrow -\epsilon < a_n - L < \epsilon \Rightarrow L - \epsilon < a_n$, and there exists a N_2 such that when $n > N_2$ then $|c_n - L| < \epsilon \Rightarrow -\epsilon < c_n - L < \epsilon \Rightarrow c_n < L + \epsilon$. If $n > \max\{N_1, N_2\}$, then $L - \epsilon < a_n \leq b_n \leq c_n < L + \epsilon \Rightarrow |b_n - L| < \epsilon \Rightarrow \lim\limits_{n \to \infty} b_n = L.$

97. $a_{n+1} \geq a_n \Rightarrow \frac{3(n+1)+1}{(n+1)+1} > \frac{3n+1}{n+1} \Rightarrow \frac{3n+4}{n+2} > \frac{3n+1}{n+1} \Rightarrow 3n^2 + 3n + 4n + 4 > 3n^2 + 6n + n + 2$
    $\Rightarrow 4 > 2$; the steps are reversible so the sequence is nondecreasing; $\frac{3n+1}{n+1} < 3 \Rightarrow 3n + 1 < 3n + 3$
    $\Rightarrow 1 < 3$; the steps are reversible so the sequence is bounded above by 3

99. $a_{n+1} \leq a_n \Rightarrow \frac{2^{n+1}3^{n+1}}{(n+1)!} \leq \frac{2^n 3^n}{n!} \Rightarrow \frac{2^{n+1}3^{n+1}}{2^n 3^n} \leq \frac{(n+1)!}{n!} \Rightarrow 2 \cdot 3 \leq n + 1$ which is true for $n \geq 5$; the steps are reversible so the sequence is decreasing after a_5, but it is not nondecreasing for all its terms; $a_1 = 6, a_2 = 18,$

$a_3 = 36$, $a_4 = 54$, $a_5 = \frac{324}{5} = 64.8 \Rightarrow$ the sequence is bounded from above by 64.8

101. $a_n = 1 - \frac{1}{n}$ converges because $\frac{1}{n} \to 0$ by Example 1; also it is a nondecreasing sequence bounded above by 1

103. $a_n = \frac{2^n - 1}{2^n} = 1 - \frac{1}{2^n}$ and $0 < \frac{1}{2^n} < \frac{1}{n}$; since $\frac{1}{n} \to 0$ (by Example 1) $\Rightarrow \frac{1}{2^n} \to 0$, the sequence converges; also it is a nondecreasing sequence bounded above by 1

105. $a_n = ((-1)^n + 1)\left(\frac{n+1}{n}\right)$ diverges because $a_n = 0$ for n odd, while for n even $a_n = 2\left(1 + \frac{1}{n}\right)$ converges to 2; it diverges by definition of divergence

107. If $\{a_n\}$ is nonincreasing with lower bound M, then $\{-a_n\}$ is a nondecreasing sequence with upper bound $-M$. By Theorem 1, $\{-a_n\}$ converges and hence $\{a_n\}$ converges. If $\{a_n\}$ has no lower bound, then $\{-a_n\}$ has no upper bound and therefore diverges. Hence, $\{a_n\}$ also diverges.

109. $a_n \geq a_{n+1} \Leftrightarrow \frac{1 + \sqrt{2n}}{\sqrt{n}} \geq \frac{1 + \sqrt{2(n+1)}}{\sqrt{n+1}} \Leftrightarrow \sqrt{n+1} + \sqrt{2n^2 + 2n} \geq \sqrt{n} + \sqrt{2n^2 + 2n} \Leftrightarrow \sqrt{n+1} \geq \sqrt{n}$ and $\frac{1 + \sqrt{2n}}{\sqrt{n}} \geq \sqrt{2}$; thus the sequence is nonincreasing and bounded below by $\sqrt{2} \Rightarrow$ it converges

111. $\frac{4^{n+1} + 3^n}{4^n} = 4 + \left(\frac{3}{4}\right)^n$ so $a_n \geq a_{n+1} \Leftrightarrow 4 + \left(\frac{3}{4}\right)^n \geq 4 + \left(\frac{3}{4}\right)^{n+1} \Leftrightarrow \left(\frac{3}{4}\right)^n \geq \left(\frac{3}{4}\right)^{n+1} \Leftrightarrow 1 \geq \frac{3}{4}$ and $4 + \left(\frac{3}{4}\right)^n \geq 4$; thus the sequence is nonincreasing and bounded below by 4 $\Rightarrow$ it converges

113. Let $0 < M < 1$ and let N be an integer greater than $\frac{M}{1-M}$. Then $n > N \Rightarrow n > \frac{M}{1-M} \Rightarrow n - nM > M$ $\Rightarrow n > M + nM \Rightarrow n > M(n+1) \Rightarrow \frac{n}{n+1} > M$.

115. The sequence $a_n = 1 + \frac{(-1)^n}{2}$ is the sequence $\frac{1}{2}, \frac{3}{2}, \frac{1}{2}, \frac{3}{2}, \ldots$. This sequence is bounded above by $\frac{3}{2}$, but it clearly does not converge, by definition of convergence.

117. Given an $\epsilon > 0$, by definition of convergence there corresponds an N such that for all $n > N$, $|L_1 - a_n| < \epsilon$ and $|L_2 - a_n| < \epsilon$. Now $|L_2 - L_1| = |L_2 - a_n + a_n - L_1| \leq |L_2 - a_n| + |a_n - L_1| < \epsilon + \epsilon = 2\epsilon$. $|L_2 - L_1| < 2\epsilon$ says that the difference between two fixed values is smaller than any positive number 2ϵ. The only nonnegative number smaller than every positive number is 0, so $|L_1 - L_2| = 0$ or $L_1 = L_2$.

119. $a_{2k} \to L \Leftrightarrow$ given an $\epsilon > 0$ there corresponds an N_1 such that $[2k > N_1 \Rightarrow |a_{2k} - L| < \epsilon]$. Similarly, $a_{2k+1} \to L \Leftrightarrow [2k + 1 > N_2 \Rightarrow |a_{2k+1} - L| < \epsilon]$. Let $N = \max\{N_1, N_2\}$. Then $n > N \Rightarrow |a_n - L| < \epsilon$ whether n is even or odd, and hence $a_n \to L$.

121. $\left|\sqrt[n]{0.5} - 1\right| < 10^{-3} \Rightarrow -\frac{1}{1000} < \left(\frac{1}{2}\right)^{1/n} - 1 < \frac{1}{1000} \Rightarrow \left(\frac{999}{1000}\right)^n < \frac{1}{2} < \left(\frac{1001}{1000}\right)^n \Rightarrow n > \frac{\ln\left(\frac{1}{2}\right)}{\ln\left(\frac{999}{1000}\right)} \Rightarrow n > 692.8$ $\Rightarrow N = 692$; $a_n = \left(\frac{1}{2}\right)^{1/n}$ and $\lim_{n \to \infty} a_n = 1$

123. $(0.9)^n < 10^{-3} \Rightarrow n \ln(0.9) < -3 \ln 10 \Rightarrow n > \frac{-3 \ln 10}{\ln(0.9)} \approx 65.54 \Rightarrow N = 65$; $a_n = \left(\frac{9}{10}\right)^n$ and $\lim_{n \to \infty} a_n = 0$

125. (a) $f(x) = x^2 - a \Rightarrow f'(x) = 2x \Rightarrow x_{n+1} = x_n - \frac{x_n^2 - a}{2x_n} \Rightarrow x_{n+1} = \frac{2x_n^2 - (x_n^2 - a)}{2x_n} = \frac{x_n^2 + a}{2x_n} = \frac{\left(x_n + \frac{a}{x_n}\right)}{2}$

(b) $x_1 = 2$, $x_2 = 1.75$, $x_3 = 1.732142857$, $x_4 = 1.73205081$, $x_5 = 1.732050808$; we are finding the positive number where $x^2 - 3 = 0$; that is, where $x^2 = 3$, $x > 0$, or where $x = \sqrt{3}$.

127. $x_1 = 1$, $x_2 = 1 + \cos(1) = 1.540302306$, $x_3 = 1.540302306 + \cos(1 + \cos(1)) = 1.570791601$,

$x_4 = 1.570791601 + \cos(1.570791601) = 1.570796327 = \frac{\pi}{2}$ to 9 decimal places. After a few steps, the

arc (x_{n-1}) and line segment $\cos(x_{n-1})$ are nearly the same as the quarter circle.

11.2 INFINITE SERIES

1. $s_n = \frac{a(1-r^n)}{(1-r)} = \frac{2\left(1-\left(\frac{1}{3}\right)^n\right)}{1-\left(\frac{1}{3}\right)} \Rightarrow \lim\limits_{n \to \infty} s_n = \frac{2}{1-\left(\frac{1}{3}\right)} = 3$

3. $s_n = \frac{a(1-r^n)}{(1-r)} = \frac{1-\left(-\frac{1}{2}\right)^n}{1-\left(-\frac{1}{2}\right)} \Rightarrow \lim\limits_{n \to \infty} s_n = \frac{1}{\left(\frac{3}{2}\right)} = \frac{2}{3}$

5. $\frac{1}{(n+1)(n+2)} = \frac{1}{n+1} - \frac{1}{n+2} \Rightarrow s_n = \left(\frac{1}{2} - \frac{1}{3}\right) + \left(\frac{1}{3} - \frac{1}{4}\right) + \dots + \left(\frac{1}{n+1} - \frac{1}{n+2}\right) = \frac{1}{2} - \frac{1}{n+2} \Rightarrow \lim\limits_{n \to \infty} s_n = \frac{1}{2}$

7. $1 - \frac{1}{4} + \frac{1}{16} - \frac{1}{64} + \dots$, the sum of this geometric series is $\frac{1}{1-\left(-\frac{1}{4}\right)} = \frac{1}{1+\left(\frac{1}{4}\right)} = \frac{4}{5}$

9. $\frac{7}{4} + \frac{7}{16} + \frac{7}{64} + \dots$, the sum of this geometric series is $\frac{\left(\frac{7}{4}\right)}{1-\left(\frac{1}{4}\right)} = \frac{7}{3}$

11. $(5+1) + \left(\frac{5}{2} + \frac{1}{3}\right) + \left(\frac{5}{4} + \frac{1}{9}\right) + \left(\frac{5}{8} + \frac{1}{27}\right) + \dots$, is the sum of two geometric series; the sum is

$\frac{5}{1-\left(\frac{1}{2}\right)} + \frac{1}{1-\left(\frac{1}{3}\right)} = 10 + \frac{3}{2} = \frac{23}{2}$

13. $(1+1) + \left(\frac{1}{2} - \frac{1}{5}\right) + \left(\frac{1}{4} + \frac{1}{25}\right) + \left(\frac{1}{8} - \frac{1}{125}\right) + \dots$, is the sum of two geometric series; the sum is

$\frac{1}{1-\left(\frac{1}{2}\right)} + \frac{1}{1+\left(\frac{1}{5}\right)} = 2 + \frac{5}{6} = \frac{17}{6}$

15. $\frac{4}{(4n-3)(4n+1)} = \frac{1}{4n-3} - \frac{1}{4n+1} \Rightarrow s_n = \left(1 - \frac{1}{5}\right) + \left(\frac{1}{5} - \frac{1}{9}\right) + \left(\frac{1}{9} - \frac{1}{13}\right) + \dots + \left(\frac{1}{4n-7} - \frac{1}{4n-3}\right)$

$+ \left(\frac{1}{4n-3} - \frac{1}{4n+1}\right) = 1 - \frac{1}{4n+1} \Rightarrow \lim\limits_{n \to \infty} s_n = \lim\limits_{n \to \infty} \left(1 - \frac{1}{4n+1}\right) = 1$

17. $\frac{40n}{(2n-1)^2(2n+1)^2} = \frac{A}{(2n-1)} + \frac{B}{(2n-1)^2} + \frac{C}{(2n+1)} + \frac{D}{(2n+1)^2}$

$= \frac{A(2n-1)(2n+1)^2 + B(2n+1)^2 + C(2n+1)(2n-1)^2 + D(2n-1)^2}{(2n-1)^2(2n+1)^2}$

$\Rightarrow A(2n-1)(2n+1)^2 + B(2n+1)^2 + C(2n+1)(2n-1)^2 + D(2n-1)^2 = 40n$

$\Rightarrow A\left(8n^3 + 4n^2 - 2n - 1\right) + B\left(4n^2 + 4n + 1\right) + C\left(8n^3 - 4n^2 - 2n + 1\right) = D\left(4n^2 - 4n + 1\right) = 40n$

$\Rightarrow (8A + 8C)n^3 + (4A + 4B - 4C + 4D)n^2 + (-2A + 4B - 2C - 4D)n + (-A + B + C + D) = 40n$

$\Rightarrow \begin{cases} 8A + 8C = 0 \\ 4A + 4B - 4C + 4D = 0 \\ -2A + 4B - 2C - 4D = 40 \\ -A + B + C + D = 0 \end{cases} \Rightarrow \begin{cases} 8A + 8C = 0 \\ A + B - C + D = 0 \\ -A + 2B - C - 2D = 20 \\ -A + B + C + D = 0 \end{cases} \Rightarrow \begin{cases} B + D = 0 \\ 2B - 2D = 20 \end{cases} \Rightarrow 4B = 20 \Rightarrow B = 5$

and $D = -5 \Rightarrow \begin{cases} A + C = 0 \\ -A + 5 + C - 5 = 0 \end{cases} \Rightarrow C = 0$ and $A = 0$. Hence, $\sum\limits_{n=1}^{k} \left[\frac{40n}{(2n-1)^2(2n+1)^2}\right]$

$= 5 \sum\limits_{n=1}^{k} \left[\frac{1}{(2n-1)^2} - \frac{1}{(2n+1)^2}\right] = 5\left(\frac{1}{1} - \frac{1}{9} + \frac{1}{9} - \frac{1}{25} + \frac{1}{25} - \dots - \frac{1}{(2(k-1)+1)^2} + \frac{1}{(2k-1)^2} - \frac{1}{(2k+1)^2}\right)$

$= 5\left(1 - \frac{1}{(2k+1)^2}\right) \Rightarrow$ the sum is $\lim\limits_{n \to \infty} 5\left(1 - \frac{1}{(2k+1)^2}\right) = 5$

19. $s_n = \left(1 - \frac{1}{\sqrt{2}}\right) + \left(\frac{1}{\sqrt{2}} - \frac{1}{\sqrt{3}}\right) + \left(\frac{1}{\sqrt{3}} - \frac{1}{\sqrt{4}}\right) + \dots + \left(\frac{1}{\sqrt{n-1}} + \frac{1}{\sqrt{n}}\right) + \left(\frac{1}{\sqrt{n}} - \frac{1}{\sqrt{n+1}}\right) = 1 - \frac{1}{\sqrt{n+1}}$

$\Rightarrow \lim\limits_{n \to \infty} s_n = \lim\limits_{n \to \infty} \left(1 - \frac{1}{\sqrt{n+1}}\right) = 1$

21. $s_n = \left(\frac{1}{\ln 3} - \frac{1}{\ln 2}\right) + \left(\frac{1}{\ln 4} - \frac{1}{\ln 3}\right) + \left(\frac{1}{\ln 5} - \frac{1}{\ln 4}\right) + \ldots + \left(\frac{1}{\ln(n+1)} - \frac{1}{\ln n}\right) + \left(\frac{1}{\ln(n+2)} - \frac{1}{\ln(n+1)}\right)$

$= -\frac{1}{\ln 2} + \frac{1}{\ln(n+2)} \Rightarrow \lim\limits_{n \to \infty} s_n = -\frac{1}{\ln 2}$

23. convergent geometric series with sum $\frac{1}{1 - \left(\frac{1}{\sqrt{2}}\right)} = \frac{\sqrt{2}}{\sqrt{2}-1} = 2 + \sqrt{2}$

25. convergent geometric series with sum $\frac{\left(\frac{3}{2}\right)}{1 - \left(-\frac{1}{2}\right)} = 1$

27. $\lim\limits_{n \to \infty} \cos(n\pi) = \lim\limits_{n \to \infty} (-1)^n \neq 0 \Rightarrow$ diverges

29. convergent geometric series with sum $\frac{1}{1 - \left(\frac{1}{e^2}\right)} = \frac{e^2}{e^2 - 1}$

31. convergent geometric series with sum $\frac{2}{1 - \left(\frac{1}{10}\right)} - 2 = \frac{20}{9} - \frac{18}{9} = \frac{2}{9}$

33. difference of two geometric series with sum $\frac{1}{1 - \left(\frac{2}{3}\right)} - \frac{1}{1 - \left(\frac{1}{3}\right)} = 3 - \frac{3}{2} = \frac{3}{2}$

35. $\lim\limits_{n \to \infty} \frac{n!}{1000^n} = \infty \neq 0 \Rightarrow$ diverges

37. $\sum\limits_{n=1}^{\infty} \ln\left(\frac{n}{n+1}\right) = \sum\limits_{n=1}^{\infty} [\ln(n) - \ln(n+1)] \Rightarrow s_n = [\ln(1) - \ln(2)] + [\ln(2) - \ln(3)] + [\ln(3) - \ln(4)] + \ldots$

$+ [\ln(n-1) - \ln(n)] + [\ln(n) - \ln(n+1)] = \ln(1) - \ln(n+1) = -\ln(n+1) \Rightarrow \lim\limits_{n \to \infty} s_n = -\infty, \Rightarrow$ diverges

39. convergent geometric series with sum $\frac{1}{1 - \left(\frac{e}{\pi}\right)} = \frac{\pi}{\pi - e}$

41. $\sum\limits_{n=0}^{\infty} (-1)^n x^n = \sum\limits_{n=0}^{\infty} (-x)^n; a = 1, r = -x;$ converges to $\frac{1}{1 - (-x)} = \frac{1}{1 + x}$ for $|x| < 1$

43. $a = 3, r = \frac{x-1}{2};$ converges to $\frac{3}{1 - \left(\frac{x-1}{2}\right)} = \frac{6}{3 - x}$ for $-1 < \frac{x-1}{2} < 1$ or $-1 < x < 3$

45. $a = 1, r = 2x;$ converges to $\frac{1}{1 - 2x}$ for $|2x| < 1$ or $|x| < \frac{1}{2}$

47. $a = 1, r = -(x+1)^n;$ converges to $\frac{1}{1 + (x+1)} = \frac{1}{2 + x}$ for $|x+1| < 1$ or $-2 < x < 0$

49. $a = 1, r = \sin x;$ converges to $\frac{1}{1 - \sin x}$ for $x \neq (2k+1)\frac{\pi}{2}$, k an integer

51. $0.\overline{23} = \sum\limits_{n=0}^{\infty} \frac{23}{100} \left(\frac{1}{10^2}\right)^n = \frac{\left(\frac{23}{100}\right)}{1 - \left(\frac{1}{100}\right)} = \frac{23}{99}$

53. $0.\overline{7} = \sum\limits_{n=0}^{\infty} \frac{7}{10} \left(\frac{1}{10}\right)^n = \frac{\left(\frac{7}{10}\right)}{1 - \left(\frac{1}{10}\right)} = \frac{7}{9}$

55. $0.0\overline{6} = \sum\limits_{n=0}^{\infty} \left(\frac{1}{10}\right) \left(\frac{6}{10}\right) \left(\frac{1}{10}\right)^n = \frac{\left(\frac{6}{100}\right)}{1 - \left(\frac{1}{10}\right)} = \frac{6}{90} = \frac{1}{15}$

57. $1.24\overline{123} = \frac{124}{100} + \sum\limits_{n=0}^{\infty} \frac{123}{10^5} \left(\frac{1}{10^3}\right)^n = \frac{124}{100} + \frac{\left(\frac{123}{10^5}\right)}{1 - \left(\frac{1}{10^3}\right)} = \frac{124}{100} + \frac{123}{10^5 - 10^2} = \frac{124}{100} + \frac{123}{99,900} = \frac{123,999}{99,900} = \frac{41,333}{33,300}$

59. (a) $\sum\limits_{n=-2}^{\infty} \frac{1}{(n+4)(n+5)}$ (b) $\sum\limits_{n=0}^{\infty} \frac{1}{(n+2)(n+3)}$ (c) $\sum\limits_{n=5}^{\infty} \frac{1}{(n-3)(n-2)}$

61. (a) one example is $\frac{1}{2} + \frac{1}{4} + \frac{1}{8} + \frac{1}{16} + \cdots = \frac{\left(\frac{1}{2}\right)}{1-\left(\frac{1}{2}\right)} = 1$

   (b) one example is $-\frac{3}{2} - \frac{3}{4} - \frac{3}{8} - \frac{3}{16} - \cdots = \frac{\left(-\frac{3}{2}\right)}{1-\left(\frac{1}{2}\right)} = -3$

   (c) one example is $1 - \frac{1}{2} - \frac{1}{4} - \frac{1}{8} - \frac{1}{16} - \cdots$; the series $\frac{k}{2} + \frac{k}{4} + \frac{k}{8} + \cdots = \frac{\left(\frac{k}{2}\right)}{1-\left(\frac{1}{2}\right)} = k$ where k is any positive or negative number.

63. Let $a_n = b_n = \left(\frac{1}{2}\right)^n$. Then $\sum\limits_{n=1}^{\infty} a_n = \sum\limits_{n=1}^{\infty} b_n = \sum\limits_{n=1}^{\infty} \left(\frac{1}{2}\right)^n = 1$, while $\sum\limits_{n=1}^{\infty} \left(\frac{a_n}{b_n}\right) = \sum\limits_{n=1}^{\infty} (1)$ diverges.

65. Let $a_n = \left(\frac{1}{4}\right)^n$ and $b_n = \left(\frac{1}{2}\right)^n$. Then $A = \sum\limits_{n=1}^{\infty} a_n = \frac{1}{3}$, $B = \sum\limits_{n=1}^{\infty} b_n = 1$ and $\sum\limits_{n=1}^{\infty} \left(\frac{a_n}{b_n}\right) = \sum\limits_{n=1}^{\infty} \left(\frac{1}{2}\right)^n = 1 \neq \frac{A}{B}$.

67. Since the sum of a finite number of terms is finite, adding or subtracting a finite number of terms from a series that diverges does not change the divergence of the series.

69. (a) $\frac{2}{1-r} = 5 \Rightarrow \frac{2}{5} = 1 - r \Rightarrow r = \frac{3}{5}$; $2 + 2\left(\frac{3}{5}\right) + 2\left(\frac{3}{5}\right)^2 + \cdots$

   (b) $\frac{\left(\frac{13}{2}\right)}{1-r} = 5 \Rightarrow \frac{13}{10} = 1 - r \Rightarrow r = -\frac{3}{10}$; $\frac{13}{2} - \frac{13}{2}\left(\frac{3}{10}\right) + \frac{13}{2}\left(\frac{3}{10}\right)^2 - \frac{13}{2}\left(\frac{3}{10}\right)^3 + \cdots$

71. $s_n = 1 + 2r + r^2 + 2r^3 + r^4 + 2r^5 + \cdots + r^{2n} + 2r^{2n+1}$, $n = 0, 1, \ldots$
   $\Rightarrow s_n = (1 + r^2 + r^4 + \cdots + r^{2n}) + (2r + 2r^3 + 2r^5 + \cdots + 2r^{2n+1}) \Rightarrow \lim\limits_{n \to \infty} s_n = \frac{1}{1-r^2} + \frac{2r}{1-r^2}$
   $= \frac{1+2r}{1-r^2}$, if $|r^2| < 1$ or $|r| < 1$

73. distance $= 4 + 2\left[(4)\left(\frac{3}{4}\right) + (4)\left(\frac{3}{4}\right)^2 + \cdots\right] = 4 + 2\left(\frac{3}{1-\left(\frac{3}{4}\right)}\right) = 28$ m

75. area $= 2^2 + \left(\sqrt{2}\right)^2 + (1)^2 + \left(\frac{1}{\sqrt{2}}\right)^2 + \cdots = 4 + 2 + 1 + \frac{1}{2} + \cdots = \frac{4}{1-\frac{1}{2}} = 8$ m^2

77. (a) $L_1 = 3$, $L_2 = 3\left(\frac{4}{3}\right)$, $L_3 = 3\left(\frac{4}{3}\right)^2, \ldots, L_n = 3\left(\frac{4}{3}\right)^{n-1} \Rightarrow \lim\limits_{n \to \infty} L_n = \lim\limits_{n \to \infty} 3\left(\frac{4}{3}\right)^{n-1} = \infty$

   (b) Using the fact that the area of an equilateral triangle of side length s is $\frac{\sqrt{3}}{4}s^2$, we see that $A_1 = \frac{\sqrt{3}}{4}$,
   $A_2 = A_1 + 3\left(\frac{\sqrt{3}}{4}\right)\left(\frac{1}{3}\right)^2 = \frac{\sqrt{3}}{4} + \frac{\sqrt{3}}{12}$, $A_3 = A_2 + 3(4)\left(\frac{\sqrt{3}}{4}\right)\left(\frac{1}{3^2}\right)^2 = \frac{\sqrt{3}}{4} + \frac{\sqrt{3}}{12} + \frac{\sqrt{3}}{27}$,
   $A_4 = A_3 + 3(4)^2\left(\frac{\sqrt{3}}{4}\right)\left(\frac{1}{3^3}\right)^2$, $A_5 = A_4 + 3(4)^3\left(\frac{\sqrt{3}}{4}\right)\left(\frac{1}{3^4}\right)^2, \ldots$,
   $A_n = \frac{\sqrt{3}}{4} + \sum\limits_{k=2}^{n} 3(4)^{k-2}\left(\frac{\sqrt{3}}{4}\right)\left(\frac{1}{3^2}\right)^{k-1} = \frac{\sqrt{3}}{4} + \sum\limits_{k=2}^{n} 3\sqrt{3}(4)^{k-3}\left(\frac{1}{9}\right)^{k-1} = \frac{\sqrt{3}}{4} + 3\sqrt{3}\left(\sum\limits_{k=2}^{n} \frac{4^{k-3}}{9^{k-1}}\right)$.
   $\lim\limits_{n \to \infty} A_n = \lim\limits_{n \to \infty} \left(\frac{\sqrt{3}}{4} + 3\sqrt{3}\left(\sum\limits_{k=2}^{n} \frac{4^{k-3}}{9^{k-1}}\right)\right) = \frac{\sqrt{3}}{4} + 3\sqrt{3}\left(\frac{\frac{1}{36}}{1-\frac{4}{9}}\right) = \frac{\sqrt{3}}{4} + 3\sqrt{3}\left(\frac{1}{20}\right) = \frac{2\sqrt{3}}{5}$

11.3 THE INTEGRAL TEST

1. converges; a geometric series with $r = \frac{1}{10} < 1$

3. diverges; by the nth-Term Test for Divergence, $\lim\limits_{n \to \infty} \frac{n}{n+1} = 1 \neq 0$

5. diverges; $\sum\limits_{n=1}^{\infty} \frac{3}{\sqrt{n}} = 3 \sum\limits_{n=1}^{\infty} \frac{1}{\sqrt{n}}$, which is a divergent p-series $(p = \frac{1}{2})$

7. converges; a geometric series with $r = \frac{1}{8} < 1$

9. diverges by the Integral Test: $\int_2^n \frac{\ln x}{x}\, dx = \frac{1}{2}(\ln^2 n - \ln 2) \Rightarrow \int_2^\infty \frac{\ln x}{x}\, dx \to \infty$

11. converges; a geometric series with $r = \frac{2}{3} < 1$

13. diverges; $\sum\limits_{n=0}^{\infty} \frac{-2}{n+1} = -2 \sum\limits_{n=0}^{\infty} \frac{1}{n+1}$, which diverges by the Integral Test

15. diverges; $\lim\limits_{n \to \infty} a_n = \lim\limits_{n \to \infty} \frac{2^n}{n+1} = \lim\limits_{n \to \infty} \frac{2^n \ln 2}{1} = \infty \neq 0$

17. diverges; $\lim\limits_{n \to \infty} \frac{\sqrt{n}}{\ln n} = \lim\limits_{n \to \infty} \frac{\left(\frac{1}{2\sqrt{n}}\right)}{\left(\frac{1}{n}\right)} = \lim\limits_{n \to \infty} \frac{\sqrt{n}}{2} = \infty \neq 0$

19. diverges; a geometric series with $r = \frac{1}{\ln 2} \approx 1.44 > 1$

21. converges by the Integral Test: $\int_3^\infty \frac{\left(\frac{1}{x}\right)}{(\ln x)\sqrt{(\ln x)^2 - 1}}\, dx;\ \begin{bmatrix} u = \ln x \\ du = \frac{1}{x}\, dx \end{bmatrix} \to \int_{\ln 3}^\infty \frac{1}{u\sqrt{u^2 - 1}}\, du$

    $= \lim\limits_{b \to \infty} \left[\sec^{-1} |u|\right]_{\ln 3}^b = \lim\limits_{b \to \infty} \left[\sec^{-1} b - \sec^{-1} (\ln 3)\right] = \lim\limits_{b \to \infty} \left[\cos^{-1} \left(\frac{1}{b}\right) - \sec^{-1} (\ln 3)\right]$
    $= \cos^{-1}(0) - \sec^{-1}(\ln 3) = \frac{\pi}{2} - \sec^{-1}(\ln 3) \approx 1.1439$

23. diverges by the nth-Term Test for divergence; $\lim\limits_{n \to \infty} n \sin \left(\frac{1}{n}\right) = \lim\limits_{n \to \infty} \frac{\sin \left(\frac{1}{n}\right)}{\left(\frac{1}{n}\right)} = \lim\limits_{x \to 0} \frac{\sin x}{x} = 1 \neq 0$

25. converges by the Integral Test: $\int_1^\infty \frac{e^x}{1 + e^{2x}}\, dx;\ \begin{bmatrix} u = e^x \\ du = e^x\, dx \end{bmatrix} \to \int_e^\infty \frac{1}{1 + u^2}\, du = \lim\limits_{n \to \infty} \left[\tan^{-1} u\right]_e^b$

    $= \lim\limits_{b \to \infty} \left(\tan^{-1} b - \tan^{-1} e\right) = \frac{\pi}{2} - \tan^{-1} e \approx 0.35$

27. converges by the Integral Test: $\int_1^\infty \frac{8 \tan^{-1} x}{1 + x^2}\, dx;\ \begin{bmatrix} u = \tan^{-1} x \\ du = \frac{dx}{1 + x^2} \end{bmatrix} \to \int_{\pi/4}^{\pi/2} 8u\, du = \left[4u^2\right]_{\pi/4}^{\pi/2} = 4\left(\frac{\pi^2}{4} - \frac{\pi^2}{16}\right) = \frac{3\pi^2}{4}$

29. converges by the Integral Test: $\int_1^\infty \operatorname{sech} x\, dx = 2 \lim\limits_{b \to \infty} \int_1^b \frac{e^x}{1 + (e^x)^2}\, dx = 2 \lim\limits_{b \to \infty} \left[\tan^{-1} e^x\right]_1^b$

    $= 2 \lim\limits_{b \to \infty} \left(\tan^{-1} e^b - \tan^{-1} e\right) = \pi - 2 \tan^{-1} e \approx 0.71$

31. $\int_1^\infty \left(\frac{a}{x+2} - \frac{1}{x+4}\right) dx = \lim_{b \to \infty} \left[a \ln|x+2| - \ln|x+4|\right]_1^b = \lim_{b \to \infty} \ln \frac{(b+2)^a}{b+4} - \ln\left(\frac{3^a}{5}\right);$

$\lim_{b \to \infty} \frac{(b+2)^a}{b+4} = a \lim_{b \to \infty} (b+2)^{a-1} = \begin{cases} \infty, a > 1 \\ 1, \quad a = 1 \end{cases} \Rightarrow$ the series converges to $\ln\left(\frac{5}{3}\right)$ if $a = 1$ and diverges to ∞ if

$a > 1$. If $a < 1$, the terms of the series eventually become negative and the Integral Test does not apply. From
that point on, however, the series behaves like a negative multiple of the harmonic series, and so it diverges.

33. (a)

$$\int_1^{n+1} \frac{1}{x} dx < 1 + \frac{1}{2} + \dots + \frac{1}{n} \qquad\qquad 1 + \frac{1}{2} + \dots + \frac{1}{n} < 1 + \int_1^n \frac{1}{x} dx$$

(b) There are $(13)(365)(24)(60)(60)\,(10^9)$ seconds in 13 billion years; by part (a) $s_n \leq 1 + \ln n$ where
$n = (13)(365)(24)(60)(60)\,(10^9) \Rightarrow s_n \leq 1 + \ln\left((13)(365)(24)(60)(60)\,(10^9)\right)$
$= 1 + \ln(13) + \ln(365) + \ln(24) + 2\ln(60) + 9\ln(10) \approx 41.55$

35. Yes. If $\sum\limits_{n=1}^\infty a_n$ is a divergent series of positive numbers, then $\left(\frac{1}{2}\right)\sum\limits_{n=1}^\infty a_n = \sum\limits_{n=1}^\infty \left(\frac{a_n}{2}\right)$ also diverges and $\frac{a_n}{2} < a_n$.

There is no "smallest" divergent series of positive numbers: for any divergent series $\sum\limits_{n=1}^\infty a_n$ of positive

numbers $\sum\limits_{n=1}^\infty \left(\frac{a_n}{2}\right)$ has smaller terms and still diverges.

37. Let $A_n = \sum\limits_{k=1}^n a_k$ and $B_n = \sum\limits_{k=1}^n 2^k a_{(2^k)}$, where $\{a_k\}$ is a nonincreasing sequence of positive terms converging to

0. Note that $\{A_n\}$ and $\{B_n\}$ are nondecreasing sequences of positive terms. Now,
$B_n = 2a_2 + 4a_4 + 8a_8 + \dots + 2^n a_{(2^n)} = 2a_2 + (2a_4 + 2a_4) + (2a_8 + 2a_8 + 2a_8 + 2a_8) + \dots$
$+ \underbrace{\left(2a_{(2^n)} + 2a_{(2^n)} + \dots + 2a_{(2^n)}\right)}_{2^{n-1} \text{ terms}} \leq 2a_1 + 2a_2 + (2a_3 + 2a_4) + (2a_5 + 2a_6 + 2a_7 + 2a_8) + \dots$

$+ \left(2a_{(2^{n-1})} + 2a_{(2^{n-1}+1)} + \dots + 2a_{(2^n)}\right) = 2A_{(2^n)} \leq 2\sum\limits_{k=1}^\infty a_k$. Therefore if $\sum a_k$ converges,

then $\{B_n\}$ is bounded above $\Rightarrow \sum 2^k a_{(2^k)}$ converges. Conversely,
$A_n = a_1 + (a_2 + a_3) + (a_4 + a_5 + a_6 + a_7) + \dots + a_n < a_1 + 2a_2 + 4a_4 + \dots + 2^n a_{(2^n)} = a_1 + B_n < a_1 + \sum\limits_{k=1}^\infty 2^k a_{(2^k)}.$

Therefore, if $\sum\limits_{k=1}^\infty 2^k a_{(2^k)}$ converges, then $\{A_n\}$ is bounded above and hence converges.

39. (a) $\int_2^\infty \frac{dx}{x(\ln x)^p}$; $\begin{bmatrix} u = \ln x \\ du = \frac{dx}{x} \end{bmatrix} \to \int_{\ln 2}^\infty u^{-p} \, du = \lim_{b \to \infty} \left[\frac{u^{-p+1}}{-p+1}\right]_{\ln 2}^b = \lim_{b \to \infty} \left(\frac{1}{1-p}\right)\left[b^{-p+1} - (\ln 2)^{-p+1}\right]$

$= \begin{cases} \frac{1}{p-1}(\ln 2)^{-p+1}, p > 1 \\ \infty, p < 1 \end{cases} \Rightarrow$ the improper integral converges if $p > 1$ and diverges

if $p < 1$. For $p = 1$: $\int_2^\infty \frac{dx}{x \ln x} = \lim_{b \to \infty} \left[\ln(\ln x)\right]_2^b = \lim_{b \to \infty} \left[\ln(\ln b) - \ln(\ln 2)\right] = \infty$, so the improper
integral diverges if $p = 1$.

(b) Since the series and the integral converge or diverge together, $\sum\limits_{n=2}^{\infty} \frac{1}{n(\ln n)^p}$ converges if and only if $p > 1$.

41. (a) From Fig. 11.8 in the text with $f(x) = \frac{1}{x}$ and $a_k = \frac{1}{k}$, we have $\int_1^{n+1} \frac{1}{x}\, dx \leq 1 + \frac{1}{2} + \frac{1}{3} + \ldots + \frac{1}{n}$

$\leq 1 + \int_1^n f(x)\, dx \Rightarrow \ln(n+1) \leq 1 + \frac{1}{2} + \frac{1}{3} + \ldots + \frac{1}{n} \leq 1 + \ln n \Rightarrow 0 \leq \ln(n+1) - \ln n$

$\leq \left(1 + \frac{1}{2} + \frac{1}{3} + \ldots + \frac{1}{n}\right) - \ln n \leq 1$. Therefore the sequence $\left\{\left(1 + \frac{1}{2} + \frac{1}{3} + \ldots + \frac{1}{n}\right) - \ln n\right\}$ is bounded above by 1 and below by 0.

(b) From the graph in Fig. 11.8(a) with $f(x) = \frac{1}{x}$, $\frac{1}{n+1} < \int_n^{n+1} \frac{1}{x}\, dx = \ln(n+1) - \ln n$

$\Rightarrow 0 > \frac{1}{n+1} - [\ln(n+1) - \ln n] = \left(1 + \frac{1}{2} + \frac{1}{3} + \ldots + \frac{1}{n+1} - \ln(n+1)\right) - \left(1 + \frac{1}{2} + \frac{1}{3} + \ldots + \frac{1}{n} - \ln n\right)$.

If we define $a_n = 1 + \frac{1}{2} = \frac{1}{3} + \frac{1}{n} - \ln n$, then $0 > a_{n+1} - a_n \Rightarrow a_{n+1} < a_n \Rightarrow \{a_n\}$ is a decreasing sequence of nonnegative terms.

11.4 COMPARISON TESTS

1. diverges by the Limit Comparison Test (part 1) when compared with $\sum\limits_{n=1}^{\infty} \frac{1}{\sqrt{n}}$, a divergent p-series:

$$\lim_{n \to \infty} \frac{\left(\frac{1}{2\sqrt{n} + \sqrt[3]{n}}\right)}{\left(\frac{1}{\sqrt{n}}\right)} = \lim_{n \to \infty} \frac{\sqrt{n}}{2\sqrt{n} + \sqrt[3]{n}} = \lim_{n \to \infty} \left(\frac{1}{2 + n^{-1/6}}\right) = \frac{1}{2}$$

3. converges by the Direct Comparison Test; $\frac{\sin^2 n}{2^n} \leq \frac{1}{2^n}$, which is the nth term of a convergent geometric series

5. diverges since $\lim\limits_{n \to \infty} \frac{2n}{3n-1} = \frac{2}{3} \neq 0$

7. converges by the Direct Comparison Test; $\left(\frac{n}{3n+1}\right)^n < \left(\frac{n}{3n}\right)^n = \left(\frac{1}{3}\right)^n$, the nth term of a convergent geometric series

9. diverges by the Direct Comparison Test; $n > \ln n \Rightarrow \ln n > \ln \ln n \Rightarrow \frac{1}{n} < \frac{1}{\ln n} < \frac{1}{\ln(\ln n)}$ and $\sum\limits_{n=3}^{\infty} \frac{1}{n}$ diverges

11. converges by the Limit Comparison Test (part 2) when compared with $\sum\limits_{n=1}^{\infty} \frac{1}{n^2}$, a convergent p-series:

$$\lim_{n \to \infty} \frac{\left[\frac{(\ln n)^2}{n^3}\right]}{\left(\frac{1}{n^2}\right)} = \lim_{n \to \infty} \frac{(\ln n)^2}{n} = \lim_{n \to \infty} \frac{2(\ln n)\left(\frac{1}{n}\right)}{1} = 2 \lim_{n \to \infty} \frac{\ln n}{n} = 0$$

13. diverges by the Limit Comparison Test (part 3) with $\frac{1}{n}$, the nth term of the divergent harmonic series:

$$\lim_{n \to \infty} \frac{\left[\frac{1}{\sqrt{n} \ln n}\right]}{\left(\frac{1}{n}\right)} = \lim_{n \to \infty} \frac{\sqrt{n}}{\ln n} = \lim_{n \to \infty} \frac{\left(\frac{1}{2\sqrt{n}}\right)}{\left(\frac{1}{n}\right)} = \lim_{n \to \infty} \frac{\sqrt{n}}{2} = \infty$$

15. diverges by the Limit Comparison Test (part 3) with $\frac{1}{n}$, the nth term of the divergent harmonic series:

$$\lim_{n \to \infty} \frac{\left(\frac{1}{1 + \ln n}\right)}{\left(\frac{1}{n}\right)} = \lim_{n \to \infty} \frac{n}{1 + \ln n} = \lim_{n \to \infty} \frac{1}{\left(\frac{1}{n}\right)} = \lim_{n \to \infty} n = \infty$$

17. diverges by the Integral Test: $\int_2^{\infty} \frac{\ln(x+1)}{x+1}\, dx = \int_{\ln 3}^{\infty} u\, du = \lim_{b \to \infty} \left[\frac{1}{2} u^2\right]_{\ln 3}^b = \lim_{b \to \infty} \frac{1}{2}(b^2 - \ln^2 3) = \infty$

19. converges by the Direct Comparison Test with $\frac{1}{n^{3/2}}$, the nth term of a convergent p-series: $n^2 - 1 > n$ for

$n \geq 2 \Rightarrow n^2(n^2 - 1) > n^3 \Rightarrow n\sqrt{n^2 - 1} > n^{3/2} \Rightarrow \frac{1}{n^{3/2}} > \frac{1}{n\sqrt{n^2 - 1}}$ or use Limit Comparison Test with $\frac{1}{n^2}$.

21. converges because $\sum_{n=1}^{\infty} \frac{1-n}{n2^n} = \sum_{n=1}^{\infty} \frac{1}{n2^n} + \sum_{n=1}^{\infty} \frac{-1}{2^n}$ which is the sum of two convergent series:

$\sum_{n=1}^{\infty} \frac{1}{n2^n}$ converges by the Direct Comparison Test since $\frac{1}{n2^n} < \frac{1}{2^n}$, and $\sum_{n=1}^{\infty} \frac{-1}{2^n}$ is a convergent geometric

series

23. converges by the Direct Comparison Test: $\frac{1}{3^{n-1}+1} < \frac{1}{3^{n-1}}$, which is the nth term of a convergent geometric

series

25. diverges by the Limit Comparison Test (part 1) with $\frac{1}{n}$, the nth term of the divergent harmonic series:

$\lim_{n \to \infty} \frac{\left(\sin \frac{1}{n}\right)}{\left(\frac{1}{n}\right)} = \lim_{x \to 0} \frac{\sin x}{x} = 1$

27. converges by the Limit Comparison Test (part 1) with $\frac{1}{n^2}$, the nth term of a convergent p-series:

$\lim_{n \to \infty} \frac{\left(\frac{10n+1}{n(n+1)(n+2)}\right)}{\left(\frac{1}{n^2}\right)} = \lim_{n \to \infty} \frac{10n^2+n}{n^2+3n+2} = \lim_{n \to \infty} \frac{20n+1}{2n+3} = \lim_{n \to \infty} \frac{20}{2} = 10$

29. converges by the Direct Comparison Test: $\frac{\tan^{-1}n}{n^{1.1}} < \frac{\frac{\pi}{2}}{n^{1.1}}$ and $\sum_{n=1}^{\infty} \frac{\frac{\pi}{2}}{n^{1.1}} = \frac{\pi}{2}\sum_{n=1}^{\infty} \frac{1}{n^{1.1}}$ is the product of a

convergent p-series and a nonzero constant

31. converges by the Limit Comparison Test (part 1) with $\frac{1}{n^2}$: $\lim_{n \to \infty} \frac{\left(\frac{\coth n}{n^2}\right)}{\left(\frac{1}{n^2}\right)} = \lim_{n \to \infty} \coth n = \lim_{n \to \infty} \frac{e^n+e^{-n}}{e^n-e^{-n}}$

$= \lim_{n \to \infty} \frac{1+e^{-2n}}{1-e^{-2n}} = 1$

33. diverges by the Limit Comparison Test (part 1) with $\frac{1}{n}$: $\lim_{n \to \infty} \frac{\left(\frac{1}{n\sqrt[n]{n}}\right)}{\left(\frac{1}{n}\right)} = \lim_{n \to \infty} \frac{1}{\sqrt[n]{n}} = 1$.

35. $\frac{1}{1+2+3+\ldots+n} = \frac{1}{\left(\frac{n(n+1)}{2}\right)} = \frac{2}{n(n+1)}$. The series converges by the Limit Comparison Test (part 1) with $\frac{1}{n^2}$:

$\lim_{n \to \infty} \frac{\left(\frac{2}{n(n+1)}\right)}{\left(\frac{1}{n^2}\right)} = \lim_{n \to \infty} \frac{2n^2}{n^2+n} = \lim_{n \to \infty} \frac{4n}{2n+1} = \lim_{n \to \infty} \frac{4}{2} = 2$.

37. (a) If $\lim_{n \to \infty} \frac{a_n}{b_n} = 0$, then there exists an integer N such that for all $n > N$, $\left|\frac{a_n}{b_n} - 0\right| < 1 \Rightarrow -1 < \frac{a_n}{b_n} < 1$

$\Rightarrow a_n < b_n$. Thus, if $\sum b_n$ converges, then $\sum a_n$ converges by the Direct Comparison Test.

(b) If $\lim_{n \to \infty} \frac{a_n}{b_n} = \infty$, then there exists an integer N such that for all $n > N$, $\frac{a_n}{b_n} > 1 \Rightarrow a_n > b_n$. Thus, if

$\sum b_n$ diverges, then $\sum a_n$ diverges by the Direct Comparison Test.

39. $\lim_{n \to \infty} \frac{a_n}{b_n} = \infty \Rightarrow$ there exists an integer N such that for all $n > N$, $\frac{a_n}{b_n} > 1 \Rightarrow a_n > b_n$. If $\sum a_n$ converges,

then $\sum b_n$ converges by the Direct Comparison Test

11.5 THE RATIO AND ROOT TESTS

1. converges by the Ratio Test: $\lim\limits_{n \to \infty} \frac{a_{n+1}}{a_n} = \lim\limits_{n \to \infty} \frac{\left[\frac{(n+1)^{\sqrt{2}}}{2^{n+1}}\right]}{\left[\frac{n^{\sqrt{2}}}{2^n}\right]} = \lim\limits_{n \to \infty} \frac{(n+1)^{\sqrt{2}}}{2^{n+1}} \cdot \frac{2^n}{n^{\sqrt{2}}}$

$= \lim\limits_{n \to \infty} \left(1 + \frac{1}{n}\right)^{\sqrt{2}} \left(\frac{1}{2}\right) = \frac{1}{2} < 1$

3. diverges by the Ratio Test: $\lim\limits_{n \to \infty} \frac{a_{n+1}}{a_n} = \lim\limits_{n \to \infty} \frac{\left(\frac{(n+1)!}{e^{n+1}}\right)}{\left(\frac{n!}{e^n}\right)} = \lim\limits_{n \to \infty} \frac{(n+1)!}{e^{n+1}} \cdot \frac{e^n}{n!} = \lim\limits_{n \to \infty} \frac{n+1}{e} = \infty$

5. converges by the Ratio Test: $\lim\limits_{n \to \infty} \frac{a_{n+1}}{a_n} = \lim\limits_{n \to \infty} \frac{\left(\frac{(n+1)^{10}}{10^{n+1}}\right)}{\left(\frac{n^{10}}{10^n}\right)} = \lim\limits_{n \to \infty} \frac{(n+1)^{10}}{10^{n+1}} \cdot \frac{10^n}{n^{10}} = \lim\limits_{n \to \infty} \left(1 + \frac{1}{n}\right)^{10} \left(\frac{1}{10}\right)$

$= \frac{1}{10} < 1$

7. converges by the Direct Comparison Test: $\frac{2+(-1)^n}{(1.25)^n} = \left(\frac{4}{5}\right)^n [2 + (-1)^n] \le \left(\frac{4}{5}\right)^n (3)$ which is the nth term of a convergent geometric series

9. diverges; $\lim\limits_{n \to \infty} a_n = \lim\limits_{n \to \infty} \left(1 - \frac{3}{n}\right)^n = \lim\limits_{n \to \infty} \left(1 + \frac{-3}{n}\right)^n = e^{-3} \approx 0.05 \ne 0$

11. converges by the Direct Comparison Test: $\frac{\ln n}{n^3} < \frac{n}{n^3} = \frac{1}{n^2}$ for $n \ge 2$, the nth term of a convergent p-series.

13. diverges by the Direct Comparison Test: $\frac{1}{n} - \frac{1}{n^2} = \frac{n-1}{n^2} > \frac{1}{2}\left(\frac{1}{n}\right)$ for $n > 2$ or by the Limit Comparison Test (part 1) with $\frac{1}{n}$.

15. diverges by the Direct Comparison Test: $\frac{\ln n}{n} > \frac{1}{n}$ for $n \ge 3$

17. converges by the Ratio Test: $\lim\limits_{n \to \infty} \frac{a_{n+1}}{a_n} = \lim\limits_{n \to \infty} \frac{(n+2)(n+3)}{(n+1)!} \cdot \frac{n!}{(n+1)(n+2)} = 0 < 1$

19. converges by the Ratio Test: $\lim\limits_{n \to \infty} \frac{a_{n+1}}{a_n} = \lim\limits_{n \to \infty} \frac{(n+4)!}{3!\,(n+1)!\,3^{n+1}} \cdot \frac{3!\,n!\,3^n}{(n+3)!} = \lim\limits_{n \to \infty} \frac{n+4}{3(n+1)} = \frac{1}{3} < 1$

21. converges by the Ratio Test: $\lim\limits_{n \to \infty} \frac{a_{n+1}}{a_n} = \lim\limits_{n \to \infty} \frac{(n+1)!}{(2n+3)!} \cdot \frac{(2n+1)!}{n!} = \lim\limits_{n \to \infty} \frac{n+1}{(2n+3)(2n+2)} = 0 < 1$

23. converges by the Root Test: $\lim\limits_{n \to \infty} \sqrt[n]{a_n} = \lim\limits_{n \to \infty} \sqrt[n]{\frac{n}{(\ln n)^n}} = \lim\limits_{n \to \infty} \frac{\sqrt[n]{n}}{\ln n} = \lim\limits_{n \to \infty} \frac{1}{\ln n} = 0 < 1$

25. converges by the Direct Comparison Test: $\frac{n!\,\ln n}{n(n+2)!} = \frac{\ln n}{n(n+1)(n+2)} < \frac{n}{n(n+1)(n+2)} = \frac{1}{(n+1)(n+2)} < \frac{1}{n^2}$ which is the nth-term of a convergent p-series

27. converges by the Ratio Test: $\lim\limits_{n \to \infty} \frac{a_{n+1}}{a_n} = \lim\limits_{n \to \infty} \frac{\left(\frac{1+\sin n}{n}\right) a_n}{a_n} = 0 < 1$

29. diverges by the Ratio Test: $\lim\limits_{n \to \infty} \frac{a_{n+1}}{a_n} = \lim\limits_{n \to \infty} \frac{\left(\frac{3n-1}{2n+1}\right) a_n}{a_n} = \lim\limits_{n \to \infty} \frac{3n-1}{2n+1} = \frac{3}{2} > 1$

31. converges by the Ratio Test: $\lim\limits_{n \to \infty} \frac{a_{n+1}}{a_n} = \lim\limits_{n \to \infty} \frac{\left(\frac{2}{n}\right) a_n}{a_n} = \lim\limits_{n \to \infty} \frac{2}{n} = 0 < 1$

33. converges by the Ratio Test: $\lim\limits_{n\to\infty} \frac{a_{n+1}}{a_n} = \lim\limits_{n\to\infty} \frac{\left(\frac{1+\ln n}{n}\right) a_n}{a_n} = \lim\limits_{n\to\infty} \frac{1+\ln n}{n} = \lim\limits_{n\to\infty} \frac{1}{n} = 0 < 1$

35. diverges by the nth-Term Test: $a_1 = \frac{1}{3}$, $a_2 = \sqrt[2]{\frac{1}{3}}$, $a_3 = \sqrt[3]{\sqrt[2]{\frac{1}{3}}} = \sqrt[6]{\frac{1}{3}}$, $a_4 = \sqrt[4]{\sqrt[3]{\sqrt[2]{\frac{1}{3}}}} = \sqrt[4!]{\frac{1}{3}}, \ldots,$

   $a_n = \sqrt[n!]{\frac{1}{3}} \Rightarrow \lim\limits_{n\to\infty} a_n = 1$ because $\left\{\sqrt[n!]{\frac{1}{3}}\right\}$ is a subsequence of $\left\{\sqrt[n]{\frac{1}{3}}\right\}$ whose limit is 1 by Table 8.1

37. converges by the Ratio Test: $\lim\limits_{n\to\infty} \frac{a_{n+1}}{a_n} = \lim\limits_{n\to\infty} \frac{2^{n+1}(n+1)!(n+1)!}{(2n+2)!} \cdot \frac{(2n)!}{2^n n! n!} = \lim\limits_{n\to\infty} \frac{2(n+1)(n+1)}{(2n+2)(2n+1)}$

   $= \lim\limits_{n\to\infty} \frac{n+1}{2n+1} = \frac{1}{2} < 1$

39. diverges by the Root Test: $\lim\limits_{n\to\infty} \sqrt[n]{a_n} \equiv \lim\limits_{n\to\infty} \sqrt[n]{\frac{(n!)^n}{(n^n)^2}} = \lim\limits_{n\to\infty} \frac{n!}{n^2} = \infty > 1$

41. converges by the Root Test: $\lim\limits_{n\to\infty} \sqrt[n]{a_n} = \lim\limits_{n\to\infty} \sqrt[n]{\frac{n^n}{2^{n^2}}} = \lim\limits_{n\to\infty} \frac{n}{2^n} = \lim\limits_{n\to\infty} \frac{1}{2^n \ln 2} = 0 < 1$

43. converges by the Ratio Test: $\lim\limits_{n\to\infty} \frac{a_{n+1}}{a_n} = \lim\limits_{n\to\infty} \frac{1\cdot 3\cdots(2n-1)(2n+1)}{4^{n+1} 2^{n+1}(n+1)!} \cdot \frac{4^n 2^n n!}{1\cdot 3\cdots(2n-1)}$

   $= \lim\limits_{n\to\infty} \frac{2n+1}{(4\cdot 2)(n+1)} = \frac{1}{4} < 1$

45. Ratio: $\lim\limits_{n\to\infty} \frac{a_{n+1}}{a_n} = \lim\limits_{n\to\infty} \frac{1}{(n+1)^p} \cdot \frac{n^p}{1} = \lim\limits_{n\to\infty} \left(\frac{n}{n+1}\right)^p = 1^p = 1 \Rightarrow$ no conclusion

   Root: $\lim\limits_{n\to\infty} \sqrt[n]{a_n} = \lim\limits_{n\to\infty} \sqrt[n]{\frac{1}{n^p}} = \lim\limits_{n\to\infty} \frac{1}{(\sqrt[n]{n})^p} = \frac{1}{(1)^p} = 1 \Rightarrow$ no conclusion

47. $a_n \le \frac{n}{2^n}$ for every n and the series $\sum\limits_{n=1}^{\infty} \frac{n}{2^n}$ converges by the Ratio Test since $\lim\limits_{n\to\infty} \frac{(n+1)}{2^{n+1}} \cdot \frac{2^n}{n} = \frac{1}{2} < 1$

   $\Rightarrow \sum\limits_{n=1}^{\infty} a_n$ converges by the Direct Comparison Test

11.6 ALTERNATING SERIES, ABSOLUTE AND CONDITIONAL CONVERGENCE

1.   converges absolutely $\Rightarrow$ converges by the Absolute Convergence Test since $\sum\limits_{n=1}^{\infty} |a_n| = \sum\limits_{n=1}^{\infty} \frac{1}{n^2}$ which is a
     convergent p-series

3.   diverges by the nth-Term Test since for $n > 10 \Rightarrow \frac{n}{10} > 1 \Rightarrow \lim\limits_{n\to\infty} \left(\frac{n}{10}\right)^n \ne 0 \Rightarrow \sum\limits_{n=1}^{\infty} (-1)^{n+1}\left(\frac{n}{10}\right)^n$ diverges

5.   converges by the Alternating Series Test because $f(x) = \ln x$ is an increasing function of $x \Rightarrow \frac{1}{\ln x}$ is decreasing
     $\Rightarrow u_n \ge u_{n+1}$ for $n \ge 1$; also $u_n \ge 0$ for $n \ge 1$ and $\lim\limits_{n\to\infty} \frac{1}{\ln n} = 0$

7.   diverges by the nth-Term Test since $\lim\limits_{n\to\infty} \frac{\ln n}{\ln n^2} = \lim\limits_{n\to\infty} \frac{\ln n}{2\ln n} = \lim\limits_{n\to\infty} \frac{1}{2} = \frac{1}{2} \ne 0$

9.   converges by the Alternating Series Test since $f(x) = \frac{\sqrt{x}+1}{x+1} \Rightarrow f'(x) = \frac{1-x-2\sqrt{x}}{2\sqrt{x}(x+1)^2} < 0 \Rightarrow f(x)$ is decreasing
     $\Rightarrow u_n \ge u_{n+1}$; also $u_n \ge 0$ for $n \ge 1$ and $\lim\limits_{n\to\infty} u_n = \lim\limits_{n\to\infty} \frac{\sqrt{n}+1}{n+1} = 0$

11. converges absolutely since $\sum\limits_{n=1}^{\infty} |a_n| = \sum\limits_{n=1}^{\infty} \left(\frac{1}{10}\right)^n$ a convergent geometric series

13. converges conditionally since $\frac{1}{\sqrt{n}} > \frac{1}{\sqrt{n+1}} > 0$ and $\lim_{n \to \infty} \frac{1}{\sqrt{n}} = 0 \Rightarrow$ convergence; but $\sum_{n=1}^{\infty} |a_n| = \sum_{n=1}^{\infty} \frac{1}{n^{1/2}}$ is a divergent p-series

15. converges absolutely since $\sum_{n=1}^{\infty} |a_n| = \sum_{n=1}^{\infty} \frac{n}{n^3+1}$ and $\frac{n}{n^3+1} < \frac{1}{n^2}$ which is the nth-term of a converging p-series

17. converges conditionally since $\frac{1}{n+3} > \frac{1}{(n+1)+3} > 0$ and $\lim_{n \to \infty} \frac{1}{n+3} = 0 \Rightarrow$ convergence; but $\sum_{n=1}^{\infty} |a_n|$

    $= \sum_{n=1}^{\infty} \frac{1}{n+3}$ diverges because $\frac{1}{n+3} \geq \frac{1}{4n}$ and $\sum_{n=1}^{\infty} \frac{1}{n}$ is a divergent series

19. diverges by the nth-Term Test since $\lim_{n \to \infty} \frac{3+n}{5+n} = 1 \neq 0$

21. converges conditionally since $f(x) = \frac{1}{x^2} + \frac{1}{x} \Rightarrow f'(x) = -\left(\frac{2}{x^3} + \frac{1}{x^2}\right) < 0 \Rightarrow f(x)$ is decreasing and hence

    $u_n > u_{n+1} > 0$ for $n \geq 1$ and $\lim_{n \to \infty} \left(\frac{1}{n^2} + \frac{1}{n}\right) = 0 \Rightarrow$ convergence; but $\sum_{n=1}^{\infty} |a_n| = \sum_{n=1}^{\infty} \frac{1+n}{n^2}$

    $= \sum_{n=1}^{\infty} \frac{1}{n^2} + \sum_{n=1}^{\infty} \frac{1}{n}$ is the sum of a convergent and divergent series, and hence diverges

23. converges absolutely by the Ratio Test: $\lim_{n \to \infty} \left(\frac{u_{n+1}}{u_n}\right) = \lim_{n \to \infty} \left[\frac{(n+1)^2 \left(\frac{2}{3}\right)^{n+1}}{n^2 \left(\frac{2}{3}\right)^n}\right] = \frac{2}{3} < 1$

25. converges absolutely by the Integral Test since $\int_1^{\infty} (\tan^{-1} x) \left(\frac{1}{1+x^2}\right) dx = \lim_{b \to \infty} \left[\frac{(\tan^{-1} x)^2}{2}\right]_1^b$

    $= \lim_{b \to \infty} \left[(\tan^{-1} b)^2 - (\tan^{-1} 1)^2\right] = \frac{1}{2}\left[\left(\frac{\pi}{2}\right)^2 - \left(\frac{\pi}{4}\right)^2\right] = \frac{3\pi^2}{32}$

27. diverges by the nth-Term Test since $\lim_{n \to \infty} \frac{n}{n+1} = 1 \neq 0$

29. converges absolutely by the Ratio Test: $\lim_{n \to \infty} \left(\frac{u_{n+1}}{u_n}\right) = \lim_{n \to \infty} \frac{(100)^{n+1}}{(n+1)!} \cdot \frac{n!}{(100)^n} = \lim_{n \to \infty} \frac{100}{n+1} = 0 < 1$

31. converges absolutely by the Direct Comparison Test since $\sum_{n=1}^{\infty} |a_n| = \sum_{n=1}^{\infty} \frac{1}{n^2+2n+1}$ and

    $\frac{1}{n^2+2n+1} < \frac{1}{n^2}$ which is the nth-term of a convergent p-series

33. converges absolutely since $\sum_{n=1}^{\infty} |a_n| = \sum_{n=1}^{\infty} \left|\frac{(-1)^n}{n\sqrt{n}}\right| = \sum_{n=1}^{\infty} \frac{1}{n^{3/2}}$ is a convergent p-series

35. converges absolutely by the Root Test: $\lim_{n \to \infty} \sqrt[n]{|a_n|} = \lim_{n \to \infty} \left(\frac{(n+1)^n}{(2n)^n}\right)^{1/n} = \lim_{n \to \infty} \frac{n+1}{2n} = \frac{1}{2} < 1$

37. diverges by the nth-Term Test since $\lim_{n \to \infty} |a_n| = \lim_{n \to \infty} \frac{(2n)!}{2^n n! n} = \lim_{n \to \infty} \frac{(n+1)(n+2)\cdots(2n)}{2^n n}$

    $= \lim_{n \to \infty} \frac{(n+1)(n+2)\cdots(n+(n-1))}{2^{n-1}} > \lim_{n \to \infty} \left(\frac{n+1}{2}\right)^{n-1} = \infty \neq 0$

39. converges conditionally since $\frac{\sqrt{n+1}-\sqrt{n}}{1} \cdot \frac{\sqrt{n+1}+\sqrt{n}}{\sqrt{n+1}+\sqrt{n}} = \frac{1}{\sqrt{n+1}+\sqrt{n}}$ and $\left\{\frac{1}{\sqrt{n+1}+\sqrt{n}}\right\}$ is a

    decreasing sequence of positive terms which converges to $0 \Rightarrow \sum_{n=1}^{\infty} \frac{(-1)^n}{\sqrt{n+1}+\sqrt{n}}$ converges; but

$\sum_{n=1}^{\infty} |a_n| = \sum_{n=1}^{\infty} \frac{1}{\sqrt{n+1}+\sqrt{n}}$ diverges by the Limit Comparison Test (part 1) with $\frac{1}{\sqrt{n}}$; a divergent p-series:

$$\lim_{n \to \infty} \left(\frac{\frac{1}{\sqrt{n+1}+\sqrt{n}}}{\frac{1}{\sqrt{n}}} \right) = \lim_{n \to \infty} \frac{\sqrt{n}}{\sqrt{n+1}+\sqrt{n}} = \lim_{n \to \infty} \frac{1}{\sqrt{1+\frac{1}{n}}+1} = \frac{1}{2}$$

41. diverges by the nth-Term Test since $\lim_{n \to \infty} \left(\sqrt{n + \sqrt{n}} - \sqrt{n} \right) = \lim_{n \to \infty} \left[\left(\sqrt{n + \sqrt{n}} - \sqrt{n} \right) \left(\frac{\sqrt{n+\sqrt{n}}+\sqrt{n}}{\sqrt{n+\sqrt{n}}+\sqrt{n}} \right) \right]$

$= \lim_{n \to \infty} \frac{\sqrt{n}}{\sqrt{n+\sqrt{n}}+\sqrt{n}} = \lim_{n \to \infty} \frac{1}{\sqrt{1+\frac{1}{\sqrt{n}}}+1} = \frac{1}{2} \neq 0$

43. converges absolutely by the Direct Comparison Test since $\text{sech}\,(n) = \frac{2}{e^n + e^{-n}} = \frac{2e^n}{e^{2n}+1} < \frac{2e^n}{e^{2n}} = \frac{2}{e^n}$ which is the nth term of a convergent geometric series

45. $|\text{error}| < \left| (-1)^6 \left(\frac{1}{5} \right) \right| = 0.2$

47. $|\text{error}| < \left| (-1)^6 \frac{(0.01)^5}{5} \right| = 2 \times 10^{-11}$

49. $\frac{1}{(2n)!} < \frac{5}{10^6} \Rightarrow (2n)! > \frac{10^6}{5} = 200{,}000 \Rightarrow n \geq 5 \Rightarrow 1 - \frac{1}{2!} + \frac{1}{4!} - \frac{1}{6!} + \frac{1}{8!} \approx 0.54030$

51. (a) $a_n \geq a_{n+1}$ fails since $\frac{1}{3} < \frac{1}{2}$

(b) Since $\sum_{n=1}^{\infty} |a_n| = \sum_{n=1}^{\infty} \left[\left(\frac{1}{3} \right)^n + \left(\frac{1}{2} \right)^n \right] = \sum_{n=1}^{\infty} \left(\frac{1}{3} \right)^n + \sum_{n=1}^{\infty} \left(\frac{1}{2} \right)^n$ is the sum of two absolutely convergent

series, we can rearrange the terms of the original series to find its sum:

$$\left(\frac{1}{3} + \frac{1}{9} + \frac{1}{27} + \dots \right) - \left(\frac{1}{2} + \frac{1}{4} + \frac{1}{8} + \dots \right) = \frac{\left(\frac{1}{3} \right)}{1-\left(\frac{1}{3} \right)} - \frac{\left(\frac{1}{2} \right)}{1-\left(\frac{1}{2} \right)} = \frac{1}{2} - 1 = -\frac{1}{2}$$

53. The unused terms are $\sum_{j=n+1}^{\infty} (-1)^{j+1} a_j = (-1)^{n+1} (a_{n+1} - a_{n+2}) + (-1)^{n+3} (a_{n+3} - a_{n+4}) + \dots$

$= (-1)^{n+1} [(a_{n+1} - a_{n+2}) + (a_{n+3} - a_{n+4}) + \dots]$. Each grouped term is positive, so the remainder has the same sign as $(-1)^{n+1}$, which is the sign of the first unused term.

55. Theorem 16 states that $\sum_{n=1}^{\infty} |a_n|$ converges $\Rightarrow \sum_{n=1}^{\infty} a_n$ converges. But this is equivalent to $\sum_{n=1}^{\infty} a_n$ diverges $\Rightarrow \sum_{n=1}^{\infty} |a_n|$

diverges.

57. (a) $\sum_{n=1}^{\infty} |a_n + b_n|$ converges by the Direct Comparison Test since $|a_n + b_n| \leq |a_n| + |b_n|$ and hence

$\sum_{n=1}^{\infty} (a_n + b_n)$ converges absolutely

(b) $\sum_{n=1}^{\infty} |b_n|$ converges $\Rightarrow \sum_{n=1}^{\infty} -b_n$ converges absolutely; since $\sum_{n=1}^{\infty} a_n$ converges absolutely and

$\sum_{n=1}^{\infty} -b_n$ converges absolutely, we have $\sum_{n=1}^{\infty} [a_n + (-b_n)] = \sum_{n=1}^{\infty} (a_n - b_n)$ converges absolutely by part (a)

(c) $\sum_{n=1}^{\infty} |a_n|$ converges $\Rightarrow |k| \sum_{n=1}^{\infty} |a_n| = \sum_{n=1}^{\infty} |ka_n|$ converges $\Rightarrow \sum ka_n$ converges absolutely

59. $s_1 = -\frac{1}{2}, s_2 = -\frac{1}{2} + 1 = \frac{1}{2}$,

$s_3 = -\frac{1}{2} + 1 - \frac{1}{4} - \frac{1}{6} - \frac{1}{8} - \frac{1}{10} - \frac{1}{12} - \frac{1}{14} - \frac{1}{16} - \frac{1}{18} - \frac{1}{20} - \frac{1}{22} \approx -0.5099$,

$s_4 = s_3 + \frac{1}{3} \approx -0.1766,$

$s_5 = s_4 - \frac{1}{24} - \frac{1}{26} - \frac{1}{28} - \frac{1}{30} - \frac{1}{32} - \frac{1}{34} - \frac{1}{36} - \frac{1}{38} - \frac{1}{40} - \frac{1}{42} - \frac{1}{44} \approx -0.512,$

$s_6 = s_5 + \frac{1}{5} \approx -0.312,$

$s_7 = s_6 - \frac{1}{46} - \frac{1}{48} - \frac{1}{50} - \frac{1}{52} - \frac{1}{54} - \frac{1}{56} - \frac{1}{58} - \frac{1}{60} - \frac{1}{62} - \frac{1}{64} - \frac{1}{66} \approx -0.51106$

61. (a) If $\sum\limits_{n=1}^{\infty} |a_n|$ converges, then $\sum\limits_{n=1}^{\infty} a_n$ converges and $\frac{1}{2}\sum\limits_{n=1}^{\infty} a_n + \frac{1}{2}\sum\limits_{n=1}^{\infty} |a_n| = \sum\limits_{n=1}^{\infty} \frac{a_n + |a_n|}{2}$

    converges where $b_n = \frac{a_n + |a_n|}{2} = \begin{cases} a_n, & \text{if } a_n \geq 0 \\ 0, & \text{if } a_n < 0 \end{cases}.$

    (b) If $\sum\limits_{n=1}^{\infty} |a_n|$ converges, then $\sum\limits_{n=1}^{\infty} a_n$ converges and $\frac{1}{2}\sum\limits_{n=1}^{\infty} a_n - \frac{1}{2}\sum\limits_{n=1}^{\infty} |a_n| = \sum\limits_{n=1}^{\infty} \frac{a_n - |a_n|}{2}$

    converges where $c_n = \frac{a_n - |a_n|}{2} = \begin{cases} 0, & \text{if } a_n \geq 0 \\ a_n, & \text{if } a_n < 0 \end{cases}.$

63. Here is an example figure when $N = 5$. Notice that $u_3 > u_2 > u_1$ and $u_3 > u_5 > u_4$, but $u_n \geq u_{n+1}$ for $n \geq 5$.

11.7 POWER SERIES

1. $\lim\limits_{n \to \infty} \left| \frac{u_{n+1}}{u_n} \right| < 1 \Rightarrow \lim\limits_{n \to \infty} \left| \frac{x^{n+1}}{x^n} \right| < 1 \Rightarrow |x| < 1 \Rightarrow -1 < x < 1$; when $x = -1$ we have $\sum\limits_{n=1}^{\infty} (-1)^n$, a divergent series; when $x = 1$ we have $\sum\limits_{n=1}^{\infty} 1$, a divergent series

    (a) the radius is 1; the interval of convergence is $-1 < x < 1$

    (b) the interval of absolute convergence is $-1 < x < 1$

    (c) there are no values for which the series converges conditionally

3. $\lim\limits_{n \to \infty} \left|\frac{u_{n+1}}{u_n}\right| < 1 \Rightarrow \lim\limits_{n \to \infty} \left|\frac{(4x+1)^{n+1}}{(4x+1)^n}\right| < 1 \Rightarrow |4x+1| < 1 \Rightarrow -1 < 4x+1 < 1 \Rightarrow -\frac{1}{2} < x < 0$; when $x = -\frac{1}{2}$ we

have $\sum\limits_{n=1}^{\infty} (-1)^n(-1)^n = \sum\limits_{n=1}^{\infty} (-1)^{2n} = \sum\limits_{n=1}^{\infty} 1^n$, a divergent series; when $x = 0$ we have $\sum\limits_{n=1}^{\infty} (-1)^n(1)^n$

$= \sum\limits_{n=1}^{\infty} (-1)^n$, a divergent series

(a) the radius is $\frac{1}{4}$; the interval of convergence is $-\frac{1}{2} < x < 0$

(b) the interval of absolute convergence is $-\frac{1}{2} < x < 0$

(c) there are no values for which the series converges conditionally

5. $\lim\limits_{n \to \infty} \left|\frac{u_{n+1}}{u_n}\right| < 1 \Rightarrow \lim\limits_{n \to \infty} \left|\frac{(x-2)^{n+1}}{10^{n+1}} \cdot \frac{10^n}{(x-2)^n}\right| < 1 \Rightarrow \frac{|x-2|}{10} < 1 \Rightarrow |x-2| < 10 \Rightarrow -10 < x-2 < 10$

$\Rightarrow -8 < x < 12$; when $x = -8$ we have $\sum\limits_{n=1}^{\infty} (-1)^n$, a divergent series; when $x = 12$ we have $\sum\limits_{n=1}^{\infty} 1$, a divergent

series

(a) the radius is 10; the interval of convergence is $-8 < x < 12$

(b) the interval of absolute convergence is $-8 < x < 12$

(c) there are no values for which the series converges conditionally

7. $\lim\limits_{n \to \infty} \left|\frac{u_{n+1}}{u_n}\right| < 1 \Rightarrow \lim\limits_{n \to \infty} \left|\frac{(n+1)x^{n+1}}{(n+3)} \cdot \frac{(n+2)}{nx^n}\right| < 1 \Rightarrow |x| \lim\limits_{n \to \infty} \frac{(n+1)(n+2)}{(n+3)(n)} < 1 \Rightarrow |x| < 1$

$\Rightarrow -1 < x < 1$; when $x = -1$ we have $\sum\limits_{n=1}^{\infty} (-1)^n \frac{n}{n+2}$, a divergent series by the nth-term Test; when $x = 1$ we

have $\sum\limits_{n=1}^{\infty} \frac{n}{n+2}$, a divergent series

(a) the radius is 1; the interval of convergence is $-1 < x < 1$

(b) the interval of absolute convergence is $-1 < x < 1$

(c) there are no values for which the series converges conditionally

9. $\lim\limits_{n \to \infty} \left|\frac{u_{n+1}}{u_n}\right| < 1 \Rightarrow \lim\limits_{n \to \infty} \left|\frac{x^{n+1}}{(n+1)\sqrt{n+1}\,3^{n+1}} \cdot \frac{n\sqrt{n}\,3^n}{x^n}\right| < 1 \Rightarrow \frac{|x|}{3}\left(\lim\limits_{n \to \infty} \frac{n}{n+1}\right)\left(\sqrt{\lim\limits_{n \to \infty} \frac{n}{n+1}}\right) < 1$

$\Rightarrow \frac{|x|}{3}(1)(1) < 1 \Rightarrow |x| < 3 \Rightarrow -3 < x < 3$; when $x = -3$ we have $\sum\limits_{n=1}^{\infty} \frac{(-1)^n}{n^{3/2}}$, an absolutely convergent series;

when $x = 3$ we have $\sum\limits_{n=1}^{\infty} \frac{1}{n^{3/2}}$, a convergent p-series

(a) the radius is 3; the interval of convergence is $-3 \le x \le 3$

(b) the interval of absolute convergence is $-3 \le x \le 3$

(c) there are no values for which the series converges conditionally

11. $\lim\limits_{n \to \infty} \left|\frac{u_{n+1}}{u_n}\right| < 1 \Rightarrow \lim\limits_{n \to \infty} \left|\frac{x^{n+1}}{(n+1)!} \cdot \frac{n!}{x^n}\right| < 1 \Rightarrow |x| \lim\limits_{n \to \infty} \left(\frac{1}{n+1}\right) < 1$ for all x

(a) the radius is ∞; the series converges for all x

(b) the series converges absolutely for all x

(c) there are no values for which the series converges conditionally

13. $\lim\limits_{n \to \infty} \left|\frac{u_{n+1}}{u_n}\right| < 1 \Rightarrow \lim\limits_{n \to \infty} \left|\frac{x^{2n+3}}{(n+1)!} \cdot \frac{n!}{x^{2n+1}}\right| < 1 \Rightarrow x^2 \lim\limits_{n \to \infty} \left(\frac{1}{n+1}\right) < 1$ for all x

(a) the radius is ∞; the series converges for all x

(b) the series converges absolutely for all x

(c) there are no values for which the series converges conditionally

15. $\lim\limits_{n \to \infty} \left| \frac{u_{n+1}}{u_n} \right| < 1 \Rightarrow \lim\limits_{n \to \infty} \left| \frac{x^{n+1}}{\sqrt{(n+1)^2+3}} \cdot \frac{\sqrt{n^2+3}}{x^n} \right| < 1 \Rightarrow |x| \sqrt{\lim\limits_{n \to \infty} \frac{n^2+3}{n^2+2n+4}} < 1 \Rightarrow |x| < 1$

$\Rightarrow -1 < x < 1$; when $x = -1$ we have $\sum\limits_{n=1}^{\infty} \frac{(-1)^n}{\sqrt{n^2+3}}$, a conditionally convergent series; when $x = 1$ we have

$\sum\limits_{n=1}^{\infty} \frac{1}{\sqrt{n^2+3}}$, a divergent series

(a) the radius is 1; the interval of convergence is $-1 \le x < 1$

(b) the interval of absolute convergence is $-1 < x < 1$

(c) the series converges conditionally at $x = -1$

17. $\lim\limits_{n \to \infty} \left| \frac{u_{n+1}}{u_n} \right| < 1 \Rightarrow \lim\limits_{n \to \infty} \left| \frac{(n+1)(x+3)^{n+1}}{5^{n+1}} \cdot \frac{5^n}{n(x+3)^n} \right| < 1 \Rightarrow \frac{|x+3|}{5} \lim\limits_{n \to \infty} \left(\frac{n+1}{n} \right) < 1 \Rightarrow \frac{|x+3|}{5} < 1$

$\Rightarrow |x+3| < 5 \Rightarrow -5 < x+3 < 5 \Rightarrow -8 < x < 2$; when $x = -8$ we have $\sum\limits_{n=1}^{\infty} \frac{n(-5)^n}{5^n} = \sum\limits_{n=1}^{\infty} (-1)^n n$, a divergent

series; when $x = 2$ we have $\sum\limits_{n=1}^{\infty} \frac{n5^n}{5^n} = \sum\limits_{n=1}^{\infty} n$, a divergent series

(a) the radius is 5; the interval of convergence is $-8 < x < 2$

(b) the interval of absolute convergence is $-8 < x < 2$

(c) there are no values for which the series converges conditionally

19. $\lim\limits_{n \to \infty} \left| \frac{u_{n+1}}{u_n} \right| < 1 \Rightarrow \lim\limits_{n \to \infty} \left| \frac{\sqrt{n+1}\, x^{n+1}}{3^{n+1}} \cdot \frac{3^n}{\sqrt{n}\, x^n} \right| < 1 \Rightarrow \frac{|x|}{3} \sqrt{\lim\limits_{n \to \infty} \left(\frac{n+1}{n} \right)} < 1 \Rightarrow \frac{|x|}{3} < 1 \Rightarrow |x| < 3$

$\Rightarrow -3 < x < 3$; when $x = -3$ we have $\sum\limits_{n=1}^{\infty} (-1)^n \sqrt{n}$, a divergent series; when $x = 3$ we have

$\sum\limits_{n=1}^{\infty} \sqrt{n}$, a divergent series

(a) the radius is 3; the interval of convergence is $-3 < x < 3$

(b) the interval of absolute convergence is $-3 < x < 3$

(c) there are no values for which the series converges conditionally

21. $\lim\limits_{n \to \infty} \left| \frac{u_{n+1}}{u_n} \right| < 1 \Rightarrow \lim\limits_{n \to \infty} \left| \frac{\left(1+\frac{1}{n+1}\right)^{n+1} x^{n+1}}{\left(1+\frac{1}{n}\right)^n x^n} \right| < 1 \Rightarrow |x| \left(\frac{\lim\limits_{t \to \infty} \left(1+\frac{1}{t}\right)^t}{\lim\limits_{n \to \infty} \left(1+\frac{1}{n}\right)^n} \right) < 1 \Rightarrow |x| \left(\frac{e}{e} \right) < 1 \Rightarrow |x| < 1$

$\Rightarrow -1 < x < 1$; when $x = -1$ we have $\sum\limits_{n=1}^{\infty} (-1)^n \left(1+\frac{1}{n}\right)^n$, a divergent series by the nth-Term Test since

$\lim\limits_{n \to \infty} \left(1+\frac{1}{n}\right)^n = e \ne 0$; when $x = 1$ we have $\sum\limits_{n=1}^{\infty} \left(1+\frac{1}{n}\right)^n$, a divergent series

(a) the radius is 1; the interval of convergence is $-1 < x < 1$

(b) the interval of absolute convergence is $-1 < x < 1$

(c) there are no values for which the series converges conditionally

23. $\lim\limits_{n \to \infty} \left| \frac{u_{n+1}}{u_n} \right| < 1 \Rightarrow \lim\limits_{n \to \infty} \left| \frac{(n+1)^{n+1} x^{n+1}}{n^n x^n} \right| < 1 \Rightarrow |x| \left(\lim\limits_{n \to \infty} \left(1+\frac{1}{n}\right)^n \right) \left(\lim\limits_{n \to \infty} (n+1) \right) < 1$

$\Rightarrow e|x| \lim\limits_{n \to \infty} (n+1) < 1 \Rightarrow$ only $x = 0$ satisfies this inequality

(a) the radius is 0; the series converges only for $x = 0$

(b) the series converges absolutely only for $x = 0$

(c) there are no values for which the series converges conditionally

25. $\lim\limits_{n \to \infty} \left| \frac{u_{n+1}}{u_n} \right| < 1 \Rightarrow \lim\limits_{n \to \infty} \left| \frac{(x+2)^{n+1}}{(n+1)2^{n+1}} \cdot \frac{n2^n}{(x+2)^n} \right| < 1 \Rightarrow \frac{|x+2|}{2} \lim\limits_{n \to \infty} \left(\frac{n}{n+1} \right) < 1 \Rightarrow \frac{|x+2|}{2} < 1 \Rightarrow |x+2| < 2$

$\Rightarrow -2 < x+2 < 2 \Rightarrow -4 < x < 0$; when $x = -4$ we have $\sum\limits_{n=1}^{\infty} \frac{-1}{n}$, a divergent series; when $x = 0$ we have

$\sum\limits_{n=1}^{\infty} \frac{(-1)^{n+1}}{n}$, the alternating harmonic series which converges conditionally

(a) the radius is 2; the interval of convergence is $-4 < x \leq 0$

(b) the interval of absolute convergence is $-4 < x < 0$

(c) the series converges conditionally at $x = 0$

27. $\lim\limits_{n \to \infty} \left| \frac{u_{n+1}}{u_n} \right| < 1 \Rightarrow \lim\limits_{n \to \infty} \left| \frac{x^{n+1}}{(n+1)(\ln(n+1))^2} \cdot \frac{n(\ln n)^2}{x^n} \right| < 1 \Rightarrow |x| \left(\lim\limits_{n \to \infty} \frac{n}{n+1} \right) \left(\lim\limits_{n \to \infty} \frac{\ln n}{\ln(n+1)} \right)^2 < 1$

$\Rightarrow |x| (1) \left(\lim\limits_{n \to \infty} \frac{\left(\frac{1}{n}\right)}{\left(\frac{1}{n+1}\right)} \right)^2 < 1 \Rightarrow |x| \left(\lim\limits_{n \to \infty} \frac{n+1}{n} \right)^2 < 1 \Rightarrow |x| < 1 \Rightarrow -1 < x < 1$; when $x = -1$ we have

$\sum\limits_{n=1}^{\infty} \frac{(-1)^n}{n(\ln n)^2}$ which converges absolutely; when $x = 1$ we have $\sum\limits_{n=1}^{\infty} \frac{1}{n(\ln n)^2}$ which converges

(a) the radius is 1; the interval of convergence is $-1 \leq x \leq 1$

(b) the interval of absolute convergence is $-1 \leq x \leq 1$

(c) there are no values for which the series converges conditionally

29. $\lim\limits_{n \to \infty} \left| \frac{u_{n+1}}{u_n} \right| < 1 \Rightarrow \lim\limits_{n \to \infty} \left| \frac{(4x-5)^{2n+3}}{(n+1)^{3/2}} \cdot \frac{n^{3/2}}{(4x-5)^{2n+1}} \right| < 1 \Rightarrow (4x-5)^2 \left(\lim\limits_{n \to \infty} \frac{n}{n+1} \right)^{3/2} < 1 \Rightarrow (4x-5)^2 < 1$

$\Rightarrow |4x-5| < 1 \Rightarrow -1 < 4x - 5 < 1 \Rightarrow 1 < x < \frac{3}{2}$; when $x = 1$ we have $\sum\limits_{n=1}^{\infty} \frac{(-1)^{2n+1}}{n^{3/2}} = \sum\limits_{n=1}^{\infty} \frac{-1}{n^{3/2}}$ which is

absolutely convergent; when $x = \frac{3}{2}$ we have $\sum\limits_{n=1}^{\infty} \frac{(1)^{2n+1}}{n^{3/2}}$, a convergent p-series

(a) the radius is $\frac{1}{4}$; the interval of convergence is $1 \leq x \leq \frac{3}{2}$

(b) the interval of absolute convergence is $1 \leq x \leq \frac{3}{2}$

(c) there are no values for which the series converges conditionally

31. $\lim\limits_{n \to \infty} \left| \frac{u_{n+1}}{u_n} \right| < 1 \Rightarrow \lim\limits_{n \to \infty} \left| \frac{(x+\pi)^{n+1}}{\sqrt{n+1}} \cdot \frac{\sqrt{n}}{(x+\pi)^n} \right| < 1 \Rightarrow |x+\pi| \lim\limits_{n \to \infty} \left| \sqrt{\frac{n}{n+1}} \right| < 1$

$\Rightarrow |x+\pi| \sqrt{\lim\limits_{n \to \infty} \left(\frac{n}{n+1} \right)} < 1 \Rightarrow |x+\pi| < 1 \Rightarrow -1 < x + \pi < 1 \Rightarrow -1 - \pi < x < 1 - \pi$;

when $x = -1 - \pi$ we have $\sum\limits_{n=1}^{\infty} \frac{(-1)^n}{\sqrt{n}} = \sum\limits_{n=1}^{\infty} \frac{(-1)^n}{n^{1/2}}$, a conditionally convergent series; when $x = 1 - \pi$ we have

$\sum\limits_{n=1}^{\infty} \frac{1^n}{\sqrt{n}} = \sum\limits_{n=1}^{\infty} \frac{1}{n^{1/2}}$, a divergent p-series

(a) the radius is 1; the interval of convergence is $(-1 - \pi) \leq x < (1 - \pi)$

(b) the interval of absolute convergence is $-1 - \pi < x < 1 - \pi$

(c) the series converges conditionally at $x = -1 - \pi$

33. $\lim\limits_{n \to \infty} \left| \frac{u_{n+1}}{u_n} \right| < 1 \Rightarrow \lim\limits_{n \to \infty} \left| \frac{(x-1)^{2n+2}}{4^{n+1}} \cdot \frac{4^n}{(x-1)^{2n}} \right| < 1 \Rightarrow \frac{(x-1)^2}{4} \lim\limits_{n \to \infty} |1| < 1 \Rightarrow (x-1)^2 < 4 \Rightarrow |x-1| < 2$

$\Rightarrow -2 < x - 1 < 2 \Rightarrow -1 < x < 3$; at $x = -1$ we have $\sum\limits_{n=0}^{\infty} \frac{(-2)^{2n}}{4^n} = \sum\limits_{n=0}^{\infty} \frac{4^n}{4^n} = \sum\limits_{n=0}^{\infty} 1$, which diverges; at $x = 3$

we have $\sum\limits_{n=0}^{\infty} \frac{2^{2n}}{4^n} = \sum\limits_{n=0}^{\infty} \frac{4^n}{4^n} = \sum\limits_{n=0}^{\infty} 1$, a divergent series; the interval of convergence is $-1 < x < 3$; the series

$\sum\limits_{n=0}^{\infty} \frac{(x-1)^{2n}}{4^n} = \sum\limits_{n=0}^{\infty} \left(\left(\frac{x-1}{2} \right)^2 \right)^n$ is a convergent geometric series when $-1 < x < 3$ and the sum is

$\frac{1}{1 - \left(\frac{x-1}{2} \right)^2} = \frac{1}{\left[\frac{4 - (x-1)^2}{4} \right]} = \frac{4}{4 - x^2 + 2x - 1} = \frac{4}{3 + 2x - x^2}$

35. $\lim\limits_{n \to \infty} \left| \frac{u_{n+1}}{u_n} \right| < 1 \Rightarrow \lim\limits_{n \to \infty} \left| \frac{(\sqrt{x}-2)^{n+1}}{2^{n+1}} \cdot \frac{2^n}{(\sqrt{x}-2)^n} \right| < 1 \Rightarrow \left| \sqrt{x} - 2 \right| < 2 \Rightarrow -2 < \sqrt{x} - 2 < 2 \Rightarrow 0 < \sqrt{x} < 4$

$\Rightarrow 0 < x < 16$; when $x = 0$ we have $\sum\limits_{n=0}^{\infty} (-1)^n$, a divergent series; when $x = 16$ we have $\sum\limits_{n=0}^{\infty} (1)^n$, a divergent

series; the interval of convergence is $0 < x < 16$; the series $\sum\limits_{n=0}^{\infty} \left(\frac{\sqrt{x}-2}{2} \right)^n$ is a convergent geometric series when

$0 < x < 16$ and its sum is $\dfrac{1}{1 - \left(\frac{\sqrt{x}-2}{2} \right)} = \dfrac{1}{\left(\frac{2-\sqrt{x}+2}{2} \right)} = \dfrac{2}{4 - \sqrt{x}}$

37. $\lim\limits_{n \to \infty} \left| \frac{u_{n+1}}{u_n} \right| < 1 \Rightarrow \lim\limits_{n \to \infty} \left| \left(\frac{x^2+1}{3} \right)^{n+1} \cdot \left(\frac{3}{x^2+1} \right)^n \right| < 1 \Rightarrow \frac{(x^2+1)}{3} \lim\limits_{n \to \infty} |1| < 1 \Rightarrow \frac{x^2+1}{3} < 1 \Rightarrow x^2 < 2$

$\Rightarrow |x| < \sqrt{2} \Rightarrow -\sqrt{2} < x < \sqrt{2}$; at $x = \pm\sqrt{2}$ we have $\sum\limits_{n=0}^{\infty} (1)^n$ which diverges; the interval of convergence is

$-\sqrt{2} < x < \sqrt{2}$; the series $\sum\limits_{n=0}^{\infty} \left(\frac{x^2+1}{3} \right)^n$ is a convergent geometric series when $-\sqrt{2} < x < \sqrt{2}$ and its sum is

$\dfrac{1}{1 - \left(\frac{x^2+1}{3} \right)} = \dfrac{1}{\left(\frac{3-x^2-1}{3} \right)} = \dfrac{3}{2-x^2}$

39. $\lim\limits_{n \to \infty} \left| \frac{(x-3)^{n+1}}{2^{n+1}} \cdot \frac{2^n}{(x-3)^n} \right| < 1 \Rightarrow |x - 3| < 2 \Rightarrow 1 < x < 5$; when $x = 1$ we have $\sum\limits_{n=1}^{\infty} (1)^n$ which diverges;

when $x = 5$ we have $\sum\limits_{n=1}^{\infty}(-1)^n$ which also diverges; the interval of convergence is $1 < x < 5$; the sum of this

convergent geometric series is $\dfrac{1}{1 + \left(\frac{x-3}{2} \right)} = \dfrac{2}{x-1}$. If $f(x) = 1 - \frac{1}{2}(x-3) + \frac{1}{4}(x-3)^2 + \ldots + \left(-\frac{1}{2} \right)^n (x-3)^n + \ldots$

$= \frac{2}{x-1}$ then $f'(x) = -\frac{1}{2} + \frac{1}{2}(x-3) + \ldots + \left(-\frac{1}{2} \right)^n n(x-3)^{n-1} + \ldots$ is convergent when $1 < x < 5$, and diverges

when $x = 1$ or 5. The sum for $f'(x)$ is $\frac{-2}{(x-1)^2}$, the derivative of $\frac{2}{x-1}$.

41. (a) Differentiate the series for $\sin x$ to get $\cos x = 1 - \frac{3x^2}{3!} + \frac{5x^4}{5!} - \frac{7x^6}{7!} + \frac{9x^8}{9!} - \frac{11x^{10}}{11!} + \ldots$

$= 1 - \frac{x^2}{2!} + \frac{x^4}{4!} - \frac{x^6}{6!} + \frac{x^8}{8!} - \frac{x^{10}}{10!} + \ldots$. The series converges for all values of x since

$\lim\limits_{n \to \infty} \left| \frac{x^{2n+2}}{(2n+2)!} \cdot \frac{(2n)!}{x^{2n}} \right| = x^2 \lim\limits_{n \to \infty} \left(\frac{1}{(2n+1)(2n+2)} \right) = 0 < 1$ for all x.

(b) $\sin 2x = 2x - \frac{2^3 x^3}{3!} + \frac{2^5 x^5}{5!} - \frac{2^7 x^7}{7!} + \frac{2^9 x^9}{9!} - \frac{2^{11} x^{11}}{11!} + \ldots = 2x - \frac{8x^3}{3!} + \frac{32x^5}{5!} - \frac{128x^7}{7!} + \frac{512x^9}{9!} - \frac{2048x^{11}}{11!} + \ldots$

(c) $2 \sin x \cos x = 2 \left[(0 \cdot 1) + (0 \cdot 0 + 1 \cdot 1)x + \left(0 \cdot \frac{-1}{2} + 1 \cdot 0 + 0 \cdot 1 \right) x^2 + \left(0 \cdot 0 - 1 \cdot \frac{1}{2} + 0 \cdot 0 - 1 \cdot \frac{1}{3!} \right) x^3 \right.$

$+ \left(0 \cdot \frac{1}{4!} + 1 \cdot 0 - 0 \cdot \frac{1}{2} - 0 \cdot \frac{1}{3!} + 0 \cdot 1 \right) x^4 + \left(0 \cdot 0 + 1 \cdot \frac{1}{4!} + 0 \cdot 0 + \frac{1}{2} \cdot \frac{1}{3!} + 0 \cdot 0 + 1 \cdot \frac{1}{5!} \right) x^5$

$+ \left. \left(0 \cdot \frac{1}{6!} + 1 \cdot 0 + 0 \cdot \frac{1}{4!} + 0 \cdot \frac{1}{3!} + 0 \cdot \frac{1}{2} + 0 \cdot \frac{1}{5!} + 0 \cdot 1 \right) x^6 + \ldots \right] = 2 \left[x - \frac{4x^3}{3!} + \frac{16x^5}{5!} - \ldots \right]$

$= 2x - \frac{2^3 x^3}{3!} + \frac{2^5 x^5}{5!} - \frac{2^7 x^7}{7!} + \frac{2^9 x^9}{9!} - \frac{2^{11} x^{11}}{11!} + \ldots$

43. (a) $\ln |\sec x| + C = \int \tan x \, dx = \int \left(x + \frac{x^3}{3} + \frac{2x^5}{15} + \frac{17x^7}{315} + \frac{62x^9}{2835} + \ldots \right) dx$

$= \frac{x^2}{2} + \frac{x^4}{12} + \frac{x^6}{45} + \frac{17x^8}{2520} + \frac{31x^{10}}{14,175} + \ldots + C; x = 0 \Rightarrow C = 0 \Rightarrow \ln |\sec x| = \frac{x^2}{2} + \frac{x^4}{12} + \frac{x^6}{45} + \frac{17x^8}{2520} + \frac{31x^{10}}{14,175} + \ldots,$

converges when $-\frac{\pi}{2} < x < \frac{\pi}{2}$

(b) $\sec^2 x = \frac{d(\tan x)}{dx} = \frac{d}{dx} \left(x + \frac{x^3}{3} + \frac{2x^5}{15} + \frac{17x^7}{315} + \frac{62x^9}{2835} + \ldots \right) = 1 + x^2 + \frac{2x^4}{3} + \frac{17x^6}{45} + \frac{62x^8}{315} + \ldots$, converges

when $-\frac{\pi}{2} < x < \frac{\pi}{2}$

(c) $\sec^2 x = (\sec x)(\sec x) = \left(1 + \frac{x^2}{2} + \frac{5x^4}{24} + \frac{61x^6}{720} + \ldots \right) \left(1 + \frac{x^2}{2} + \frac{5x^4}{24} + \frac{61x^6}{720} + \ldots \right)$

$= 1 + \left(\frac{1}{2} + \frac{1}{2} \right) x^2 + \left(\frac{5}{24} + \frac{1}{4} + \frac{5}{24} \right) x^4 + \left(\frac{61}{720} + \frac{5}{48} + \frac{5}{48} + \frac{61}{720} \right) x^6 + \ldots$

$= 1 + x^2 + \frac{2x^4}{3} + \frac{17x^6}{45} + \frac{62x^8}{315} + \ldots, -\frac{\pi}{2} < x < \frac{\pi}{2}$

45. (a) If $f(x) = \sum\limits_{n=0}^{\infty} a_n x^n$, then $f^{(k)}(x) = \sum\limits_{n=k}^{\infty} n(n-1)(n-2)\cdots(n-(k-1)) a_n x^{n-k}$ and $f^{(k)}(0) = k! a_k$

  $\Rightarrow a_k = \frac{f^{(k)}(0)}{k!}$; likewise if $f(x) = \sum\limits_{n=0}^{\infty} b_n x^n$, then $b_k = \frac{f^{(k)}(0)}{k!} \Rightarrow a_k = b_k$ for every nonnegative integer k

   (b) If $f(x) = \sum\limits_{n=0}^{\infty} a_n x^n = 0$ for all x, then $f^{(k)}(x) = 0$ for all x $\Rightarrow$ from part (a) that $a_k = 0$ for every

   nonnegative integer k

47. The series $\sum\limits_{n=1}^{\infty} \frac{x^n}{n}$ converges conditionally at the left-hand endpoint of its interval of convergence $[-1, 1)$; the

   series $\sum\limits_{n=1}^{\infty} \frac{x^n}{(n^2)}$ converges absolutely at the left-hand endpoint of its interval of convergence $[-1, 1]$

11.8 TAYLOR AND MACLAURIN SERIES

1. $f(x) = \ln x, f'(x) = \frac{1}{x}, f''(x) = -\frac{1}{x^2}, f'''(x) = \frac{2}{x^3}; f(1) = \ln 1 = 0, f'(1) = 1, f''(1) = -1, \ f'''(1) = 2 \Rightarrow P_0(x) = 0,$
   $P_1(x) = (x-1), P_2(x) = (x-1) - \frac{1}{2}(x-1)^2, P_3(x) = (x-1) - \frac{1}{2}(x-1)^2 + \frac{1}{3}(x-1)^3$

3. $f(x) = \frac{1}{x} = x^{-1}, f'(x) = -x^{-2}, f''(x) = 2x^{-3}, f'''(x) = -6x^{-4}; f(2) = \frac{1}{2}, f'(2) = -\frac{1}{4}, f''(2) = \frac{1}{4}, f'''(x) = -\frac{3}{8}$
   $\Rightarrow P_0(x) = \frac{1}{2}, P_1(x) = \frac{1}{2} - \frac{1}{4}(x-2), P_2(x) = \frac{1}{2} - \frac{1}{4}(x-2) + \frac{1}{8}(x-2)^2,$
   $P_3(x) = \frac{1}{2} - \frac{1}{4}(x-2) + \frac{1}{8}(x-2)^2 - \frac{1}{16}(x-2)^3$

5. $f(x) = \sin x, f'(x) = \cos x, f''(x) = -\sin x, f'''(x) = -\cos x; f\left(\frac{\pi}{4}\right) = \sin\frac{\pi}{4} = \frac{\sqrt{2}}{2}, f'\left(\frac{\pi}{4}\right) = \cos\frac{\pi}{4} = \frac{\sqrt{2}}{2},$
   $f''\left(\frac{\pi}{4}\right) = -\sin\frac{\pi}{4} = -\frac{\sqrt{2}}{2}, f'''\left(\frac{\pi}{4}\right) = -\cos\frac{\pi}{4} = -\frac{\sqrt{2}}{2} \Rightarrow P_0 = \frac{\sqrt{2}}{2}, P_1(x) = \frac{\sqrt{2}}{2} + \frac{\sqrt{2}}{2}\left(x - \frac{\pi}{4}\right),$
   $P_2(x) = \frac{\sqrt{2}}{2} + \frac{\sqrt{2}}{2}\left(x - \frac{\pi}{4}\right) - \frac{\sqrt{2}}{4}\left(x - \frac{\pi}{4}\right)^2, P_3(x) = \frac{\sqrt{2}}{2} + \frac{\sqrt{2}}{2}\left(x - \frac{\pi}{4}\right) - \frac{\sqrt{2}}{4}\left(x - \frac{\pi}{4}\right)^2 - \frac{\sqrt{2}}{12}\left(x - \frac{\pi}{4}\right)^3$

7. $f(x) = \sqrt{x} = x^{1/2}, f'(x) = \left(\frac{1}{2}\right)x^{-1/2}, f''(x) = \left(-\frac{1}{4}\right)x^{-3/2}, f'''(x) = \left(\frac{3}{8}\right)x^{-5/2}; f(4) = \sqrt{4} = 2,$
   $f'(4) = \left(\frac{1}{2}\right)4^{-1/2} = \frac{1}{4}, f''(4) = \left(-\frac{1}{4}\right)4^{-3/2} = -\frac{1}{32}, f'''(4) = \left(\frac{3}{8}\right)4^{-5/2} = \frac{3}{256} \Rightarrow P_0(x) = 2, P_1(x) = 2 + \frac{1}{4}(x-4),$
   $P_2(x) = 2 + \frac{1}{4}(x-4) - \frac{1}{64}(x-4)^2, P_3(x) = 2 + \frac{1}{4}(x-4) - \frac{1}{64}(x-4)^2 + \frac{1}{512}(x-4)^3$

9. $e^x = \sum\limits_{n=0}^{\infty} \frac{x^n}{n!} \Rightarrow e^{-x} = \sum\limits_{n=0}^{\infty} \frac{(-x)^n}{n!} = 1 - x + \frac{x^2}{2!} - \frac{x^3}{3!} + \frac{x^4}{4!} - \cdots$

11. $f(x) = (1+x)^{-1} \Rightarrow f'(x) = -(1+x)^{-2}, f''(x) = 2(1+x)^{-3}, f'''(x) = -3!(1+x)^{-4} \Rightarrow \ \ldots \ f^{(k)}(x)$
   $= (-1)^k k!(1+x)^{-k-1}; f(0) = 1, f'(0) = -1, f''(0) = 2, f'''(0) = -3!, \ldots, f^{(k)}(0) = (-1)^k k!$
   $\Rightarrow \frac{1}{1+x} = 1 - x + x^2 - x^3 + \cdots = \sum\limits_{n=0}^{\infty}(-x)^n = \sum\limits_{n=0}^{\infty}(-1)^n x^n$

13. $\sin x = \sum\limits_{n=0}^{\infty} \frac{(-1)^n x^{2n+1}}{(2n+1)!} \Rightarrow \sin 3x = \sum\limits_{n=0}^{\infty} \frac{(-1)^n(3x)^{2n+1}}{(2n+1)!} = \sum\limits_{n=0}^{\infty} \frac{(-1)^n 3^{2n+1} x^{2n+1}}{(2n+1)!} = 3x - \frac{3^3 x^3}{3!} + \frac{3^5 x^5}{5!} - \cdots$

15. $7\cos(-x) = 7\cos x = 7\sum\limits_{n=0}^{\infty} \frac{(-1)^n x^{2n}}{(2n)!} = 7 - \frac{7x^2}{2!} + \frac{7x^4}{4!} - \frac{7x^6}{6!} + \cdots$, since the cosine is an even function

17. $\cosh x = \frac{e^x + e^{-x}}{2} = \frac{1}{2}\left[\left(1 + x^2 + \frac{x^2}{2!} + \frac{x^3}{3!} + \frac{x^4}{4!} + \cdots\right) + \left(1 - x + \frac{x^2}{2!} - \frac{x^3}{3!} + \frac{x^4}{4!} - \cdots\right)\right] = 1 + \frac{x^2}{2!} + \frac{x^4}{4!} + \frac{x^6}{6!} + \cdots$
   $= \sum\limits_{n=0}^{\infty} \frac{x^{2n}}{(2n)!}$

19. $f(x) = x^4 - 2x^3 - 5x + 4 \Rightarrow f'(x) = 4x^3 - 6x^2 - 5, f''(x) = 12x^2 - 12x, f'''(x) = 24x - 12, f^{(4)}(x) = 24$
$\Rightarrow f^{(n)}(x) = 0$ if $n \geq 5$; $f(0) = 4, f'(0) = -5, f''(0) = 0, f'''(0) = -12, f^{(4)}(0) = 24, f^{(n)}(0) = 0$ if $n \geq 5$
$\Rightarrow x^4 - 2x^3 - 5x + 4 = 4 - 5x - \frac{12}{3!}x^3 + \frac{24}{4!}x^4 = x^4 - 2x^3 - 5x + 4$ itself

21. $f(x) = x^3 - 2x + 4 \Rightarrow f'(x) = 3x^2 - 2, f''(x) = 6x, f'''(x) = 6 \Rightarrow f^{(n)}(x) = 0$ if $n \geq 4$; $f(2) = 8, f'(2) = 10$,
$f''(2) = 12, f'''(2) = 6, f^{(n)}(2) = 0$ if $n \geq 4 \Rightarrow x^3 - 2x + 4 = 8 + 10(x - 2) + \frac{12}{2!}(x - 2)^2 + \frac{6}{3!}(x - 2)^3$
$= 8 + 10(x - 2) + 6(x - 2)^2 + (x - 2)^3$

23. $f(x) = x^4 + x^2 + 1 \Rightarrow f'(x) = 4x^3 + 2x, f''(x) = 12x^2 + 2, f'''(x) = 24x, f^{(4)}(x) = 24, f^{(n)}(x) = 0$ if $n \geq 5$;
$f(-2) = 21, f'(-2) = -36, f''(-2) = 50, f'''(-2) = -48, f^{(4)}(-2) = 24, f^{(n)}(-2) = 0$ if $n \geq 5 \Rightarrow x^4 + x^2 + 1$
$= 21 - 36(x + 2) + \frac{50}{2!}(x + 2)^2 - \frac{48}{3!}(x + 2)^3 + \frac{24}{4!}(x + 2)^4 = 21 - 36(x + 2) + 25(x + 2)^2 - 8(x + 2)^3 + (x + 2)^4$

25. $f(x) = x^{-2} \Rightarrow f'(x) = -2x^{-3}, f''(x) = 3!\,x^{-4}, f'''(x) = -4!\,x^{-5} \Rightarrow f^{(n)}(x) = (-1)^n(n + 1)!\,x^{-n-2}$;
$f(1) = 1, f'(1) = -2, f''(1) = 3!, f'''(1) = -4!, f^{(n)}(1) = (-1)^n(n + 1)! \Rightarrow \frac{1}{x^2}$
$= 1 - 2(x - 1) + 3(x - 1)^2 - 4(x - 1)^3 + \ldots = \sum_{n=0}^{\infty} (-1)^n(n + 1)(x - 1)^n$

27. $f(x) = e^x \Rightarrow f'(x) = e^x, f''(x) = e^x \Rightarrow f^{(n)}(x) = e^x$; $f(2) = e^2, f'(2) = e^2, \ldots f^{(n)}(2) = e^2$
$\Rightarrow e^x = e^2 + e^2(x - 2) + \frac{e^2}{2}(x - 2)^2 + \frac{e^3}{3!}(x - 2)^3 + \ldots = \sum_{n=0}^{\infty} \frac{e^2}{n!}(x - 2)^n$

29. If $e^x = \sum_{n=0}^{\infty} \frac{f^{(n)}(a)}{n!}(x - a)^n$ and $f(x) = e^x$, we have $f^{(n)}(a) = e^a$ f or all $n = 0, 1, 2, 3, \ldots$
$\Rightarrow e^x = e^a\left[\frac{(x - a)^0}{0!} + \frac{(x - a)^1}{1!} + \frac{(x - a)^2}{2!} + \ldots\right] = e^a\left[1 + (x - a) + \frac{(x - a)^2}{2!} + \ldots\right]$ at $x = a$

31. $f(x) = f(a) + f'(a)(x - a) + \frac{f''(a)}{2}(x - a)^2 + \frac{f'''(a)}{3!}(x - a)^3 + \ldots \Rightarrow f'(x)$
$= f'(a) + f''(a)(x - a) + \frac{f'''(a)}{3!}3(x - a)^2 + \ldots \Rightarrow f''(x) = f''(a) + f'''(a)(x - a) + \frac{f^{(4)}(a)}{4!}4 \cdot 3(x - a)^2 + \ldots$
$\Rightarrow f^{(n)}(x) = f^{(n)}(a) + f^{(n+1)}(a)(x - a) + \frac{f^{(n+2)}(a)}{2}(x - a)^2 + \ldots$
$\Rightarrow f(a) = f(a) + 0, f'(a) = f'(a) + 0, \ldots, f^{(n)}(a) = f^{(n)}(a) + 0$

33. $f(x) = \ln(\cos x) \Rightarrow f'(x) = -\tan x$ and $f''(x) = -\sec^2 x$; $f(0) = 0, f'(0) = 0, f''(0) = -1$
$\Rightarrow L(x) = 0$ and $Q(x) = -\frac{x^2}{2}$

35. $f(x) = (1 - x^2)^{-1/2} \Rightarrow f'(x) = x(1 - x^2)^{-3/2}$ and $f''(x) = (1 - x^2)^{-3/2} + 3x^2(1 - x^2)^{-5/2}$; $f(0) = 1$,
$f'(0) = 0, f''(0) = 1 \Rightarrow L(x) = 1$ and $Q(x) = 1 + \frac{x^2}{2}$

37. $f(x) = \sin x \Rightarrow f'(x) = \cos x$ and $f''(x) = -\sin x$; $f(0) = 0, f'(0) = 1, f''(0) = 0 \Rightarrow L(x) = x$ and $Q(x) = x$

11.9 CONVERGENCE OF TAYLOR SERIES; ERROR ESTIMATES

1. $e^x = 1 + x + \frac{x^2}{2!} + \ldots = \sum_{n=0}^{\infty} \frac{x^n}{n!} \Rightarrow e^{-5x} = 1 + (-5x) + \frac{(-5x)^2}{2!} + \ldots = 1 - 5x + \frac{5^2 x^2}{2!} - \frac{5^3 x^3}{3!} + \ldots = \sum_{n=0}^{\infty} \frac{(-1)^n 5^n x^n}{n!}$

3. $\sin x = x - \frac{x^3}{3!} + \frac{x^5}{5!} - \ldots = \sum_{n=0}^{\infty} \frac{(-1)^n x^{2n+1}}{(2n+1)!} \Rightarrow 5\sin(-x) = 5\left[(-x) - \frac{(-x)^3}{3!} + \frac{(-x)^5}{5!} - \ldots\right]$
$= \sum_{n=0}^{\infty} \frac{5(-1)^{n+1} x^{2n+1}}{(2n+1)!}$

5. $\cos x = \sum\limits_{n=0}^{\infty} \frac{(-1)^n x^{2n}}{(2n)!} \Rightarrow \cos\sqrt{x+1} = \sum\limits_{n=0}^{\infty} \frac{(-1)^n \left[(x+1)^{1/2}\right]^{2n}}{(2n)!} = \sum\limits_{n=0}^{\infty} \frac{(-1)^n (x+1)^n}{(2n)!} = 1 - \frac{x+1}{2!} + \frac{(x+1)^2}{4!} - \frac{(x+1)^3}{6!} + \cdots$

7. $e^x = \sum\limits_{n=0}^{\infty} \frac{x^n}{n!} \Rightarrow xe^x = x\left(\sum\limits_{n=0}^{\infty} \frac{x^n}{n!}\right) = \sum\limits_{n=0}^{\infty} \frac{x^{n+1}}{n!} = x + x^2 + \frac{x^3}{2!} + \frac{x^4}{3!} + \frac{x^5}{4!} + \cdots$

9. $\cos x = \sum\limits_{n=0}^{\infty} \frac{(-1)^n x^{2n}}{(2n)!} \Rightarrow \frac{x^2}{2} - 1 + \cos x = \frac{x^2}{2} - 1 + \sum\limits_{n=0}^{\infty} \frac{(-1)^n x^{2n}}{(2n)!} = \frac{x^2}{2} - 1 + 1 - \frac{x^2}{2} + \frac{x^4}{4!} - \frac{x^6}{6!} + \frac{x^8}{8!} - \frac{x^{10}}{10!} + \cdots$

$= \frac{x^4}{4!} - \frac{x^6}{6!} + \frac{x^8}{8!} - \frac{x^{10}}{10!} + \cdots = \sum\limits_{n=2}^{\infty} \frac{(-1)^n x^{2n}}{(2n)!}$

11. $\cos x = \sum\limits_{n=0}^{\infty} \frac{(-1)^n x^{2n}}{(2n)!} \Rightarrow x\cos\pi x = x\sum\limits_{n=0}^{\infty} \frac{(-1)^n (\pi x)^{2n}}{(2n)!} = \sum\limits_{n=0}^{\infty} \frac{(-1)^n \pi^{2n} x^{2n+1}}{(2n)!} = x - \frac{\pi^2 x^3}{2!} + \frac{\pi^4 x^5}{4!} - \frac{\pi^6 x^7}{6!} + \cdots$

13. $\cos^2 x = \frac{1}{2} + \frac{\cos 2x}{2} = \frac{1}{2} + \frac{1}{2}\sum\limits_{n=0}^{\infty} \frac{(-1)^n (2x)^{2n}}{(2n)!} = \frac{1}{2} + \frac{1}{2}\left[1 - \frac{(2x)^2}{2!} + \frac{(2x)^4}{4!} - \frac{(2x)^6}{6!} + \frac{(2x)^8}{8!} - \cdots\right]$

$= 1 - \frac{(2x)^2}{2\cdot 2!} + \frac{(2x)^4}{2\cdot 4!} - \frac{(2x)^6}{2\cdot 6!} + \frac{(2x)^8}{2\cdot 8!} - \cdots = 1 + \sum\limits_{n=1}^{\infty} \frac{(-1)^n (2x)^{2n}}{2\cdot(2n)!} = 1 + \sum\limits_{n=1}^{\infty} \frac{(-1)^n 2^{2n-1} x^{2n}}{(2n)!}$

15. $\frac{x^2}{1-2x} = x^2\left(\frac{1}{1-2x}\right) = x^2\sum\limits_{n=0}^{\infty} (2x)^n = \sum\limits_{n=0}^{\infty} 2^n x^{n+2} = x^2 + 2x^3 + 2^2 x^4 + 2^3 x^5 + \cdots$

17. $\frac{1}{1-x} = \sum\limits_{n=0}^{\infty} x^n = 1 + x + x^2 + x^3 + \cdots \Rightarrow \frac{d}{dx}\left(\frac{1}{1-x}\right) = \frac{1}{(1-x)^2} = 1 + 2x + 3x^2 + \cdots = \sum\limits_{n=1}^{\infty} nx^{n-1}$

$= \sum\limits_{n=0}^{\infty} (n+1)x^n$

19. By the Alternating Series Estimation Theorem, the error is less than $\frac{|x|^5}{5!} \Rightarrow |x|^5 < (5!)\,(5\times 10^{-4})$

$\Rightarrow |x|^5 < 600\times 10^{-4} \Rightarrow |x| < \sqrt[5]{6\times 10^{-2}} \approx 0.56968$

21. If $\sin x = x$ and $|x| < 10^{-3}$, then the error is less than $\frac{(10^{-3})^3}{3!} \approx 1.67\times 10^{-10}$, by Alternating Series Estimation Theorem;
The Alternating Series Estimation Theorem says $R_2(x)$ has the same sign as $-\frac{x^3}{3!}$. Moreover, $x < \sin x$
$\Rightarrow 0 < \sin x - x = R_2(x) \Rightarrow x < 0 \Rightarrow -10^{-3} < x < 0.$

23. $|R_2(x)| = \left|\frac{e^c x^3}{3!}\right| < \frac{3^{(0.1)}(0.1)^3}{3!} < 1.87\times 10^{-4}$, where c is between 0 and x

25. $|R_4(x)| < \left|\frac{\cosh c}{5!}x^5\right| = \left|\frac{e^c + e^{-c}}{2}\frac{x^5}{5!}\right| < \frac{1.65 + \frac{1}{1.65}}{2}\cdot\frac{(0.5)^5}{5!} = (1.13)\frac{(0.5)^5}{5!} \approx 0.000294$

27. $|R_1| = \left|\frac{1}{(1+c)^2}\frac{x^2}{2!}\right| < \frac{x^2}{2} = \left|\frac{x}{2}\right||x| < .01\,|x| = (1\%)\,|x| \Rightarrow \left|\frac{x}{2}\right| < .01 \Rightarrow 0 < |x| < .02$

29. (a) $\sin x = x - \frac{x^3}{3!} + \frac{x^5}{5!} - \frac{x^7}{7!} + \cdots \Rightarrow \frac{\sin x}{x} = 1 - \frac{x^2}{3!} + \frac{x^4}{5!} - \frac{x^6}{7!} + \cdots$, $s_1 = 1$ and $s_2 = 1 - \frac{x^2}{6}$; if L is the sum of the
series representing $\frac{\sin x}{x}$, then by the Alternating Series Estimation Theorem, $L - s_1 = \frac{\sin x}{x} - 1 < 0$ and
$L - s_2 = \frac{\sin x}{x} - \left(1 - \frac{x^2}{6}\right) > 0$. Therefore $1 - \frac{x^2}{6} < \frac{\sin x}{x} < 1$

(b) The graph of $y = \frac{\sin x}{x}$, $x \neq 0$, is bounded below by the graph of $y = 1 - \frac{x^2}{6}$ and above by the graph of $y = 1$ as derived in part (a).

31. $\sin x$ when $x = 0.1$; the sum is $\sin(0.1) \approx 0.099833417$

33. $\tan^{-1} x$ when $x = \frac{\pi}{3}$; the sum is $\tan^{-1}\left(\frac{\pi}{3}\right) \approx 0.808448$

35. $e^x \sin x = 0 + x + x^2 + x^3\left(-\frac{1}{3!} + \frac{1}{2!}\right) + x^4\left(-\frac{1}{3!} + \frac{1}{3!}\right) + x^5\left(\frac{1}{5!} - \frac{1}{2!}\frac{1}{3!} + \frac{1}{4!}\right) + x^6\left(\frac{1}{5!} - \frac{1}{3!}\frac{1}{3!} + \frac{1}{5!}\right) + \cdots$
$= x + x^2 + \frac{1}{3}x^3 - \frac{1}{30}x^5 - \frac{1}{90}x^6 + \cdots$

37. $\sin^2 x = \left(\frac{1 - \cos 2x}{2}\right) = \frac{1}{2} - \frac{1}{2}\cos 2x = \frac{1}{2} - \frac{1}{2}\left(1 - \frac{(2x)^2}{2!} + \frac{(2x)^4}{4!} - \frac{(2x)^6}{6!} + \cdots\right) = \frac{2x^2}{2!} - \frac{2^3 x^4}{4!} + \frac{2^5 x^6}{6!} - \cdots$
$\Rightarrow \frac{d}{dx}(\sin^2 x) = \frac{d}{dx}\left(\frac{2x^2}{2!} - \frac{2^3 x^4}{4!} + \frac{2^5 x^6}{6!} - \cdots\right) = 2x - \frac{(2x)^3}{3!} + \frac{(2x)^5}{5!} - \frac{(2x)^7}{7!} + \cdots \Rightarrow 2\sin x \cos x$
$= 2x - \frac{(2x)^3}{3!} + \frac{(2x)^5}{5!} - \frac{(2x)^7}{7!} + \cdots = \sin 2x$, which checks

39. A special case of Taylor's Theorem is $f(b) = f(a) + f'(c)(b - a)$, where c is between a and b $\Rightarrow$ $f(b) - f(a) = f'(c)(b - a)$, the Mean Value Theorem.

41. (a) $f'' \leq 0$, $f'(a) = 0$ and $x = a$ interior to the interval I $\Rightarrow$ $f(x) - f(a) = \frac{f''(c_2)}{2}(x - a)^2 \leq 0$ throughout I $\Rightarrow f(x) \leq f(a)$ throughout I $\Rightarrow$ f has a local maximum at $x = a$
(b) similar reasoning gives $f(x) - f(a) = \frac{f''(c_2)}{2}(x - a)^2 \geq 0$ throughout I $\Rightarrow$ $f(x) \geq f(a)$ throughout I $\Rightarrow$ f has a local minimum at $x = a$

43. (a) $f(x) = (1 + x)^k \Rightarrow f'(x) = k(1 + x)^{k-1} \Rightarrow f''(x) = k(k - 1)(1 + x)^{k-2}$; $f(0) = 1$, $f'(0) = k$, and $f''(0) = k(k - 1)$ $\Rightarrow Q(x) = 1 + kx + \frac{k(k-1)}{2}x^2$
(b) $|R_2(x)| = \left|\frac{3 \cdot 2 \cdot 1}{3!}x^3\right| < \frac{1}{100} \Rightarrow |x^3| < \frac{1}{100} \Rightarrow 0 < x < \frac{1}{100^{1/3}}$ or $0 < x < .21544$

45. If $f(x) = \sum_{n=0}^{\infty} a_n x^n$, then $f^{(k)}(x) = \sum_{n=k}^{\infty} n(n - 1)(n - 2)\cdots(n - k + 1)a_n x^{n-k}$ and $f^{(k)}(0) = k! a_k$
$\Rightarrow a_k = \frac{f^{(k)}(0)}{k!}$ for k a nonnegative integer. Therefore, the coefficients of $f(x)$ are identical with the corresponding coefficients in the Maclaurin series of $f(x)$ and the statement follows.

47. (a) Suppose $f(x)$ is a continuous periodic function with period p. Let x_0 be an arbitrary real number. Then f assumes a minimum m_1 and a maximum m_2 in the interval $[x_0, x_0 + p]$; i.e., $m_1 \leq f(x) \leq m_2$ for all x in $[x_0, x_0 + p]$. Since f is periodic it has exactly the same values on all other intervals $[x_0 + p, x_0 + 2p]$, $[x_0 + 2p, x_0 + 3p], \ldots$, and $[x_0 - p, x_0], [x_0 - 2p, x_0 - p], \ldots$, and so forth. That is, for all real numbers $-\infty < x < \infty$ we have $m_1 \leq f(x) \leq m_2$. Now choose $M = \max\{|m_1|, |m_2|\}$. Then $-M \leq -|m_1| \leq m_1 \leq f(x) \leq m_2 \leq |m_2| \leq M \Rightarrow |f(x)| \leq M$ for all x.
(b) The dominate term in the nth order Taylor polynomial generated by $\cos x$ about $x = a$ is $\frac{\sin(a)}{n!}(x - a)^n$ or $\frac{\cos(a)}{n!}(x - a)^n$. In both cases, as $|x|$ increases the absolute value of these dominate terms tends to ∞,

causing the graph of $P_n(x)$ to move away from cos x.

49. (a) $e^{-i\pi} = \cos(-\pi) + i\sin(-\pi) = -1 + i(0) = -1$

   (b) $e^{i\pi/4} = \cos\left(\frac{\pi}{4}\right) + i\sin\left(\frac{\pi}{4}\right) = \frac{1}{\sqrt{2}} + \frac{i}{\sqrt{2}} = \left(\frac{1}{\sqrt{2}}\right)(1+i)$

   (c) $e^{-i\pi/2} = \cos\left(-\frac{\pi}{2}\right) + i\sin\left(-\frac{\pi}{2}\right) = 0 + i(-1) = -i$

51. $e^x = 1 + x + \frac{x^2}{2!} + \frac{x^3}{3!} + \frac{x^4}{4!} + \ldots \Rightarrow e^{i\theta} = 1 + i\theta + \frac{(i\theta)^2}{2!} + \frac{(i\theta)^3}{3!} + \frac{(i\theta)^4}{4!} + \ldots$ and

   $e^{-i\theta} = 1 - i\theta + \frac{(-i\theta)^2}{2!} + \frac{(-i\theta)^3}{3!} + \frac{(-i\theta)^4}{4!} + \ldots = 1 - i\theta + \frac{(i\theta)^2}{2!} - \frac{(i\theta)^3}{3!} + \frac{(i\theta)^4}{4!} - \ldots$

   $\Rightarrow \frac{e^{i\theta} + e^{-i\theta}}{2} = \frac{\left(1 + i\theta + \frac{(i\theta)^2}{2!} + \frac{(i\theta)^3}{3!} + \frac{(i\theta)^4}{4!} + \ldots\right) + \left(1 - i\theta + \frac{(i\theta)^2}{2!} - \frac{(i\theta)^3}{3!} + \frac{(i\theta)^4}{4!} - \ldots\right)}{2}$

   $= 1 - \frac{\theta^2}{2!} + \frac{\theta^4}{4!} - \frac{\theta^6}{6!} + \ldots = \cos\theta;$

   $\frac{e^{i\theta} - e^{-i\theta}}{2i} = \frac{\left(1 + i\theta + \frac{(i\theta)^2}{2!} + \frac{(i\theta)^3}{3!} + \frac{(i\theta)^4}{4!} + \ldots\right) - \left(1 - i\theta + \frac{(i\theta)^2}{2!} - \frac{(i\theta)^3}{3!} + \frac{(i\theta)^4}{4!} - \ldots\right)}{2i}$

   $= \theta - \frac{\theta^3}{3!} + \frac{\theta^5}{5!} - \frac{\theta^7}{7!} + \ldots = \sin\theta$

53. $e^x \sin x = \left(1 + x + \frac{x^2}{2!} + \frac{x^3}{3!} + \frac{x^4}{4!} + \ldots\right)\left(x - \frac{x^3}{3!} + \frac{x^5}{5!} - \frac{x^7}{7!} + \ldots\right)$

   $= (1)x + (1)x^2 + \left(-\frac{1}{6} + \frac{1}{2}\right)x^3 + \left(-\frac{1}{6} + \frac{1}{6}\right)x^4 + \left(\frac{1}{120} - \frac{1}{12} + \frac{1}{24}\right)x^5 + \ldots = x + x^2 + \frac{1}{3}x^3 - \frac{1}{30}x^5 + \ldots;$

   $e^x \cdot e^{ix} = e^{(1+i)x} = e^x(\cos x + i\sin x) = e^x\cos x + i(e^x\sin x) \Rightarrow e^x\sin x$ is the series of the imaginary part

   of $e^{(1+i)x}$ which we calculate next; $e^{(1+i)x} = \sum_{n=0}^{\infty} \frac{(x+ix)^n}{n!} = 1 + (x+ix) + \frac{(x+ix)^2}{2!} + \frac{(x+ix)^3}{3!} + \frac{(x+ix)^4}{4!} + \ldots$

   $= 1 + x + ix + \frac{1}{2!}(2ix^2) + \frac{1}{3!}(2ix^3 - 2x^3) + \frac{1}{4!}(-4x^4) + \frac{1}{5!}(-4x^5 - 4ix^5) + \frac{1}{6!}(-8ix^6) + \ldots \Rightarrow$ the imaginary part

   of $e^{(1+i)x}$ is $x + \frac{2}{2!}x^2 + \frac{2}{3!}x^3 - \frac{4}{5!}x^5 - \frac{8}{6!}x^6 + \ldots = x + x^2 + \frac{1}{3}x^3 - \frac{1}{30}x^5 - \frac{1}{90}x^6 + \ldots$ in agreement with our

   product calculation. The series for $e^x\sin x$ converges for all values of x.

55. (a) $e^{i\theta_1}e^{i\theta_2} = (\cos\theta_1 + i\sin\theta_1)(\cos\theta_2 + i\sin\theta_2) = (\cos\theta_1\cos\theta_2 - \sin\theta_1\sin\theta_2) + i(\sin\theta_1\cos\theta_2 + \sin\theta_2\cos\theta_1)$

   $= \cos(\theta_1 + \theta_2) + i\sin(\theta_1 + \theta_2) = e^{i(\theta_1+\theta_2)}$

   (b) $e^{-i\theta} = \cos(-\theta) + i\sin(-\theta) = \cos\theta - i\sin\theta = (\cos\theta - i\sin\theta)\left(\frac{\cos\theta + i\sin\theta}{\cos\theta + i\sin\theta}\right) = \frac{1}{\cos\theta + i\sin\theta} = \frac{1}{e^{i\theta}}$

11.10 APPLICATIONS OF POWER SERIES

1. $(1+x)^{1/2} = 1 + \frac{1}{2}x + \frac{\left(\frac{1}{2}\right)\left(-\frac{1}{2}\right)x^2}{2!} + \frac{\left(\frac{1}{2}\right)\left(-\frac{1}{2}\right)\left(-\frac{3}{2}\right)x^3}{3!} + \ldots = 1 + \frac{1}{2}x - \frac{1}{8}x^2 + \frac{1}{16}x^3 - \ldots$

3. $(1-x)^{-1/2} = 1 - \frac{1}{2}(-x) + \frac{\left(-\frac{1}{2}\right)\left(-\frac{3}{2}\right)(-x)^2}{2!} + \frac{\left(-\frac{1}{2}\right)\left(-\frac{3}{2}\right)\left(-\frac{5}{2}\right)(-x)^3}{3!} + \ldots = 1 + \frac{1}{2}x + \frac{3}{8}x^2 + \frac{5}{16}x^3 + \ldots$

5. $\left(1+\frac{x}{2}\right)^{-2} = 1 - 2\left(\frac{x}{2}\right) + \frac{(-2)(-3)\left(\frac{x}{2}\right)^2}{2!} + \frac{(-2)(-3)(-4)\left(\frac{x}{2}\right)^3}{3!} + \ldots = 1 - x + \frac{3}{4}x^2 - \frac{1}{2}x^3$

7. $(1+x^3)^{-1/2} = 1 - \frac{1}{2}x^3 + \frac{\left(-\frac{1}{2}\right)\left(-\frac{3}{2}\right)(x^3)^2}{2!} + \frac{\left(-\frac{1}{2}\right)\left(-\frac{3}{2}\right)\left(-\frac{5}{2}\right)(x^3)^3}{3!} + \ldots = 1 - \frac{1}{2}x^3 + \frac{3}{8}x^6 - \frac{5}{16}x^9 + \ldots$

9. $\left(1+\frac{1}{x}\right)^{1/2} = 1 + \frac{1}{2}\left(\frac{1}{x}\right) + \frac{\left(\frac{1}{2}\right)\left(-\frac{1}{2}\right)\left(\frac{1}{x}\right)^2}{2!} + \frac{\left(\frac{1}{2}\right)\left(-\frac{1}{2}\right)\left(-\frac{3}{2}\right)\left(\frac{1}{x}\right)^3}{3!} + \ldots = 1 + \frac{1}{2x} - \frac{1}{8x^2} + \frac{1}{16x^3}$

11. $(1+x)^4 = 1 + 4x + \frac{(4)(3)x^2}{2!} + \frac{(4)(3)(2)x^3}{3!} + \frac{(4)(3)(2)x^4}{4!} = 1 + 4x + 6x^2 + 4x^3 + x^4$

13. $(1-2x)^3 = 1 + 3(-2x) + \frac{(3)(2)(-2x)^2}{2!} + \frac{(3)(2)(1)(-2x)^3}{3!} = 1 - 6x + 12x^2 - 8x^3$

15. Assume the solution has the form $y = a_0 + a_1x + a_2x^2 + \ldots + a_{n-1}x^{n-1} + a_nx^n + \ldots$

$\Rightarrow \frac{dy}{dx} = a_1 + 2a_2x + \ldots + na_nx^{n-1} + \ldots$

$\Rightarrow \frac{dy}{dx} + y = (a_1 + a_0) + (2a_2 + a_1)x + (3a_3 + a_2)x^2 + \ldots + (na_n + a_{n-1})x^{n-1} + \ldots = 0$

$\Rightarrow a_1 + a_0 = 0, 2a_2 + a_1 = 0, 3a_3 + a_2 = 0$ and in general $na_n + a_{n-1} = 0$. Since $y = 1$ when $x = 0$ we have

$a_0 = 1$. Therefore $a_1 = -1, a_2 = \frac{-a_1}{2 \cdot 1} = \frac{1}{2}, a_3 = \frac{-a_2}{3} = -\frac{1}{3 \cdot 2}, \ldots, a_n = \frac{-a_{n-1}}{n} = \frac{(-1)^n}{n!}$

$\Rightarrow y = 1 - x + \frac{1}{2}x^2 - \frac{1}{3!}x^3 + \ldots + \frac{(-1)^n}{n!}x^n + \ldots = \sum_{n=0}^{\infty} \frac{(-1)^n x^n}{n!} = e^{-x}$

17. Assume the solution has the form $y = a_0 + a_1x + a_2x^2 + \ldots + a_{n-1}x^{n-1} + a_nx^n + \ldots$

$\Rightarrow \frac{dy}{dx} = a_1 + 2a_2x + \ldots + na_nx^{n-1} + \ldots$

$\Rightarrow \frac{dy}{dx} - y = (a_1 - a_0) + (2a_2 - a_1)x + (3a_3 - a_2)x^2 + \ldots + (na_n - a_{n-1})x^{n-1} + \ldots = 1$

$\Rightarrow a_1 - a_0 = 1, 2a_2 - a_1 = 0, 3a_3 - a_2 = 0$ and in general $na_n - a_{n-1} = 0$. Since $y = 0$ when $x = 0$ we have

$a_0 = 0$. Therefore $a_1 = 1, a_2 = \frac{a_1}{2} = \frac{1}{2}, a_3 = \frac{a_2}{3} = \frac{1}{3 \cdot 2}, a_4 = \frac{a_3}{4} = \frac{1}{4 \cdot 3 \cdot 2}, \ldots, a_n = \frac{a_{n-1}}{n} = \frac{1}{n!}$

$\Rightarrow y = 0 + 1x + \frac{1}{2}x^2 + \frac{1}{3 \cdot 2}x^3 + \frac{1}{4 \cdot 3 \cdot 2}x^4 + \ldots + \frac{1}{n!}x^n + \ldots$

$= \left(1 + 1x + \frac{1}{2}x^2 + \frac{1}{3 \cdot 2}x^3 + \frac{1}{4 \cdot 3 \cdot 2}x^4 + \ldots + \frac{1}{n!}x^n + \ldots\right) - 1 = \sum_{n=0}^{\infty} \frac{x^n}{n!} - 1 = e^x - 1$

19. Assume the solution has the form $y = a_0 + a_1x + a_2x^2 + \ldots + a_{n-1}x^{n-1} + a_nx^n + \ldots$

$\Rightarrow \frac{dy}{dx} = a_1 + 2a_2x + \ldots + na_nx^{n-1} + \ldots$

$\Rightarrow \frac{dy}{dx} - y = (a_1 - a_0) + (2a_2 - a_1)x + (3a_3 - a_2)x^2 + \ldots + (na_n - a_{n-1})x^{n-1} + \ldots = x$

$\Rightarrow a_1 - a_0 = 0, 2a_2 - a_1 = 1, 3a_3 - a_2 = 0$ and in general $na_n - a_{n-1} = 0$. Since $y = 0$ when $x = 0$ we have

$a_0 = 0$. Therefore $a_1 = 0, a_2 = \frac{1 + a_1}{2} = \frac{1}{2}, a_3 = \frac{a_2}{3} = \frac{1}{3 \cdot 2}, a_4 = \frac{a_3}{4} = \frac{1}{4 \cdot 3 \cdot 2}, \ldots, a_n = \frac{a_{n-1}}{n} = \frac{1}{n!}$

$\Rightarrow y = 0 + 0x + \frac{1}{2}x^2 + \frac{1}{3 \cdot 2}x^3 + \frac{1}{4 \cdot 3 \cdot 2}x^4 + + \ldots + \frac{1}{n!}x^n + \ldots$

$= \left(1 + 1x + \frac{1}{2}x^2 + \frac{1}{3 \cdot 2}x^3 + \frac{1}{4 \cdot 3 \cdot 2}x^4 + \ldots + \frac{1}{n!}x^n + \ldots\right) - 1 - x = \sum_{n=0}^{\infty} \frac{x^n}{n!} - 1 - x = e^x - x - 1$

21. $y' - xy = a_1 + (2a_2 - a_0)x + (3a_3 - a_1)x + \ldots + (na_n - a_{n-2})x^{n-1} + \ldots = 0 \Rightarrow a_1 = 0, 2a_2 - a_0 = 0, 3a_3 - a_1 = 0,$

$4a_4 - a_2 = 0$ and in general $na_n - a_{n-2} = 0$. Since $y = 1$ when $x = 0$, we have $a_0 = 1$. Therefore $a_2 = \frac{a_0}{2} = \frac{1}{2},$

$a_3 = \frac{a_1}{3} = 0, a_4 = \frac{a_2}{4} = \frac{1}{2 \cdot 4}, a_5 = \frac{a_3}{5} = 0, \ldots, a_{2n} = \frac{1}{2 \cdot 4 \cdot 6 \cdots 2n}$ and $a_{2n+1} = 0$

$\Rightarrow y = 1 + \frac{1}{2}x^2 + \frac{1}{2 \cdot 4}x^4 + \frac{1}{2 \cdot 4 \cdot 6}x^6 + \ldots + \frac{1}{2 \cdot 4 \cdot 6 \cdots 2n}x^{2n} + \ldots = \sum_{n=0}^{\infty} \frac{x^{2n}}{2^n n!} = \sum_{n=0}^{\infty} \frac{\left(\frac{x^2}{2}\right)^n}{n!} = e^{x^2/2}$

23. $(1 - x)y' - y = (a_1 - a_0) + (2a_2 - a_1 - a_1)x + (3a_3 - 2a_2 - a_2)x^2 + (4a_4 - 3a_3 - a_3)x^3 + \ldots$

$+ (na_n - (n-1)a_{n-1} - a_{n-1})x^{n-1} + \ldots = 0 \Rightarrow a_1 - a_0 = 0, 2a_2 - 2a_1 = 0, 3a_3 - 3a_2 = 0$ and in

general $(na_n - na_{n-1}) = 0$. Since $y = 2$ when $x = 0$, we have $a_0 = 2$. Therefore

$a_1 = 2, a_2 = 2, \ldots, a_n = 2 \Rightarrow y = 2 + 2x + 2x^2 + \ldots = \sum_{n=0}^{\infty} 2x^n = \frac{2}{1 - x}$

25. $y = a_0 + a_1x + a_2x^2 + \ldots + a_nx^n + \ldots \Rightarrow y'' = 2a_2 + 3 \cdot 2a_3x + \ldots + n(n-1)a_nx^{n-2} + \ldots \Rightarrow y'' - y$

$= (2a_2 - a_0) + (3 \cdot 2a_3 - a_1)x + (4 \cdot 3a_4 - a_2)x^2 + \ldots + (n(n-1)a_n - a_{n-2})x^{n-2} + \ldots = 0 \Rightarrow 2a_2 - a_0 = 0,$

$3 \cdot 2a_3 - a_1 = 0, 4 \cdot 3a_4 - a_2 = 0$ and in general $n(n-1)a_n - a_{n-2} = 0$. Since $y' = 1$ and $y = 0$ when $x = 0,$

we have $a_0 = 0$ and $a_1 = 1$. Therefore $a_2 = 0, a_3 = \frac{1}{3 \cdot 2}, a_4 = 0, a_5 = \frac{1}{5 \cdot 4 \cdot 3 \cdot 2}, \ldots, a_{2n+1} = \frac{1}{(2n+1)!}$ and

$a_{2n} = 0 \Rightarrow y = x + \frac{1}{3!}x^3 + \frac{1}{5!}x^5 + \ldots = \sum_{n=0}^{\infty} \frac{x^{2n+1}}{(2n+1)!} = \sinh x$

27. $y = a_0 + a_1x + a_2x^2 + \ldots + a_nx^n + \ldots \Rightarrow y'' = 2a_2 + 3 \cdot 2a_3x + \ldots + n(n-1)a_nx^{n-2} + \ldots \Rightarrow y'' + y$

$= (2a_2 + a_0) + (3 \cdot 2a_3 + a_1)x + (4 \cdot 3a_4 + a_2)x^2 + \ldots + (n(n-1)a_n + a_{n-2})x^{n-2} + \ldots = x \Rightarrow 2a_2 + a_0 = 0,$

$3 \cdot 2a_3 + a_1 = 1, 4 \cdot 3a_4 + a_2 = 0$ and in general $n(n-1)a_n + a_{n-2} = 0$. Since $y' = 1$ and $y = 2$ when $x = 0,$

we have $a_0 = 2$ and $a_1 = 1$. Therefore $a_2 = -1$, $a_3 = 0$, $a_4 = \frac{1}{4 \cdot 3}$, $a_5 = 0, \ldots$, $a_{2n} = -2 \cdot \frac{(-1)^{n+1}}{(2n)!}$ and

$a_{2n+1} = 0 \Rightarrow y = 2 + x - x^2 + 2 \cdot \frac{x^4}{4!} + \ldots = 2 + x - 2 \sum_{n=1}^{\infty} \frac{(-1)^{n+1} x^{2n}}{(2n)!} = x + \cos 2x$

29. $y = a_0 + a_1(x - 2) + a_2(x - 2)^2 + \ldots + a_n(x - 2)^n + \ldots$

$\Rightarrow y'' = 2a_2 + 3 \cdot 2a_3(x - 2) + \ldots + n(n-1)a_n(x-2)^{n-2} + \ldots \Rightarrow y'' - y$

$= (2a_2 - a_0) + (3 \cdot 2a_3 - a_1)(x - 2) + (4 \cdot 3a_4 - a_2)(x - 2)^2 + \ldots + (n(n-1)a_n - a_{n-2})(x-2)^{n-2} + \ldots = -x$

$= -(x - 2) - 2 \Rightarrow 2a_2 - a_0 = -2$, $3 \cdot 2a_3 - a_1 = -1$, and $n(n-1)a_n - a_{n-2} = 0$ for $n > 3$. Since $y = 0$ when $x = 2$,

we have $a_0 = 0$, and since $y' = -2$ when $x = 2$, we have $a_1 = -2$. Therefore $a_2 = -1$, $a_3 = -\frac{1}{2}$, $a_4 = \frac{1}{4 \cdot 3}(-1) = \frac{-2}{4 \cdot 3 \cdot 2 \cdot 1}$,

$a_5 = \frac{1}{5 \cdot 4}\left(-\frac{1}{2}\right) = \frac{-3}{5 \cdot 4 \cdot 3 \cdot 2 \cdot 1}, \ldots$, $a_{2n} = \frac{-2}{(2n)!}$, and $a_{2n+1} = \frac{-3}{(2n+1)!}$. Since $a_1 = -2$, we have $a_1(x - 2) = (-2)(x - 2)$ and

$(-2)(x - 2) = (-3 + 1)(x - 2) = (-3)(x - 2) + (1)(x - 2) = x - 2 - 3(x - 2)$.

$\Rightarrow y = x - 2 - 3(x - 2) - \frac{2}{2!}(x-2)^2 - \frac{3}{3!}(x-2)^3 - \frac{2}{4!}(x-2)^4 - \frac{3}{5!}(x-2)^5 - \ldots$

$\Rightarrow y = x - 2 - \frac{2}{2!}(x-2)^2 - \frac{2}{4!}(x-2)^4 - \ldots - 3(x-2)x - \frac{3}{3!}(x-2)^3 - \frac{3}{5!}(x-2)^5 - \ldots$

$\Rightarrow y = x - 2\sum_{n=0}^{\infty} \frac{(x-2)^{2n}}{(2n)!} - 3\sum_{n=0}^{\infty} \frac{(x-2)^{2n+1}}{(2n+1)!}$

31. $y'' + x^2 y = 2a_2 + 6a_3 x + (4 \cdot 3a_4 + a_0)x^2 + \ldots + (n(n-1)a_n + a_{n-4})x^{n-2} + \ldots = x \Rightarrow 2a_2 = 0$, $6a_3 = 1$,

$4 \cdot 3a_4 + a_0 = 0$, $5 \cdot 4a_5 + a_1 = 0$, and in general $n(n-1)a_n + a_{n-4} = 0$. Since $y' = b$ and $y = a$ when $x = 0$,

we have $a_0 = a$ and $a_1 = b$. Therefore $a_2 = 0$, $a_3 = \frac{1}{2 \cdot 3}$, $a_4 = -\frac{a}{3 \cdot 4}$, $a_5 = -\frac{b}{4 \cdot 5}$, $a_6 = 0$, $a_7 = \frac{-1}{2 \cdot 3 \cdot 6 \cdot 7}$

$\Rightarrow y = a + bx + \frac{1}{2 \cdot 3}x^3 - \frac{a}{3 \cdot 4}x^4 - \frac{b}{4 \cdot 5}x^5 - \frac{1}{2 \cdot 3 \cdot 6 \cdot 7}x^7 + \frac{ax^8}{3 \cdot 4 \cdot 7 \cdot 8} + \frac{bx^9}{4 \cdot 5 \cdot 8 \cdot 9} + \ldots$

33. $\int_0^{0.2} \sin x^2 \, dx = \int_0^{0.2} \left(x^2 - \frac{x^6}{3!} + \frac{x^{10}}{5!} - \ldots\right) dx = \left[\frac{x^3}{3} - \frac{x^7}{7 \cdot 3!} + \ldots\right]_0^{0.2} \approx \left[\frac{x^3}{3}\right]_0^{0.2} \approx 0.00267$ with error

$|E| \leq \frac{(.2)^7}{7 \cdot 3!} \approx 0.0000003$

35. $\int_0^{0.1} \frac{1}{\sqrt{1 + x^4}} \, dx = \int_0^{0.1} \left(1 - \frac{x^4}{2} + \frac{3x^8}{8} - \ldots\right) dx = \left[x - \frac{x^5}{10} + \ldots\right]_0^{0.1} \approx [x]_0^{0.1} \approx 0.1$ with error

$|E| \leq \frac{(0.1)^5}{10} = 0.000001$

37. $\int_0^{0.1} \frac{\sin x}{x} \, dx = \int_0^{0.1} \left(1 - \frac{x^2}{3!} + \frac{x^4}{5!} - \frac{x^6}{7!} + \ldots\right) dx = \left[x - \frac{x^3}{3 \cdot 3!} + \frac{x^5}{5 \cdot 5!} - \frac{x^7}{7 \cdot 7!} + \ldots\right]_0^{0.1} \approx \left[x - \frac{x^3}{3 \cdot 3!} + \frac{x^5}{5 \cdot 5!}\right]_0^{0.1}$

≈ 0.0999444611, $|E| \leq \frac{(0.1)^7}{7 \cdot 7!} \approx 2.8 \times 10^{-12}$

39. $(1 + x^4)^{1/2} = (1)^{1/2} + \frac{\left(\frac{1}{2}\right)}{1}(1)^{-1/2}(x^4) + \frac{\left(\frac{1}{2}\right)\left(-\frac{1}{2}\right)}{2!}(1)^{-3/2}(x^4)^2 + \frac{\left(\frac{1}{2}\right)\left(-\frac{1}{2}\right)\left(-\frac{3}{2}\right)}{3!}(1)^{-5/2}(x^4)^3$

$+ \frac{\left(\frac{1}{2}\right)\left(-\frac{1}{2}\right)\left(-\frac{3}{2}\right)\left(-\frac{5}{2}\right)}{4!}(1)^{-7/2}(x^4)^4 + \ldots = 1 + \frac{x^4}{2} - \frac{x^8}{8} + \frac{x^{12}}{16} - \frac{5x^{16}}{128} + \ldots$

$\Rightarrow \int_0^{0.1}\left(1 + \frac{x^4}{2} - \frac{x^8}{8} + \frac{x^{12}}{16} - \frac{5x^{16}}{128} + \ldots\right) dx \approx \left[x + \frac{x^5}{10}\right]_0^{0.1} \approx 0.100001$, $|E| \leq \frac{(0.1)^9}{72} \approx 1.39 \times 10^{-11}$

41. $\int_0^1 \cos t^2 \, dt = \int_0^1 \left(1 - \frac{t^4}{2} + \frac{t^8}{4!} - \frac{t^{12}}{6!} + \ldots\right) dt = \left[t - \frac{t^5}{10} + \frac{t^9}{9 \cdot 4!} - \frac{t^{13}}{13 \cdot 6!} + \ldots\right]_0^1 \Rightarrow |\text{error}| < \frac{1}{13 \cdot 6!} \approx .00011$

43. $F(x) = \int_0^x \left(t^2 - \frac{t^6}{3!} + \frac{t^{10}}{5!} - \frac{t^{14}}{7!} + \ldots\right) dt = \left[\frac{t^3}{3} - \frac{t^7}{7 \cdot 3!} + \frac{t^{11}}{11 \cdot 5!} - \frac{t^{15}}{15 \cdot 7!} + \ldots\right]_0^x \approx \frac{x^3}{3} - \frac{x^7}{7 \cdot 3!} + \frac{x^{11}}{11 \cdot 5!}$

$\Rightarrow |\text{error}| < \frac{1}{15 \cdot 7!} \approx 0.000013$

45. (a) $F(x) = \int_0^x \left(t - \frac{t^3}{3} + \frac{t^5}{5} - \frac{t^7}{7} + \ldots\right) dt = \left[\frac{t^2}{2} - \frac{t^4}{12} + \frac{t^6}{30} - \ldots\right]_0^x \approx \frac{x^2}{2} - \frac{x^4}{12} \Rightarrow |\text{error}| < \frac{(0.5)^6}{30} \approx .00052$

(b) $|\text{error}| < \frac{1}{33 \cdot 34} \approx .00089$ when $F(x) \approx \frac{x^2}{2} - \frac{x^4}{3 \cdot 4} + \frac{x^6}{5 \cdot 6} - \frac{x^8}{7 \cdot 8} + \ldots + (-1)^{15}\frac{x^{32}}{31 \cdot 32}$

47. $\frac{1}{x^2}\left(e^x - (1+x)\right) = \frac{1}{x^2}\left(\left(1 + x + \frac{x^2}{2} + \frac{x^3}{3!} + \dots\right) - 1 - x\right) = \frac{1}{2} + \frac{x}{3!} + \frac{x^2}{4!} + \dots \Rightarrow \lim_{x \to 0} \frac{e^x - (1+x)}{x^2}$

$= \lim_{x \to 0} \left(\frac{1}{2} + \frac{x}{3!} + \frac{x^2}{4!} + \dots\right) = \frac{1}{2}$

49. $\frac{1}{t^4}\left(1 - \cos t - \frac{t^2}{2}\right) = \frac{1}{t^4}\left[1 - \frac{t^2}{2} - \left(1 - \frac{t^2}{2} + \frac{t^4}{4!} - \frac{t^6}{6!} + \dots\right)\right] = -\frac{1}{4!} + \frac{t^2}{6!} - \frac{t^4}{8!} + \dots \Rightarrow \lim_{t \to 0} \frac{1 - \cos t - \left(\frac{t^2}{2}\right)}{t^4}$

$= \lim_{t \to 0} \left(-\frac{1}{4!} + \frac{t^2}{6!} - \frac{t^4}{8!} + \dots\right) = -\frac{1}{24}$

51. $\frac{1}{y^3}\left(y - \tan^{-1} y\right) = \frac{1}{y^3}\left[y - \left(y - \frac{y^3}{3} + \frac{y^5}{5} - \dots\right)\right] = \frac{1}{3} - \frac{y^2}{5} + \frac{y^4}{7} - \dots \Rightarrow \lim_{y \to 0} \frac{y - \tan^{-1} y}{y^3} = \lim_{y \to 0} \left(\frac{1}{3} - \frac{y^2}{5} + \frac{y^4}{7} - \dots\right)$

$= \frac{1}{3}$

53. $x^2 \left(-1 + e^{-1/x^2}\right) = x^2\left(-1 + 1 - \frac{1}{x^2} + \frac{1}{2x^4} - \frac{1}{6x^6} + \dots\right) = -1 + \frac{1}{2x^2} - \frac{1}{6x^4} + \dots \Rightarrow \lim_{x \to \infty} x^2\left(e^{-1/x^2} - 1\right)$

$= \lim_{x \to \infty} \left(-1 + \frac{1}{2x^2} - \frac{1}{6x^4} + \dots\right) = -1$

55. $\frac{\ln(1+x^2)}{1 - \cos x} = \frac{\left(x^2 - \frac{x^4}{2} + \frac{x^6}{3} - \dots\right)}{1 - \left(1 - \frac{x^2}{2!} + \frac{x^4}{4!} - \dots\right)} = \frac{\left(1 - \frac{x^2}{2} + \frac{x^4}{3} - \dots\right)}{\left(\frac{1}{2!} - \frac{x^2}{4!} + \dots\right)} \Rightarrow \lim_{x \to 0} \frac{\ln(1+x^2)}{1 - \cos x} = \lim_{x \to 0} \frac{\left(1 - \frac{x^2}{2} + \frac{x^4}{3} - \dots\right)}{\left(\frac{1}{2!} - \frac{x^2}{4!} + \dots\right)} = 2! = 2$

57. $\ln\left(\frac{1+x}{1-x}\right) = \ln(1+x) - \ln(1-x) = \left(x - \frac{x^2}{2} + \frac{x^3}{3} - \frac{x^4}{4} + \dots\right) - \left(-x - \frac{x^2}{2} - \frac{x^3}{3} - \frac{x^4}{4} - \dots\right) = 2\left(x + \frac{x^3}{3} + \frac{x^5}{5} + \dots\right)$

59. $\tan^{-1} x = x - \frac{x^3}{3} + \frac{x^5}{5} - \frac{x^7}{7} + \frac{x^9}{9} - \dots + \frac{(-1)^{n-1}x^{2n-1}}{2n-1} + \dots \Rightarrow |\text{error}| = \left|\frac{(-1)^{n-1}x^{2n-1}}{2n-1}\right| = \frac{1}{2n-1}$ when $x = 1$;

$\frac{1}{2n-1} < \frac{1}{10^3} \Rightarrow n > \frac{1001}{2} = 500.5 \Rightarrow$ the first term not used is the 501st $\Rightarrow$ we must use 500 terms

61. $\tan^{-1} x = x - \frac{x^3}{3} + \frac{x^5}{5} - \frac{x^7}{7} + \frac{x^9}{9} - \dots + \frac{(-1)^{n-1}x^{2n-1}}{2n-1} + \dots$ and when the series representing $48 \tan^{-1}\left(\frac{1}{18}\right)$ has an

error less than $\frac{1}{3} \cdot 10^{-6}$, then the series representing the sum

$48 \tan^{-1}\left(\frac{1}{18}\right) + 32 \tan^{-1}\left(\frac{1}{57}\right) - 20 \tan^{-1}\left(\frac{1}{239}\right)$ also has an error of magnitude less than 10^{-6}; thus

$|\text{error}| = 48 \frac{\left(\frac{1}{18}\right)^{2n-1}}{2n-1} < \frac{1}{3 \cdot 10^6} \Rightarrow n \geq 4$ using a calculator $\Rightarrow$ 4 terms

63. (a) $(1 - x^2)^{-1/2} \approx 1 + \frac{x^2}{2} + \frac{3x^4}{8} + \frac{5x^6}{16} \Rightarrow \sin^{-1} x \approx x + \frac{x^3}{6} + \frac{3x^5}{40} + \frac{5x^7}{112}$; Using the Ratio Test:

$\lim_{n \to \infty} \left|\frac{1 \cdot 3 \cdot 5 \cdots (2n-1)(2n+1)x^{2n+3}}{2 \cdot 4 \cdot 6 \cdots (2n)(2n+2)(2n+3)} \cdot \frac{2 \cdot 4 \cdot 6 \cdots (2n)(2n+1)}{1 \cdot 3 \cdot 5 \cdots (2n-1)x^{2n+1}}\right| < 1 \Rightarrow x^2 \lim_{n \to \infty} \left|\frac{(2n+1)(2n+1)}{(2n+2)(2n+3)}\right| < 1$

$\Rightarrow |x| < 1 \Rightarrow$ the radius of convergence is 1. See Exercise 69.

(b) $\frac{d}{dx}\left(\cos^{-1} x\right) = -(1 - x^2)^{-1/2} \Rightarrow \cos^{-1} x = \frac{\pi}{2} - \sin^{-1} x \approx \frac{\pi}{2} - \left(x + \frac{x^3}{6} + \frac{3x^5}{40} + \frac{5x^7}{112}\right) \approx \frac{\pi}{2} - x - \frac{x^3}{6} - \frac{3x^5}{40} - \frac{5x^7}{112}$

65. $\frac{-1}{1+x} = -\frac{1}{1 - (-x)} = -1 + x - x^2 + x^3 - \dots \Rightarrow \frac{d}{dx}\left(\frac{-1}{1+x}\right) = \frac{1}{1+x^2} = \frac{d}{dx}\left(-1 + x - x^2 + x^3 - \dots\right)$

$= 1 - 2x + 3x^2 - 4x^3 + \dots$

67. Wallis' formula gives the approximation $\pi \approx 4\left[\frac{2 \cdot 4 \cdot 4 \cdot 6 \cdot 8 \cdots (2n-2) \cdot (2n)}{3 \cdot 3 \cdot 5 \cdot 5 \cdot 7 \cdot 7 \cdots (2n-1) \cdot (2n-1)}\right]$ to produce the table

n	$\sim \pi$
10	3.221088998
20	3.181104886
30	3.167880758
80	3.151425420
90	3.150331383
93	3.150049112
94	3.149959030
95	3.149870848
100	3.149456425

At n = 1929 we obtain the first approximation accurate to 3 decimals: 3.141999845. At n = 30,000 we still do not obtain accuracy to 4 decimals: 3.141617732, so the convergence to π is very slow. Here is a <u>Maple</u> CAS procedure to produce these approximations:

```
pie :=
  proc(n)
  local i,j;
    a(2) := evalf(8/9);
    for i from 3 to n do a(i) := evalf(2*(2*i−2)*i/(2*i−1)^2*a(i−1)) od;
    [[j,4*a(j)] $ (j = n−5 .. n)]
  end
```

69. $(1 - x^2)^{-1/2} = (1 + (-x^2))^{-1/2} = (1)^{-1/2} + \left(-\frac{1}{2}\right)(1)^{-3/2}(-x^2) + \frac{\left(-\frac{1}{2}\right)\left(-\frac{3}{2}\right)(1)^{-5/2}(-x^2)^2}{2!}$

$+ \frac{\left(-\frac{1}{2}\right)\left(-\frac{3}{2}\right)\left(-\frac{5}{2}\right)(1)^{-7/2}(-x^2)^3}{3!} + \dots = 1 + \frac{x^2}{2} + \frac{1 \cdot 3 x^4}{2^2 \cdot 2!} + \frac{1 \cdot 3 \cdot 5 x^6}{2^3 \cdot 3!} + \dots = 1 + \sum_{n=1}^{\infty} \frac{1 \cdot 3 \cdot 5 \cdots (2n-1) x^{2n}}{2^n \cdot n!}$

$\Rightarrow \sin^{-1} x = \int_0^x (1 - t^2)^{-1/2}\, dt = \int_0^x \left(1 + \sum_{n=1}^{\infty} \frac{1 \cdot 3 \cdot 5 \cdots (2n-1) x^{2n}}{2^n \cdot n!}\right) dt = x + \sum_{n=1}^{\infty} \frac{1 \cdot 3 \cdot 5 \cdots (2n-1) x^{2n+1}}{2 \cdot 4 \cdots (2n)(2n+1)},$

where $|x| < 1$

71. (a) $\tan\left(\tan^{-1}(n+1) - \tan^{-1}(n-1)\right) = \frac{\tan(\tan^{-1}(n+1)) - \tan(\tan^{-1}(n-1))}{1 + \tan(\tan^{-1}(n+1))\tan(\tan^{-1}(n-1))} = \frac{(n+1) - (n-1)}{1 + (n+1)(n-1)} = \frac{2}{n^2}$

(b) $\sum_{n=1}^{N} \tan^{-1}\left(\frac{2}{n^2}\right) = \sum_{n=1}^{N}\left[\tan^{-1}(n+1) - \tan^{-1}(n-1)\right] = (\tan^{-1} 2 - \tan^{-1} 0) + (\tan^{-1} 3 - \tan^{-1} 1)$

$+ (\tan^{-1} 4 - \tan^{-1} 2) + \dots + (\tan^{-1}(N+1) - \tan^{-1}(N-1)) = \tan^{-1}(N+1) + \tan^{-1} N - \frac{\pi}{4}$

(c) $\sum_{n=1}^{\infty} \tan^{-1}\left(\frac{2}{n^2}\right) = \lim_{n \to \infty}\left[\tan^{-1}(N+1) + \tan^{-1} N - \frac{\pi}{4}\right] = \frac{\pi}{2} + \frac{\pi}{2} - \frac{\pi}{4} = \frac{3\pi}{4}$

11.11 FOURIER SERIES

1. $a_0 = \frac{1}{2\pi}\int_0^{2\pi} 1\, dx = 1,\ a_k = \frac{1}{\pi}\int_0^{2\pi} \cos kx\, dx = \frac{1}{\pi}\left[\frac{\sin kx}{k}\right]_0^{2\pi} = 0,\ b_k = \frac{1}{\pi}\int_0^{2\pi} \sin kx\, dx = \frac{1}{\pi}\left[-\frac{\cos kx}{k}\right]_0^{2\pi} = 0.$

Thus, the Fourier series for f(x) is 1.

3. $a_0 = \frac{1}{2\pi}\left[\int_0^\pi x\,dx + \int_\pi^{2\pi}(x - 2\pi)\,dx\right] = \frac{1}{2\pi}\left[\frac{1}{2}\pi^2 + \frac{1}{2}(4\pi^2 - \pi^2) - 2\pi^2\right] = 0$. Note,

$\int_\pi^{2\pi}(x - 2\pi)\cos kx\,dx = -\int_0^\pi u\cos ku\,du$ (Let $u = 2\pi - x$). So $a_k = \frac{1}{\pi}\left[\int_0^\pi x\cos kx\,dx + \int_\pi^{2\pi}(x - 2\pi)\cos kx\,dx\right] = 0$.

Note, $\int_\pi^{2\pi}(x - 2\pi)\sin kx\,dx = \int_0^\pi u\sin ku\,du$ (Let $u = 2\pi - x$). So $b_k = \frac{1}{\pi}\left[\int_0^\pi x\sin kx\,dx + \int_\pi^{2\pi}(x - 2\pi)\sin kx\,dx\right]$

$= \frac{2}{\pi}\int_0^\pi x\sin kx\,dx = \frac{2}{\pi}\left[-\frac{x}{k}\cos kx + \frac{1}{k^2}\sin kx\right]_0^\pi = -\frac{2}{k}\cos k\pi = \frac{2}{k}(-1)^{k+1}$.

Thus, the Fourier series for $f(x)$ is $\sum_{k=1}^\infty (-1)^{k+1}\frac{2\sin kx}{k}$.

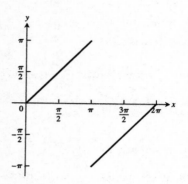

5. $a_0 = \frac{1}{2\pi}\int_0^{2\pi} e^x\,dx = \frac{1}{2\pi}(e^{2\pi} - 1)$, $a_k = \frac{1}{\pi}\int_0^{2\pi} e^x\cos kx\,dx = \frac{1}{\pi}\left[\frac{e^x}{1+k^2}(\cos kx + k\sin kx)\right]_0^{2\pi} = \frac{e^{2\pi} - 1}{\pi(1+k^2)}$,

$b_k = \frac{1}{\pi}\int_0^{2\pi} e^x\sin kx\,dx = \frac{1}{\pi}\left[\frac{e^x}{1+k^2}(\sin kx - k\cos kx)\right]_0^{2\pi} = \frac{k(1 - e^{2\pi})}{\pi(1+k^2)}$.

Thus, the Fourier series for $f(x)$ is $\frac{1}{2\pi}(e^{2\pi} - 1) + \frac{e^{2\pi} - 1}{\pi}\sum_{k=1}^\infty\left(\frac{\cos kx}{1+k^2} - \frac{k\sin kx}{1+k^2}\right)$.

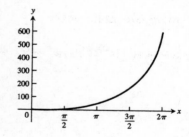

7. $a_0 = \frac{1}{2\pi}\int_0^{2\pi} f(x)\,dx = \frac{1}{2\pi}\int_0^{2\pi}\cos x\,dx = 0$, $a_k = \frac{1}{\pi}\int_0^{2\pi}\cos x\cos kx\,dx = \begin{cases} \frac{1}{\pi}\left[\frac{\sin(k-1)x}{2(k-1)} + \frac{\sin(k+1)x}{2(k+1)}\right]_0^\pi, & k \neq 1 \\ \frac{1}{\pi}\left[\frac{1}{2}x + \frac{1}{4}\sin 2x\right]_0^\pi, & k = 1 \end{cases}$

$= \begin{cases} 0, & k \neq 1 \\ \frac{1}{2}, & k = 1 \end{cases}$.

$b_k = \frac{1}{\pi}\int_0^{2\pi}\cos x\sin kx\,dx = \begin{cases} -\frac{1}{\pi}\left[\frac{\cos(k-1)x}{2(k-1)} + \frac{\cos(k+1)x}{2(k+1)}\right]_0^\pi, & k \neq 1 \\ -\frac{1}{4\pi}\cos 2x\Big|_0^\pi, & k = 1 \end{cases} = \begin{cases} 0, & k\text{ odd} \\ \frac{2k}{\pi(k^2 - 1)}, & k\text{ even} \end{cases}$.

Thus, the Fourier series for $f(x)$ is $\frac{1}{2}\cos x + \sum_{k\text{ even}}\frac{2k}{\pi(k^2 - 1)}\sin kx$.

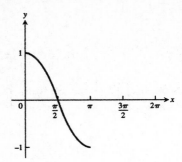

9. $\int_0^{2\pi} \cos px \, dx = \frac{1}{p} \sin px \Big|_0^{2\pi} = 0$ if $p \neq 0$.

11. $\int_0^{2\pi} \cos px \cos qx \, dx = \int_0^{2\pi} \frac{1}{2}[\cos(p+q)x + \cos(p-q)x]dx = \frac{1}{2}\left[\frac{1}{p+q}\sin(p+q)x + \frac{1}{p-q}\sin(p-q)x\right]_0^{2\pi} = 0$ if $p \neq q$.

If $p = q$ then $\int_0^{2\pi} \cos px \cos qx \, dx = \int_0^{2\pi} \cos^2 px \, dx = \int_0^{2\pi} \frac{1}{2}(1 + \cos 2px) \, dx = \frac{1}{2}\left(x + \frac{1}{2p}\sin 2px\right)\Big|_0^{2\pi} = \pi$.

13. $\int_0^{2\pi} \sin px \cos qx \, dx = \int_0^{2\pi} \frac{1}{2}[\sin(p+q)x + \sin(p-q)x]dx = -\frac{1}{2}\left[\frac{1}{p+q}\cos(p+q)x + \frac{1}{p-q}\cos(p-q)x\right]_0^{2\pi}$

$= -\frac{1}{2}\left[(1-1)\frac{1}{p+q} + (1-1)\frac{1}{p-q}\right] = 0$. If $p = q$ then $\int_0^{2\pi} \sin px \cos qx \, dx = \int_0^{2\pi} \sin px \cos px \, dx = \int_0^{2\pi} \frac{1}{2}\sin 2px \, dx$

$= -\frac{1}{4\pi}\cos 2px \Big|_0^{2\pi} = -\frac{1}{4\pi}(1-1) = 0$.

15. (a) $f(x)$ is piecewise continuous on $[0, 2\pi]$ and $f'(x) = 1$ for all $x \neq \pi \Rightarrow f'(x)$ is piecewise continuous on $[0, 2\pi]$. Then by Theorem 24, the Fourier series for $f(x)$ converges to $f(x)$ for all $x \neq \pi$ and converges to $\frac{1}{2}(f(\pi^+) + f(\pi^-))$

$= \frac{1}{2}(-\pi + \pi) = 0$ at $x = \pi$.

(b) The Fourier series for $f(x)$ is $\sum_{k=1}^{\infty} (-1)^{k+1} \frac{2\sin kx}{k}$. If we differentiate this series term by term we get the series

$\sum_{k=1}^{\infty} (-1)^{k+1} 2 \cos kx$, which diverges by the nth term test for divergence for any x since $\lim_{k\to\infty} (-1)^{k+1} 2 \cos kx \neq 0$.

CHAPTER 11 PRACTICE EXERCISES

1. converges to 1, since $\lim_{n\to\infty} a_n = \lim_{n\to\infty} \left(1 + \frac{(-1)^n}{n}\right) = 1$

3. converges to -1, since $\lim_{n\to\infty} a_n = \lim_{n\to\infty} \left(\frac{1-2^n}{2^n}\right) = \lim_{n\to\infty} \left(\frac{1}{2^n} - 1\right) = -1$

5. diverges, since $\left\{\sin \frac{n\pi}{2}\right\} = \{0, 1, 0, -1, 0, 1, \dots\}$

7. converges to 0, since $\lim_{n\to\infty} a_n = \lim_{n\to\infty} \frac{\ln n^2}{n} = 2\lim_{n\to\infty} \frac{\left(\frac{1}{n}\right)}{1} = 0$

9. converges to 1, since $\lim_{n\to\infty} a_n = \lim_{n\to\infty} \left(\frac{n + \ln n}{n}\right) = \lim_{n\to\infty} \frac{1 + \left(\frac{1}{n}\right)}{1} = 1$

11. converges to e^{-5}, since $\lim_{n\to\infty} a_n = \lim_{n\to\infty} \left(\frac{n-5}{n}\right)^n = \lim_{n\to\infty} \left(1 + \frac{(-5)}{n}\right)^n = e^{-5}$ by Theorem 5

13. converges to 3, since $\lim_{n\to\infty} a_n = \lim_{n\to\infty} \left(\frac{3^n}{n}\right)^{1/n} = \lim_{n\to\infty} \frac{3}{n^{1/n}} = \frac{3}{1} = 3$ by Theorem 5

15. converges to ln 2, since $\lim\limits_{n \to \infty} a_n = \lim\limits_{n \to \infty} n(2^{1/n} - 1) = \lim\limits_{n \to \infty} \frac{2^{1/n} - 1}{\left(\frac{1}{n}\right)} = \lim\limits_{n \to \infty} \frac{\left[\frac{(-2^{1/n} \ln 2)}{n^2}\right]}{\left(\frac{-1}{n^2}\right)} = \lim\limits_{n \to \infty} 2^{1/n} \ln 2$

$= 2^0 \cdot \ln 2 = \ln 2$

17. diverges, since $\lim\limits_{n \to \infty} a_n = \lim\limits_{n \to \infty} \frac{(n+1)!}{n!} = \lim\limits_{n \to \infty} (n+1) = \infty$

19. $\frac{1}{(2n-3)(2n-1)} = \frac{\left(\frac{1}{2}\right)}{2n-3} - \frac{\left(\frac{1}{2}\right)}{2n-1} \Rightarrow s_n = \left[\frac{\left(\frac{1}{2}\right)}{3} - \frac{\left(\frac{1}{2}\right)}{5}\right] + \left[\frac{\left(\frac{1}{2}\right)}{5} - \frac{\left(\frac{1}{2}\right)}{7}\right] + \cdots + \left[\frac{\left(\frac{1}{2}\right)}{2n-3} - \frac{\left(\frac{1}{2}\right)}{2n-1}\right] = \frac{\left(\frac{1}{2}\right)}{3} - \frac{\left(\frac{1}{2}\right)}{2n-1}$

$\Rightarrow \lim\limits_{n \to \infty} s_n = \lim\limits_{n \to \infty} \left[\frac{1}{6} - \frac{\left(\frac{1}{2}\right)}{2n-1}\right] = \frac{1}{6}$

21. $\frac{9}{(3n-1)(3n+2)} = \frac{3}{3n-1} - \frac{3}{3n+2} \Rightarrow s_n = \left(\frac{3}{2} - \frac{3}{5}\right) + \left(\frac{3}{5} - \frac{3}{8}\right) + \left(\frac{3}{8} - \frac{3}{11}\right) + \cdots + \left(\frac{3}{3n-1} - \frac{3}{3n+2}\right)$

$= \frac{3}{2} - \frac{3}{3n+2} \Rightarrow \lim\limits_{n \to \infty} s_n = \lim\limits_{n \to \infty} \left(\frac{3}{2} - \frac{3}{3n+2}\right) = \frac{3}{2}$

23. $\sum\limits_{n=0}^{\infty} e^{-n} = \sum\limits_{n=0}^{\infty} \frac{1}{e^n}$, a convergent geometric series with $r = \frac{1}{e}$ and $a = 1 \Rightarrow$ the sum is $\frac{1}{1 - \left(\frac{1}{e}\right)} = \frac{e}{e-1}$

25. diverges, a p-series with $p = \frac{1}{2}$

27. Since $f(x) = \frac{1}{x^{1/2}} \Rightarrow f'(x) = -\frac{1}{2x^{3/2}} < 0 \Rightarrow f(x)$ is decreasing $\Rightarrow a_{n+1} < a_n$, and $\lim\limits_{n \to \infty} a_n = \lim\limits_{n \to \infty} \frac{1}{\sqrt{n}} = 0$, the

series $\sum\limits_{n=1}^{\infty} \frac{(-1)^n}{\sqrt{n}}$ converges by the Alternating Series Test. Since $\sum\limits_{n=1}^{\infty} \frac{1}{\sqrt{n}}$ diverges, the given series converges

conditionally.

29. The given series does not converge absolutely by the Direct Comparison Test since $\frac{1}{\ln(n+1)} > \frac{1}{n+1}$, which is

the nth term of a divergent series. Since $f(x) = \frac{1}{\ln(x+1)} \Rightarrow f'(x) = -\frac{1}{(\ln(x+1))^2(x+1)} < 0 \Rightarrow f(x)$ is

decreasing $\Rightarrow a_{n+1} < a_n$, and $\lim\limits_{n \to \infty} a_n = \lim\limits_{n \to \infty} \frac{1}{\ln(n+1)} = 0$, the given series converges conditionally by the

Alternating Series Test.

31. converges absolutely by the Direct Comparison Test since $\frac{\ln n}{n^3} < \frac{n}{n^3} = \frac{1}{n^2}$, the nth term of a convergent p-series

33. $\lim\limits_{n \to \infty} \frac{\left(\frac{1}{n\sqrt{n^2+1}}\right)}{\left(\frac{1}{n^2}\right)} = \sqrt{\lim\limits_{n \to \infty} \frac{n^2}{n^2+1}} = \sqrt{1} = 1 \Rightarrow$ converges absolutely by the Limit Comparison Test

35. converges absolutely by the Ratio Test since $\lim\limits_{n \to \infty} \left[\frac{n+2}{(n+1)!} \cdot \frac{n!}{n+1}\right] = \lim\limits_{n \to \infty} \frac{n+2}{(n+1)^2} = 0 < 1$

37. converges absolutely by the Ratio Test since $\lim\limits_{n \to \infty} \left[\frac{3^{n+1}}{(n+1)!} \cdot \frac{n!}{3^n}\right] = \lim\limits_{n \to \infty} \frac{3}{n+1} = 0 < 1$

39. converges absolutely by the Limit Comparison Test since $\lim\limits_{n \to \infty} \frac{\left(\frac{1}{n^{3/2}}\right)}{\left(\frac{1}{\sqrt{n(n+1)(n+2)}}\right)} = \sqrt{\lim\limits_{n \to \infty} \frac{n(n+1)(n+2)}{n^3}} = 1$

41. $\lim\limits_{n \to \infty} \left|\frac{u_{n+1}}{u_n}\right| < 1 \Rightarrow \lim\limits_{n \to \infty} \left|\frac{(x+4)^{n+1}}{(n+1)3^{n+1}} \cdot \frac{n3^n}{(x+4)^n}\right| < 1 \Rightarrow \frac{|x+4|}{3} \lim\limits_{n \to \infty} \left(\frac{n}{n+1}\right) < 1 \Rightarrow \frac{|x+4|}{3} < 1$

$\Rightarrow |x+4| < 3 \Rightarrow -3 < x+4 < 3 \Rightarrow -7 < x < -1;$ at $x = -7$ we have $\sum\limits_{n=1}^{\infty} \frac{(-1)^n 3^n}{n3^n} = \sum\limits_{n=1}^{\infty} \frac{(-1)^n}{n}$, the alternating harmonic series, which converges conditionally; at $x = -1$ we have $\sum\limits_{n=1}^{\infty} \frac{3^n}{n3^n} = \sum\limits_{n=1}^{\infty} \frac{1}{n}$, the divergent

harmonic series

(a) the radius is 3; the interval of convergence is $-7 \leq x < -1$

(b) the interval of absolute convergence is $-7 < x < -1$

(c) the series converges conditionally at $x = -7$

43. $\lim\limits_{n \to \infty} \left| \frac{u_{n+1}}{u_n} \right| < 1 \Rightarrow \lim\limits_{n \to \infty} \left| \frac{(3x-1)^{n+1}}{(n+1)^2} \cdot \frac{n^2}{(3x-1)^n} \right| < 1 \Rightarrow |3x-1| \lim\limits_{n \to \infty} \frac{n^2}{(n+1)^2} < 1 \Rightarrow |3x-1| < 1$

$\Rightarrow -1 < 3x - 1 < 1 \Rightarrow 0 < 3x < 2 \Rightarrow 0 < x < \frac{2}{3}$; at $x = 0$ we have $\sum\limits_{n=1}^{\infty} \frac{(-1)^{n-1}(-1)^n}{n^2} = \sum\limits_{n=1}^{\infty} \frac{(-1)^{2n-1}}{n^2}$

$= -\sum\limits_{n=1}^{\infty} \frac{1}{n^2}$, a nonzero constant multiple of a convergent p-series, which is absolutely convergent; at $x = \frac{2}{3}$ we

have $\sum\limits_{n=1}^{\infty} \frac{(-1)^{n-1}(1)^n}{n^2} = \sum\limits_{n=1}^{\infty} \frac{(-1)^{n-1}}{n^2}$, which converges absolutely

(a) the radius is $\frac{1}{3}$; the interval of convergence is $0 \leq x \leq \frac{2}{3}$

(b) the interval of absolute convergence is $0 \leq x \leq \frac{2}{3}$

(c) there are no values for which the series converges conditionally

45. $\lim\limits_{n \to \infty} \left| \frac{u_{n+1}}{u_n} \right| < 1 \Rightarrow \lim\limits_{n \to \infty} \left| \frac{x^{n+1}}{(n+1)^{n+1}} \cdot \frac{n^n}{x^n} \right| < 1 \Rightarrow |x| \lim\limits_{n \to \infty} \left| \left(\frac{n}{n+1}\right)^n \left(\frac{1}{n+1}\right) \right| < 1 \Rightarrow \frac{|x|}{e} \lim\limits_{n \to \infty} \left(\frac{1}{n+1}\right) < 1$

$\Rightarrow \frac{|x|}{e} \cdot 0 < 1$, which holds for all x

(a) the radius is ∞; the series converges for all x

(b) the series converges absolutely for all x

(c) there are no values for which the series converges conditionally

47. $\lim\limits_{n \to \infty} \left| \frac{u_{n+1}}{u_n} \right| < 1 \Rightarrow \lim\limits_{n \to \infty} \left| \frac{(n+2)x^{2n+1}}{3^{n+1}} \cdot \frac{3^n}{(n+1)x^{2n-1}} \right| < 1 \Rightarrow \frac{x^2}{3} \lim\limits_{n \to \infty} \left(\frac{n+2}{n+1}\right) < 1 \Rightarrow -\sqrt{3} < x < \sqrt{3}$;

the series $\sum\limits_{n=1}^{\infty} -\frac{n+1}{\sqrt{3}}$ and $\sum\limits_{n=1}^{\infty} \frac{n+1}{\sqrt{3}}$, obtained with $x = \pm\sqrt{3}$, both diverge

(a) the radius is $\sqrt{3}$; the interval of convergence is $-\sqrt{3} < x < \sqrt{3}$

(b) the interval of absolute convergence is $-\sqrt{3} < x < \sqrt{3}$

(c) there are no values for which the series converges conditionally

49. $\lim\limits_{n \to \infty} \left| \frac{u_{n+1}}{u_n} \right| < 1 \Rightarrow \lim\limits_{n \to \infty} \left| \frac{\text{csch}\,(n+1)x^{n+1}}{\text{csch}\,(n)x^n} \right| < 1 \Rightarrow |x| \lim\limits_{n \to \infty} \left| \frac{\left(\frac{2}{e^{n+1} - e^{-n-1}}\right)}{\left(\frac{2}{e^n - e^{-n}}\right)} \right| < 1$

$\Rightarrow |x| \lim\limits_{n \to \infty} \left| \frac{e^{-1} - e^{-2n-1}}{1 - e^{-2n-2}} \right| < 1 \Rightarrow \frac{|x|}{e} < 1 \Rightarrow -e < x < e$; the series $\sum\limits_{n=1}^{\infty} (\pm e)^n \,\text{csch}\,n$, obtained with $x = \pm e$,

both diverge since $\lim\limits_{n \to \infty} (\pm e)^n \,\text{csch}\,n \neq 0$

(a) the radius is e; the interval of convergence is $-e < x < e$

(b) the interval of absolute convergence is $-e < x < e$

(c) there are no values for which the series converges conditionally

51. The given series has the form $1 - x + x^2 - x^3 + \ldots + (-x)^n + \ldots = \frac{1}{1+x}$, where $x = \frac{1}{4}$; the sum is $\frac{1}{1+\left(\frac{1}{4}\right)} = \frac{4}{5}$

53. The given series has the form $x - \frac{x^3}{3!} + \frac{x^5}{5!} - \ldots + (-1)^n \frac{x^{2n+1}}{(2n+1)!} + \ldots = \sin x$, where $x = \pi$; the sum is $\sin \pi = 0$

55. The given series has the form $1 + x + \frac{x^2}{2!} + \frac{x^2}{3!} + \ldots + \frac{x^n}{n!} + \ldots = e^x$, where $x = \ln 2$; the sum is $e^{\ln(2)} = 2$

57. Consider $\frac{1}{1-2x}$ as the sum of a convergent geometric series with $a = 1$ and $r = 2x \Rightarrow \frac{1}{1-2x}$

$= 1 + (2x) + (2x)^2 + (2x)^3 + \ldots = \sum_{n=0}^{\infty} (2x)^n = \sum_{n=0}^{\infty} 2^n x^n$ where $|2x| < 1 \Rightarrow |x| < \frac{1}{2}$

59. $\sin x = \sum_{n=0}^{\infty} \frac{(-1)^n x^{2n+1}}{(2n+1)!} \Rightarrow \sin \pi x = \sum_{n=0}^{\infty} \frac{(-1)^n (\pi x)^{2n+1}}{(2n+1)!} = \sum_{n=0}^{\infty} \frac{(-1)^n \pi^{2n+1} x^{2n+1}}{(2n+1)!}$

61. $\cos x = \sum_{n=0}^{\infty} \frac{(-1)^n x^{2n}}{(2n)!} \Rightarrow \cos\left(x^{5/2}\right) = \sum_{n=0}^{\infty} \frac{(-1)^n \left(x^{5/2}\right)^{2n}}{(2n)!} = \sum_{n=0}^{\infty} \frac{(-1)^n x^{5n}}{(2n)!}$

63. $e^x = \sum_{n=0}^{\infty} \frac{x^n}{n!} \Rightarrow e^{(\pi x/2)} = \sum_{n=0}^{\infty} \frac{\left(\frac{\pi x}{2}\right)^n}{n!} = \sum_{n=0}^{\infty} \frac{\pi^n x^n}{2^n n!}$

65. $f(x) = \sqrt{3 + x^2} = (3 + x^2)^{1/2} \Rightarrow f'(x) = x(3 + x^2)^{-1/2} \Rightarrow f''(x) = -x^2(3 + x^2)^{-3/2} + (3 + x^2)^{-1/2}$

$\Rightarrow f'''(x) = 3x^3(3 + x^2)^{-5/2} - 3x(3 + x^2)^{-3/2}; f(-1) = 2, f'(-1) = -\frac{1}{2}, f''(-1) = -\frac{1}{8} + \frac{1}{2} = \frac{3}{8},$

$f'''(-1) = -\frac{3}{32} + \frac{3}{8} = \frac{9}{32} \Rightarrow \sqrt{3 + x^2} = 2 - \frac{(x+1)}{2 \cdot 1!} + \frac{3(x+1)^2}{2^3 \cdot 2!} + \frac{9(x+1)^3}{2^5 \cdot 3!} + \ldots$

67. $f(x) = \frac{1}{x+1} = (x + 1)^{-1} \Rightarrow f'(x) = -(x + 1)^{-2} \Rightarrow f''(x) = 2(x + 1)^{-3} \Rightarrow f'''(x) = -6(x + 1)^{-4}; f(3) = \frac{1}{4},$

$f'(3) = -\frac{1}{4^2}, f''(3) = \frac{2}{4^3}, f'''(2) = \frac{-6}{4^4} \Rightarrow \frac{1}{x+1} = \frac{1}{4} - \frac{1}{4^2}(x - 3) + \frac{1}{4^3}(x - 3)^2 - \frac{1}{4^4}(x - 3)^3 + \ldots$

69. Assume the solution has the form $y = a_0 + a_1 x + a_2 x^2 + \ldots + a_{n-1} x^{n-1} + a_n x^n + \ldots$

$\Rightarrow \frac{dy}{dx} = a_1 + 2a_2 x + \ldots + na_n x^{n-1} + \ldots \Rightarrow \frac{dy}{dx} + y$

$= (a_1 + a_0) + (2a_2 + a_1)x + (3a_3 + a_2)x^2 + \ldots + (na_n + a_{n-1})x^{n-1} + \ldots = 0 \Rightarrow a_1 + a_0 = 0, 2a_2 + a_1 = 0,$

$3a_3 + a_2 = 0$ and in general $na_n + a_{n-1} = 0$. Since $y = -1$ when $x = 0$ we have $a_0 = -1$. Therefore $a_1 = 1$,

$a_2 = \frac{-a_1}{2 \cdot 1} = -\frac{1}{2}, a_3 = \frac{-a_2}{3} = \frac{1}{3 \cdot 2}, a_4 = \frac{-a_3}{4} = -\frac{1}{4 \cdot 3 \cdot 2}, \ldots, a_n = \frac{-a_{n-1}}{n} = \frac{-1}{n} \frac{(-1)^n}{(n-1)!} = \frac{(-1)^{n+1}}{n!}$

$\Rightarrow y = -1 + x - \frac{1}{2}x^2 + \frac{1}{3 \cdot 2}x^3 - \ldots + \frac{(-1)^{n+1}}{n!}x^n + \ldots = -\sum_{n=0}^{\infty} \frac{(-1)^n x^n}{n!} = -e^{-x}$

71. Assume the solution has the form $y = a_0 + a_1 x + a_2 x^2 + \ldots + a_{n-1} x^{n-1} + a_n x^n + \ldots$

$\Rightarrow \frac{dy}{dx} = a_1 + 2a_2 x + \ldots + na_n x^{n-1} + \ldots \Rightarrow \frac{dy}{dx} + 2y$

$= (a_1 + 2a_0) + (2a_2 + 2a_1)x + (3a_3 + 2a_2)x^2 + \ldots + (na_n + 2a_{n-1})x^{n-1} + \ldots = 0$. Since $y = 3$ when $x = 0$ we

have $a_0 = 3$. Therefore $a_1 = -2a_0 = -2(3) = -3(2), a_2 = -\frac{2}{2}a_1 = -\frac{2}{2}(-2 \cdot 3) = 3\left(\frac{2^2}{2}\right), a_3 = -\frac{2}{3}a_2$

$= -\frac{2}{3}\left[3\left(\frac{2^2}{2}\right)\right] = -3\left(\frac{2^3}{3 \cdot 2}\right), \ldots, a_n = \left(-\frac{2}{n}\right)a_{n-1} = \left(-\frac{2}{n}\right)\left(3\left(\frac{(-1)^{n-1}2^{n-1}}{(n-1)!}\right)\right) = 3\left(\frac{(-1)^n 2^n}{n!}\right)$

$\Rightarrow y = 3 - 3(2x) + 3\frac{(2)^2}{2}x^2 - 3\frac{(2)^3}{3 \cdot 2}x^3 + \ldots + 3\frac{(-1)^n 2^n}{n!}x^n + \ldots$

$= 3\left[1 - (2x) + \frac{(2x)^2}{2!} - \frac{(2x)^3}{3!} + \ldots + \frac{(-1)^n (2x)^n}{n!} + \ldots\right] = 3\sum_{n=0}^{\infty} \frac{(-1)^n (2x)^n}{n!} = 3e^{-2x}$

73. Assume the solution has the form $y = a_0 + a_1 x + a_2 x^2 + \ldots + a_{n-1} x^{n-1} + a_n x^n + \ldots$

$\Rightarrow \frac{dy}{dx} = a_1 + 2a_2 x + \ldots + na_n x^{n-1} + \ldots \Rightarrow \frac{dy}{dx} - y$

$= (a_1 - a_0) + (2a_2 - a_1)x + (3a_3 - a_2)x^2 + \ldots + (na_n - a_{n-1})x^{n-1} + \ldots = 3x \Rightarrow a_1 - a_0 = 0, 2a_2 - a_1 = 3,$

$3a_3 - a_2 = 0$ and in general $na_n - a_{n-1} = 0$ for $n > 2$. Since $y = -1$ when $x = 0$ we have $a_0 = -1$. Therefore

$a_1 = -1, a_2 = \frac{3 + a_1}{2} = \frac{2}{2}, a_3 = \frac{a_2}{3} = \frac{2}{3 \cdot 2}, a_4 = \frac{a_3}{4} = \frac{2}{4 \cdot 3 \cdot 2}, \ldots, a_n = \frac{a_{n-1}}{n} = \frac{2}{n!}$

$\Rightarrow y = -1 - x + \left(\frac{2}{2}\right)x^2 + \frac{3}{3 \cdot 2}x^3 + \frac{2}{4 \cdot 3 \cdot 2}x^4 + \ldots + \frac{2}{n!}x^n + \ldots$

$= 2\left(1 + x + \frac{1}{2}x^2 + \frac{1}{3 \cdot 2}x^3 + \frac{1}{4 \cdot 3 \cdot 2}x^4 + \ldots + \frac{1}{n!}x^n + \ldots\right) - 3 - 3x = 2\sum_{n=0}^{\infty} \frac{x^n}{n!} - 3 - 3x = 2e^x - 3x - 3$

75. Assume the solution has the form $y = a_0 + a_1x + a_2x^2 + \ldots + a_{n-1}x^{n-1} + a_nx^n + \ldots$

$\Rightarrow \frac{dy}{dx} = a_1 + 2a_2x + \ldots + na_nx^{n-1} + \ldots \Rightarrow \frac{dy}{dx} - y$

$= (a_1 - a_0) + (2a_2 - a_1)x + (3a_3 - a_2)x^2 + \ldots + (na_n - a_{n-1})x^{n-1} + \ldots = x \Rightarrow a_1 - a_0 = 0, 2a_2 - a_1 = 1,$

$3a_3 - a_2 = 0$ and in general $na_n - a_{n-1} = 0$ for $n > 2$. Since $y = 1$ when $x = 0$ we have $a_0 = 1$. Therefore

$a_1 = 1, a_2 = \frac{1+a_1}{2} = \frac{2}{2}, a_3 = \frac{a_2}{3} = \frac{2}{3\cdot2}, a_4 = \frac{a_3}{4} = \frac{2}{4\cdot3\cdot2}, \ldots, a_n = \frac{a_{n-1}}{n} = \frac{2}{n!}$

$\Rightarrow y = 1 + x + \left(\frac{2}{2}\right)x^2 + \frac{2}{3\cdot2}x^3 + \frac{2}{4\cdot2\cdot2}x^4 + \ldots + \frac{2}{n!}x^n + \ldots$

$= 2\left(1 + x + \frac{1}{2}x^2 + \frac{1}{3\cdot2}x^3 + \frac{1}{4\cdot3\cdot2}x^4 + \ldots + \frac{1}{n!}x^n + \ldots\right) - 1 - x = 2\sum_{n=0}^{\infty}\frac{x^n}{n!} - 1 - x = 2e^x - x - 1$

77. $\int_0^{1/2} \exp(-x^3)\,dx = \int_0^{1/2}\left(1 - x^3 + \frac{x^6}{2!} - \frac{x^9}{3!} + \frac{x^{12}}{4!} + \ldots\right)dx = \left[x - \frac{x^4}{4} + \frac{x^7}{7\cdot2!} - \frac{x^{10}}{10\cdot3!} + \frac{x^{13}}{13\cdot4!} - \ldots\right]_0^{1/2}$

$\approx \frac{1}{2} - \frac{1}{2^4\cdot4} + \frac{1}{2^7\cdot7\cdot2!} - \frac{1}{2^{10}\cdot10\cdot3!} + \frac{1}{2^{13}\cdot13\cdot4!} - \frac{1}{2^{16}\cdot16\cdot5!} \approx 0.484917143$

79. $\int_1^{1/2} \frac{\tan^{-1}x}{x}\,dx = \int_1^{1/2}\left(1 - \frac{x^2}{3} + \frac{x^4}{5} - \frac{x^6}{7} + \frac{x^8}{9} - \frac{x^{10}}{11} + \ldots\right)dx = \left[x - \frac{x^3}{9} + \frac{x^5}{25} - \frac{x^7}{49} + \frac{x^9}{81} - \frac{x^{11}}{121} + \ldots\right]_0^{1/2}$

$\approx \frac{1}{2} - \frac{1}{9\cdot2^3} + \frac{1}{5^2\cdot2^5} - \frac{1}{7^2\cdot2^7} + \frac{1}{9^2\cdot2^9} - \frac{1}{11^2\cdot2^{11}} + \frac{1}{13^2\cdot2^{13}} - \frac{1}{15^2\cdot2^{15}} + \frac{1}{17^2\cdot2^{17}} - \frac{1}{19^2\cdot2^{19}} + \frac{1}{21^2\cdot2^{21}}$

≈ 0.4872223583

81. $\lim_{x \to 0} \frac{7\sin x}{e^{2x} - 1} = \lim_{x \to 0} \frac{7\left(x - \frac{x^3}{3!} + \frac{x^5}{5!} - \ldots\right)}{\left(2x + \frac{2^2x^2}{2!} + \frac{2^3x^3}{3!} + \ldots\right)} = \lim_{x \to 0} \frac{7\left(1 - \frac{x^2}{3!} + \frac{x^4}{5!} - \ldots\right)}{\left(2 + \frac{2^2x}{2!} + \frac{2^3x^2}{3!} + \ldots\right)} = \frac{7}{2}$

83. $\lim_{t \to 0}\left(\frac{1}{2 - 2\cos t} - \frac{1}{t^2}\right) = \lim_{t \to 0}\frac{t^2 - 2 + 2\cos t}{2t^2(1 - \cos t)} = \lim_{t \to 0}\frac{t^2 - 2 + 2\left(1 - \frac{t^2}{2} + \frac{t^4}{4!} - \ldots\right)}{2t^2\left(1 - 1 + \frac{t^2}{2} - \frac{t^4}{4!} + \ldots\right)} = \lim_{t \to 0}\frac{2\left(\frac{t^4}{4!} - \frac{t^6}{6!} + \ldots\right)}{\left(t^4 - \frac{2t^6}{4!} + \ldots\right)}$

$= \lim_{t \to 0}\frac{2\left(\frac{1}{4!} - \frac{t^2}{6!} + \ldots\right)}{\left(1 - \frac{2t^2}{4!} + \ldots\right)} = \frac{1}{12}$

85. $\lim_{z \to 0}\frac{1 - \cos^2 z}{\ln(1 - z) + \sin z} = \lim_{z \to 0}\frac{1 - \left(1 - z^2 + \frac{z^4}{3} - \ldots\right)}{\left(-z - \frac{z^2}{2} - \frac{z^3}{3} - \ldots\right) + \left(z - \frac{z^3}{3!} + \frac{z^5}{5!} - \ldots\right)} = \lim_{z \to 0}\frac{\left(z^2 - \frac{z^4}{3} + \ldots\right)}{\left(-\frac{z^2}{2} - \frac{2z^3}{3} - \frac{z^4}{4} - \ldots\right)}$

$= \lim_{z \to 0}\frac{\left(1 - \frac{z^2}{3} + \ldots\right)}{\left(-\frac{1}{2} - \frac{2z}{3} - \frac{z^2}{4} - \ldots\right)} = -2$

87. $\lim_{x \to 0}\left(\frac{\sin 3x}{x^3} + \frac{r}{x^2} + s\right) = \lim_{x \to 0}\left[\frac{\left(3x - \frac{(3x)^3}{6} + \frac{(3x)^5}{120} - \ldots\right)}{x^3} + \frac{r}{x^2} + s\right] = \lim_{x \to 0}\left(\frac{3}{x^2} - \frac{9}{2} + \frac{81x^2}{40} + \ldots + \frac{r}{x^2} + s\right) = 0$

$\Rightarrow \frac{r}{x^2} + \frac{3}{x^2} = 0$ and $s - \frac{9}{2} = 0 \Rightarrow r = -3$ and $s = \frac{9}{2}$

89. (a) $\sum_{n=1}^{\infty}\left(\sin\frac{1}{2n} - \sin\frac{1}{2n+1}\right) = \left(\sin\frac{1}{2} - \sin\frac{1}{3}\right) + \left(\sin\frac{1}{4} - \sin\frac{1}{5}\right) + \left(\sin\frac{1}{6} - \sin\frac{1}{7}\right) + \ldots + \left(\sin\frac{1}{2n} - \sin\frac{1}{2n+1}\right)$

$+ \ldots = \sum_{n=2}^{\infty}(-1)^n\sin\frac{1}{n}; f(x) = \sin\frac{1}{x} \Rightarrow f'(x) = \frac{-\cos\left(\frac{1}{x}\right)}{x^2} < 0$ if $x \geq 2 \Rightarrow \sin\frac{1}{n+1} < \sin\frac{1}{n}$, and

$\lim_{n \to \infty}\sin\frac{1}{n} = 0 \Rightarrow \sum_{n=2}^{\infty}(-1)^n\sin\frac{1}{n}$ converges by the Alternating Series Test

(b) $|error| < \left|\sin\frac{1}{42}\right| \approx 0.02381$ and the sum is an underestimate because the remainder is positive

91. $\lim_{n \to \infty}\left|\frac{2\cdot5\cdot8\cdots(3n-1)(3n+2)x^{n+1}}{2\cdot4\cdot6\cdots(2n)(2n+2)} \cdot \frac{2\cdot4\cdot6\cdots(2n)}{2\cdot5\cdot8\cdots(3n-1)x^n}\right| < 1 \Rightarrow |x|\lim_{n \to \infty}\left|\frac{3n+2}{2n+2}\right| < 1 \Rightarrow |x| < \frac{2}{3}$

$\Rightarrow$ the radius of convergence is $\frac{2}{3}$

93. $\sum\limits_{k=2}^{n} \ln\left(1 - \frac{1}{k^2}\right) = \sum\limits_{k=2}^{n} \left[\ln\left(1 + \frac{1}{k}\right) + \ln\left(1 - \frac{1}{k}\right)\right] = \sum\limits_{k=2}^{n} \left[\ln(k+1) - \ln k + \ln(k-1) - \ln k\right]$

$= [\ln 3 - \ln 2 + \ln 1 - \ln 2] + [\ln 4 - \ln 3 + \ln 2 - \ln 3] + [\ln 5 - \ln 4 + \ln 3 - \ln 4] + [\ln 6 - \ln 5 + \ln 4 - \ln 5]$

$+ \ldots + [\ln(n+1) - \ln n + \ln(n-1) - \ln n] = [\ln 1 - \ln 2] + [\ln(n+1) - \ln n] \qquad$ after cancellation

$\Rightarrow \sum\limits_{k=2}^{n} \ln\left(1 - \frac{1}{k^2}\right) = \ln\left(\frac{n+1}{2n}\right) \Rightarrow \sum\limits_{k=2}^{\infty} \ln\left(1 - \frac{1}{k^2}\right) = \lim\limits_{n \to \infty} \ln\left(\frac{n+1}{2n}\right) = \ln \frac{1}{2}$ is the sum

95. (a) $\lim\limits_{n \to \infty} \left| \frac{1 \cdot 4 \cdot 7 \cdots (3n-2)(3n+1)x^{3n+3}}{(3n+3)!} \cdot \frac{(3n)!}{1 \cdot 4 \cdot 7 \cdots (3n-2)x^{3n}} \right| < 1 \Rightarrow |x^3| \lim\limits_{n \to \infty} \frac{(3n+1)}{(3n+1)(3n+2)(3n+3)}$

$= |x^3| \cdot 0 < 1 \Rightarrow$ the radius of convergence is ∞

(b) $y = 1 + \sum\limits_{n=1}^{\infty} \frac{1 \cdot 4 \cdot 7 \cdots (3n-2)}{(3n)!} x^{3n} \Rightarrow \frac{dy}{dx} = \sum\limits_{n=1}^{\infty} \frac{1 \cdot 4 \cdot 7 \cdots (3n-2)}{(3n-1)!} x^{3n-1}$

$\Rightarrow \frac{d^2y}{dx^2} = \sum\limits_{n=1}^{\infty} \frac{1 \cdot 4 \cdot 7 \cdots (3n-2)}{(3n-2)!} x^{3n-2} = x + \sum\limits_{n=2}^{\infty} \frac{1 \cdot 4 \cdot 7 \cdots (3n-5)}{(3n-3)!} x^{3n-2}$

$= x \left(1 + \sum\limits_{n=1}^{\infty} \frac{1 \cdot 4 \cdot 7 \cdots (3n-2)}{(3n)!} x^{3n} \right) = xy + 0 \Rightarrow a = 1$ and $b = 0$

97. Yes, the series $\sum\limits_{n=1}^{\infty} a_n b_n$ converges as we now show. Since $\sum\limits_{n=1}^{\infty} a_n$ converges it follows that $a_n \to 0 \Rightarrow a_n < 1$

for $n >$ some index $N \Rightarrow a_n b_n < b_n$ for $n > N \Rightarrow \sum\limits_{n=1}^{\infty} a_n b_n$ converges by the Direct Comparison Test with $\sum\limits_{n=1}^{\infty} b_n$

99. $\sum\limits_{n=1}^{\infty} (x_{n+1} - x_n) = \lim\limits_{n \to \infty} \sum\limits_{k=1}^{\infty} (x_{k+1} - x_k) = \lim\limits_{n \to \infty} (x_{n+1} - x_1) = \lim\limits_{n \to \infty} (x_{n+1}) - x_1 \Rightarrow$ both the series and

sequence must either converge or diverge.

101. Newton's method gives $x_{n+1} = x_n - \frac{(x_n - 1)^{40}}{40(x_n - 1)^{39}} = \frac{39}{40} x_n + \frac{1}{40}$, and if the sequence $\{x_n\}$ has the limit L, then

$L = \frac{39}{40} L + \frac{1}{40} \Rightarrow L = 1$ and $\{x_n\}$ converges since $\left| \frac{f(x)f''(x)}{[f'(x)]^2} \right| = \frac{39}{40} < 1$

103. $a_n = \frac{1}{\ln n}$ for $n \geq 2 \Rightarrow a_2 \geq a_3 \geq a_4 \geq \ldots$, and $\frac{1}{\ln 2} + \frac{1}{\ln 4} + \frac{1}{\ln 8} + \ldots = \frac{1}{\ln 2} + \frac{1}{2\ln 2} + \frac{1}{3\ln 2} + \ldots$

$= \frac{1}{\ln 2} \left(1 + \frac{1}{2} + \frac{1}{3} + \ldots \right)$ which diverges so that $1 + \sum\limits_{n=2}^{\infty} \frac{1}{n \ln n}$ diverges by the Integral Test.

105. $a_0 = \frac{1}{2\pi} \int_0^{2\pi} f(x)\,dx = \frac{1}{2\pi} \int_\pi^{2\pi} 1\,dx = \frac{1}{2}$, $a_k = \frac{1}{\pi} \int_0^{2\pi} f(x) \cos kx\,dx = \frac{1}{\pi} \int_\pi^{2\pi} \cos kx\,dx = 0$.

$b_k = \frac{1}{\pi} \int_0^{2\pi} f(x) \sin kx\,dx = \frac{1}{\pi} \int_\pi^{2\pi} \sin kx\,dx = -\frac{\cos kx}{\pi k} \Big|_\pi^{2\pi} = -\frac{1}{\pi k} \left(1 - (-1)^k \right) = \begin{cases} -\frac{2}{\pi k}, & k \text{ odd} \\ 0, & k \text{ even} \end{cases}$.

Thus, the Fourier series of $f(x)$ is $\frac{1}{2} - \sum\limits_{k \text{ odd}} \frac{2}{\pi k} \sin kx$

107. $a_0 = \frac{1}{2\pi} \left[\int_0^\pi (\pi - x)\,dx + \int_\pi^{2\pi} (x - 2\pi)\,dx \right] = \frac{1}{2\pi} \left[\int_0^\pi (\pi - x)\,dx - \int_0^\pi (\pi - u)\,du \right] = 0$ where we used the

substitution $u = x - \pi$ in the second integral. We have $a_k = \frac{1}{\pi} \left[\int_0^\pi (\pi - x) \cos kx\,dx + \int_\pi^{2\pi} (x - 2\pi) \cos kx\,dx \right]$. Using the substitution $u = x - \pi$ in the second integral gives $\int_\pi^{2\pi} (x - 2\pi) \cos kx \, dx = \int_0^\pi -(\pi - u) \cos(ku + k\pi) \, du$

$$= \begin{cases} \int_0^\pi (\pi - u) \cos ku \, du, & k \text{ odd} \\ \int_0^\pi -(\pi - u) \cos ku \, du, & k \text{ even} \end{cases}.$$

Thus, $a_k = \begin{cases} \frac{2}{\pi} \int_0^\pi (\pi - x) \cos kx \, dx, & k \text{ odd} \\ 0, & k \text{ even} \end{cases}$.

Now, since k is odd, letting $v = \pi - x \Rightarrow \frac{2}{\pi} \int_0^\pi (\pi - x) \cos kx \, dx = -\frac{2}{\pi} \int_0^\pi v \cos kv \, dv = -\frac{2}{\pi} \left(-\frac{2}{k^2} \right) = \frac{4}{\pi k^2}$, k odd. (See

Exercise 106). So, $a_k = \begin{cases} \frac{4}{\pi k^2}, & k \text{ odd} \\ 0, & k \text{ even} \end{cases}$.

Using similar techniques we see that $b_k = \begin{cases} \frac{2}{\pi} \int_0^\pi (\pi - u) \sin ku \, du, & k \text{ odd} \\ 0, & k \text{ even} \end{cases} = \begin{cases} \frac{2}{k}, & k \text{ odd} \\ 0, & k \text{ even} \end{cases}$.

Thus, the Fourier series of $f(x)$ is $\sum_{k \text{ odd}} \left(\frac{4}{\pi k^2} \cos kx + \frac{2}{k} \sin kx \right)$.

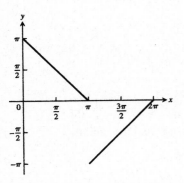

CHAPTER 11 ADDITIONAL AND ADVANCED EXERCISES

1. converges since $\frac{1}{(3n-2)^{(2n+1)/2}} < \frac{1}{(3n-2)^{3/2}}$ and $\sum_{n=1}^\infty \frac{1}{(3n-2)^{3/2}}$ converges by the Limit Comparison Test:

$$\lim_{n \to \infty} \frac{\left(\frac{1}{n^{3/2}} \right)}{\left(\frac{1}{(3n-2)^{3/2}} \right)} = \lim_{n \to \infty} \left(\frac{3n-2}{n} \right)^{3/2} = 3^{3/2}$$

3. diverges by the nth-Term Test since $\lim_{n \to \infty} a_n = \lim_{n \to \infty} (-1)^n \tanh n = \lim_{b \to \infty} (-1)^n \left(\frac{1 - e^{-2n}}{1 + e^{-2n}} \right) = \lim_{n \to \infty} (-1)^n$

   does not exist

5. converges by the Direct Comparison Test: $a_1 = 1 = \frac{12}{(1)(3)(2)^2}$, $a_2 = \frac{1 \cdot 2}{3 \cdot 4} = \frac{12}{(2)(4)(3)^2}$, $a_3 = \left(\frac{2 \cdot 3}{4 \cdot 5} \right) \left(\frac{1 \cdot 2}{3 \cdot 4} \right)$

   $= \frac{12}{(3)(5)(4)^2}$, $a_4 = \left(\frac{3 \cdot 4}{5 \cdot 6} \right) \left(\frac{2 \cdot 3}{4 \cdot 5} \right) \left(\frac{1 \cdot 2}{3 \cdot 4} \right) = \frac{12}{(4)(6)(5)^2}, \ldots \Rightarrow 1 + \sum_{n=1}^\infty \frac{12}{(n+1)(n+3)(n+2)^2}$ represents the

   given series and $\frac{12}{(n+1)(n+3)(n+2)^2} < \frac{12}{n^4}$, which is the nth-term of a convergent p-series

7. diverges by the nth-Term Test since if $a_n \to L$ as $n \to \infty$, then $L = \frac{1}{1+L} \Rightarrow L^2 + L - 1 = 0 \Rightarrow L = \frac{-1 \pm \sqrt{5}}{2}$

   $\neq 0$

9. $f(x) = \cos x$ with $a = \frac{\pi}{3} \Rightarrow f\left(\frac{\pi}{3} \right) = 0.5, f'\left(\frac{\pi}{3} \right) = -\frac{\sqrt{3}}{2}, f''\left(\frac{\pi}{3} \right) = -0.5, f'''\left(\frac{\pi}{3} \right) = \frac{\sqrt{3}}{2}, f^{(4)}\left(\frac{\pi}{3} \right) = 0.5$;

   $\cos x = \frac{1}{2} - \frac{\sqrt{3}}{2} \left(x - \frac{\pi}{3} \right) - \frac{1}{4} \left(x - \frac{\pi}{3} \right)^2 + \frac{\sqrt{3}}{12} \left(x - \frac{\pi}{3} \right)^3 + \cdots$

11. $e^x = 1 + x + \frac{x^2}{2!} + \frac{x^3}{3!} + \cdots$ with $a = 0$

13. $f(x) = \cos x$ with $a = 22\pi \Rightarrow f(22\pi) = 1, f'(22\pi) = 0, f''(22\pi) = -1, f'''(22\pi) = 0, f^{(4)}(22\pi) = 1,$
$f^{(5)}(22\pi) = 0, f^{(6)}(22\pi) = -1; \cos x = 1 - \frac{1}{2}(x - 22\pi)^2 + \frac{1}{4!}(x - 22\pi)^4 - \frac{1}{6!}(x - 22\pi)^6 + \ldots$

15. Yes, the sequence converges: $c_n = (a^n + b^n)^{1/n} \Rightarrow c_n = b\left(\left(\frac{a}{b}\right)^n + 1\right)^{1/n} \Rightarrow \lim\limits_{n \to \infty} c_n = \ln b + \lim\limits_{n \to \infty} \frac{\ln\left(\left(\frac{a}{b}\right)^n + 1\right)}{n}$
$= \ln b + \lim\limits_{n \to \infty} \frac{\left(\frac{a}{b}\right)^n \ln\left(\frac{a}{b}\right)}{\left(\frac{a}{b}\right)^n + 1} = \ln b + \frac{0 \cdot \ln\left(\frac{a}{b}\right)}{0 + 1} = \ln b$ since $0 < a < b$. Thus, $\lim\limits_{n \to \infty} c_n = e^{\ln b} = b$.

17. $s_n = \sum\limits_{k=0}^{n-1} \int_k^{k+1} \frac{dx}{1 + x^2} \Rightarrow s_n = \int_0^1 \frac{dx}{1 + x^2} + \int_1^2 \frac{dx}{1 + x^2} + \ldots + \int_{n-1}^n \frac{dx}{1 + x^2} \Rightarrow s_n = \int_0^n \frac{dx}{1 + x^2}$
$\Rightarrow \lim\limits_{n \to \infty} s_n = \lim\limits_{n \to \infty} (\tan^{-1} n - \tan^{-1} 0) = \frac{\pi}{2}$

19. (a) Each A_{n+1} fits into the corresponding upper triangular region, whose vertices are:
$(n, f(n) - f(n+1)), (n+1, f(n+1))$ and $(n, f(n))$ along the line whose slope is $f(n + 1) - f(n)$.

All the A_n's fit into the first upper triangular region whose area is $\frac{f(1) - f(2)}{2} \Rightarrow \sum\limits_{n=1}^{\infty} A_n < \frac{f(1) - f(2)}{2}$

(b) If $A_k = \frac{f(k + 1) + f(k)}{2} - \int_k^{k+1} f(x)\, dx$, then

$\sum\limits_{k=1}^{n-1} A_k = \frac{f(1) + f(2) + f(2) + f(3) + f(3) + \ldots + f(n-1) + f(n)}{2} - \int_1^2 f(x)\, dx - \int_2^3 f(x)\, dx - \ldots - \int_{n-1}^n f(x)\, dx$

$= \frac{f(1) + f(n)}{2} + \sum\limits_{k=2}^{n-1} f(k) - \int_1^n f(x)\, dx \Rightarrow \sum\limits_{k=1}^{n-1} A_k = \sum\limits_{k=1}^n f(k) - \frac{f(1) + f(n)}{2} - \int_1^n f(x)\, dx < \frac{f(1) - f(2)}{2}$, from

part (a). The sequence $\left\{\sum\limits_{k=1}^{n-1} A_k\right\}$ is bounded above and increasing, so it converges and the limit in
question must exist.

(c) Let $L = \lim\limits_{n \to \infty} \left[\sum\limits_{k=1}^n f(k) - \int_1^n f(x)\, dx - \frac{1}{2}(f(1) + f(n))\right]$, which exists by part (b). Since f is positive and

decreasing $\lim\limits_{n \to \infty} f(n) = M \geq 0$ exists. Thus $\lim\limits_{n \to \infty} \left[\sum\limits_{k=1}^n f(k) - \int_1^n f(x)\, dx\right] = L + \frac{1}{2}(f(1) + M)$.

21. (a) No, the limit does not appear to depend on the value of the constant a
(b) Yes, the limit depends on the value of b

(c) $s = \left(1 - \frac{\cos\left(\frac{a}{n}\right)}{n}\right)^n \Rightarrow \ln s = \frac{\ln\left(1 - \frac{\cos\left(\frac{a}{n}\right)}{n}\right)}{\left(\frac{1}{n}\right)} \Rightarrow \lim\limits_{n \to \infty} \ln s = \frac{\left(\frac{1}{1 - \frac{\cos\left(\frac{a}{n}\right)}{n}}\right)\left(\frac{-\frac{a}{n}\sin\left(\frac{a}{n}\right) + \cos\left(\frac{a}{n}\right)}{n^2}\right)}{\left(-\frac{1}{n^2}\right)}$

$= \lim\limits_{n \to \infty} \frac{\frac{a}{n}\sin\left(\frac{a}{n}\right) - \cos\left(\frac{a}{n}\right)}{1 - \frac{\cos\left(\frac{a}{n}\right)}{n}} = \frac{0 - 1}{1 - 0} = -1 \Rightarrow \lim\limits_{n \to \infty} s = e^{-1} \approx 0.3678794412$; similarly,

$\lim\limits_{n \to \infty} \left(1 - \frac{\cos\left(\frac{a}{n}\right)}{bn}\right)^n = e^{-1/b}$

23. $\lim\limits_{n \to \infty} \left|\frac{u_{n+1}}{u_n}\right| < 1 \Rightarrow \lim\limits_{n \to \infty} \left|\frac{b^{n+1} x^{n+1}}{\ln(n+1)} \cdot \frac{\ln n}{b^n x^n}\right| < 1 \Rightarrow |bx| < 1 \Rightarrow -\frac{1}{b} < x < \frac{1}{b} = 5 \Rightarrow b = \pm\frac{1}{5}$

25. $\lim\limits_{x \to 0} \frac{\sin(ax) - \sin x - x}{x^3} = \lim\limits_{x \to 0} \frac{\left(ax - \frac{a^3 x^3}{3!} + \ldots\right) - \left(x - \frac{x^3}{3!} + \ldots\right) - x}{x^3}$
$= \lim\limits_{x \to 0} \left[\frac{a - 2}{x^2} - \frac{a^3}{3!} + \frac{1}{3!} - \left(\frac{a^5}{5!} - \frac{1}{5!}\right)x^2 + \ldots\right]$ is finite if $a - 2 = 0 \Rightarrow a = 2$;
$\lim\limits_{x \to 0} \frac{\sin 2x - \sin x - x}{x^3} = -\frac{2^3}{3!} + \frac{1}{3!} = -\frac{7}{6}$

27. (a) $\frac{u_n}{u_{n+1}} = \frac{(n+1)^2}{n^2} = 1 + \frac{2}{n} + \frac{1}{n^2} \Rightarrow C = 2 > 1$ and $\sum\limits_{n=1}^{\infty} \frac{1}{n^2}$ converges

(b) $\frac{u_n}{u_{n+1}} = \frac{n+1}{n} = 1 + \frac{1}{n} + \frac{0}{n^2} \Rightarrow C = 1 \le 1$ and $\sum_{n=1}^{\infty} \frac{1}{n}$ diverges

29. (a) $\sum_{n=1}^{\infty} a_n = L \Rightarrow a_n^2 \le a_n \sum_{n=1}^{\infty} a_n = a_n L \Rightarrow \sum_{n=1}^{\infty} a_n^2$ converges by the Direct Comparison Test

(b) converges by the Limit Comparison Test: $\lim_{n \to \infty} \frac{\left(\frac{a_n}{1-a_n}\right)}{a_n} = \lim_{n \to \infty} \frac{1}{1-a_n} = 1$ since $\sum_{n=1}^{\infty} a_n$ converges and

therefore $\lim_{x \to \infty} a_n = 0$

31. $(1-x)^{-1} = 1 + \sum_{n=1}^{\infty} x^n$ where $|x| < 1 \Rightarrow \frac{1}{(1-x)^2} = \frac{d}{dx}(1-x)^{-1} = \sum_{n=1}^{\infty} nx^{n-1}$ and when $x = \frac{1}{2}$ we have

$4 = 1 + 2\left(\frac{1}{2}\right) + 3\left(\frac{1}{2}\right)^2 + 4\left(\frac{1}{2}\right)^3 + \ldots + n\left(\frac{1}{2}\right)^{n-1} + \ldots$

33. The sequence $\{x_n\}$ converges to $\frac{\pi}{2}$ from below so $\epsilon_n = \frac{\pi}{2} - x_n > 0$ for each n. By the Alternating Series Estimation Theorem $\epsilon_{n+1} \approx \frac{1}{3!}(\epsilon_n)^3$ with $|\text{error}| < \frac{1}{5!}(\epsilon_n)^5$, and since the remainder is negative this is an overestimate $\Rightarrow 0 < \epsilon_{n+1} < \frac{1}{6}(\epsilon_n)^3$.

35. (a) $\frac{1}{(1-x)^2} = \frac{d}{dx}\left(\frac{1}{1-x}\right) = \frac{d}{dx}(1 + x + x^2 + x^3 + \ldots) = 1 + 2x + 3x^2 + 4x^3 + \ldots = \sum_{n=1}^{\infty} nx^{n-1}$

(b) from part (a) we have $\sum_{n=1}^{\infty} n\left(\frac{5}{6}\right)^{n-1}\left(\frac{1}{6}\right) = \left(\frac{1}{6}\right)\left[\frac{1}{1-\left(\frac{5}{6}\right)}\right]^2 = 6$

(c) from part (a) we have $\sum_{n=1}^{\infty} np^{n-1}q = \frac{q}{(1-p)^2} = \frac{q}{q^2} = \frac{1}{q}$

37. (a) $R_n = C_0 e^{-kt_0} + C_0 e^{-2kt_0} + \ldots + C_0 e^{-nkt_0} = \frac{C_0 e^{-kt_0}(1 - e^{-nkt_0})}{1 - e^{-kt_0}} \Rightarrow R = \lim_{n \to \infty} R_n = \frac{C_0 e^{-kt_0}}{1 - e^{-kt_0}} = \frac{C_0}{e^{kt_0} - 1}$

(b) $R_n = \frac{e^{-1}(1 - e^{-n})}{1 - e^{-1}} \Rightarrow R_1 = e^{-1} \approx 0.36787944$ and $R_{10} = \frac{e^{-1}(1 - e^{-10})}{1 - e^{-1}} \approx 0.58195028$;

$R = \frac{1}{e-1} \approx 0.58197671; R - R_{10} \approx 0.00002643 \Rightarrow \frac{R - R_{10}}{R} < 0.0001$

(c) $R_n = \frac{e^{-.1}(1 - e^{-.1n})}{1 - e^{-.1}}, \frac{R}{2} = \frac{1}{2}\left(\frac{1}{e^{.1} - 1}\right) \approx 4.7541659; R_n > \frac{R}{2} \Rightarrow \frac{1 - e^{-.1n}}{e^{.1} - 1} > \left(\frac{1}{2}\right)\left(\frac{1}{e^{.1} - 1}\right)$

$\Rightarrow 1 - e^{-n/10} > \frac{1}{2} \Rightarrow e^{-n/10} < \frac{1}{2} \Rightarrow -\frac{n}{10} < \ln\left(\frac{1}{2}\right) \Rightarrow \frac{n}{10} > -\ln\left(\frac{1}{2}\right) \Rightarrow n > 6.93 \Rightarrow n = 7$

39. The convergence of $\sum_{n=1}^{\infty} |a_n|$ implies that $\lim_{n \to \infty} |a_n| = 0$. Let $N > 0$ be such that $|a_n| < \frac{1}{2} \Rightarrow 1 - |a_n| > \frac{1}{2}$

$\Rightarrow \frac{|a_n|}{1 - |a_n|} < 2|a_n|$ for all $n > N$. Now $|\ln(1 + a_n)| = \left|a_n - \frac{a_n^2}{2} + \frac{a_n^3}{3} - \frac{a_n^4}{4} + \ldots\right| \le |a_n| + \left|\frac{a_n^2}{2}\right| + \left|\frac{a_n^3}{3}\right| + \left|\frac{a_n^4}{4}\right| + \ldots$

$< |a_n| + |a_n|^2 + |a_n|^3 + |a_n|^4 + \ldots = \frac{|a_n|}{1 - |a_n|} < 2|a_n|$. Therefore $\sum_{n=1}^{\infty} \ln(1 + a_n)$ converges by the Direct

Comparison Test since $\sum_{n=1}^{\infty} |a_n|$ converges.

41. (a) $s_{2n+1} = \frac{c_1}{1} + \frac{c_2}{2} + \frac{c_3}{3} + \ldots + \frac{c_{2n+1}}{2n+1} = \frac{t_1}{1} + \frac{t_2 - t_1}{2} + \frac{t_3 - t_2}{3} + \ldots + \frac{t_{2n+1} - t_{2n}}{2n+1}$

$= t_1\left(1 - \frac{1}{2}\right) + t_2\left(\frac{1}{2} - \frac{1}{3}\right) + \ldots + t_{2n}\left(\frac{1}{2n} - \frac{1}{2n+1}\right) + \frac{t_{2n+1}}{2n+1} = \sum_{k=1}^{2n} \frac{t_k}{k(k+1)} + \frac{t_{2n+1}}{2n+1}$.

(b) $\{c_n\} = \{(-1)^n\} \Rightarrow \sum_{n=1}^{\infty} \frac{(-1)^n}{n}$ converges

(c) $\{c_n\} = \{1, -1, -1, 1, 1, -1, -1, 1, 1, \ldots\} \Rightarrow$ the series $1 - \frac{1}{2} - \frac{1}{3} + \frac{1}{4} + \frac{1}{5} - \frac{1}{6} - \frac{1}{7} + \ldots$ converges

NOTES:

CHAPTER 12 VECTORS AND THE GEOMETRY OF SPACE

12.1 THREE-DIMENSIONAL COORDINATE SYSTEMS

1. The line through the point $(2, 3, 0)$ parallel to the z-axis

3. The x-axis

5. The circle $x^2 + y^2 = 4$ in the xy-plane

7. The circle $x^2 + z^2 = 4$ in the xz-plane

9. The circle $y^2 + z^2 = 1$ in the yz-plane

11. The circle $x^2 + y^2 = 16$ in the xy-plane

13. (a) The first quadrant of the xy-plane   (b) The fourth quadrant of the xy-plane

15. (a) The solid ball of radius 1 centered at the origin
    (b) The exterior of the sphere of radius 1 centered at the origin

17. (a) The closed upper hemisphere of radius 1 centered at the origin
    (b) The solid upper hemisphere of radius 1 centered at the origin

19. (a) $x = 3$ (b) $y = -1$ (c) $z = -2$

21. (a) $z = 1$ (b) $x = 3$ (c) $y = -1$

23. (a) $x^2 + (y - 2)^2 = 4, z = 0$ (b) $(y - 2)^2 + z^2 = 4, x = 0$ (c) $x^2 + z^2 = 4, y = 2$

25. (a) $y = 3, z = -1$ (b) $x = 1, z = -1$ (c) $x = 1, y = 3$

27. $x^2 + y^2 + z^2 = 25, z = 3 \Rightarrow x^2 + y^2 = 16$ in the plane $z = 3$

29. $0 \leq z \leq 1$

31. $z \leq 0$

33. (a) $(x - 1)^2 + (y - 1)^2 + (z - 1)^2 < 1$ (b) $(x - 1)^2 + (y - 1)^2 + (z - 1)^2 > 1$

35. $|P_1 P_2| = \sqrt{(3 - 1)^2 + (3 - 1)^2 + (0 - 1)^2} = \sqrt{9} = 3$

37. $|P_1 P_2| = \sqrt{(4 - 1)^2 + (-2 - 4)^2 + (7 - 5)^2} = \sqrt{49} = 7$

39. $|P_1 P_2| = \sqrt{(2 - 0)^2 + (-2 - 0)^2 + (-2 - 0)^2} = \sqrt{3 \cdot 4} = 2\sqrt{3}$

41. center $(-2, 0, 2)$, radius $2\sqrt{2}$

43. center $\left(\sqrt{2}, \sqrt{2}, -\sqrt{2}\right)$, radius $\sqrt{2}$

45. $(x - 1)^2 + (y - 2)^2 + (z - 3)^2 = 14$

47. $(x + 2)^2 + y^2 + z^2 = 3$

49. $x^2 + y^2 + z^2 + 4x - 4z = 0 \Rightarrow (x^2 + 4x + 4) + y^2 + (z^2 - 4z + 4) = 4 + 4$

$\Rightarrow (x + 2)^2 + (y - 0)^2 + (z - 2)^2 = \left(\sqrt{8}\right)^2 \Rightarrow$ the center is at $(-2, 0, 2)$ and the radius is $\sqrt{8}$

51. $2x^2 + 2y^2 + 2z^2 + x + y + z = 9 \Rightarrow x^2 + \frac{1}{2}x + y^2 + \frac{1}{2}y + z^2 + \frac{1}{2}z = \frac{9}{2}$

$\Rightarrow \left(x^2 + \frac{1}{2}x + \frac{1}{16}\right) + \left(y^2 + \frac{1}{2}y + \frac{1}{16}\right) + \left(z^2 + \frac{1}{2}z + \frac{1}{16}\right) = \frac{9}{2} + \frac{3}{16} \Rightarrow \left(x + \frac{1}{4}\right)^2 + \left(y + \frac{1}{4}\right)^2 + \left(z + \frac{1}{4}\right)^2 = \left(\frac{5\sqrt{3}}{4}\right)^2$

$\Rightarrow$ the center is at $\left(-\frac{1}{4}, -\frac{1}{4}, -\frac{1}{4}\right)$ and the radius is $\frac{5\sqrt{3}}{4}$

53. (a) the distance between (x, y, z) and $(x, 0, 0)$ is $\sqrt{y^2 + z^2}$

(b) the distance between (x, y, z) and $(0, y, 0)$ is $\sqrt{x^2 + z^2}$

(c) the distance between (x, y, z) and $(0, 0, z)$ is $\sqrt{x^2 + y^2}$

55. $|AB| = \sqrt{(1 - (-1))^2 + (-1 - 2)^2 + (3 - 1)^2} = \sqrt{4 + 9 + 4} = \sqrt{17}$

$|BC| = \sqrt{(3 - 1)^2 + (4 - (-1))^2 + (5 - 3)^2} = \sqrt{4 + 25 + 4} = \sqrt{33}$

$|CA| = \sqrt{(-1 - 3)^2 + (2 - 4)^2 + (1 - 5)^2} = \sqrt{16 + 4 + 16} = \sqrt{36} = 6$

Thus the perimeter of triangle ABC is $\sqrt{17} + \sqrt{33} + 6$.

12.2 VECTORS

1. (a) $\langle 3(3), 3(-2) \rangle = \langle 9, -6 \rangle$

(b) $\sqrt{9^2 + (-6)^2} = \sqrt{117} = 3\sqrt{13}$

3. (a) $\langle 3 + (-2), -2 + 5 \rangle = \langle 1, 3 \rangle$

(b) $\sqrt{1^2 + 3^2} = \sqrt{10}$

5. (a) $2\mathbf{u} = \langle 2(3), 2(-2) \rangle = \langle 6, -4 \rangle$

$3\mathbf{v} = \langle 3(-2), 3(5) \rangle = \langle -6, 15 \rangle$

$2\mathbf{u} - 3\mathbf{v} = \langle 6 - (-4), -4 - 15 \rangle = \langle 12, -19 \rangle$

(b) $\sqrt{12^2 + (-19)^2} = \sqrt{505}$

7. (a) $\frac{3}{5}\mathbf{u} = \left\langle \frac{3}{5}(3), \frac{3}{5}(-2) \right\rangle = \left\langle \frac{9}{5}, -\frac{6}{5} \right\rangle$

$\frac{4}{5}\mathbf{v} = \left\langle \frac{4}{5}(-2), \frac{4}{5}(5) \right\rangle = \left\langle -\frac{8}{5}, 4 \right\rangle$

$\frac{3}{5}\mathbf{u} + \frac{4}{5}\mathbf{v} = \left\langle \frac{9}{5} + \left(-\frac{8}{5}\right), -\frac{6}{5} + 4 \right\rangle = \left\langle \frac{1}{5}, \frac{14}{5} \right\rangle$

(b) $\sqrt{\left(\frac{1}{5}\right)^2 + \left(\frac{14}{5}\right)^2} = \frac{\sqrt{197}}{5}$

9. $\langle 2 - 1, -1 - 3 \rangle = \langle 1, -4 \rangle$

11. $\langle 0 - 2, 0 - 3 \rangle = \langle -2, -3 \rangle$

13. $\left\langle \cos \frac{2\pi}{3}, \sin \frac{2\pi}{3} \right\rangle = \left\langle -\frac{1}{2}, \frac{\sqrt{3}}{2} \right\rangle$

15. This is the unit vector which makes an angle of $120° + 90° = 210°$ with the positive x-axis;
$\langle \cos 210°, \sin 210° \rangle = \left\langle -\frac{\sqrt{3}}{2}, -\frac{1}{2} \right\rangle$

17. $\vec{P_1P_2} = (2 - 5)\mathbf{i} + (9 - 7)\mathbf{j} + (-2 - (-1))\mathbf{k} = -3\mathbf{i} + 2\mathbf{j} - \mathbf{k}$

19. $\vec{AB} = (-10 - (-7))\mathbf{i} + (8 - (-8))\mathbf{j} + (1 - 1)\mathbf{k} = -3\mathbf{i} + 16\mathbf{j}$

21. $5\mathbf{u} - \mathbf{v} = 5\langle 1, 1, -1 \rangle - \langle 2, 0, 3 \rangle = \langle 5, 5, -5 \rangle - \langle 2, 0, 3 \rangle = \langle 5 - 2, 5 - 0, -5 - 3 \rangle = \langle 3, 5, -8 \rangle = 3\mathbf{i} + 5\mathbf{j} - 8\mathbf{k}$

23. The vector $\mathbf{v}$ is horizontal and 1 in. long. The vectors $\mathbf{u}$ and $\mathbf{w}$ are $\frac{11}{16}$ in. long. $\mathbf{w}$ is vertical and $\mathbf{u}$ makes a 45° angle with the horizontal. All vectors must be drawn to scale.

(a)

(b)

(c)

(d)

25. length $= |2\mathbf{i} + \mathbf{j} - 2\mathbf{k}| = \sqrt{2^2 + 1^2 + (-2)^2} = 3$, the direction is $\frac{2}{3}\mathbf{i} + \frac{1}{3}\mathbf{j} - \frac{2}{3}\mathbf{k} \Rightarrow 2\mathbf{i} + \mathbf{j} - 2\mathbf{k} = 3\left(\frac{2}{3}\mathbf{i} + \frac{1}{3}\mathbf{j} - \frac{2}{3}\mathbf{k}\right)$

27. length $= |5\mathbf{k}| = \sqrt{25} = 5$, the direction is $\mathbf{k} \Rightarrow 5\mathbf{k} = 5(\mathbf{k})$

29. length $= \left|\frac{1}{\sqrt{6}}\mathbf{i} - \frac{1}{\sqrt{6}}\mathbf{j} - \frac{1}{\sqrt{6}}\mathbf{k}\right| = \sqrt{3\left(\frac{1}{\sqrt{6}}\right)^2} = \sqrt{\frac{1}{2}}$, the direction is $\frac{1}{\sqrt{3}}\mathbf{i} - \frac{1}{\sqrt{3}}\mathbf{j} - \frac{1}{\sqrt{3}}\mathbf{k}$
$\Rightarrow \frac{1}{\sqrt{6}}\mathbf{i} - \frac{1}{\sqrt{6}}\mathbf{j} - \frac{1}{\sqrt{6}}\mathbf{k} = \sqrt{\frac{1}{2}}\left(\frac{1}{\sqrt{3}}\mathbf{i} - \frac{1}{\sqrt{3}}\mathbf{j} - \frac{1}{\sqrt{3}}\mathbf{k}\right)$

31. (a) $2\mathbf{i}$ (b) $-\sqrt{3}\mathbf{k}$ (c) $\frac{3}{10}\mathbf{j} + \frac{2}{5}\mathbf{k}$ (d) $6\mathbf{i} - 2\mathbf{j} + 3\mathbf{k}$

33. $|\mathbf{v}| = \sqrt{12^2 + 5^2} = \sqrt{169} = 13$; $\frac{\mathbf{v}}{|\mathbf{v}|} = \frac{1}{13}\mathbf{v} = \frac{1}{13}(12\mathbf{i} - 5\mathbf{k}) \Rightarrow$ the desired vector is $\frac{7}{13}(12\mathbf{i} - 5\mathbf{k})$

35. (a) $3\mathbf{i} + 4\mathbf{j} - 5\mathbf{k} = 5\sqrt{2}\left(\frac{3}{5\sqrt{2}}\mathbf{i} + \frac{4}{5\sqrt{2}}\mathbf{j} - \frac{1}{\sqrt{2}}\mathbf{k}\right) \Rightarrow$ the direction is $\frac{3}{5\sqrt{2}}\mathbf{i} + \frac{4}{5\sqrt{2}}\mathbf{j} - \frac{1}{\sqrt{2}}\mathbf{k}$
(b) the midpoint is $\left(\frac{1}{2}, 3, \frac{5}{2}\right)$

37. (a) $-\mathbf{i}-\mathbf{j}-\mathbf{k}=\sqrt{3}\left(-\frac{1}{\sqrt{3}}\mathbf{i}-\frac{1}{\sqrt{3}}\mathbf{j}-\frac{1}{\sqrt{3}}\mathbf{k}\right) \Rightarrow$ the direction is $-\frac{1}{\sqrt{3}}\mathbf{i}-\frac{1}{\sqrt{3}}\mathbf{j}-\frac{1}{\sqrt{3}}\mathbf{k}$

    (b) the midpoint is $\left(\frac{5}{2},\frac{7}{2},\frac{9}{2}\right)$

39. $\overrightarrow{AB}=(5-a)\mathbf{i}+(1-b)\mathbf{j}+(3-c)\mathbf{k}=\mathbf{i}+4\mathbf{j}-2\mathbf{k} \Rightarrow 5-a=1,\ 1-b=4,$ and $3-c=-2 \Rightarrow a=4,\ b=-3,$ and
    $c=5 \Rightarrow$ A is the point $(4,-3,5)$

41. $2\mathbf{i}+\mathbf{j}=a(\mathbf{i}+\mathbf{j})+b(\mathbf{i}-\mathbf{j})=(a+b)\mathbf{i}+(a-b)\mathbf{j} \Rightarrow a+b=2$ and $a-b=1 \Rightarrow 2a=3 \Rightarrow a=\frac{3}{2}$ and
    $b=a-1=\frac{1}{2}$

43. If $|x|$ is the magnitude of the x-component, then $\cos 30°=\frac{|x|}{|F|} \Rightarrow |x|=|F|\cos 30°=(10)\left(\frac{\sqrt{3}}{2}\right)=5\sqrt{3}$ lb
    $\Rightarrow \mathbf{F}_x=5\sqrt{3}\,\mathbf{i}$;
    if $|y|$ is the magnitude of the y-component, then $\sin 30°=\frac{|y|}{|F|} \Rightarrow |y|=|F|\sin 30°=(10)\left(\frac{1}{2}\right)=5$ lb $\Rightarrow \mathbf{F}_y=5\mathbf{j}$.

45. 25° west of north is $90°+25°=115°$ north of east. $800\langle\cos 155°,\sin 115°\rangle\approx\langle -338.095,725.046\rangle$

47. (a) The tree is located at the tip of the vector $\overrightarrow{OP}=(5\cos 60°)\mathbf{i}+(5\sin 60°)\mathbf{j}=\frac{5}{2}\mathbf{i}+\frac{5\sqrt{3}}{2}\mathbf{j} \Rightarrow P=\left(\frac{5}{2},\frac{5\sqrt{3}}{2}\right)$

    (b) The telephone pole is located at the point Q, which is the tip of the vector $\overrightarrow{OP}+\overrightarrow{PQ}$
    $=\left(\frac{5}{2}\mathbf{i}+\frac{5\sqrt{3}}{2}\mathbf{j}\right)+(10\cos 315°)\mathbf{i}+(10\sin 315°)\mathbf{j}=\left(\frac{5}{2}+\frac{10\sqrt{2}}{2}\right)\mathbf{i}+\left(\frac{5\sqrt{3}}{2}-\frac{10\sqrt{2}}{2}\right)\mathbf{j}$
    $\Rightarrow Q=\left(\frac{5+10\sqrt{2}}{2},\frac{5\sqrt{3}-10\sqrt{2}}{2}\right)$

49. (a) the midpoint of AB is $M\left(\frac{5}{2},\frac{5}{2},0\right)$ and $\overrightarrow{CM}=\left(\frac{5}{2}-1\right)\mathbf{i}+\left(\frac{5}{2}-1\right)\mathbf{j}+(0-3)\mathbf{k}=\frac{3}{2}\mathbf{i}+\frac{3}{2}\mathbf{j}-3\mathbf{k}$

    (b) the desired vector is $\left(\frac{2}{3}\right)\overrightarrow{CM}=\frac{2}{3}\left(\frac{3}{2}\mathbf{i}+\frac{3}{2}\mathbf{j}-3\mathbf{k}\right)=\mathbf{i}+\mathbf{j}-2\mathbf{k}$

    (c) the vector whose sum is the vector from the origin to C and the result of part (b) will terminate
    at the center of mass $\Rightarrow$ the terminal point of $(\mathbf{i}+\mathbf{j}+3\mathbf{k})+(\mathbf{i}+\mathbf{j}-2\mathbf{k})=2\mathbf{i}+2\mathbf{j}+\mathbf{k}$ is the point
    $(2,2,1)$, which is the location of the center of mass

51. Without loss of generality we identify the vertices of the quadrilateral such that $A(0,0,0)$, $B(x_b,0,0)$,
    $C(x_c,y_c,0)$ and $D(x_d,y_d,z_d) \Rightarrow$ the midpoint of AB is $M_{AB}\left(\frac{x_b}{2},0,0\right)$, the midpoint of BC is
    $M_{BC}\left(\frac{x_b+x_c}{2},\frac{y_c}{2},0\right)$, the midpoint of CD is $M_{CD}\left(\frac{x_c+x_d}{2},\frac{y_c+y_d}{2},\frac{z_d}{2}\right)$ and the midpoint of AD is
    $M_{AD}\left(\frac{x_d}{2},\frac{y_d}{2},\frac{z_d}{2}\right) \Rightarrow$ the midpoint of $M_{AB}M_{CD}$ is $\left(\frac{\frac{x_b}{2}+\frac{x_c+x_d}{2}}{2},\frac{y_c+y_d}{4},\frac{z_d}{4}\right)$ which is the same as the midpoint
    of $M_{AD}M_{BC}=\left(\frac{\frac{x_b+x_c}{2}+\frac{x_d}{2}}{2},\frac{y_c+y_d}{4},\frac{z_d}{4}\right)$.

53. Without loss of generality we can coordinatize the vertices of the triangle such that $A(0,0)$, $B(b,0)$ and
    $C(x_c,y_c) \Rightarrow$ a is located at $\left(\frac{b+x_c}{2},\frac{y_c}{2}\right)$, b is at $\left(\frac{x_c}{2},\frac{y_c}{2}\right)$ and c is at $\left(\frac{b}{2},0\right)$. Therefore, $\overrightarrow{Aa}=\left(\frac{b}{2}+\frac{x_c}{2}\right)\mathbf{i}+\left(\frac{y_c}{2}\right)\mathbf{j}$,
    $\overrightarrow{Bb}=\left(\frac{x_c}{2}-b\right)\mathbf{i}+\left(\frac{y_c}{2}\right)\mathbf{j}$, and $\overrightarrow{Cc}=\left(\frac{b}{2}-x_c\right)\mathbf{i}+(-y_c)\mathbf{j} \Rightarrow \overrightarrow{Aa}+\overrightarrow{Bb}+\overrightarrow{Cc}=\mathbf{0}$.

12.3 THE DOT PRODUCT

NOTE: In Exercises 1-7 below we calculate proj$_v$ **u** as the vector $\left(\frac{|u|\cos\theta}{|v|}\right)$ **v**, so the scalar multiplier of **v** is the number in column 5 divided by the number in column 2.

	v · u	**|v|**	**|u|**	**cos θ**	**|u| cos θ**	**proj$_v$ u**
1.	-25	5	5	-1	-5	$-2\mathbf{i}+4\mathbf{j}-\sqrt{5}\mathbf{k}$
3.	25	15	5	$\frac{1}{3}$	$\frac{5}{3}$	$\frac{1}{9}(10\mathbf{i}+11\mathbf{j}-2\mathbf{k})$
5.	2	$\sqrt{34}$	$\sqrt{3}$	$\frac{2}{\sqrt{3}\sqrt{34}}$	$\frac{2}{\sqrt{34}}$	$\frac{1}{17}(5\mathbf{j}-3\mathbf{k})$
7.	$10+\sqrt{17}$	$\sqrt{26}$	$\sqrt{21}$	$\frac{10+\sqrt{17}}{\sqrt{546}}$	$\frac{10+\sqrt{17}}{\sqrt{26}}$	$\frac{10+\sqrt{17}}{\sqrt{26}}(-5\mathbf{i}+\mathbf{j})$

9. $\theta = \cos^{-1}\left(\frac{\mathbf{u\cdot v}}{|\mathbf{u}||\mathbf{v}|}\right) = \cos^{-1}\left(\frac{(2)(1)+(1)(2)+(0)(-1)}{\sqrt{2^2+1^2+0^2}\sqrt{1^2+2^2+(-1)^2}}\right) = \cos^{-1}\left(\frac{4}{\sqrt{5}\sqrt{6}}\right) = \cos^{-1}\left(\frac{4}{\sqrt{30}}\right) \approx 0.75$ rad

11. $\theta = \cos^{-1}\left(\frac{\mathbf{u\cdot v}}{|\mathbf{u}||\mathbf{v}|}\right) = \cos^{-1}\left(\frac{\left(\sqrt{3}\right)\left(\sqrt{3}\right)+(-7)(1)+(0)(-2)}{\sqrt{\left(\sqrt{3}\right)^2+(-7)^2+0^2}\sqrt{\left(\sqrt{3}\right)^2+(1)^2+(-2)^2}}\right) = \cos^{-1}\left(\frac{3-7}{\sqrt{52}\sqrt{8}}\right)$

$= \cos^{-1}\left(\frac{-1}{\sqrt{26}}\right) \approx 1.77$ rad

13. $\vec{AB} = \langle 3,1 \rangle$, $\vec{BC} = \langle -1,-3 \rangle$, and $\vec{AC} = \langle 2,-2 \rangle$. $\vec{BA} = \langle -3,-1 \rangle$, $\vec{CB} = \langle 1,3 \rangle$, $\vec{CA} = \langle -2,2 \rangle$.

$\left|\vec{AB}\right| = \left|\vec{BA}\right| = \sqrt{10}$, $\left|\vec{BC}\right| = \left|\vec{CB}\right| = \sqrt{10}$, $\left|\vec{AC}\right| = \left|\vec{CA}\right| = 2\sqrt{2}$,

Angle at A $= \cos^{-1}\left(\frac{\vec{AB}\cdot\vec{AC}}{\left|\vec{AB}\right|\left|\vec{AC}\right|}\right) = \cos^{-1}\left(\frac{3(2)+1(-2)}{\left(\sqrt{10}\right)\left(2\sqrt{2}\right)}\right) = \cos^{-1}\left(\frac{1}{\sqrt{5}}\right) \approx 63.435°$

Angle at B $= \cos^{-1}\left(\frac{\vec{BC}\cdot\vec{BA}}{\left|\vec{BC}\right|\left|\vec{BA}\right|}\right) = \cos^{-1}\left(\frac{(-1)(-3)+(-3)(-1)}{\left(\sqrt{10}\right)\left(\sqrt{10}\right)}\right) = \cos^{-1}\left(\frac{3}{5}\right) \approx 53.130°$, and

Angle at C $= \cos^{-1}\left(\frac{\vec{CB}\cdot\vec{CA}}{\left|\vec{CB}\right|\left|\vec{CA}\right|}\right) = \cos^{-1}\left(\frac{1(-2)+3(2)}{\left(\sqrt{10}\right)\left(2\sqrt{2}\right)}\right) = \cos^{-1}\left(\frac{1}{\sqrt{5}}\right) \approx 63.435°$

15. (a) $\cos\alpha = \frac{\mathbf{i\cdot v}}{|\mathbf{i}||\mathbf{v}|} = \frac{a}{|\mathbf{v}|}$, $\cos\beta = \frac{\mathbf{j\cdot v}}{|\mathbf{j}||\mathbf{v}|} = \frac{b}{|\mathbf{v}|}$, $\cos\gamma = \frac{\mathbf{k\cdot v}}{|\mathbf{k}||\mathbf{v}|} = \frac{c}{|\mathbf{v}|}$ and

$\cos^2\alpha + \cos^2\beta + \cos^2\gamma = \left(\frac{a}{|\mathbf{v}|}\right)^2 + \left(\frac{b}{|\mathbf{v}|}\right)^2 + \left(\frac{c}{|\mathbf{v}|}\right)^2 = \frac{a^2+b^2+c^2}{|\mathbf{v}||\mathbf{v}|} = \frac{|\mathbf{v}||\mathbf{v}|}{|\mathbf{v}||\mathbf{v}|} = 1$

(b) $|\mathbf{v}| = 1 \Rightarrow \cos\alpha = \frac{a}{|\mathbf{v}|} = a$, $\cos\beta = \frac{b}{|\mathbf{v}|} = b$ and $\cos\gamma = \frac{c}{|\mathbf{v}|} = c$ are the direction cosines of **v**

17. $\mathbf{u} = \left(\frac{\mathbf{v\cdot u}}{\mathbf{v\cdot v}}\mathbf{v}\right) + \left(\mathbf{u} - \frac{\mathbf{v\cdot u}}{\mathbf{v\cdot v}}\mathbf{v}\right) = \frac{3}{2}(\mathbf{i}+\mathbf{j}) + \left[(3\mathbf{j}+4\mathbf{k}) - \frac{3}{2}(\mathbf{i}+\mathbf{j})\right] = \left(\frac{3}{2}\mathbf{i}+\frac{3}{2}\mathbf{j}\right) + \left(-\frac{3}{2}\mathbf{i}+\frac{3}{2}\mathbf{j}+4\mathbf{k}\right)$, where

$\mathbf{v\cdot u} = 3$ and $\mathbf{v\cdot v} = 2$

19. $\mathbf{u} = \left(\frac{\mathbf{v\cdot u}}{\mathbf{v\cdot v}}\mathbf{v}\right) + \left(\mathbf{u} - \frac{\mathbf{v\cdot u}}{\mathbf{v\cdot v}}\mathbf{v}\right) = \frac{14}{3}(\mathbf{i}+2\mathbf{j}-\mathbf{k}) + \left[(8\mathbf{i}+4\mathbf{j}-12\mathbf{k}) - \left(\frac{14}{3}\mathbf{i}+\frac{28}{3}\mathbf{j}-\frac{14}{3}\mathbf{k}\right)\right]$

$= \left(\frac{14}{3}\mathbf{i}+\frac{28}{3}\mathbf{j}-\frac{14}{3}\mathbf{k}\right) + \left(\frac{10}{3}\mathbf{i}-\frac{16}{3}\mathbf{j}-\frac{22}{3}\mathbf{k}\right)$, where $\mathbf{v\cdot u} = 28$ and $\mathbf{v\cdot v} = 6$

21. The sum of two vectors of equal length is *always* orthogonal to their difference, as we can see from the equation

$(\mathbf{v}_1 + \mathbf{v}_2) \cdot (\mathbf{v}_1 - \mathbf{v}_2) = \mathbf{v}_1 \cdot \mathbf{v}_1 + \mathbf{v}_2 \cdot \mathbf{v}_1 - \mathbf{v}_1 \cdot \mathbf{v}_2 - \mathbf{v}_2 \cdot \mathbf{v}_2 = |\mathbf{v}_1|^2 - |\mathbf{v}_2|^2 = 0$

23. Let **u** and **v** be the sides of a rhombus $\Rightarrow$ the diagonals are $\mathbf{d}_1 = \mathbf{u} + \mathbf{v}$ and $\mathbf{d}_2 = -\mathbf{u} + \mathbf{v}$
$\Rightarrow \mathbf{d}_1 \cdot \mathbf{d}_2 = (\mathbf{u} + \mathbf{v}) \cdot (-\mathbf{u} + \mathbf{v}) = -\mathbf{u} \cdot \mathbf{u} + \mathbf{u} \cdot \mathbf{v} - \mathbf{v} \cdot \mathbf{u} + \mathbf{v} \cdot \mathbf{v} = |\mathbf{v}|^2 - |\mathbf{u}|^2 = 0$ because $|\mathbf{u}| = |\mathbf{v}|$, since a rhombus has equal sides.

25. Clearly the diagonals of a rectangle are equal in length. What is not as obvious is the statement that equal diagonals happen only in a rectangle. We show this is true by letting the adjacent sides of a parallelogram be the vectors $(v_1\mathbf{i} + v_2\mathbf{j})$ and $(u_1\mathbf{i} + u_2\mathbf{j})$. The equal diagonals of the parallelogram are
$\mathbf{d}_1 = (v_1\mathbf{i} + v_2\mathbf{j}) + (u_1\mathbf{i} + u_2\mathbf{j})$ and $\mathbf{d}_2 = (v_1\mathbf{i} + v_2\mathbf{j}) - (u_1\mathbf{i} + u_2\mathbf{j})$. Hence $|\mathbf{d}_1| = |\mathbf{d}_2| = |(v_1\mathbf{i} + v_2\mathbf{j}) + (u_1\mathbf{i} + u_2\mathbf{j})|$
$= |(v_1\mathbf{i} + v_2\mathbf{j}) - (u_1\mathbf{i} + u_2\mathbf{j})| \Rightarrow |(v_1 + u_1)\mathbf{i} + (v_2 + u_2)\mathbf{j}| = |(v_1 - u_1)\mathbf{i} + (v_2 - u_2)\mathbf{j}|$
$\Rightarrow \sqrt{(v_1 + u_1)^2 + (v_2 + u_2)^2} = \sqrt{(v_1 - u_1)^2 + (v_2 - u_2)^2} \Rightarrow v_1^2 + 2v_1u_1 + u_1^2 + v_2^2 + 2v_2u_2 + u_2^2$
$= v_1^2 - 2v_1u_1 + u_1^2 + v_2^2 - 2v_2u_2 + u_2^2 \Rightarrow 2(v_1u_1 + v_2u_2) = -2(v_1u_1 + v_2u_2) \Rightarrow v_1u_1 + v_2u_2 = 0$
$\Rightarrow (v_1\mathbf{i} + v_2\mathbf{j}) \cdot (u_1\mathbf{i} + u_2\mathbf{j}) = 0 \Rightarrow$ the vectors $(v_1\mathbf{i} + v_2\mathbf{j})$ and $(u_1\mathbf{i} + u_2\mathbf{j})$ are perpendicular and the parallelogram must be a rectangle.

27. horizontal component: $1200 \cos(8°) \approx 1188$ ft/s; vertical component: $1200 \sin(8°) \approx 167$ ft/s

29. (a) Since $|\cos \theta| \le 1$, we have $|\mathbf{u} \cdot \mathbf{v}| = |\mathbf{u}|\,|\mathbf{v}|\,|\cos \theta| \le |\mathbf{u}|\,|\mathbf{v}|\,(1) = |\mathbf{u}|\,|\mathbf{v}|$.
    (b) We have equality precisely when $|\cos \theta| = 1$ or when one or both of **u** and **v** is **0**. In the case of nonzero vectors, we have equality when $\theta = 0$ or π, i.e., when the vectors are parallel.

31. $\mathbf{v} \cdot \mathbf{u}_1 = (a\mathbf{u}_1 + b\mathbf{u}_2) \cdot \mathbf{u}_1 = a\mathbf{u}_1 \cdot \mathbf{u}_1 + b\mathbf{u}_2 \cdot \mathbf{u}_1 = a\,|\mathbf{u}_1|^2 + b(\mathbf{u}_2 \cdot \mathbf{u}_1) = a(1)^2 + b(0) = a$

33. $P(x_1, y_1) = P\left(x_1, \frac{c}{b} - \frac{a}{b}x_1\right)$ and $Q(x_2, y_2) = Q\left(x_2, \frac{c}{b} - \frac{a}{b}x_2\right)$ are any two points P and Q on the line with $b \ne 0$
$\Rightarrow \overrightarrow{PQ} = (x_2 - x_1)\mathbf{i} + \frac{a}{b}(x_1 - x_2)\mathbf{j} \Rightarrow \overrightarrow{PQ} \cdot \mathbf{v} = \left[(x_2 - x_1)\mathbf{i} + \frac{a}{b}(x_1 - x_2)\mathbf{j}\right] \cdot (a\mathbf{i} + b\mathbf{j}) = a(x_2 - x_1) + b\left(\frac{a}{b}\right)(x_1 - x_2)$
$= 0 \Rightarrow \mathbf{v}$ is perpendicular to $\overrightarrow{PQ}$ for $b \ne 0$. If $b = 0$, then $\mathbf{v} = a\mathbf{i}$ is perpendicular to the vertical line $ax = c$.
Alternatively, the slope of **v** is $\frac{b}{a}$ and the slope of the line $ax + by = c$ is $-\frac{a}{b}$, so the slopes are negative reciprocals $\Rightarrow$ the vector **v** and the line are perpendicular.

35. $\mathbf{v} = \mathbf{i} + 2\mathbf{j}$ is perpendicular to the line $x + 2y = c$;
    $P(2, 1)$ on the line $\Rightarrow 2 + 2 = c \Rightarrow x + 2y = 4$

37. $\mathbf{v} = -2\mathbf{i} + \mathbf{j}$ is perpendicular to the line $-2x + y = c$;
    $P(-2, -7)$ on the line $\Rightarrow (-2)(-2) - 7 = c$
    $\Rightarrow -2x + y = -3$

39. $\mathbf{v} = \mathbf{i} - \mathbf{j}$ is parallel to the line $-x - y = c$;
$P(-2, 1)$ on the line $\Rightarrow -(-2) - 1 = c \Rightarrow -x - y = 1$
or $x + y = -1$.

41. $\mathbf{v} = -\mathbf{i} - 2\mathbf{j}$ is parallel to the line $-2x + y = c$;
$P(1, 2)$ on the line $\Rightarrow -2(1) + 2 = c \Rightarrow -2x - y = 0$
or $2x - y = 0$.

43. $P(0, 0)$, $Q(1, 1)$ and $\mathbf{F} = 5\mathbf{j} \Rightarrow \vec{PQ} = \mathbf{i} + \mathbf{j}$ and $W = \mathbf{F} \cdot \vec{PQ} = (5\mathbf{j}) \cdot (\mathbf{i} + \mathbf{j}) = 5 \, \text{N} \cdot \text{m} = 5 \, \text{J}$

45. $W = |\mathbf{F}| \left| \vec{PQ} \right| \cos \theta = (200)(20)(\cos 30°) = 2000\sqrt{3} = 3464.10 \, \text{N} \cdot \text{m} = 3464.10 \, \text{J}$

In Exercises 47-51 we use the fact that $\mathbf{n} = a\mathbf{i} + b\mathbf{j}$ is normal to the line $ax + by = c$.

47. $\mathbf{n}_1 = 3\mathbf{i} + \mathbf{j}$ and $\mathbf{n}_2 = 2\mathbf{i} - \mathbf{j} \Rightarrow \theta = \cos^{-1}\left(\frac{\mathbf{n}_1 \cdot \mathbf{n}_2}{|\mathbf{n}_1| \, |\mathbf{n}_2|}\right) = \cos^{-1}\left(\frac{6-1}{\sqrt{10}\sqrt{5}}\right) = \cos^{-1}\left(\frac{1}{\sqrt{2}}\right) = \frac{\pi}{4}$

49. $\mathbf{n}_1 = \sqrt{3}\mathbf{i} - \mathbf{j}$ and $\mathbf{n}_2 = \mathbf{i} - \sqrt{3}\mathbf{j} \Rightarrow \theta = \cos^{-1}\left(\frac{\mathbf{n}_1 \cdot \mathbf{n}_2}{|\mathbf{n}_1| \, |\mathbf{n}_2|}\right) = \cos^{-1}\left(\frac{\sqrt{3}+\sqrt{3}}{\sqrt{4}\sqrt{4}}\right) = \cos^{-1}\left(\frac{\sqrt{3}}{2}\right) = \frac{\pi}{6}$

51. $\mathbf{n}_1 = 3\mathbf{i} - 4\mathbf{j}$ and $\mathbf{n}_2 = \mathbf{i} - \mathbf{j} \Rightarrow \theta = \cos^{-1}\left(\frac{\mathbf{n}_1 \cdot \mathbf{n}_2}{|\mathbf{n}_1| \, |\mathbf{n}_2|}\right) = \cos^{-1}\left(\frac{3+4}{\sqrt{25}\sqrt{2}}\right) = \cos^{-1}\left(\frac{7}{5\sqrt{2}}\right) \approx 0.14 \, \text{rad}$

53. The angle between the corresponding normals is equal to the angle between the corresponding tangents. The
points of intersection are $\left(-\frac{\sqrt{3}}{2}, \frac{3}{4}\right)$ and $\left(\frac{\sqrt{3}}{2}, \frac{3}{4}\right)$. At $\left(-\frac{\sqrt{3}}{2}, \frac{3}{4}\right)$ the tangent line for $f(x) = x^2$ is
$y - \frac{3}{4} = f'\left(-\frac{\sqrt{3}}{2}\right)\left(x - \left(-\frac{\sqrt{3}}{2}\right)\right) \Rightarrow y = -\sqrt{3}\left(x + \frac{\sqrt{3}}{2}\right) + \frac{3}{4} \Rightarrow y = -\sqrt{3}x - \frac{3}{4}$, and the tangent line for
$f(x) = \left(\frac{3}{2}\right) - x^2$ is $y - \frac{3}{4} = f'\left(-\frac{\sqrt{3}}{2}\right)\left(x - \left(-\frac{\sqrt{3}}{2}\right)\right) \Rightarrow y = \sqrt{3}\left(x + \frac{\sqrt{3}}{2}\right) + \frac{3}{4} = \sqrt{3}x + \frac{9}{4}$. The corresponding
normals are $\mathbf{n}_1 = \sqrt{3}\mathbf{i} + \mathbf{j}$ and $\mathbf{n}_2 = -\sqrt{3}\mathbf{i} + \mathbf{j}$. The angle at $\left(-\frac{\sqrt{3}}{2}, \frac{3}{4}\right)$ is $\theta = \cos^{-1}\left(\frac{\mathbf{n}_1 \cdot \mathbf{n}_2}{|\mathbf{n}_1| \, |\mathbf{n}_2|}\right)$
$= \cos^{-1}\left(\frac{-3+1}{\sqrt{4}\sqrt{4}}\right) = \cos^{-1}\left(-\frac{1}{2}\right) = \frac{2\pi}{3}$, the angle is $\frac{\pi}{3}$ and $\frac{2\pi}{3}$. At $\left(\frac{\sqrt{3}}{2}, \frac{3}{4}\right)$ the tangent line for $f(x) = x^2$ is
$y = \sqrt{3}\left(x + \frac{\sqrt{3}}{2}\right) + \frac{3}{4} = \sqrt{3}x + \frac{9}{4}$ and the tangent line for $f(x) = \frac{3}{2} - x^2$ is $y = -\sqrt{3}\left(x + \frac{\sqrt{3}}{2}\right) + \frac{3}{4}$
$= -\sqrt{3}x - \frac{3}{4}$. The corresponding normals are $\mathbf{n}_1 = -\sqrt{3}\mathbf{i} + \mathbf{j}$ and $\mathbf{n}_2 = \sqrt{3}\mathbf{i} + \mathbf{j}$. The angle at $\left(\frac{\sqrt{3}}{2}, \frac{3}{4}\right)$ is
$\theta = \cos^{-1}\left(\frac{\mathbf{n}_1 \cdot \mathbf{n}_2}{|\mathbf{n}_1| \, |\mathbf{n}_2|}\right) = \cos^{-1}\left(\frac{-3+1}{\sqrt{4}\sqrt{4}}\right) = \cos^{-1}\left(-\frac{1}{2}\right) = \frac{2\pi}{3}$, the angle is $\frac{\pi}{3}$ and $\frac{2\pi}{3}$.

55. The curves intersect when $y = x^3 = (y^2)^3 = y^6 \Rightarrow y = 0$ or $y = 1$. The points of intersection are $(0, 0)$ and
$(1, 1)$. Note that $y \geq 0$ since $y = y^6$. At $(0, 0)$ the tangent line for $y = x^3$ is $y = 0$ and the tangent line for
$y = \sqrt{x}$ is $x = 0$. Therefore, the angle of intersection at $(0, 0)$ is $\frac{\pi}{2}$. At $(1, 1)$ the tangent line for $y = x^3$ is
$y = 3x - 2$ and the tangent line for $y = \sqrt{x}$ is $y = \frac{1}{2}x + \frac{1}{2}$. The corresponding normal vectors are

$\mathbf{n}_1 = -3\mathbf{i} + \mathbf{j}$ and $\mathbf{n}_2 = -\frac{1}{2}\mathbf{i} + \mathbf{j} \Rightarrow \theta = \cos^{-1}\left(\frac{\mathbf{n}_1 \cdot \mathbf{n}_2}{|\mathbf{n}_1|\,|\mathbf{n}_2|}\right) = \cos^{-1}\left(\frac{1}{\sqrt{2}}\right) = \frac{\pi}{4}$, the angle is $\frac{\pi}{4}$ and $\frac{3\pi}{4}$.

12.4 THE CROSS PRODUCT

1. $\mathbf{u} \times \mathbf{v} = \begin{vmatrix} \mathbf{i} & \mathbf{j} & \mathbf{k} \\ 2 & -2 & -1 \\ 1 & 0 & -1 \end{vmatrix} = 3\left(\frac{2}{3}\mathbf{i} + \frac{1}{3}\mathbf{j} + \frac{2}{3}\mathbf{k}\right) \Rightarrow$ length $= 3$ and the direction is $\frac{2}{3}\mathbf{i} + \frac{1}{3}\mathbf{j} + \frac{2}{3}\mathbf{k}$;

   $\mathbf{v} \times \mathbf{u} = -(\mathbf{u} \times \mathbf{v}) = -3\left(\frac{2}{3}\mathbf{i} + \frac{1}{3}\mathbf{j} + \frac{2}{3}\mathbf{k}\right) \Rightarrow$ length $= 3$ and the direction is $-\frac{2}{3}\mathbf{i} - \frac{1}{3}\mathbf{j} - \frac{2}{3}\mathbf{k}$

3. $\mathbf{u} \times \mathbf{v} = \begin{vmatrix} \mathbf{i} & \mathbf{j} & \mathbf{k} \\ 2 & -2 & 4 \\ -1 & 1 & -2 \end{vmatrix} = \mathbf{0} \Rightarrow$ length $= 0$ and has no direction

   $\mathbf{v} \times \mathbf{u} = -(\mathbf{u} \times \mathbf{v}) = \mathbf{0} \Rightarrow$ length $= 0$ and has no direction

5. $\mathbf{u} \times \mathbf{v} = \begin{vmatrix} \mathbf{i} & \mathbf{j} & \mathbf{k} \\ 2 & 0 & 0 \\ 0 & -3 & 0 \end{vmatrix} = -6(\mathbf{k}) \Rightarrow$ length $= 6$ and the direction is $-\mathbf{k}$

   $\mathbf{v} \times \mathbf{u} = -(\mathbf{u} \times \mathbf{v}) = 6(\mathbf{k}) \Rightarrow$ length $= 6$ and the direction is $\mathbf{k}$

7. $\mathbf{u} \times \mathbf{v} = \begin{vmatrix} \mathbf{i} & \mathbf{j} & \mathbf{k} \\ -8 & -2 & -4 \\ 2 & 2 & 1 \end{vmatrix} = 6\mathbf{i} - 12\mathbf{k} \Rightarrow$ length $= 6\sqrt{5}$ and the direction is $\frac{1}{\sqrt{5}}\mathbf{i} - \frac{2}{\sqrt{5}}\mathbf{k}$

   $\mathbf{v} \times \mathbf{u} = -(\mathbf{u} \times \mathbf{v}) = -(6\mathbf{i} - 12\mathbf{k}) \Rightarrow$ length $= 6\sqrt{5}$ and the direction is $-\frac{1}{\sqrt{5}}\mathbf{i} + \frac{2}{\sqrt{5}}\mathbf{k}$

9. $\mathbf{u} \times \mathbf{v} = \begin{vmatrix} \mathbf{i} & \mathbf{j} & \mathbf{k} \\ 1 & 0 & 0 \\ 0 & 1 & 0 \end{vmatrix} = \mathbf{k}$
                          11. $\mathbf{u} \times \mathbf{v} = \begin{vmatrix} \mathbf{i} & \mathbf{j} & \mathbf{k} \\ 1 & 0 & -1 \\ 0 & 1 & 1 \end{vmatrix} = \mathbf{i} - \mathbf{j} + \mathbf{k}$

13. $\mathbf{u} \times \mathbf{v} = \begin{vmatrix} \mathbf{i} & \mathbf{j} & \mathbf{k} \\ 1 & 1 & 0 \\ 1 & -1 & 0 \end{vmatrix} = -2\mathbf{k}$

15. (a) $\vec{PQ} \times \vec{PR} = \begin{vmatrix} \mathbf{i} & \mathbf{j} & \mathbf{k} \\ 1 & 1 & -3 \\ -1 & 3 & -1 \end{vmatrix} = 8\mathbf{i} + 4\mathbf{j} + 4\mathbf{k} \Rightarrow$ Area $= \frac{1}{2}\left|\vec{PQ} \times \vec{PR}\right| = \frac{1}{2}\sqrt{64 + 16 + 16} = 2\sqrt{6}$

   (b) $\mathbf{u} = \pm \frac{\vec{PQ} \times \vec{PR}}{\left|\vec{PQ} \times \vec{PR}\right|} = \pm \frac{1}{\sqrt{6}}(2\mathbf{i} + \mathbf{j} + \mathbf{k})$

17. (a) $\vec{PQ} \times \vec{PR} = \begin{vmatrix} \mathbf{i} & \mathbf{j} & \mathbf{k} \\ 1 & 1 & 1 \\ 1 & 1 & 0 \end{vmatrix} = -\mathbf{i} + \mathbf{j} \Rightarrow$ Area $= \frac{1}{2}\left|\vec{PQ} \times \vec{PR}\right| = \frac{1}{2}\sqrt{1+1} = \frac{\sqrt{2}}{2}$

   (b) $\mathbf{u} = \pm \frac{\vec{PQ} \times \vec{PR}}{\left|\vec{PQ} \times \vec{PR}\right|} = \pm \frac{1}{\sqrt{2}}(-\mathbf{i} + \mathbf{j}) = \pm \frac{1}{\sqrt{2}}(\mathbf{i} - \mathbf{j})$

19. If $\mathbf{u} = a_1\mathbf{i} + a_2\mathbf{j} + a_3\mathbf{k}$, $\mathbf{v} = b_1\mathbf{i} + b_2\mathbf{j} + b_3\mathbf{k}$, and $\mathbf{w} = c_1\mathbf{i} + c_2\mathbf{j} + c_3\mathbf{k}$, then $\mathbf{u} \cdot (\mathbf{v} \times \mathbf{w}) = \begin{vmatrix} a_1 & a_2 & a_3 \\ b_1 & b_2 & b_3 \\ c_1 & c_2 & c_3 \end{vmatrix}$,

   $\mathbf{v} \cdot (\mathbf{w} \times \mathbf{u}) = \begin{vmatrix} b_1 & b_2 & b_3 \\ c_1 & c_2 & c_3 \\ a_1 & a_2 & a_3 \end{vmatrix}$ and $\mathbf{w} \cdot (\mathbf{u} \times \mathbf{v}) = \begin{vmatrix} c_1 & c_2 & c_3 \\ a_1 & a_2 & a_3 \\ b_1 & b_2 & b_3 \end{vmatrix}$ which all have the same value, since the

   interchanging of two pair of rows in a determinant does not change its value $\Rightarrow$ the volume is

   $|(\mathbf{u} \times \mathbf{v}) \cdot \mathbf{w}| = $ abs $\begin{vmatrix} 2 & 0 & 0 \\ 0 & 2 & 0 \\ 0 & 0 & 2 \end{vmatrix} = 8$

21. $|(\mathbf{u} \times \mathbf{v}) \cdot \mathbf{w}| = $ abs $\begin{vmatrix} 2 & 1 & 0 \\ 2 & -1 & 1 \\ 1 & 0 & 2 \end{vmatrix} = |-7| = 7$ (for details about verification, see Exercise 19)

23. (a) $\mathbf{u} \cdot \mathbf{v} = -6$, $\mathbf{u} \cdot \mathbf{w} = -81$, $\mathbf{v} \cdot \mathbf{w} = 18 \Rightarrow$ none

   (b) $\mathbf{u} \times \mathbf{v} = \begin{vmatrix} \mathbf{i} & \mathbf{j} & \mathbf{k} \\ 5 & -1 & 1 \\ 0 & 1 & -5 \end{vmatrix} \neq \mathbf{0}$, $\mathbf{u} \times \mathbf{w} = \begin{vmatrix} \mathbf{i} & \mathbf{j} & \mathbf{k} \\ 5 & -1 & 1 \\ -15 & 3 & -3 \end{vmatrix} = \mathbf{0}$, $\mathbf{v} \times \mathbf{w} = \begin{vmatrix} \mathbf{i} & \mathbf{j} & \mathbf{k} \\ 0 & 1 & -5 \\ -15 & 3 & -3 \end{vmatrix} \neq \mathbf{0}$

   $\Rightarrow \mathbf{u}$ and $\mathbf{w}$ are parallel

25. $\left|\vec{PQ} \times \mathbf{F}\right| = \left|\vec{PQ}\right| |\mathbf{F}| \sin(60°) = \frac{2}{3} \cdot 30 \cdot \frac{\sqrt{3}}{2}$ ft $\cdot$ lb $= 10\sqrt{3}$ ft $\cdot$ lb

27. (a) true, $|\mathbf{u}| = \sqrt{a_1^2 + a_2^2 + a_3^2} = \sqrt{\mathbf{u} \cdot \mathbf{u}}$

   (b) not always true, $\mathbf{u} \cdot \mathbf{u} = |\mathbf{u}|^2$

   (c) true, $\mathbf{u} \times \mathbf{0} = \begin{vmatrix} \mathbf{i} & \mathbf{j} & \mathbf{k} \\ a_1 & a_2 & a_3 \\ 0 & 0 & 0 \end{vmatrix} = 0\mathbf{i} + 0\mathbf{j} + 0\mathbf{k} = \mathbf{0}$ and $\mathbf{0} \times \mathbf{u} = \begin{vmatrix} \mathbf{i} & \mathbf{j} & \mathbf{k} \\ 0 & 0 & 0 \\ a_1 & a_2 & a_3 \end{vmatrix} = 0\mathbf{i} + 0\mathbf{j} + 0\mathbf{k} = \mathbf{0}$

   (d) true, $\mathbf{u} \times (-\mathbf{u}) = \begin{vmatrix} \mathbf{i} & \mathbf{j} & \mathbf{k} \\ a_1 & a_2 & a_3 \\ -a_1 & -a_2 & -a_3 \end{vmatrix} = (-a_2a_3 + a_2a_3)\mathbf{i} - (-a_1a_3 + a_1a_3)\mathbf{j} + (-a_1a_2 + a_1a_2)\mathbf{k} = \mathbf{0}$

   (e) not always true, $\mathbf{i} \times \mathbf{j} = \mathbf{k} \neq -\mathbf{k} = \mathbf{j} \times \mathbf{i}$ for example

   (f) true, distributive property of the cross product

   (g) true, $(\mathbf{u} \times \mathbf{v}) \cdot \mathbf{v} = \mathbf{u} \cdot (\mathbf{v} \times \mathbf{v}) = \mathbf{u} \cdot \mathbf{0} = 0$

   (h) true, the volume of a parallelpiped with $\mathbf{u}$, $\mathbf{v}$, and $\mathbf{w}$ along the three edges is $(\mathbf{u} \times \mathbf{v}) \cdot \mathbf{w} = (\mathbf{v} \times \mathbf{w}) \cdot \mathbf{u} = \mathbf{u} \cdot (\mathbf{v} \times \mathbf{w})$, since the dot product is commutative.

29. (a) $\text{proj}_v\,\mathbf{u} = \left(\dfrac{\mathbf{u}\cdot\mathbf{v}}{|\mathbf{v}||\mathbf{v}|}\right)\mathbf{v}$ (b) $\pm(\mathbf{u}\times\mathbf{v})$ (c) $\pm((\mathbf{u}\times\mathbf{v})\times\mathbf{w})$ (d) $|(\mathbf{u}\times\mathbf{v})\cdot\mathbf{w}|$

31. (a) yes, $\mathbf{u}\times\mathbf{v}$ and $\mathbf{w}$ are both vectors   (b) no, $\mathbf{u}$ is a vector but $\mathbf{v}\cdot\mathbf{w}$ is a scalar
    (c) yes, $\mathbf{u}$ and $\mathbf{u}\times\mathbf{w}$ are both vectors   (d) no, $\mathbf{u}$ is a vector but $\mathbf{v}\cdot\mathbf{w}$ is a scalar

33. No, $\mathbf{v}$ need not equal $\mathbf{w}$. For example, $\mathbf{i}+\mathbf{j}\neq -\mathbf{i}+\mathbf{j}$, but $\mathbf{i}\times(\mathbf{i}+\mathbf{j}) = \mathbf{i}\times\mathbf{i}+\mathbf{i}\times\mathbf{j} = \mathbf{0}+\mathbf{k} = \mathbf{k}$ and
    $\mathbf{i}\times(-\mathbf{i}+\mathbf{j}) = -\mathbf{i}\times\mathbf{i}+\mathbf{i}\times\mathbf{j} = \mathbf{0}+\mathbf{k} = \mathbf{k}$.

35. $\overrightarrow{AB} = -\mathbf{i}+\mathbf{j}$ and $\overrightarrow{AD} = -\mathbf{i}-\mathbf{j} \Rightarrow \overrightarrow{AB}\times\overrightarrow{AD} = \begin{vmatrix} \mathbf{i} & \mathbf{j} & \mathbf{k} \\ -1 & 1 & 0 \\ -1 & -1 & 0 \end{vmatrix} = 2\mathbf{k} \Rightarrow \text{area} = \left|\overrightarrow{AB}\times\overrightarrow{AD}\right| = 2$

37. $\overrightarrow{AB} = 3\mathbf{i}-2\mathbf{j}$ and $\overrightarrow{AD} = 5\mathbf{i}+\mathbf{j} \Rightarrow \overrightarrow{AB}\times\overrightarrow{AD} = \begin{vmatrix} \mathbf{i} & \mathbf{j} & \mathbf{k} \\ 3 & -2 & 0 \\ 5 & 1 & 0 \end{vmatrix} = 13\mathbf{k} \Rightarrow \text{area} = \left|\overrightarrow{AB}\times\overrightarrow{AD}\right| = 13$

39. $\overrightarrow{AB} = -2\mathbf{i}+3\mathbf{j}$ and $\overrightarrow{AC} = 3\mathbf{i}+\mathbf{j} \Rightarrow \overrightarrow{AB}\times\overrightarrow{AC} = \begin{vmatrix} \mathbf{i} & \mathbf{j} & \mathbf{k} \\ -2 & 3 & 0 \\ 3 & 1 & 0 \end{vmatrix} = -11\mathbf{k} \Rightarrow \text{area} = \tfrac{1}{2}\left|\overrightarrow{AB}\times\overrightarrow{AC}\right| = \tfrac{11}{2}$

41. $\overrightarrow{AB} = 6\mathbf{i}-5\mathbf{j}$ and $\overrightarrow{AC} = 11\mathbf{i}-5\mathbf{j} \Rightarrow \overrightarrow{AB}\times\overrightarrow{AC} = \begin{vmatrix} \mathbf{i} & \mathbf{j} & \mathbf{k} \\ 6 & -5 & 0 \\ 11 & -5 & 0 \end{vmatrix} = 25\mathbf{k} \Rightarrow \text{area} = \tfrac{1}{2}\left|\overrightarrow{AB}\times\overrightarrow{AC}\right| = \tfrac{25}{2}$

43. If $\mathbf{A} = a_1\mathbf{i}+a_2\mathbf{j}$ and $\mathbf{B} = b_1\mathbf{i}+b_2\mathbf{j}$, then $\mathbf{A}\times\mathbf{B} = \begin{vmatrix} \mathbf{i} & \mathbf{j} & \mathbf{k} \\ a_1 & a_2 & 0 \\ b_1 & b_2 & 0 \end{vmatrix} = \begin{vmatrix} a_1 & a_2 \\ b_1 & b_2 \end{vmatrix}\mathbf{k}$ and the triangle's area is

    $\tfrac{1}{2}\left|\mathbf{A}\times\mathbf{B}\right| = \pm\tfrac{1}{2}\begin{vmatrix} a_1 & a_2 \\ b_1 & b_2 \end{vmatrix}$. The applicable sign is $(+)$ if the acute angle from $\mathbf{A}$ to $\mathbf{B}$ runs counterclockwise

    in the xy-plane, and $(-)$ if it runs clockwise, because the area must be a nonnegative number.

12.5 LINES AND PLANES IN SPACE

1. The direction $\mathbf{i}+\mathbf{j}+\mathbf{k}$ and $P(3,-4,-1) \Rightarrow x = 3+t,\, y = -4+t,\, z = -1+t$

3. The direction $\overrightarrow{PQ} = 5\mathbf{i}+5\mathbf{j}-5\mathbf{k}$ and $P(-2,0,3) \Rightarrow x = -2+5t,\, y = 5t,\, z = 3-5t$

5. The direction $2\mathbf{j}+\mathbf{k}$ and $P(0,0,0) \Rightarrow x = 0,\, y = 2t,\, z = t$

7. The direction $\mathbf{k}$ and $P(1,1,1) \Rightarrow x = 1,\, y = 1,\, z = 1+t$

9. The direction $\mathbf{i}+2\mathbf{j}+2\mathbf{k}$ and $P(0,-7,0) \Rightarrow x = t,\, y = -7+2t,\, z = 2t$

11. The direction $\mathbf{i}$ and $P(0,0,0) \Rightarrow x = t,\, y = 0,\, z = 0$

13. The direction $\overrightarrow{PQ} = \mathbf{i} + \mathbf{j} + \frac{3}{2}\mathbf{k}$ and $P(0,0,0) \Rightarrow x = t$,
$y = t, z = \frac{3}{2}t$, where $0 \le t \le 1$

15. The direction $\overrightarrow{PQ} = \mathbf{j}$ and $P(1,1,0) \Rightarrow x = 1, y = 1 + t$,
$z = 0$, where $-1 \le t \le 0$

17. The direction $\overrightarrow{PQ} = -2\mathbf{j}$ and $P(0,1,1) \Rightarrow x = 0$,
$y = 1 - 2t, z = 1$, where $0 \le t \le 1$

19. The direction $\overrightarrow{PQ} = -2\mathbf{i} + 2\mathbf{j} - 2\mathbf{k}$ and $P(2,0,2)$
$\Rightarrow x = 2 - 2t, y = 2t, z = 2 - 2t$, where $0 \le t \le 1$

21. $3(x - 0) + (-2)(y - 2) + (-1)(z + 1) = 0 \Rightarrow 3x - 2y - z = -3$

23. $\overrightarrow{PQ} = \mathbf{i} - \mathbf{j} + 3\mathbf{k}, \overrightarrow{PS} = -\mathbf{i} - 3\mathbf{j} + 2\mathbf{k} \Rightarrow \overrightarrow{PQ} \times \overrightarrow{PS} = \begin{vmatrix} \mathbf{i} & \mathbf{j} & \mathbf{k} \\ 1 & -1 & 3 \\ -1 & -3 & 2 \end{vmatrix} = 7\mathbf{i} - 5\mathbf{j} - 4\mathbf{k}$ is normal to the plane

$\Rightarrow 7(x - 2) + (-5)(y - 0) + (-4)(z - 2) = 0 \Rightarrow 7x - 5y - 4z = 6$

25. $\mathbf{n} = \mathbf{i} + 3\mathbf{j} + 4\mathbf{k}, P(2,4,5) = (1)(x - 2) + (3)(y - 4) + (4)(z - 5) = 0 \Rightarrow x + 3y + 4z = 34$

27. $\begin{cases} x = 2t + 1 = s + 2 \\ y = 3t + 2 = 2s + 4 \end{cases} \Rightarrow \begin{cases} 2t - s = 1 \\ 3t - 2s = 2 \end{cases} \Rightarrow \begin{cases} 4t - 2s = 2 \\ 3t - 2s = 2 \end{cases} \Rightarrow t = 0$ and $s = -1$; then $z = 4t + 3 = -4s - 1$

$\Rightarrow 4(0) + 3 = (-4)(-1) - 1$ is satisfied $\Rightarrow$ the lines do intersect when $t = 0$ and $s = -1 \Rightarrow$ the point of
intersection is $x = 1, y = 2$, and $z = 3$ or $P(1,2,3)$. A vector normal to the plane determined by these lines is

$\mathbf{n_1} \times \mathbf{n_2} = \begin{vmatrix} \mathbf{i} & \mathbf{j} & \mathbf{k} \\ 2 & 3 & 4 \\ 1 & 2 & -4 \end{vmatrix} = -20\mathbf{i} + 12\mathbf{j} + \mathbf{k}$, where $\mathbf{n_1}$ and $\mathbf{n_2}$ are directions of the lines $\Rightarrow$ the plane

containing the lines is represented by $(-20)(x - 1) + (12)(y - 2) + (1)(z - 3) = 0 \Rightarrow -20x + 12y + z = 7$.

29. The cross product of $\mathbf{i} + \mathbf{j} - \mathbf{k}$ and $-4\mathbf{i} + 2\mathbf{j} - 2\mathbf{k}$ has the same direction as the normal to the plane

$$\Rightarrow \mathbf{n} = \begin{vmatrix} \mathbf{i} & \mathbf{j} & \mathbf{k} \\ 1 & 1 & -1 \\ -4 & 2 & -2 \end{vmatrix} = 6\mathbf{j} + 6\mathbf{k}. \text{ Select a point on either line, such as } P(-1, 2, 1). \text{ Since the lines are given}$$

to intersect, the desired plane is $0(x + 1) + 6(y - 2) + 6(z - 1) = 0 \Rightarrow 6y + 6z = 18 \Rightarrow y + z = 3$.

31. $\mathbf{n}_1 \times \mathbf{n}_2 = \begin{vmatrix} \mathbf{i} & \mathbf{j} & \mathbf{k} \\ 2 & 1 & -1 \\ 1 & 2 & 1 \end{vmatrix} = 3\mathbf{i} - 3\mathbf{j} + 3\mathbf{k}$ is a vector in the direction of the line of intersection of the planes

$\Rightarrow 3(x - 2) + (-3)(y - 1) + 3(z + 1) = 0 \Rightarrow 3x - 3y + 3z = 0 \Rightarrow x - y + z = 0$ is the desired plane containing $P_0(2, 1, -1)$

33. $S(0, 0, 12), P(0, 0, 0)$ and $\mathbf{v} = 4\mathbf{i} - 2\mathbf{j} + 2\mathbf{k} \Rightarrow \overrightarrow{PS} \times \mathbf{v} = \begin{vmatrix} \mathbf{i} & \mathbf{j} & \mathbf{k} \\ 0 & 0 & 12 \\ 4 & -2 & 2 \end{vmatrix} = 24\mathbf{i} + 48\mathbf{j} = 24(\mathbf{i} + 2\mathbf{j})$

$\Rightarrow d = \frac{|\overrightarrow{PS} \times \mathbf{v}|}{|\mathbf{v}|} = \frac{24\sqrt{1+4}}{\sqrt{16+4+4}} = \frac{24\sqrt{5}}{\sqrt{24}} = \sqrt{5 \cdot 24} = 2\sqrt{30}$ is the distance from S to the line

35. $S(2, 1, 3), P(2, 1, 3)$ and $\mathbf{v} = 2\mathbf{i} + 6\mathbf{j} \Rightarrow \overrightarrow{PS} \times \mathbf{v} = \mathbf{0} \Rightarrow d = \frac{|\overrightarrow{PS} \times \mathbf{v}|}{|\mathbf{v}|} = \frac{0}{\sqrt{40}} = 0$ is the distance from S to the line

(i.e., the point S lies on the line)

37. $S(3, -1, 4), P(4, 3, -5)$ and $\mathbf{v} = -\mathbf{i} + 2\mathbf{j} + 3\mathbf{k} \Rightarrow \overrightarrow{PS} \times \mathbf{v} = \begin{vmatrix} \mathbf{i} & \mathbf{j} & \mathbf{k} \\ -1 & -4 & 9 \\ -1 & 2 & 3 \end{vmatrix} = -30\mathbf{i} - 6\mathbf{j} - 6\mathbf{k}$

$\Rightarrow d = \frac{|\overrightarrow{PS} \times \mathbf{v}|}{|\mathbf{v}|} = \frac{\sqrt{900 + 36 + 36}}{\sqrt{1+4+9}} = \frac{\sqrt{972}}{\sqrt{14}} = \frac{\sqrt{486}}{\sqrt{7}} = \frac{\sqrt{81 \cdot 6}}{\sqrt{7}} = \frac{9\sqrt{42}}{7}$ is the distance from S to the line

39. $S(2, -3, 4), x + 2y + 2z = 13$ and $P(13, 0, 0)$ is on the plane $\Rightarrow \overrightarrow{PS} = -11\mathbf{i} - 3\mathbf{j} + 4\mathbf{k}$ and $\mathbf{n} = \mathbf{i} + 2\mathbf{j} + 2\mathbf{k}$

$\Rightarrow d = \left| \overrightarrow{PS} \cdot \frac{\mathbf{n}}{|\mathbf{n}|} \right| = \left| \frac{-11 - 6 + 8}{\sqrt{1+4+4}} \right| = \left| \frac{-9}{\sqrt{9}} \right| = 3$

41. $S(0, 1, 1), 4y + 3z = -12$ and $P(0, -3, 0)$ is on the plane $\Rightarrow \overrightarrow{PS} = 4\mathbf{j} + \mathbf{k}$ and $\mathbf{n} = 4\mathbf{j} + 3\mathbf{k}$

$\Rightarrow d = \left| \overrightarrow{PS} \cdot \frac{\mathbf{n}}{|\mathbf{n}|} \right| = \left| \frac{16 + 3}{\sqrt{16 + 9}} \right| = \frac{19}{5}$

43. $S(0, -1, 0), 2x + y + 2z = 4$ and $P(2, 0, 0)$ is on the plane $\Rightarrow \overrightarrow{PS} = -2\mathbf{i} - \mathbf{j}$ and $\mathbf{n} = 2\mathbf{i} + \mathbf{j} + 2\mathbf{k}$

$\Rightarrow d = \left| \overrightarrow{PS} \cdot \frac{\mathbf{n}}{|\mathbf{n}|} \right| = \left| \frac{-4 - 1 + 0}{\sqrt{4 + 1 + 4}} \right| = \frac{5}{3}$

45. The point $P(1, 0, 0)$ is on the first plane and $S(10, 0, 0)$ is a point on the second plane $\Rightarrow \overrightarrow{PS} = 9\mathbf{i}$, and

$\mathbf{n} = \mathbf{i} + 2\mathbf{j} + 6\mathbf{k}$ is normal to the first plane $\Rightarrow$ the distance from S to the first plane is $d = \left| \overrightarrow{PS} \cdot \frac{\mathbf{n}}{|\mathbf{n}|} \right|$

$= \left| \frac{9}{\sqrt{1 + 4 + 36}} \right| = \frac{9}{\sqrt{41}}$, which is also the distance between the planes.

47. $\mathbf{n}_1 = \mathbf{i} + \mathbf{j}$ and $\mathbf{n}_2 = 2\mathbf{i} + \mathbf{j} - 2\mathbf{k} \Rightarrow \theta = \cos^{-1}\left(\frac{\mathbf{n}_1 \cdot \mathbf{n}_2}{|\mathbf{n}_1| \, |\mathbf{n}_2|}\right) = \cos^{-1}\left(\frac{2+1}{\sqrt{2}\sqrt{9}}\right) = \cos^{-1}\left(\frac{1}{\sqrt{2}}\right) = \frac{\pi}{4}$

49. $\mathbf{n}_1 = 2\mathbf{i} + 2\mathbf{j} + 2\mathbf{k}$ and $\mathbf{n}_2 = 2\mathbf{i} - 2\mathbf{j} - \mathbf{k} \Rightarrow \theta = \cos^{-1}\left(\frac{\mathbf{n}_1 \cdot \mathbf{n}_2}{|\mathbf{n}_1| \, |\mathbf{n}_2|}\right) = \cos^{-1}\left(\frac{4 - 4 - 2}{\sqrt{12}\sqrt{9}}\right) = \cos^{-1}\left(\frac{-1}{3\sqrt{3}}\right) \approx 1.76 \text{ rad}$

51. $\mathbf{n}_1 = 2\mathbf{i} + 2\mathbf{j} - \mathbf{k}$ and $\mathbf{n}_2 = \mathbf{i} + 2\mathbf{j} + \mathbf{k} \Rightarrow \theta = \cos^{-1}\left(\frac{\mathbf{n}_1 \cdot \mathbf{n}_2}{|\mathbf{n}_1||\mathbf{n}_2|}\right) = \cos^{-1}\left(\frac{2+4-1}{\sqrt{9}\sqrt{6}}\right) = \cos^{-1}\left(\frac{5}{3\sqrt{6}}\right) \approx 0.82$ rad

53. $2x - y + 3z = 6 \Rightarrow 2(1-t) - (3t) + 3(1+t) = 6 \Rightarrow -2t + 5 = 6 \Rightarrow t = -\frac{1}{2} \Rightarrow x = \frac{3}{2}, y = -\frac{3}{2}$ and $z = \frac{1}{2}$
    $\Rightarrow \left(\frac{3}{2}, -\frac{3}{2}, \frac{1}{2}\right)$ is the point

55. $x + y + z = 2 \Rightarrow (1 + 2t) + (1 + 5t) + (3t) = 2 \Rightarrow 10t + 2 = 2 \Rightarrow t = 0 \Rightarrow x = 1, y = 1$ and $z = 0$
    $\Rightarrow (1, 1, 0)$ is the point

57. $\mathbf{n}_1 = \mathbf{i} + \mathbf{j} + \mathbf{k}$ and $\mathbf{n}_2 = \mathbf{i} + \mathbf{j} \Rightarrow \mathbf{n}_1 \times \mathbf{n}_2 = \begin{vmatrix} \mathbf{i} & \mathbf{j} & \mathbf{k} \\ 1 & 1 & 1 \\ 1 & 1 & 0 \end{vmatrix} = -\mathbf{i} + \mathbf{j}$, the direction of the desired line; $(1, 1, -1)$
    is on both planes $\Rightarrow$ the desired line is $x = 1 - t, y = 1 + t, z = -1$

59. $\mathbf{n}_1 = \mathbf{i} - 2\mathbf{j} + 4\mathbf{k}$ and $\mathbf{n}_2 = \mathbf{i} + \mathbf{j} - 2\mathbf{k} \Rightarrow \mathbf{n}_1 \times \mathbf{n}_2 = \begin{vmatrix} \mathbf{i} & \mathbf{j} & \mathbf{k} \\ 1 & -2 & 4 \\ 1 & 1 & -2 \end{vmatrix} = 6\mathbf{j} + 3\mathbf{k}$, the direction of the
    desired line; $(4, 3, 1)$ is on both planes $\Rightarrow$ the desired line is $x = 4, y = 3 + 6t, z = 1 + 3t$

61. <u>L1 & L2:</u> $x = 3 + 2t = 1 + 4s$ and $y = -1 + 4t = 1 + 2s \Rightarrow \begin{cases} 2t - 4s = -2 \\ 4t - 2s = 2 \end{cases} \Rightarrow \begin{cases} 2t - 4s = -2 \\ 2t - s = 1 \end{cases}$
    $\Rightarrow -3s = -3 \Rightarrow s = 1$ and $t = 1 \Rightarrow$ on L1, $z = 1$ and on L2, $z = 1 \Rightarrow$ L1 and L2 intersect at $(5, 3, 1)$.

    <u>L2 & L3:</u> The direction of L2 is $\frac{1}{6}(4\mathbf{i} + 2\mathbf{j} + 4\mathbf{k}) = \frac{1}{3}(2\mathbf{i} + \mathbf{j} + 2\mathbf{k})$ which is the same as the direction
    $\frac{1}{3}(2\mathbf{i} + \mathbf{j} + 2\mathbf{k})$ of L3; hence L2 and L3 are parallel.

    <u>L1 & L3:</u> $x = 3 + 2t = 3 + 2r$ and $y = -1 + 4t = 2 + r \Rightarrow \begin{cases} 2t - 2r = 0 \\ 4t - r = 3 \end{cases} \Rightarrow \begin{cases} t - r = 0 \\ 4t - r = 3 \end{cases} \Rightarrow 3t = 3$
    $\Rightarrow t = 1$ and $r = 1 \Rightarrow$ on L1, $z = 2$ while on L3, $z = 0 \Rightarrow$ L1 and L2 do not intersect. The direction of L1
    is $\frac{1}{\sqrt{21}}(2\mathbf{i} + 4\mathbf{j} - \mathbf{k})$ while the direction of L3 is $\frac{1}{3}(2\mathbf{i} + \mathbf{j} + 2\mathbf{k})$ and neither is a multiple of the other; hence
    L1 and L3 are skew.

63. $x = 2 + 2t, y = -4 - t, z = 7 + 3t; x = -2 - t, y = -2 + \frac{1}{2}t, z = 1 - \frac{3}{2}t$

65. $x = 0 \Rightarrow t = -\frac{1}{2}, y = -\frac{1}{2}, z = -\frac{3}{2} \Rightarrow \left(0, -\frac{1}{2}, -\frac{3}{2}\right); y = 0 \Rightarrow t = -1, x = -1, z = -3 \Rightarrow (-1, 0, -3); z = 0$
    $\Rightarrow t = 0, x = 1, y = -1 \Rightarrow (1, -1, 0)$

67. With substitution of the line into the plane we have $2(1 - 2t) + (2 + 5t) - (-3t) = 8 \Rightarrow 2 - 4t + 2 + 5t + 3t = 8$
    $\Rightarrow 4t + 4 = 8 \Rightarrow t = 1 \Rightarrow$ the point $(-1, 7, -3)$ is contained in both the line and plane, so they are not parallel.

69. There are many possible answers. One is found as follows: eliminate t to get $t = x - 1 = 2 - y = \frac{z-3}{2}$
    $\Rightarrow x - 1 = 2 - y$ and $2 - y = \frac{z-3}{2} \Rightarrow x + y = 3$ and $2y + z = 7$ are two such planes.

71. The points $(a, 0, 0)$, $(0, b, 0)$ and $(0, 0, c)$ are the x, y, and z intercepts of the plane. Since a, b, and c are all
    nonzero, the plane must intersect all three coordinate axes and cannot pass through the origin. Thus,
    $\frac{x}{a} + \frac{y}{b} + \frac{z}{c} = 1$ describes all planes <u>except</u> those through the origin or parallel to a coordinate axis.

73. (a) $\overrightarrow{EP} = c\overrightarrow{EP_1} \Rightarrow -x_0\mathbf{i} + y\mathbf{j} + z\mathbf{k} = c\left[(x_1 - x_0)\mathbf{i} + y_1\mathbf{j} + z_1\mathbf{k}\right] \Rightarrow -x_0 = c(x_1 - x_0), y = cy_1$ and $z = cz_1$,
    where c is a positive real number

(b) At $x_1 = 0 \Rightarrow c = 1 \Rightarrow y = y_1$ and $z = z_1$; at $x_1 = x_0 \Rightarrow x_0 = 0, y = 0, z = 0$; $\lim\limits_{x_0 \to \infty} c = \lim\limits_{x_0 \to \infty} \frac{-x_0}{x_1 - x_0}$

$= \lim\limits_{x_0 \to \infty} \frac{-1}{-1} = 1 \Rightarrow c \to 1$ so that $y \to y_1$ and $z \to z_1$

12.6 CYLINDERS AND QUADRIC SURFACES

1. d, ellipsoid

3. a, cylinder

5. l, hyperbolic paraboloid

7. b, cylinder

9. k, hyperbolic paraboloid

11. h, cone

13. $x^2 + y^2 = 4$

15. $z = y^2 - 1$

17. $x^2 + 4z^2 = 16$

19. $z^2 - y^2 = 1$

21. $9x^2 + y^2 + z^2 = 9$

23. $4x^2 + 9y^2 + 4z^2 = 36$

25. $x^2 + 4y^2 = z$

27. $z = 8 - x^2 - y^2$

29. $x = 4 - 4y^2 - z^2$

31. $x^2 + y^2 = z^2$

33. $4x^2 + 9z^2 = 9y^2$

35. $x^2 + y^2 - z^2 = 1$

37. $\frac{y^2}{4} + \frac{z^2}{9} - \frac{x^2}{4} = 1$

39. $z^2 - x^2 - y^2 = 1$

41. $x^2 - y^2 - \frac{z^2}{4} = 1$

43. $y^2 - x^2 = z$

45. $x^2 + y^2 + z^2 = 4$

47. $z = 1 + y^2 - x^2$

49. $y = -(x^2 + z^2)$

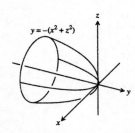

51. $16x^2 + 4y^2 = 1$

53. $x^2 + y^2 - z^2 = 4$

55. $x^2 + z^2 = y$

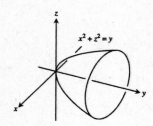

57. $x^2 + z^2 = 1$

59. $16y^2 + 9z^2 = 4x^2$

61. $9x^2 + 4y^2 + z^2 = 36$

63. $x^2 + y^2 - 16z^2 = 16$

65. $z = -(x^2 + y^2)$

67. $x^2 - 4y^2 = 1$

69. $4y^2 + z^2 - 4x^2 = 4$

71. $x^2 + y^2 = z$

73. $yz = 1$

75. $9x^2 + 16y^2 = 4z^2$

77. (a) If $x^2 + \frac{y^2}{4} + \frac{z^2}{9} = 1$ and $z = c$, then $x^2 + \frac{y^2}{4} = \frac{9-c^2}{9} \Rightarrow \frac{x^2}{\left(\frac{9-c^2}{9}\right)} + \frac{y^2}{\left[\frac{4(9-c^2)}{9}\right]} = 1 \Rightarrow A = ab\pi$

$= \pi\left(\frac{\sqrt{9-c^2}}{3}\right)\left(\frac{2\sqrt{9-c^2}}{3}\right) = \frac{2\pi(9-c^2)}{9}$

(b) From part (a), each slice has the area $\frac{2\pi(9-z^2)}{9}$, where $-3 \le z \le 3$. Thus $V = 2\int_0^3 \frac{2\pi}{9}(9-z^2)\,dz$

$= \frac{4\pi}{9}\int_0^3 (9-z^2)\,dz = \frac{4\pi}{9}\left[9z - \frac{z^3}{3}\right]_0^3 = \frac{4\pi}{9}(27-9) = 8\pi$

(c) $\frac{x^2}{a^2} + \frac{y^2}{b^2} + \frac{z^2}{c^2} = 1 \Rightarrow \frac{x^2}{\left[\frac{a^2(c^2-z^2)}{c^2}\right]} + \frac{y^2}{\left[\frac{b^2(c^2-z^2)}{c^2}\right]} = 1 \Rightarrow A = \pi \left(\frac{a\sqrt{c^2-z^2}}{c}\right)\left(\frac{b\sqrt{c^2-z^2}}{c}\right)$

$\Rightarrow V = 2\int_0^c \frac{\pi ab}{c^2}(c^2 - z^2)\, dz = \frac{2\pi ab}{c^2}\left[c^2 z - \frac{z^3}{3}\right]_0^c = \frac{2\pi ab}{c^2}\left(\frac{2}{3}c^3\right) = \frac{4\pi abc}{3}$. Note that if $r = a = b = c$,

then $V = \frac{4\pi r^3}{3}$, which is the volume of a sphere.

79. We calculate the volume by the slicing method, taking slices parallel to the xy-plane. For fixed z, $\frac{x^2}{a^2} + \frac{y^2}{b^2} = \frac{z}{c}$

gives the ellipse $\frac{x^2}{\left(\frac{za^2}{c}\right)} + \frac{y^2}{\left(\frac{zb^2}{c}\right)} = 1$. The area of this ellipse is $\pi\left(a\sqrt{\frac{z}{c}}\right)\left(b\sqrt{\frac{z}{c}}\right) = \frac{\pi abz}{c}$ (see Exercise 77a). Hence

the volume is given by $V = \int_0^h \frac{\pi abz}{c}\, dz = \left[\frac{\pi abz^2}{2c}\right]_0^h = \frac{\pi abh^2}{c}$. Now the area of the elliptic base when z = h is

$A = \frac{\pi abh}{c}$, as determined previously. Thus, $V = \frac{\pi abh^2}{c} = \frac{1}{2}\left(\frac{\pi abh}{c}\right)h = \frac{1}{2}$ (base)(altitude), as claimed.

81. $y = y_1 \Rightarrow \frac{z}{c} = \frac{y_1^2}{b^2} - \frac{x^2}{a^2}$, a parabola in the plane $y = y_1 \Rightarrow$ vertex when $\frac{dz}{dx} = 0$ or $c\frac{dz}{dx} = -\frac{2x}{a^2} = 0 \Rightarrow x = 0$

$\Rightarrow$ Vertex $\left(0, y_1, \frac{cy_1^2}{b^2}\right)$; writing the parabola as $x^2 = -\frac{a^2}{c}z + \frac{a^2 y_1^2}{b^2}$ we see that $4p = -\frac{a^2}{c} \Rightarrow p = -\frac{a^2}{4c}$

$\Rightarrow$ Focus $\left(0, y_1, \frac{cy_1^2}{b^2} - \frac{a^2}{4c}\right)$

83. No, it is not mere coincidence. A plane parallel to one of the coordinate planes will set one of the variables
x, y, or z equal to a constant in the general equation $Ax^2 + By^2 + Cz^2 + Dxy + Eyz + Fxz + Gx + Hy + Jz + K$
$= 0$ for a quadric surface. The resulting equation then has the general form for a conic in that parallel plane.
For example, setting $y = y_1$ results in the equation $Ax^2 + Cz^2 + D'x + E'z + Fxz + Gx + Jz + K' = 0$ where
$D' = Dy_1$, $E' = Ey_1$, and $K' = K + By_1^2 + Hy_1$, which is the general form of a conic section in the plane $y = y_1$
by Section 10.3.

85. $z = y^2$

87. $z = x^2 + y^2$

CHAPTER 12 PRACTICE EXERCISES

1. (a) $3\langle -3, 4\rangle - 4\langle 2, -5\rangle = \langle -9 - 8, 12 + 20\rangle = \langle -17, 32\rangle$

   (b) $\sqrt{17^2 + 32^2} = \sqrt{1313}$

3. (a) $\langle -2(-3), -2(4)\rangle = \langle 6, -8\rangle$

   (b) $\sqrt{6^2 + (-8)^2} = 10$

5. $\frac{\pi}{6}$ radians below the negative x-axis: $\left\langle -\frac{\sqrt{3}}{2}, -\frac{1}{2} \right\rangle$ [assuming counterclockwise].

7. $2\left(\frac{1}{\sqrt{4^2 + 1^2}} \right)(4\mathbf{i} - \mathbf{j}) = \left(\frac{8}{\sqrt{17}}\mathbf{i} - \frac{2}{\sqrt{17}}\mathbf{j} \right)$

9. length $= \left| \sqrt{2}\mathbf{i} + \sqrt{2}\mathbf{j} \right| = \sqrt{2 + 2} = 2$, $\sqrt{2}\mathbf{i} + \sqrt{2}\mathbf{j} = 2 \left(\frac{1}{\sqrt{2}}\mathbf{i} + \frac{1}{\sqrt{2}}\mathbf{j} \right) \Rightarrow$ the direction is $\frac{1}{\sqrt{2}}\mathbf{i} + \frac{1}{\sqrt{2}}\mathbf{j}$

11. $t = \frac{\pi}{2} \Rightarrow \mathbf{v} = (-2 \sin \frac{\pi}{2})\mathbf{i} + \left(2 \cos \frac{\pi}{2} \right)\mathbf{j} = -2\mathbf{i}$; length $= |-2\mathbf{i}| = \sqrt{4 + 0} = 2$; $-2\mathbf{i} = 2(-\mathbf{i}) \Rightarrow$ the direction is $-\mathbf{i}$

13. length $= |2\mathbf{i} - 3\mathbf{j} + 6\mathbf{k}| = \sqrt{4 + 9 + 36} = 7$, $2\mathbf{i} - 3\mathbf{j} + 6\mathbf{k} = 7 \left(\frac{2}{7}\mathbf{i} - \frac{3}{7}\mathbf{j} + \frac{6}{7}\mathbf{k} \right) \Rightarrow$ the direction is $\frac{2}{7}\mathbf{i} - \frac{3}{7}\mathbf{j} + \frac{6}{7}\mathbf{k}$

15. $2 \frac{\mathbf{v}}{|\mathbf{v}|} = 2 \cdot \frac{4\mathbf{i} - \mathbf{j} + 4\mathbf{k}}{\sqrt{4^2 + (-1)^2 + 4^2}} = 2 \cdot \frac{4\mathbf{i} - \mathbf{j} + 4\mathbf{k}}{\sqrt{33}} = \frac{8}{\sqrt{33}}\mathbf{i} - \frac{2}{\sqrt{33}}\mathbf{j} + \frac{8}{\sqrt{33}}\mathbf{k}$

17. $|\mathbf{v}| = \sqrt{1 + 1} = \sqrt{2}$, $|\mathbf{u}| = \sqrt{4 + 1 + 4} = 3$, $\mathbf{v} \cdot \mathbf{u} = 3$, $\mathbf{u} \cdot \mathbf{v} = 3$, $\mathbf{v} \times \mathbf{u} = \begin{vmatrix} \mathbf{i} & \mathbf{j} & \mathbf{k} \\ 1 & 1 & 0 \\ 2 & 1 & -2 \end{vmatrix} = -2\mathbf{i} + 2\mathbf{j} - \mathbf{k}$,

$\mathbf{u} \times \mathbf{v} = -(\mathbf{v} \times \mathbf{u}) = 2\mathbf{i} - 2\mathbf{j} + \mathbf{k}$, $|\mathbf{v} \times \mathbf{u}| = \sqrt{4 + 4 + 1} = 3$, $\theta = \cos^{-1} \left(\frac{\mathbf{v} \cdot \mathbf{u}}{|\mathbf{v}||\mathbf{u}|} \right) = \cos^{-1} \left(\frac{1}{\sqrt{2}} \right) = \frac{\pi}{4}$,

$|\mathbf{u}| \cos \theta = \frac{3}{\sqrt{2}}$, $\text{proj}_{\mathbf{v}} \mathbf{u} = \left(\frac{\mathbf{v} \cdot \mathbf{u}}{|\mathbf{v}||\mathbf{v}|} \right) \mathbf{v} = \frac{3}{2}(\mathbf{i} + \mathbf{j})$

19. $\mathbf{u} = \left(\frac{\mathbf{v} \cdot \mathbf{u}}{|\mathbf{v}||\mathbf{v}|} \right) \mathbf{v} + \left[\mathbf{u} - \left(\frac{\mathbf{v} \cdot \mathbf{u}}{|\mathbf{v}||\mathbf{v}|} \right) \mathbf{v} \right] = \frac{4}{3}(2\mathbf{i} + \mathbf{j} - \mathbf{k}) + \left[(\mathbf{i} + \mathbf{j} - 5\mathbf{k}) - \frac{4}{3}(2\mathbf{i} + \mathbf{j} - \mathbf{k}) \right] = \frac{4}{3}(2\mathbf{i} + \mathbf{j} - \mathbf{k}) - \frac{1}{3}(5\mathbf{i} + \mathbf{j} + 11\mathbf{k})$,

where $\mathbf{v} \cdot \mathbf{u} = 8$ and $\mathbf{v} \cdot \mathbf{v} = 6$

21. $\mathbf{u} \times \mathbf{v} = \begin{vmatrix} \mathbf{i} & \mathbf{j} & \mathbf{k} \\ 1 & 0 & 0 \\ 1 & 1 & 0 \end{vmatrix} = \mathbf{k}$

23. Let $\mathbf{v} = v_1\mathbf{i} + v_2\mathbf{j} + v_3\mathbf{k}$ and $\mathbf{w} = w_1\mathbf{i} + w_2\mathbf{j} + w_3\mathbf{k}$. Then $|\mathbf{v} - 2\mathbf{w}|^2 = |(v_1\mathbf{i} + v_2\mathbf{j} + v_3\mathbf{k}) - 2(w_1\mathbf{i} + w_2\mathbf{j} + w_3\mathbf{k})|^2$

$= |(v_1 - 2w_1)\mathbf{i} + (v_2 - 2w_2)\mathbf{j} + (v_3 - 2w_3)\mathbf{k}|^2 = \left(\sqrt{(v_1 - 2w_1)^2 + (v_2 - 2w_2)^2 + (v_3 - 2w_3)^2} \right)^2$

$= (v_1^2 + v_2^2 + v_3^2) - 4(v_1w_1 + v_2w_2 + v_3w_3) + 4(w_1^2 + w_2^2 + w_3^2) = |\mathbf{v}|^2 - 4\mathbf{v} \cdot \mathbf{w} + 4|\mathbf{w}|^2$

$= |\mathbf{v}|^2 - 4|\mathbf{v}||\mathbf{w}| \cos \theta + 4|\mathbf{w}|^2 = 4 - 4(2)(3) \left(\cos \frac{\pi}{3} \right) + 36 = 40 - 24 \left(\frac{1}{2} \right) = 40 - 12 = 28 \Rightarrow |\mathbf{v} - 2\mathbf{w}| = \sqrt{28}$

$= 2\sqrt{7}$

25. (a) area $= |\mathbf{u} \times \mathbf{v}| = \text{abs} \begin{vmatrix} \mathbf{i} & \mathbf{j} & \mathbf{k} \\ 1 & 1 & -1 \\ 2 & 1 & 1 \end{vmatrix} = |2\mathbf{i} - 3\mathbf{j} - \mathbf{k}| = \sqrt{4 + 9 + 1} = \sqrt{14}$

(b) volume $= \mathbf{u} \cdot (\mathbf{v} \times \mathbf{w}) = \begin{vmatrix} 1 & 1 & -1 \\ 2 & 1 & 1 \\ -1 & -2 & 3 \end{vmatrix} = 1(3 + 2) + 1(-1 - 6) - 1(-4 + 1) = 1$

27. The desired vector is $\mathbf{n} \times \mathbf{v}$ or $\mathbf{v} \times \mathbf{n}$ since $\mathbf{n} \times \mathbf{v}$ is perpendicular to both $\mathbf{n}$ and $\mathbf{v}$ and, therefore, also parallel to the plane.

29. The line L passes through the point $P(0, 0, -1)$ parallel to $\mathbf{v} = -\mathbf{i} + \mathbf{j} + \mathbf{k}$. With $\overrightarrow{PS} = 2\mathbf{i} + 2\mathbf{j} + \mathbf{k}$ and

$$\overrightarrow{PS} \times \mathbf{v} = \begin{vmatrix} \mathbf{i} & \mathbf{j} & \mathbf{k} \\ 2 & 2 & 1 \\ -1 & 1 & 1 \end{vmatrix} = (2-1)\mathbf{i} + (-1-2)\mathbf{j} + (2+2)\mathbf{k} = \mathbf{i} - 3\mathbf{j} + 4\mathbf{k}, \text{ we find the distance}$$

$$d = \frac{|\overrightarrow{PS} \times \mathbf{v}|}{|\mathbf{v}|} = \frac{\sqrt{1+9+16}}{\sqrt{1+1+1}} = \frac{\sqrt{26}}{\sqrt{3}} = \frac{\sqrt{78}}{3}.$$

31. Parametric equations for the line are $x = 1 - 3t, y = 2, z = 3 + 7t$.

33. The point $P(4, 0, 0)$ lies on the plane $x - y = 4$, and $\overrightarrow{PS} = (6-4)\mathbf{i} + 0\mathbf{j} + (-6+0)\mathbf{k} = 2\mathbf{i} - 6\mathbf{k}$ with $\mathbf{n} = \mathbf{i} - \mathbf{j}$

$$\Rightarrow d = \frac{|\mathbf{n} \cdot \overrightarrow{PS}|}{|\mathbf{n}|} = \left| \frac{2+0+0}{\sqrt{1+1+0}} \right| = \frac{2}{\sqrt{2}} = \sqrt{2}.$$

35. $P(3, -2, 1)$ and $\mathbf{n} = 2\mathbf{i} + \mathbf{j} - \mathbf{k} \Rightarrow (2)(x-3) + (1)(y-(-2)) + (1)(z-1) = 0 \Rightarrow 2x + y + z = 5$

37. $P(1, -1, 2), Q(2, 1, 3)$ and $R(-1, 2, -1) \Rightarrow \overrightarrow{PQ} = \mathbf{i} + 2\mathbf{j} + \mathbf{k}, \overrightarrow{PR} = -2\mathbf{i} + 3\mathbf{j} - 3\mathbf{k}$ and $\overrightarrow{PQ} \times \overrightarrow{PR}$

$$= \begin{vmatrix} \mathbf{i} & \mathbf{j} & \mathbf{k} \\ 1 & 2 & 1 \\ -2 & 3 & -3 \end{vmatrix} = -9\mathbf{i} + \mathbf{j} + 7\mathbf{k} \text{ is normal to the plane} \Rightarrow (-9)(x-1) + (1)(y+1) + (7)(z-2) = 0$$

$$\Rightarrow -9x + y + 7z = 4$$

39. $\left(0, -\frac{1}{2}, -\frac{3}{2} \right)$, since $t = -\frac{1}{2}$, $y = -\frac{1}{2}$ and $z = -\frac{3}{2}$ when $x = 0$; $(-1, 0, -3)$, since $t = -1$, $x = -1$ and $z = -3$

when $y = 0$; $(1, -1, 0)$, since $t = 0$, $x = 1$ and $y = -1$ when $z = 0$

41. $\mathbf{n}_1 = \mathbf{i}$ and $\mathbf{n}_2 = \mathbf{i} + \mathbf{j} + \sqrt{2}\mathbf{k} \Rightarrow$ the desired angle is $\cos^{-1} \left(\frac{\mathbf{n}_1 \cdot \mathbf{n}_2}{|\mathbf{n}_1| \, |\mathbf{n}_2|} \right) = \cos^{-1} \left(\frac{1}{2} \right) = \frac{\pi}{3}$

43. The direction of the line is $\mathbf{n}_1 \times \mathbf{n}_2 = \begin{vmatrix} \mathbf{i} & \mathbf{j} & \mathbf{k} \\ 1 & 2 & 1 \\ 1 & -1 & 2 \end{vmatrix} = 5\mathbf{i} - \mathbf{j} - 3\mathbf{k}$. Since the point $(-5, 3, 0)$ is on

both planes, the desired line is $x = -5 + 5t, y = 3 - t, z = -3t$.

45. (a) The corresponding normals are $\mathbf{n}_1 = 3\mathbf{i} + 6\mathbf{k}$ and $\mathbf{n}_2 = 2\mathbf{i} + 2\mathbf{j} - \mathbf{k}$ and since $\mathbf{n}_1 \cdot \mathbf{n}_2$

$= (3)(2) + (0)(2) + (6)(-1) = 6 + 0 - 6 = 0$, we have that the planes are orthogonal

(b) The line of intersection is parallel to $\mathbf{n}_1 \times \mathbf{n}_2 = \begin{vmatrix} \mathbf{i} & \mathbf{j} & \mathbf{k} \\ 3 & 0 & 6 \\ 2 & 2 & -1 \end{vmatrix} = -12\mathbf{i} + 15\mathbf{j} + 6\mathbf{k}$. Now to find a point in

the intersection, solve $\begin{cases} 3x + 6z = 1 \\ 2x + 2y - z = 3 \end{cases} \Rightarrow \begin{cases} 3x + 6z = 1 \\ 12x + 12y - 6z = 18 \end{cases} \Rightarrow 15x + 12y = 19 \Rightarrow x = 0$ and $y = \frac{19}{12}$

$\Rightarrow \left(0, \frac{19}{12}, \frac{1}{6} \right)$ is a point on the line we seek. Therefore, the line is $x = -12t, y = \frac{19}{12} + 15t$ and $z = \frac{1}{6} + 6t$.

47. Yes; $\mathbf{v} \cdot \mathbf{n} = (2\mathbf{i} - 4\mathbf{j} + \mathbf{k}) \cdot (2\mathbf{i} + \mathbf{j} + 0\mathbf{k}) = 2 \cdot 2 - 4 \cdot 1 + 1 \cdot 0 = 0 \Rightarrow$ the vector is orthogonal to the plane's normal

$\Rightarrow \mathbf{v}$ is parallel to the plane

49. A normal to the plane is $\mathbf{n} = \overrightarrow{AB} \times \overrightarrow{AC} = \begin{vmatrix} \mathbf{i} & \mathbf{j} & \mathbf{k} \\ 2 & 0 & -1 \\ 2 & -1 & 0 \end{vmatrix} = -\mathbf{i} - 2\mathbf{j} - 2\mathbf{k} \Rightarrow$ the distance is $d = \left| \frac{\overrightarrow{AP} \cdot \mathbf{n}}{\mathbf{n}} \right|$

$= \left| \frac{(\mathbf{i} + 4\mathbf{j}) \cdot (-\mathbf{i} - 2\mathbf{j} - 2\mathbf{k})}{\sqrt{1+4+4}} \right| = \left| \frac{-1 - 8 + 0}{3} \right| = 3$

51. $\mathbf{n} = 2\mathbf{i} - \mathbf{j} - \mathbf{k}$ is normal to the plane $\Rightarrow \mathbf{n} \times \mathbf{v} = \begin{vmatrix} \mathbf{i} & \mathbf{j} & \mathbf{k} \\ 2 & -1 & -1 \\ 1 & 1 & 1 \end{vmatrix} = 0\mathbf{i} - 3\mathbf{j} + 3\mathbf{k} = -3\mathbf{j} + 3\mathbf{k}$ is orthogonal

to $\mathbf{v}$ and parallel to the plane

53. A vector parallel to the line of intersection is $\mathbf{v} = \mathbf{n}_1 \times \mathbf{n}_2 = \begin{vmatrix} \mathbf{i} & \mathbf{j} & \mathbf{k} \\ 1 & 2 & 1 \\ 1 & -1 & 2 \end{vmatrix} = 5\mathbf{i} - \mathbf{j} - 3\mathbf{k}$

$\Rightarrow |\mathbf{v}| = \sqrt{25 + 1 + 9} = \sqrt{35} \Rightarrow 2\left(\frac{\mathbf{v}}{|\mathbf{v}|}\right) = \frac{2}{\sqrt{35}}(5\mathbf{i} - \mathbf{j} - 3\mathbf{k})$ is the desired vector.

55. The line is represented by $x = 3 + 2t$, $y = 2 - t$, and $z = 1 + 2t$. It meets the plane $2x - y + 2z = -2$ when $2(3 + 2t) - (2 - t) + 2(1 + 2t) = -2 \Rightarrow t = -\frac{8}{9} \Rightarrow$ the point is $\left(\frac{11}{9}, \frac{26}{9}, -\frac{7}{9}\right)$.

57. The intersection occurs when $(3 + 2t) + 3(2t) - t = -4 \Rightarrow t = -1 \Rightarrow$ the point is $(1, -2, -1)$. The required line

must be perpendicular to both the given line and to the normal, and hence is parallel to $\begin{vmatrix} \mathbf{i} & \mathbf{j} & \mathbf{k} \\ 2 & 2 & 1 \\ 1 & 3 & -1 \end{vmatrix}$

$= -5\mathbf{i} + 3\mathbf{j} + 4\mathbf{k} \Rightarrow$ the line is represented by $x = 1 - 5t$, $y = -2 + 3t$, and $z = -1 + 4t$.

59. The vector $\overrightarrow{AB} \times \overrightarrow{CD} = \begin{vmatrix} \mathbf{i} & \mathbf{j} & \mathbf{k} \\ 3 & -2 & 4 \\ \frac{26}{5} & 0 & -\frac{26}{5} \end{vmatrix} = \frac{26}{5}(2\mathbf{i} + 7\mathbf{j} + 2\mathbf{k})$ is normal to the plane and $A(-2, 0, -3)$ lies on the

plane $\Rightarrow 2(x + 2) + 7(y - 0) + 2(z - (-3)) = 0 \Rightarrow 2x + 7y + 2z + 10 = 0$ is an equation of the plane.

61. The vector $\overrightarrow{PQ} \times \overrightarrow{PR} = \begin{vmatrix} \mathbf{i} & \mathbf{j} & \mathbf{k} \\ 2 & -1 & 3 \\ -3 & 0 & 1 \end{vmatrix} = -\mathbf{i} - 11\mathbf{j} - 3\mathbf{k}$ is normal to the plane.

(a) No, the plane is not orthogonal to $\overrightarrow{PQ} \times \overrightarrow{PR}$.

(b) No, these equations represent a line, not a plane.

(c) No, the plane $(x + 2) + 11(y - 1) - 3z = 0$ has normal $\mathbf{i} + 11\mathbf{j} - 3\mathbf{k}$ which is not parallel to $\overrightarrow{PQ} \times \overrightarrow{PR}$.

(d) No, this vector equation is equivalent to the equations $3y + 3z = 3$, $3x - 2z = -6$, and $3x + 2y = -4$
$\Rightarrow x = -\frac{4}{3} - \frac{2}{3}t$, $y = t$, $z = 1 - t$, which represents a line, not a plane.

(e) Yes, this is a plane containing the point $R(-2, 1, 0)$ with normal $\overrightarrow{PQ} \times \overrightarrow{PR}$.

63. $\overrightarrow{AB} = -2\mathbf{i} + \mathbf{j} + \mathbf{k}$, $\overrightarrow{CD} = \mathbf{i} + 4\mathbf{j} - \mathbf{k}$, and $\overrightarrow{AC} = 2\mathbf{i} + \mathbf{j} \Rightarrow \mathbf{n} = \begin{vmatrix} \mathbf{i} & \mathbf{j} & \mathbf{k} \\ -2 & 1 & 1 \\ 1 & 4 & -1 \end{vmatrix} = -5\mathbf{i} - \mathbf{j} - 9\mathbf{k} \Rightarrow$ the distance is

$d = \left| \frac{(2\mathbf{i} + \mathbf{j}) \cdot (-5\mathbf{i} - \mathbf{j} - 9\mathbf{k})}{\sqrt{25 + 1 + 81}} \right| = \frac{11}{\sqrt{107}}$

65. $x^2 + y^2 + z^2 = 4$

67. $4x^2 + 4y^2 + z^2 = 4$

69. $z = -(x^2 + y^2)$

71. $x^2 + y^2 = z^2$

73. $x^2 + y^2 - z^2 = 4$

75. $y^2 - x^2 - z^2 = 1$

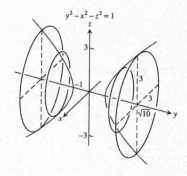

CHAPTER 12 ADDITIONAL AND ADVANCED EXERCISES

1. Information from ship A indicates the submarine is now on the line L_1: $x = 4 + 2t$, $y = 3t$, $z = -\frac{1}{3}t$; information from ship B indicates the submarine is now on the line L_2: $x = 18s$, $y = 5 - 6s$, $z = -s$. The current position of the sub is $\left(6, 3, -\frac{1}{3}\right)$ and occurs when the lines intersect at $t = 1$ and $s = \frac{1}{3}$. The straight line path of the submarine contains both points $P\left(2, -1, -\frac{1}{3}\right)$ and $Q\left(6, 3, -\frac{1}{3}\right)$; the line representing this path is L: $x = 2 + 4t$, $y = -1 + 4t$, $z = -\frac{1}{3}$. The submarine traveled the distance between P and Q in 4 minutes $\Rightarrow$ a speed of $\frac{\left|\overrightarrow{PQ}\right|}{4} = \frac{\sqrt{32}}{4} = \sqrt{2}$ thousand ft/min. In 20 minutes the submarine will move $20\sqrt{2}$ thousand ft from Q along the line L $\Rightarrow$ $20\sqrt{2} = \sqrt{(2 + 4t - 6)^2 + (-1 + 4t - 3)^2 + 0^2}$ $\Rightarrow$ $800 = 16(t-1)^2 + 16(t-1)^2 = 32(t-1)^2$ $\Rightarrow$ $(t-1)^2 = \frac{800}{32} = 25$ $\Rightarrow$ $t = 6$ $\Rightarrow$ the submarine will be located at $\left(26, 23, -\frac{1}{3}\right)$ in 20 minutes.

3. Torque $= \left|\overrightarrow{PQ} \times \mathbf{F}\right|$ $\Rightarrow$ 15 ft-lb $= \left|\overrightarrow{PQ}\right| |\mathbf{F}| \sin\frac{\pi}{2} = \frac{3}{4}$ ft $\cdot |\mathbf{F}|$ $\Rightarrow$ $|\mathbf{F}| = 20$ lb

5. (a) If $P(x, y, z)$ is a point in the plane determined by the three points $P_1(x_1, y_1, z_1)$, $P_2(x_2, y_2, z_2)$ and $P_3(x_3, y_3, z_3)$, then the vectors $\overrightarrow{PP_1}$, $\overrightarrow{PP_2}$ and $\overrightarrow{PP_3}$ all lie in the plane. Thus $\overrightarrow{PP_1} \cdot (\overrightarrow{PP_2} \times \overrightarrow{PP_3}) = 0$

$$\Rightarrow \begin{vmatrix} x_1 - x & y_1 - y & z_1 - z \\ x_2 - x & y_2 - y & z_2 - z \\ x_3 - x & y_3 - y & z_3 - z \end{vmatrix} = 0$$ by the determinant formula for the triple scalar product in Section 10.4.

(b) Subtract row 1 from rows 2, 3, and 4 and evaluate the resulting determinant (which has the same value as the given determinant) by cofactor expansion about column 4. This expansion is exactly the determinant in part (a) so we have all points $P(x, y, z)$ in the plane determined by $P_1(x_1, y_1, z_1)$, $P_2(x_2, y_2, z_2)$, and $P_3(x_3, y_3, z_3)$.

7. (a) $\overrightarrow{BD} = \overrightarrow{AD} - \overrightarrow{AB}$

   (b) $\overrightarrow{AP} = \overrightarrow{AB} + \frac{1}{2}\overrightarrow{BD} = \frac{1}{2}\left(\overrightarrow{AB} + \overrightarrow{AD}\right)$

   (c) $\overrightarrow{AC} = \overrightarrow{AB} + \overrightarrow{AD}$, so by part (b), $\overrightarrow{AP} = \frac{1}{2}\overrightarrow{AC}$

9. If $Q(x, y)$ is a point on the line $ax + by = c$, then $\overrightarrow{P_1Q} = (x - x_1)\mathbf{i} + (y - y_1)\mathbf{j}$, and $\mathbf{n} = a\mathbf{i} + b\mathbf{j}$ is normal to the

   line. The distance is $\left|\text{proj}_{\mathbf{n}}\, \overrightarrow{P_1Q}\right| = \left|\frac{[(x - x_1)\mathbf{i} + (y - y_1)\mathbf{j}]\cdot(a\mathbf{i} + b\mathbf{j})}{\sqrt{a^2 + b^2}}\right| = \frac{|a(x - x_1) + b(y - y_1)|}{\sqrt{a^2 + b^2}}$

   $= \frac{|ax_1 + by_1 - c|}{\sqrt{a^2 + b^2}}$, since $c = ax + by$.

11. (a) If (x_1, y_1, z_1) is on the plane $Ax + By + Cz = D_1$, then the distance d between the planes is

     $d = \frac{|Ax_1 + By_1 + Cz_1 - D_2|}{\sqrt{A^2 + B^2 + C^2}} = \frac{|D_1 - D_2|}{|A\mathbf{i} + B\mathbf{j} + C\mathbf{k}|}$, since $Ax_1 + By_1 + Cz_1 = D_1$, by Exercise 10(a).

   (b) $d = \frac{|12 - 6|}{\sqrt{4 + 9 + 1}} = \frac{6}{\sqrt{14}}$

   (c) $\frac{|2(3) + (-1)(2) + 2(-1) + 4|}{\sqrt{14}} = \frac{|2(3) + (-1)(2) + 2(-1) - D|}{\sqrt{14}} \Rightarrow D = 8$ or $-4 \Rightarrow$ the desired plane is

     $2x - y + 2x = 8$

   (d) Choose the point $(2, 0, 1)$ on the plane. Then $\frac{|3 - D|}{\sqrt{6}} = 5 \Rightarrow D = 3 \pm 5\sqrt{6} \Rightarrow$ the desired planes are

     $x - 2y + z = 3 + 5\sqrt{6}$ and $x - 2y + z = 3 - 5\sqrt{6}$.

13. $\mathbf{n} = \mathbf{i} + 2\mathbf{j} + 6\mathbf{k}$ is normal to the plane $x + 2y + 6z = 6$; $\mathbf{v} \times \mathbf{n} = \begin{vmatrix} \mathbf{i} & \mathbf{j} & \mathbf{k} \\ 1 & 1 & 1 \\ 1 & 2 & 6 \end{vmatrix} = 4\mathbf{i} - 5\mathbf{j} + \mathbf{k}$ is parallel to the

plane and perpendicular to the plane of $\mathbf{v}$ and $\mathbf{n} \Rightarrow \mathbf{w} = \mathbf{n} \times (\mathbf{v} \times \mathbf{n}) = \begin{vmatrix} \mathbf{i} & \mathbf{j} & \mathbf{k} \\ 1 & 2 & 6 \\ 4 & -5 & 1 \end{vmatrix} = 32\mathbf{i} + 23\mathbf{j} - 13\mathbf{k}$ is a

vector parallel to the plane $x + 2y + 6z = 6$ in the direction of the projection vector $\text{proj}_P\, \mathbf{v}$. Therefore,

$\text{proj}_P\, \mathbf{v} = \text{proj}_{\mathbf{w}}\, \mathbf{v} = \left(\mathbf{v} \cdot \frac{\mathbf{w}}{|\mathbf{w}|}\right) \frac{\mathbf{w}}{|\mathbf{w}|} = \left(\frac{\mathbf{v}\cdot\mathbf{w}}{|\mathbf{w}|^2}\right)\mathbf{w} = \left(\frac{32 + 23 - 13}{32^2 + 23^2 + 13^2}\right)\mathbf{w} = \frac{42}{1722}\mathbf{w} = \frac{1}{41}\mathbf{w} = \frac{32}{41}\mathbf{i} + \frac{23}{41}\mathbf{j} - \frac{13}{41}\mathbf{k}$

15. (a) $\mathbf{u} \times \mathbf{v} = 2\mathbf{i} \times 2\mathbf{j} = 4\mathbf{k} \Rightarrow (\mathbf{u} \times \mathbf{v}) \times \mathbf{C} = \mathbf{0}$; $(\mathbf{u} \cdot \mathbf{w})\mathbf{v} - (\mathbf{v} \cdot \mathbf{w})\mathbf{u} = 0\mathbf{v} - 0\mathbf{u} = \mathbf{0}$; $\mathbf{v} \times \mathbf{w} = 4\mathbf{i} \Rightarrow \mathbf{u} \times (\mathbf{v} \times \mathbf{w}) = \mathbf{0}$;

     $(\mathbf{u} \cdot \mathbf{w})\mathbf{v} - (\mathbf{u} \cdot \mathbf{v})\mathbf{w} = 0\mathbf{v} - 0\mathbf{w} = \mathbf{0}$

   (b) $\mathbf{u} \times \mathbf{v} = \begin{vmatrix} \mathbf{i} & \mathbf{j} & \mathbf{k} \\ 1 & -1 & 1 \\ 2 & 1 & -2 \end{vmatrix} = \mathbf{i} + 4\mathbf{j} + 3\mathbf{k} \Rightarrow (\mathbf{u} \times \mathbf{v}) \times \mathbf{w} = \begin{vmatrix} \mathbf{i} & \mathbf{j} & \mathbf{k} \\ 1 & 4 & 3 \\ -1 & 2 & -1 \end{vmatrix} = -10\mathbf{i} - 2\mathbf{j} + 6\mathbf{k}$;

     $(\mathbf{u} \cdot \mathbf{w})\mathbf{v} - (\mathbf{v} \cdot \mathbf{w})\mathbf{u} = -4(2\mathbf{i} + \mathbf{j} - 2\mathbf{k}) - 2(\mathbf{i} - \mathbf{j} + \mathbf{k}) = -10\mathbf{i} - 2\mathbf{j} + 6\mathbf{k}$;

     $\mathbf{v} \times \mathbf{w} = \begin{vmatrix} \mathbf{i} & \mathbf{j} & \mathbf{k} \\ 2 & 1 & -2 \\ -1 & 2 & -1 \end{vmatrix} = 3\mathbf{i} + 4\mathbf{j} + 5\mathbf{k} \Rightarrow \mathbf{u} \times (\mathbf{v} \times \mathbf{w}) = \begin{vmatrix} \mathbf{i} & \mathbf{j} & \mathbf{k} \\ 1 & -1 & 1 \\ 3 & 4 & 5 \end{vmatrix} = -9\mathbf{i} - 2\mathbf{j} + 7\mathbf{k}$;

     $(\mathbf{u} \cdot \mathbf{w})\mathbf{v} - (\mathbf{u} \cdot \mathbf{v})\mathbf{w} = -4(2\mathbf{i} + \mathbf{j} - 2\mathbf{k}) - (-1)(-\mathbf{i} + 2\mathbf{j} - \mathbf{k}) = -9\mathbf{i} - 2\mathbf{j} + 7\mathbf{k}$

   (c) $\mathbf{u} \times \mathbf{v} = \begin{vmatrix} \mathbf{i} & \mathbf{j} & \mathbf{k} \\ 2 & 1 & 0 \\ 2 & -1 & 1 \end{vmatrix} = \mathbf{i} - 2\mathbf{j} - 4\mathbf{k} \Rightarrow (\mathbf{u} \times \mathbf{v}) \times \mathbf{w} = \begin{vmatrix} \mathbf{i} & \mathbf{j} & \mathbf{k} \\ 1 & -2 & -4 \\ 1 & 0 & 2 \end{vmatrix} = -4\mathbf{i} - 6\mathbf{j} + 2\mathbf{k}$;

     $(\mathbf{u} \cdot \mathbf{w})\mathbf{v} - (\mathbf{v} \cdot \mathbf{w})\mathbf{u} = 2(2\mathbf{i} - \mathbf{j} + \mathbf{k}) - 4(2\mathbf{i} + \mathbf{j}) = -4\mathbf{i} - 6\mathbf{j} + 2\mathbf{k}$;

     $\mathbf{v} \times \mathbf{w} = \begin{vmatrix} \mathbf{i} & \mathbf{j} & \mathbf{k} \\ 2 & -1 & 1 \\ 1 & 0 & 2 \end{vmatrix} = -2\mathbf{i} - 3\mathbf{j} + \mathbf{k} \Rightarrow \mathbf{u} \times (\mathbf{v} \times \mathbf{w}) = \begin{vmatrix} \mathbf{i} & \mathbf{j} & \mathbf{k} \\ 2 & 1 & 0 \\ -2 & -3 & 1 \end{vmatrix} = \mathbf{i} - 2\mathbf{j} - 4\mathbf{k}$;

     $(\mathbf{u} \cdot \mathbf{w})\mathbf{v} - (\mathbf{u} \cdot \mathbf{v})\mathbf{w} = 2(2\mathbf{i} - \mathbf{j} + \mathbf{k}) - 3(\mathbf{i} + 2\mathbf{k}) = \mathbf{i} - 2\mathbf{j} - 4\mathbf{k}$

(d) $\mathbf{u} \times \mathbf{v} = \begin{vmatrix} \mathbf{i} & \mathbf{j} & \mathbf{k} \\ 1 & 1 & -2 \\ -1 & 0 & -1 \end{vmatrix} = -\mathbf{i} + 3\mathbf{j} + \mathbf{k} \Rightarrow (\mathbf{u} \times \mathbf{v}) \times \mathbf{w} = \begin{vmatrix} \mathbf{i} & \mathbf{j} & \mathbf{k} \\ -1 & 3 & 1 \\ 2 & 4 & -2 \end{vmatrix} = -10\mathbf{i} - 10\mathbf{k};$

$(\mathbf{u} \cdot \mathbf{w})\mathbf{v} - (\mathbf{v} \cdot \mathbf{w})\mathbf{u} = 10(-\mathbf{i} - \mathbf{k}) - 0(\mathbf{i} + \mathbf{j} - 2\mathbf{k}) = -10\mathbf{i} - 10\mathbf{k};$

$\mathbf{v} \times \mathbf{w} = \begin{vmatrix} \mathbf{i} & \mathbf{j} & \mathbf{k} \\ -1 & 0 & -1 \\ 2 & 4 & -2 \end{vmatrix} = 4\mathbf{i} - 4\mathbf{j} - 4\mathbf{k} \Rightarrow \mathbf{u} \times (\mathbf{v} \times \mathbf{w}) = \begin{vmatrix} \mathbf{i} & \mathbf{j} & \mathbf{k} \\ 1 & 1 & -2 \\ 4 & -4 & -4 \end{vmatrix} = -12\mathbf{i} - 4\mathbf{j} - 8\mathbf{k};$

$(\mathbf{u} \cdot \mathbf{w})\mathbf{v} - (\mathbf{u} \cdot \mathbf{v})\mathbf{w} = 10(-\mathbf{i} - \mathbf{k}) - 1(2\mathbf{i} + 4\mathbf{j} - 2\mathbf{k}) = -12\mathbf{i} - 4\mathbf{j} - 8\mathbf{k}$

17. The formula is always true; $\mathbf{u} \times [\mathbf{u} \times (\mathbf{u} \times \mathbf{v})] \cdot \mathbf{w} = \mathbf{u} \times [(\mathbf{u} \cdot \mathbf{v})\mathbf{u} - (\mathbf{u} \cdot \mathbf{u})\mathbf{v}] \cdot \mathbf{w}$
$= [(\mathbf{u} \cdot \mathbf{v})\mathbf{u} \times \mathbf{u} - (\mathbf{u} \cdot \mathbf{u})\mathbf{u} \times \mathbf{v}] \cdot \mathbf{w} = -|\mathbf{u}|^2 \mathbf{u} \times \mathbf{v} \cdot \mathbf{w} = -|\mathbf{u}|^2 \mathbf{u} \cdot \mathbf{v} \times \mathbf{w}$

19. If $\mathbf{A} = a\mathbf{i} + b\mathbf{j}$ and $\mathbf{B} = c\mathbf{i} + d\mathbf{j}$, then $\mathbf{A} \cdot \mathbf{B} = |\mathbf{A}| \, |\mathbf{B}| \cos \theta \Rightarrow ac + bd = \sqrt{a^2 + b^2} \sqrt{c^2 + d^2} \cos \theta$
$\Rightarrow (ac + bd)^2 = (a^2 + b^2)(c^2 + d^2) \cos^2 \theta \Rightarrow (ac + bd)^2 \le (a^2 + b^2)(c^2 + d^2)$, since $\cos^2 \theta \le 1$.

21. $|\mathbf{A} + \mathbf{B}|^2 = (\mathbf{A} + \mathbf{B}) \cdot (\mathbf{A} + \mathbf{B}) = \mathbf{A} \cdot \mathbf{A} + 2\mathbf{A} \cdot \mathbf{B} + \mathbf{B} \cdot \mathbf{B} \le |\mathbf{A}|^2 + 2\,|\mathbf{A}|\,|\mathbf{B}| + |\mathbf{B}|^2 = (|\mathbf{A}| + |\mathbf{B}|)^2 \Rightarrow |\mathbf{A} + \mathbf{B}|$
$\le |\mathbf{A}| + |\mathbf{B}|$

23. $(a\mathbf{B} + b\mathbf{A}) \cdot (b\mathbf{A} - a\mathbf{B}) = a\mathbf{B} \cdot b\mathbf{A} + b\mathbf{A} \cdot b\mathbf{A} - a\mathbf{B} \cdot a\mathbf{B} - b\mathbf{A} \cdot a\mathbf{B} = b\mathbf{A} \cdot a\mathbf{B} + b^2\mathbf{A} \cdot \mathbf{A} - a^2\mathbf{B} \cdot \mathbf{B} - b\mathbf{A} \cdot a\mathbf{B}$
$= b^2 a^2 - a^2 b^2 = 0$, where $a = |\mathbf{A}|$ and $b = |\mathbf{B}|$

25. (a) The vector from $(0, d)$ to $(kd, 0)$ is $\mathbf{r}_k = kd\mathbf{i} - d\mathbf{j} \Rightarrow \dfrac{1}{|\mathbf{r}_k|^3} = \dfrac{1}{d^3 (k^2 + 1)^{3/2}} \Rightarrow \dfrac{\mathbf{r}_k}{|\mathbf{r}_k|^3} = \dfrac{k\mathbf{i} - \mathbf{j}}{d^2 (k^2 + 1)^{3/2}}$. The

total force on the mass $(0, d)$ due to the masses Q_k for $k = -n, -n+1, \ldots, n-1, n$ is

$\mathbf{F} = \dfrac{GMm}{d^2}(-\mathbf{j}) + \dfrac{GMm}{2d^2}\left(\dfrac{\mathbf{i} - \mathbf{j}}{\sqrt{2}}\right) + \dfrac{GMm}{5d^2}\left(\dfrac{2\mathbf{i} - \mathbf{j}}{\sqrt{5}}\right) + \cdots + \dfrac{GMm}{(n^2+1)d^2}\left(\dfrac{n\mathbf{i} - \mathbf{j}}{\sqrt{n^2+1}}\right) + \dfrac{GMm}{2d^2}\left(\dfrac{-\mathbf{i} - \mathbf{j}}{\sqrt{2}}\right)$

$+ \dfrac{GMm}{5d^2}\left(\dfrac{-2\mathbf{i} - \mathbf{j}}{\sqrt{5}}\right) + \cdots + \dfrac{GMm}{(n^2+1)d^2}\left(\dfrac{-n\mathbf{i} - \mathbf{j}}{\sqrt{n^2+1}}\right)$

The $\mathbf{i}$ components cancel, giving

$\mathbf{F} = \dfrac{GMm}{d^2}\left(-1 - \dfrac{2}{2\sqrt{2}} - \dfrac{2}{5\sqrt{5}} - \cdots - \dfrac{2}{(n^2+1)(n^2+1)^{1/2}}\right)\mathbf{j} \Rightarrow$ the magnitude of the force is

$|\mathbf{F}| = \dfrac{GMm}{d^2}\left(1 + \displaystyle\sum_{i=1}^{n} \dfrac{2}{(i^2+1)^{3/2}}\right).$

(b) Yes, it is finite: $\displaystyle\lim_{n \to \infty} |\mathbf{F}| = \dfrac{GMm}{d^2}\left(1 + \displaystyle\sum_{i=1}^{\infty} \dfrac{2}{(i^2+1)^{3/2}}\right)$ is finite since $\displaystyle\sum_{i=1}^{\infty} \dfrac{2}{(i^2+1)^{3/2}}$ converges.

NOTES:

CHAPTER 13 VECTOR-VALUED FUNCTIONS AND MOTION IN SPACE

13.1 VECTOR FUNCTIONS

1. $x = t + 1$ and $y = t^2 - 1 \Rightarrow y = (x-1)^2 - 1 = x^2 - 2x$; $\mathbf{v} = \frac{d\mathbf{r}}{dt} = \mathbf{i} + 2t\mathbf{j} \Rightarrow \mathbf{a} = \frac{d\mathbf{v}}{dt} = 2\mathbf{j} \Rightarrow \mathbf{v} = \mathbf{i} + 2\mathbf{j}$ and $\mathbf{a} = 2\mathbf{j}$ at $t = 1$

3. $x = e^t$ and $y = \frac{2}{9} e^{2t} \Rightarrow y = \frac{2}{9} x^2$; $\mathbf{v} = \frac{d\mathbf{r}}{dt} = e^t\mathbf{i} + \frac{4}{9} e^{2t}\mathbf{j} \Rightarrow \mathbf{a} = e^t\mathbf{i} + \frac{8}{9} e^{2t}\mathbf{j} \Rightarrow \mathbf{v} = 3\mathbf{i} + 4\mathbf{j}$ and $\mathbf{a} = 3\mathbf{i} + 8\mathbf{j}$ at $t = \ln 3$

5. $\mathbf{v} = \frac{d\mathbf{r}}{dt} = (\cos t)\mathbf{i} - (\sin t)\mathbf{j}$ and $\mathbf{a} = \frac{d\mathbf{v}}{dt} = -(\sin t)\mathbf{i} - (\cos t)\mathbf{j}$

   $\Rightarrow$ for $t = \frac{\pi}{4}$, $\mathbf{v}\left(\frac{\pi}{4}\right) = \frac{\sqrt{2}}{2}\mathbf{i} - \frac{\sqrt{2}}{2}\mathbf{j}$ and

   $\mathbf{a}\left(\frac{\pi}{4}\right) = -\frac{\sqrt{2}}{2}\mathbf{i} - \frac{\sqrt{2}}{2}\mathbf{j}$; for $t = \frac{\pi}{2}$, $\mathbf{v}\left(\frac{\pi}{2}\right) = -\mathbf{j}$ and

   $\mathbf{a}\left(\frac{\pi}{2}\right) = -\mathbf{i}$

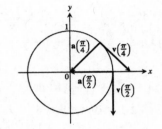

7. $\mathbf{v} = \frac{d\mathbf{r}}{dt} = (1 - \cos t)\mathbf{i} + (\sin t)\mathbf{j}$ and $\mathbf{a} = \frac{d\mathbf{v}}{dt}$

   $= (\sin t)\mathbf{i} + (\cos t)\mathbf{j} \Rightarrow$ for $t = \pi$, $\mathbf{v}(\pi) = 2\mathbf{i}$ and $\mathbf{a}(\pi) = -\mathbf{j}$;

   for $t = \frac{3\pi}{2}$, $\mathbf{v}\left(\frac{3\pi}{2}\right) = \mathbf{i} - \mathbf{j}$ and $\mathbf{a}\left(\frac{3\pi}{2}\right) = -\mathbf{i}$

9. $\mathbf{r} = (t+1)\mathbf{i} + (t^2 - 1)\mathbf{j} + 2t\mathbf{k} \Rightarrow \mathbf{v} = \frac{d\mathbf{r}}{dt} = \mathbf{i} + 2t\mathbf{j} + 2\mathbf{k} \Rightarrow \mathbf{a} = \frac{d^2\mathbf{r}}{dt^2} = 2\mathbf{j}$; Speed: $|\mathbf{v}(1)| = \sqrt{1^2 + (2(1))^2 + 2^2} = 3$;

   Direction: $\frac{\mathbf{v}(1)}{|\mathbf{v}(1)|} = \frac{\mathbf{i} + 2(1)\mathbf{j} + 2\mathbf{k}}{3} = \frac{1}{3}\mathbf{i} + \frac{2}{3}\mathbf{j} + \frac{2}{3}\mathbf{k} \Rightarrow \mathbf{v}(1) = 3\left(\frac{1}{3}\mathbf{i} + \frac{2}{3}\mathbf{j} + \frac{2}{3}\mathbf{k}\right)$

11. $\mathbf{r} = (2\cos t)\mathbf{i} + (3\sin t)\mathbf{j} + 4t\mathbf{k} \Rightarrow \mathbf{v} = \frac{d\mathbf{r}}{dt} = (-2\sin t)\mathbf{i} + (3\cos t)\mathbf{j} + 4\mathbf{k} \Rightarrow \mathbf{a} = \frac{d^2\mathbf{r}}{dt^2} = (-2\cos t)\mathbf{i} - (3\sin t)\mathbf{j}$;

   Speed: $\left|\mathbf{v}\left(\frac{\pi}{2}\right)\right| = \sqrt{\left(-2\sin\frac{\pi}{2}\right)^2 + \left(3\cos\frac{\pi}{2}\right)^2 + 4^2} = 2\sqrt{5}$; Direction: $\frac{\mathbf{v}\left(\frac{\pi}{2}\right)}{\left|\mathbf{v}\left(\frac{\pi}{2}\right)\right|}$

   $= \left(-\frac{2}{2\sqrt{5}}\sin\frac{\pi}{2}\right)\mathbf{i} + \left(\frac{3}{2\sqrt{5}}\cos\frac{\pi}{2}\right)\mathbf{j} + \frac{4}{2\sqrt{5}}\mathbf{k} = -\frac{1}{\sqrt{5}}\mathbf{i} + \frac{2}{\sqrt{5}}\mathbf{k} \Rightarrow \mathbf{v}\left(\frac{\pi}{2}\right) = 2\sqrt{5}\left(-\frac{1}{\sqrt{5}}\mathbf{i} + \frac{2}{\sqrt{5}}\mathbf{k}\right)$

13. $\mathbf{r} = (2\ln(t+1))\mathbf{i} + t^2\mathbf{j} + \frac{t^2}{2}\mathbf{k} \Rightarrow \mathbf{v} = \frac{d\mathbf{r}}{dt} = \left(\frac{2}{t+1}\right)\mathbf{i} + 2t\mathbf{j} + t\mathbf{k} \Rightarrow \mathbf{a} = \frac{d^2\mathbf{r}}{dt^2} = \left[\frac{-2}{(t+1)^2}\right]\mathbf{i} + 2\mathbf{j} + \mathbf{k}$;

   Speed: $|\mathbf{v}(1)| = \sqrt{\left(\frac{2}{1+1}\right)^2 + (2(1))^2 + 1^2} = \sqrt{6}$; Direction: $\frac{\mathbf{v}(1)}{|\mathbf{v}(1)|} = \frac{\left(\frac{2}{1+1}\right)\mathbf{i} + 2(1)\mathbf{j} + (1)\mathbf{k}}{\sqrt{6}}$

   $= \frac{1}{\sqrt{6}}\mathbf{i} + \frac{2}{\sqrt{6}}\mathbf{j} + \frac{1}{\sqrt{6}}\mathbf{k} \Rightarrow \mathbf{v}(1) = \sqrt{6}\left(\frac{1}{\sqrt{6}}\mathbf{i} + \frac{2}{\sqrt{6}}\mathbf{j} + \frac{1}{\sqrt{6}}\mathbf{k}\right)$

15. $\mathbf{v} = 3\mathbf{i} + \sqrt{3}\mathbf{j} + 2t\mathbf{k}$ and $\mathbf{a} = 2\mathbf{k} \Rightarrow \mathbf{v}(0) = 3\mathbf{i} + \sqrt{3}\mathbf{j}$ and $\mathbf{a}(0) = 2\mathbf{k} \Rightarrow |\mathbf{v}(0)| = \sqrt{3^2 + \left(\sqrt{3}\right)^2 + 0^2} = \sqrt{12}$ and

   $|\mathbf{a}(0)| = \sqrt{2^2} = 2$; $\mathbf{v}(0) \cdot \mathbf{a}(0) = 0 \Rightarrow \cos\theta = 0 \Rightarrow \theta = \frac{\pi}{2}$

17. $\mathbf{v} = \left(\frac{2t}{t^2+1}\right)\mathbf{i} + \left(\frac{1}{t^2+1}\right)\mathbf{j} + t(t^2+1)^{-1/2}\mathbf{k}$ and $\mathbf{a} = \left[\frac{-2t^2+2}{(t^2+1)^2}\right]\mathbf{i} - \left[\frac{2t}{(t^2+1)^2}\right]\mathbf{j} + \left[\frac{1}{(t^2+1)^{3/2}}\right]\mathbf{k} \Rightarrow \mathbf{v}(0) = \mathbf{j}$ and

   $\mathbf{a}(0) = 2\mathbf{i} + \mathbf{k} \Rightarrow |\mathbf{v}(0)| = 1$ and $|\mathbf{a}(0)| = \sqrt{2^2 + 1^2} = \sqrt{5}$; $\mathbf{v}(0) \cdot \mathbf{a}(0) = 0 \Rightarrow \cos\theta = 0 \Rightarrow \theta = \frac{\pi}{2}$

19. $\mathbf{v} = (1 - \cos t)\mathbf{i} + (\sin t)\mathbf{j}$ and $\mathbf{a} = (\sin t)\mathbf{i} + (\cos t)\mathbf{j} \Rightarrow \mathbf{v} \cdot \mathbf{a} = (\sin t)(1 - \cos t) + (\sin t)(\cos t) = \sin t$. Thus,
$\mathbf{v} \cdot \mathbf{a} = 0 \Rightarrow \sin t = 0 \Rightarrow t = 0, \pi,$ or 2π

21. $\int_0^1 [t^3\mathbf{i} + 7\mathbf{j} + (t+1)\mathbf{k}] \, dt = \left[\frac{t^4}{4}\right]_0^1 \mathbf{i} + [7t]_0^1 \mathbf{j} + \left[\frac{t^2}{2} + t\right]_0^1 \mathbf{k} = \frac{1}{4}\mathbf{i} + 7\mathbf{j} + \frac{3}{2}\mathbf{k}$

23. $\int_{-\pi/4}^{\pi/4} [(\sin t)\mathbf{i} + (1 + \cos t)\mathbf{j} + (\sec^2 t)\,\mathbf{k}] \, dt = [-\cos t]_{-\pi/4}^{\pi/4}\mathbf{i} + [t + \sin t]_{-\pi/4}^{\pi/4}\mathbf{j} + [\tan t]_{-\pi/4}^{\pi/4}\mathbf{k}$
$= \left(\frac{\pi + 2\sqrt{2}}{2}\right)\mathbf{j} + 2\mathbf{k}$

25. $\int_1^4 \left(\frac{1}{t}\mathbf{i} + \frac{1}{5-t}\mathbf{j} + \frac{1}{2t}\mathbf{k}\right) dt = = [\ln t]_1^4 \mathbf{i} + [-\ln(5-t)]_1^4 \mathbf{j} + \left[\frac{1}{2}\ln t\right]_1^4 \mathbf{k} = (\ln 4)\mathbf{i} + (\ln 4)\mathbf{j} + (\ln 2)\mathbf{k}$

27. $\mathbf{r} = \int (-t\mathbf{i} - t\mathbf{j} - t\mathbf{k}) \, dt = -\frac{t^2}{2}\mathbf{i} - \frac{t^2}{2}\mathbf{j} - \frac{t^2}{2}\mathbf{k} + \mathbf{C}$; $\mathbf{r}(0) = 0\mathbf{i} - 0\mathbf{j} - 0\mathbf{k} + \mathbf{C} = \mathbf{i} + 2\mathbf{j} + 3\mathbf{k} \Rightarrow \mathbf{C} = \mathbf{i} + 2\mathbf{j} + 3\mathbf{k}$
$\Rightarrow \mathbf{r} = \left(-\frac{t^2}{2} + 1\right)\mathbf{i} + \left(-\frac{t^2}{2} + 2\right)\mathbf{j} + \left(-\frac{t^2}{2} + 3\right)\mathbf{k}$

29. $\mathbf{r} = \int \left[\left(\frac{3}{2}(t+1)^{1/2}\right)\mathbf{i} + e^{-t}\mathbf{j} + \left(\frac{1}{t+1}\right)\mathbf{k}\right] dt = (t+1)^{3/2}\mathbf{i} - e^{-t}\mathbf{j} + \ln(t+1)\mathbf{k} + \mathbf{C}$;
$\mathbf{r}(0) = (0+1)^{3/2}\mathbf{i} - e^{-0}\mathbf{j} + \ln(0+1)\mathbf{k} + \mathbf{C} = \mathbf{k} \Rightarrow \mathbf{C} = -\mathbf{i} + \mathbf{j} + \mathbf{k}$
$\Rightarrow \mathbf{r} = [(t+1)^{3/2} - 1]\mathbf{i} + (1 - e^{-t})\mathbf{j} + [1 + \ln(t+1)]\mathbf{k}$

31. $\frac{d\mathbf{r}}{dt} = \int (-32\mathbf{k}) \, dt = -32t\mathbf{k} + \mathbf{C}_1$; $\frac{d\mathbf{r}}{dt}(0) = 8\mathbf{i} + 8\mathbf{j} \Rightarrow -32(0)\mathbf{k} + \mathbf{C}_1 = 8\mathbf{i} + 8\mathbf{j} \Rightarrow \mathbf{C}_1 = 8\mathbf{i} + 8\mathbf{j}$
$\Rightarrow \frac{d\mathbf{r}}{dt} = 8\mathbf{i} + 8\mathbf{j} - 32t\mathbf{k}$; $\mathbf{r} = \int (8\mathbf{i} + 8\mathbf{j} - 32t\mathbf{k}) \, dt = 8t\mathbf{i} + 8t\mathbf{j} - 16t^2\mathbf{k} + \mathbf{C}_2$; $\mathbf{r}(0) = 100\mathbf{k}$
$\Rightarrow 8(0)\mathbf{i} + 8(0)\mathbf{j} - 16(0)^2\mathbf{k} + \mathbf{C}_2 = 100\mathbf{k} \Rightarrow \mathbf{C}_2 = 100\mathbf{k} \Rightarrow \mathbf{r} = 8t\mathbf{i} + 8t\mathbf{j} + (100 - 16t^2)\mathbf{k}$

33. $\mathbf{r}(t) = (\sin t)\mathbf{i} + (t^2 - \cos t)\mathbf{j} + e^t\mathbf{k} \Rightarrow \mathbf{v}(t) = (\cos t)\mathbf{i} + (2t + \sin t)\mathbf{j} + e^t\mathbf{k}$; $t_0 = 0 \Rightarrow \mathbf{v}(t_0) = \mathbf{i} + \mathbf{k}$ and
$\mathbf{r}(t_0) = P_0 = (0, -1, 1) \Rightarrow x = 0 + t = t, y = -1,$ and $z = 1 + t$ are parametric equations of the tangent line

35. $\mathbf{r}(t) = (a \sin t)\mathbf{i} + (a \cos t)\mathbf{j} + bt\mathbf{k} \Rightarrow \mathbf{v}(t) = (a \cos t)\mathbf{i} - (a \sin t)\mathbf{j} + b\mathbf{k}$; $t_0 = 2\pi \Rightarrow \mathbf{v}(t_0) = a\mathbf{i} + b\mathbf{k}$ and
$\mathbf{r}(t_0) = P_0 = (0, a, 2b\pi) \Rightarrow x = 0 + at = at, y = a,$ and $z = 2\pi b + bt$ are parametric equations of the tangent line

37. (a) $\mathbf{v}(t) = -(\sin t)\mathbf{i} + (\cos t)\mathbf{j} \Rightarrow \mathbf{a}(t) = -(\cos t)\mathbf{i} - (\sin t)\mathbf{j}$;
   (i) $|\mathbf{v}(t)| = \sqrt{(-\sin t)^2 + (\cos t)^2} = 1 \Rightarrow$ constant speed;
   (ii) $\mathbf{v} \cdot \mathbf{a} = (\sin t)(\cos t) - (\cos t)(\sin t) = 0 \Rightarrow$ yes, orthogonal;
   (iii) counterclockwise movement;
   (iv) yes, $\mathbf{r}(0) = \mathbf{i} + 0\mathbf{j}$
   (b) $\mathbf{v}(t) = -(2 \sin 2t)\mathbf{i} + (2 \cos 2t)\mathbf{j} \Rightarrow \mathbf{a}(t) = -(4 \cos 2t)\mathbf{i} - (4 \sin 2t)\mathbf{j}$;
   (i) $|\mathbf{v}(t)| = \sqrt{4 \sin^2 2t + 4 \cos^2 2t} = 2 \Rightarrow$ constant speed;
   (ii) $\mathbf{v} \cdot \mathbf{a} = 8 \sin 2t \cos 2t - 8 \cos 2t \sin 2t = 0 \Rightarrow$ yes, orthogonal;
   (iii) counterclockwise movement;
   (iv) yes, $\mathbf{r}(0) = \mathbf{i} + 0\mathbf{j}$
   (c) $\mathbf{v}(t) = -\sin\left(t - \frac{\pi}{2}\right)\mathbf{i} + \cos\left(t - \frac{\pi}{2}\right)\mathbf{j} \Rightarrow \mathbf{a}(t) = -\cos\left(t - \frac{\pi}{2}\right)\mathbf{i} - \sin\left(t - \frac{\pi}{2}\right)\mathbf{j}$;
   (i) $|\mathbf{v}(t)| = \sqrt{\sin^2\left(t - \frac{\pi}{2}\right) + \cos^2\left(t - \frac{\pi}{2}\right)} = 1 \Rightarrow$ constant speed;
   (ii) $\mathbf{v} \cdot \mathbf{a} = \sin\left(t - \frac{\pi}{2}\right) \cos\left(t - \frac{\pi}{2}\right) - \cos\left(t - \frac{\pi}{2}\right) \sin\left(t - \frac{\pi}{2}\right) = 0 \Rightarrow$ yes, orthogonal;
   (iii) counterclockwise movement;
   (iv) no, $\mathbf{r}(0) = 0\mathbf{i} - \mathbf{j}$ instead of $\mathbf{i} + 0\mathbf{j}$
   (d) $\mathbf{v}(t) = -(\sin t)\mathbf{i} - (\cos t)\mathbf{j} \Rightarrow \mathbf{a}(t) = -(\cos t)\mathbf{i} + (\sin t)\mathbf{j}$;
   (i) $|\mathbf{v}(t)| = \sqrt{(-\sin t)^2 + (-\cos t)^2} = 1 \Rightarrow$ constant speed;

    (ii) $\mathbf{v} \cdot \mathbf{a} = (\sin t)(\cos t) - (\cos t)(\sin t) = 0 \Rightarrow$ yes, orthogonal;

    (iii) clockwise movement;

    (iv) yes, $\mathbf{r}(0) = \mathbf{i} - 0\mathbf{j}$

(e) $\mathbf{v}(t) = -(2t \sin t)\mathbf{i} + (2t \cos t)\mathbf{j} \Rightarrow \mathbf{a}(t) = -(2 \sin t + 2t \cos t)\mathbf{i} + (2 \cos t - 2t \sin t)\mathbf{j}$;

    (i) $|\mathbf{v}(t)| = \sqrt{[-(2t \sin t)]^2 + (2t \cos t)^2} = \sqrt{4t^2(\sin^2 t + \cos^2 t)} = 2|t| = 2t, t \geq 0$

         $\Rightarrow$ variable speed;

    (ii) $\mathbf{v} \cdot \mathbf{a} = 4(t \sin^2 t + t^2 \sin t \cos t) + 4(t \cos^2 t - t^2 \cos t \sin t) = 4t \neq 0$ in general

        $\Rightarrow$ not orthogonal in general;

    (iii) counterclockwise movement;

    (iv) yes, $\mathbf{r}(0) = \mathbf{i} + 0\mathbf{j}$

39. $\frac{d\mathbf{v}}{dt} = \mathbf{a} = 3\mathbf{i} - \mathbf{j} + \mathbf{k} \Rightarrow \mathbf{v}(t) = 3t\mathbf{i} - t\mathbf{j} + t\mathbf{k} + \mathbf{C}_1$; the particle travels in the direction of the vector

$(4-1)\mathbf{i} + (1-2)\mathbf{j} + (4-3)\mathbf{k} = 3\mathbf{i} - \mathbf{j} + \mathbf{k}$ (since it travels in a straight line), and at time $t = 0$ it has speed

$2 \Rightarrow \mathbf{v}(0) = \frac{2}{\sqrt{9+1+1}}(3\mathbf{i} - \mathbf{j} + \mathbf{k}) = \mathbf{C}_1 \Rightarrow \frac{d\mathbf{r}}{dt} = \mathbf{v}(t) = \left(3t + \frac{6}{\sqrt{11}}\right)\mathbf{i} - \left(t + \frac{2}{\sqrt{11}}\right)\mathbf{j} + \left(t + \frac{2}{\sqrt{11}}\right)\mathbf{k}$

$\Rightarrow \mathbf{r}(t) = \left(\frac{3}{2}t^2 + \frac{6}{\sqrt{11}}t\right)\mathbf{i} - \left(\frac{1}{2}t^2 + \frac{2}{\sqrt{11}}t\right)\mathbf{j} + \left(\frac{1}{2}t^2 + \frac{2}{\sqrt{11}}t\right)\mathbf{k} + \mathbf{C}_2; \mathbf{r}(0) = \mathbf{i} + 2\mathbf{j} + 3\mathbf{k} = \mathbf{C}_2$

$\Rightarrow \mathbf{r}(t) = \left(\frac{3}{2}t^2 + \frac{6}{\sqrt{11}}t + 1\right)\mathbf{i} - \left(\frac{1}{2}t^2 + \frac{2}{\sqrt{11}}t - 2\right)\mathbf{j} + \left(\frac{1}{2}t^2 + \frac{2}{\sqrt{11}}t + 3\right)\mathbf{k}$

$= \left(\frac{1}{2}t^2 + \frac{2}{\sqrt{11}}t\right)(3\mathbf{i} - \mathbf{j} + \mathbf{k}) + (\mathbf{i} + 2\mathbf{j} + 3\mathbf{k})$

41. The velocity vector is tangent to the graph of $y^2 = 2x$ at the point $(2, 2)$, has length 5, and a positive $\mathbf{i}$

component. Now, $y^2 = 2x \Rightarrow 2y\frac{dy}{dx} = 2 \Rightarrow \frac{dy}{dx}\Big|_{(2,2)} = \frac{2}{2 \cdot 2} = \frac{1}{2} \Rightarrow$ the tangent vector lies in the direction of the

vector $\mathbf{i} + \frac{1}{2}\mathbf{j} \Rightarrow$ the velocity vector is $\mathbf{v} = \frac{5}{\sqrt{1 + \frac{1}{4}}}\left(\mathbf{i} + \frac{1}{2}\mathbf{j}\right) = \frac{5}{\left(\frac{\sqrt{5}}{2}\right)}\left(\mathbf{i} + \frac{1}{2}\mathbf{j}\right) = 2\sqrt{5}\mathbf{i} + \sqrt{5}\mathbf{j}$

43. $\mathbf{v} = (-3 \sin t)\mathbf{j} + (2 \cos t)\mathbf{k}$ and $\mathbf{a} = (-3 \cos t)\mathbf{j} - (2 \sin t)\mathbf{k}$; $|\mathbf{v}|^2 = 9 \sin^2 t + 4 \cos^2 t \Rightarrow \frac{d}{dt}\left(|\mathbf{v}|^2\right)$

$= 18 \sin t \cos t - 8 \cos t \sin t = 10 \sin t \cos t$; $\frac{d}{dt}\left(|\mathbf{v}|^2\right) = 0 \Rightarrow 10 \sin t \cos t = 0 \Rightarrow \sin t = 0$ or $\cos t = 0$

$\Rightarrow t = 0, \pi$ or $t = \frac{\pi}{2}, \frac{3\pi}{2}$. When $t = 0, \pi$, $|\mathbf{v}|^2 = 4 \Rightarrow |\mathbf{v}| = \sqrt{4} = 2$; when $t = \frac{\pi}{2}, \frac{3\pi}{2}$, $|\mathbf{v}| = \sqrt{9} = 3$.

Therefore max $|\mathbf{v}|$ is 3 when $t = \frac{\pi}{2}, \frac{3\pi}{2}$, and min $|\mathbf{v}| = 2$ when $t = 0, \pi$. Next, $|\mathbf{a}|^2 = 9 \cos^2 t + 4 \sin^2 t$

$\Rightarrow \frac{d}{dt}\left(|\mathbf{a}|^2\right) = -18 \cos t \sin t + 8 \sin t \cos t = -10 \sin t \cos t$; $\frac{d}{dt}\left(|\mathbf{a}|^2\right) = 0 \Rightarrow -10 \sin t \cos t = 0 \Rightarrow \sin t = 0$ or

$\cos t = 0 \Rightarrow t = 0, \pi$ or $t = \frac{\pi}{2}, \frac{3\pi}{2}$. When $t = 0, \pi$, $|\mathbf{a}|^2 = 9 \Rightarrow |\mathbf{a}| = 3$; when $t = \frac{\pi}{2}, \frac{3\pi}{2}$, $|\mathbf{a}|^2 = 4 \Rightarrow |\mathbf{a}| = 2$.

Therefore, max $|\mathbf{a}| = 3$ when $t = 0, \pi$, and min $|\mathbf{a}| = 2$ when $t = \frac{\pi}{2}, \frac{3\pi}{2}$.

45. $\frac{d}{dt}(\mathbf{v} \cdot \mathbf{v}) = \mathbf{v} \cdot \frac{d\mathbf{v}}{dt} + \frac{d\mathbf{v}}{dt} \cdot \mathbf{v} = 2\mathbf{v} \cdot \frac{d\mathbf{v}}{dt} = 2 \cdot 0 = 0 \Rightarrow \mathbf{v} \cdot \mathbf{v}$ is a constant $\Rightarrow |\mathbf{v}| = \sqrt{\mathbf{v} \cdot \mathbf{v}}$ is constant

47. $\frac{d}{dt}\left[\mathbf{r} \cdot \left(\frac{d\mathbf{r}}{dt} \times \frac{d^2\mathbf{r}}{dt^2}\right)\right] = \frac{d\mathbf{r}}{dt} \cdot \left(\frac{d\mathbf{r}}{dt} \times \frac{d^2\mathbf{r}}{dt^2}\right) + \mathbf{r} \cdot \left(\frac{d^2\mathbf{r}}{dt^2} \times \frac{d^2\mathbf{r}}{dt^2}\right) + \mathbf{r} \cdot \left(\frac{d\mathbf{r}}{dt} \times \frac{d^3\mathbf{r}}{dt^3}\right) = \mathbf{r} \cdot \left(\frac{d\mathbf{r}}{dt} \times \frac{d^3\mathbf{r}}{dt^3}\right)$, since $\mathbf{A} \cdot (\mathbf{A} \times \mathbf{B}) = 0$

and $\mathbf{A} \cdot (\mathbf{B} \times \mathbf{B}) = 0$ for any vectors $\mathbf{A}$ and $\mathbf{B}$

49. (a) $\mathbf{u} = f(t)\mathbf{i} + g(t)\mathbf{j} + h(t)\mathbf{k} \Rightarrow c\mathbf{u} = cf(t)\mathbf{i} + cg(t)\mathbf{j} + ch(t)\mathbf{k} \Rightarrow \frac{d}{dt}(c\mathbf{u}) = c\frac{df}{dt}\mathbf{i} + c\frac{dg}{dt}\mathbf{j} + c\frac{dh}{dt}\mathbf{k}$

      $= c\left(\frac{df}{dt}\mathbf{i} + \frac{dg}{dt}\mathbf{j} + \frac{dh}{dt}\mathbf{k}\right) = c\frac{d\mathbf{u}}{dt}$

    (b) $f\mathbf{u} = ff(t)\mathbf{i} + fg(t)\mathbf{j} + fh(t)\mathbf{k} \Rightarrow \frac{d}{dt}(f\mathbf{u}) = \left[\frac{df}{dt}f(t) + f\frac{df}{dt}\right]\mathbf{i} + \left[\frac{df}{dt}g(t) + f\frac{dg}{dt}\right]\mathbf{j} + \left[\frac{df}{dt}h(t) + f\frac{dh}{dt}\right]\mathbf{k}$

      $= \frac{df}{dt}[f(t)\mathbf{i} + g(t)\mathbf{j} + h(t)\mathbf{k}] + f\left[\frac{df}{dt}\mathbf{i} + \frac{dg}{dt}\mathbf{j} + \frac{dh}{dt}\mathbf{k}\right] = \frac{df}{dt}\mathbf{u} + f\frac{d\mathbf{u}}{dt}$

51. Suppose $\mathbf{r}$ is continuous at $t = t_0$. Then $\lim\limits_{t \to t_0} \mathbf{r}(t) = \mathbf{r}(t_0) \Leftrightarrow \lim\limits_{t \to t_0} [f(t)\mathbf{i} + g(t)\mathbf{j} + h(t)\mathbf{k}]$

$= f(t_0)\mathbf{i} + g(t_0)\mathbf{j} + h(t_0)\mathbf{k} \Leftrightarrow \lim\limits_{t \to t_0} f(t) = f(t_0),\ \lim\limits_{t \to t_0} g(t) = g(t_0),$ and $\lim\limits_{t \to t_0} h(t) = h(t_0) \Leftrightarrow$ f, g, and h are

continuous at $t = t_0$.

53. $\mathbf{r}'(t_0)$ exists $\Rightarrow f'(t_0)\mathbf{i} + g'(t_0)\mathbf{j} + h'(t_0)\mathbf{k}$ exists $\Rightarrow f'(t_0),\ g'(t_0),\ h'(t_0)$ all exist $\Rightarrow$ f, g, and h are continuous at
$t = t_0 \Rightarrow \mathbf{r}(t)$ is continuous at $t = t_0$

55. (a) Let u and $\mathbf{r}$ be continuous on $[a, b]$. Then $\lim\limits_{t \to t_0} u(t)\mathbf{r}(t) = \lim\limits_{t \to t_0} [u(t)f(t)\mathbf{i} + u(t)g(t)\mathbf{j} + u(t)h(t)\mathbf{k}]$

$= u(t_0)f(t_0)\mathbf{i} + u(t_0)g(t_0)\mathbf{j} + u(t_0)h(t_0)\mathbf{k} = u(t_0)\mathbf{r}(t_0) \Rightarrow u\mathbf{r}$ is continuous for every t_0 in $[a, b]$.

(b) Let u and $\mathbf{r}$ be differentiable. Then $\frac{d}{dt}(u\mathbf{r}) = \frac{d}{dt}[u(t)f(t)\mathbf{i} + u(t)g(t)\mathbf{j} + u(t)h(t)\mathbf{k}]$

$= \left(\frac{du}{dt} f(t) + u(t) \frac{df}{dt}\right)\mathbf{i} + \left(\frac{du}{dt} g(t) + u(t) \frac{dg}{dt}\right)\mathbf{j} + \left(\frac{du}{dt} h(t) + u(t) \frac{dh}{dt}\right)\mathbf{k}$

$= [f(t)\mathbf{i} + g(t)\mathbf{j} + h(t)\mathbf{k}]\frac{du}{dt} + u(t)\left(\frac{df}{dt}\mathbf{i} + \frac{dg}{dt}\mathbf{j} + \frac{dh}{dt}\mathbf{k}\right) = \mathbf{r}\frac{du}{dt} + u\frac{d\mathbf{r}}{dt}$

57. $\frac{d}{dt}\int_a^t \mathbf{r}(\tau)\,d\tau = \frac{d}{dt}\int_a^t [f(\tau)\mathbf{i} + g(\tau)\mathbf{j} + h(\tau)\mathbf{k}]\,d\tau = \frac{d}{dt}\int_a^t f(\tau)\,d\tau\,\mathbf{i} + \frac{d}{dt}\int_a^t g(\tau)\,d\tau\,\mathbf{j} + \frac{d}{dt}\int_a^t h(\tau)\,d\tau\,\mathbf{k}$

$= f(t)\mathbf{i} + g(t)\mathbf{j} + h(t)\mathbf{k} = \mathbf{r}(t)$. Since $\frac{d}{dt}\int_a^t \mathbf{r}(\tau)\,d\tau = \mathbf{r}(t)$, we have that $\int_a^t \mathbf{r}(\tau)\,d\tau$ is an antiderivative of

$\mathbf{r}$. If $\mathbf{R}$ is any antiderivative of $\mathbf{r}$, then $\mathbf{R}(t) = \int_a^t \mathbf{r}(\tau)\,d\tau + \mathbf{C}$ by Exercise 56(b). Then $\mathbf{R}(a) = \int_a^a \mathbf{r}(\tau)\,d\tau + \mathbf{C}$

$= \mathbf{0} + \mathbf{C} \Rightarrow \mathbf{C} = \mathbf{R}(a) \Rightarrow \int_a^t \mathbf{r}(\tau)\,d\tau = \mathbf{R}(t) - \mathbf{C} = \mathbf{R}(t) - \mathbf{R}(a) \Rightarrow \int_a^b \mathbf{r}(\tau)\,d\tau = \mathbf{R}(b) - \mathbf{R}(a)$.

13.2 MODELING PROJECTILE MOTION

1. $x = (v_0 \cos \alpha)t \Rightarrow (21\text{ km})\left(\frac{1000\text{ m}}{1\text{ km}}\right) = (840\text{ m/s})(\cos 60°)t \Rightarrow t = \frac{21{,}000\text{ m}}{(840\text{ m/s})(\cos 60°)} = 50$ seconds

3. (a) $t = \frac{2v_0 \sin \alpha}{g} = \frac{2(500\text{ m/s})(\sin 45°)}{9.8\text{ m/s}^2} \approx 72.2$ seconds; $R = \frac{v_0^2}{g} \sin 2\alpha = \frac{(500\text{ m/s})^2}{9.8\text{ m/s}^2}(\sin 90°) \approx 25{,}510.2$ m

(b) $x = (v_0 \cos \alpha)t \Rightarrow 5000\text{ m} = (500\text{ m/s})(\cos 45°)t \Rightarrow t = \frac{5000\text{ m}}{(500\text{ m/s})(\cos 45°)} \approx 14.14$ s; thus,

$y = (v_0 \sin \alpha)t - \frac{1}{2}gt^2 \Rightarrow y \approx (500\text{ m/s})(\sin 45°)(14.14\text{ s}) - \frac{1}{2}(9.8\text{ m/s}^2)(14.14\text{ s})^2 \approx 4020$ m

(c) $y_{max} = \frac{(v_0 \sin \alpha)^2}{2g} = \frac{((500\text{ m/s})(\sin 45°))^2}{2(9.8\text{ m/s}^2)} \approx 6378$ m

5. $x = x_0 + (v_0 \cos \alpha)t = 0 + (44 \cos 45°)t = 22\sqrt{2}t$ and $y = y_0 + (v_0 \sin \alpha)t - \frac{1}{2}gt^2 = 6.5 + (44 \sin 45°)t - 16t^2$

$= 6.5 + 22\sqrt{2}t - 16t^2$; the shot lands when $y = 0 \Rightarrow t = \frac{22\sqrt{2} \pm \sqrt{968 + 416}}{32} \approx 2.135$ sec since $t > 0$; thus

$x = 22\sqrt{2}t \approx \left(22\sqrt{2}\right)(2.135) \approx 66.43$ ft

7. (a) $R = \frac{v_0^2}{g} \sin 2\alpha \Rightarrow 10\text{ m} = \left(\frac{v_0^2}{9.8\text{ m/s}^2}\right)(\sin 90°) \Rightarrow v_0^2 = 98\text{ m}^2\text{s}^2 \Rightarrow v_0 \approx 9.9$ m/s;

(b) $6\text{m} \approx \frac{(9.9\text{ m/s})^2}{9.8\text{ m/s}^2}(\sin 2\alpha) \Rightarrow \sin 2\alpha \approx 0.59999 \Rightarrow 2\alpha \approx 36.87°$ or $143.12° \Rightarrow \alpha \approx 18.4°$ or $71.6°$

9. $R = \frac{v_0^2}{g} \sin 2\alpha \Rightarrow 3(248.8)\text{ ft} = \left(\frac{v_0^2}{32\text{ ft/sec}^2}\right)(\sin 18°) \Rightarrow v_0^2 \approx 77{,}292.84\text{ ft}^2/\text{sec}^2 \Rightarrow v_0 \approx 278.02\text{ ft/sec} \approx 190$ mph

11. $x = (v_0 \cos \alpha)t \Rightarrow 135\text{ ft} = (90\text{ ft/sec})(\cos 30°)t \Rightarrow t \approx 1.732\text{ sec}; y = (v_0 \sin \alpha)t - \frac{1}{2}gt^2$

$\Rightarrow y \approx (90\text{ ft/sec})(\sin 30°)(1.732\text{ sec}) - \frac{1}{2}(32\text{ ft/sec}^2)(1.732\text{ sec})^2 \Rightarrow y \approx 29.94\text{ ft} \Rightarrow$ the golf ball will clip
the leaves at the top

13. We do part b first.

   (b) $x = (v_0 \cos \alpha)t \Rightarrow 315 \text{ ft} = (v_0 \cos 20°)t \Rightarrow v_0 = \frac{315}{t \cos 20°}$; also $y = (v_0 \sin \alpha)t - \frac{1}{2} gt^2$

     $\Rightarrow 34 \text{ ft} = \left(\frac{315}{t \cos 20°}\right)(t \sin 20°) - \frac{1}{2}(32)t^2 \Rightarrow 34 = 315 \tan 20° - 16t^2 \Rightarrow t^2 \approx 5.04 \text{ sec}^2 \Rightarrow t \approx 2.25 \text{ sec}$

   (a) $v_0 = \frac{315}{(2.25)(\cos 20°)} \approx 149 \text{ ft/sec}$

15. $R = \frac{v_0^2}{g} \sin 2\alpha \Rightarrow 16,000 \text{ m} = \frac{(400 \text{ m/s})^2}{9.8 \text{ m/s}^2} \sin 2\alpha \Rightarrow \sin 2\alpha = 0.98 \Rightarrow 2\alpha \approx 78.5° \text{ or } 2\alpha \approx 101.5° \Rightarrow \alpha \approx 39.3°$
   or $50.7°$

17. $x = x_0 + (v_0 \cos \alpha)t = 0 + (v_0 \cos 40°)t \approx 0.766 v_0 t$ and $y = y_0 + (v_0 \sin \alpha)t - \frac{1}{2} gt^2 = 6.5 + (v_0 \sin 40°)t - 16t^2$
   $\approx 6.5 + 0.643 v_0 t - 16t^2$; now the shot went 73.833 ft $\Rightarrow 73.833 = 0.766 v_0 t \Rightarrow t \approx \frac{96.383}{v_0}$ sec; the shot lands

   when $y = 0 \Rightarrow 0 = 6.5 + (0.643)(96.383) - 16\left(\frac{96.383}{v_0}\right)^2 \Rightarrow 0 \approx 68.474 - \frac{148,635}{v_0^2} \Rightarrow v_0 \approx \sqrt{\frac{148,635}{68.474}}$
   ≈ 46.6 ft/sec, the shot's initial speed

19. $\frac{d\mathbf{r}}{dt} = \int (-g\mathbf{j}) \, dt = -gt\mathbf{j} + \mathbf{C}_1$ and $\frac{d\mathbf{r}}{dt}(0) = (v_0 \cos \alpha)\mathbf{i} + (v_0 \sin \alpha)\mathbf{j} \Rightarrow -g(0)\mathbf{j} + \mathbf{C}_1 = (v_0 \cos \alpha)\mathbf{i} + (v_0 \sin \alpha)\mathbf{j}$

   $\Rightarrow \mathbf{C}_1 = (v_0 \cos \alpha)\mathbf{i} + (v_0 \sin \alpha)\mathbf{j} \Rightarrow \frac{d\mathbf{r}}{dt} = (v_0 \cos \alpha)\mathbf{i} + (v_0 \sin \alpha - gt)\mathbf{j} \, ; \mathbf{r} = \int [(v_0 \cos \alpha)\mathbf{i} + (v_0 \sin \alpha - gt)\mathbf{j}] \, dt$

   $= (v_0 t \cos \alpha)\mathbf{i} + \left(v_0 t \sin \alpha - \frac{1}{2} gt^2\right)\mathbf{j} + \mathbf{C}_2$ and $\mathbf{r}(0) = x_0\mathbf{i} + y_0\mathbf{j} \Rightarrow [v_0(0) \cos \alpha]\mathbf{i} + \left[v_0(0) \sin \alpha - \frac{1}{2} g(0)^2\right]\mathbf{j} + \mathbf{C}_2$

   $= x_0\mathbf{i} + y_0\mathbf{j} \Rightarrow \mathbf{C}_2 = x_0\mathbf{i} + y_0\mathbf{j} \Rightarrow \mathbf{r} = (x_0 + v_0 t \cos \alpha)\mathbf{i} + \left(y_0 + v_0 t \sin \alpha - \frac{1}{2} gt^2\right)\mathbf{j} \Rightarrow x = x_0 + v_0 t \cos \alpha$ and

   $y = y_0 + v_0 t \sin \alpha - \frac{1}{2} gt^2$

21. The horizontal distance from Rebollo to the center of the cauldron is 90 ft $\Rightarrow$ the horizontal distance to the

   nearest rim is $x = 90 - \frac{1}{2}(12) = 84 \Rightarrow 84 = x_0 + (v_0 \cos \alpha)t \approx 0 + \left(\frac{90g}{v_0 \sin \alpha}\right)t \Rightarrow 84 = \frac{(90)(32)}{\sqrt{(68)(64)}} t$

   $\Rightarrow t = 1.92$ sec. The vertical distance at this time is $y = y_0 + (v_0 \sin \alpha)t - \frac{1}{2} gt^2$

   $\approx 6 + \sqrt{(68)(64)}\,(1.92) - 16(1.92)^2 \approx 73.7$ ft $\Rightarrow$ the arrow clears the rim by 3.7 ft

23. Flight time = 1 sec and the measure of the angle of elevation is about 64° (using a protractor) so that

   $t = \frac{2v_0 \sin \alpha}{g} \Rightarrow 1 = \frac{2v_0 \sin 64°}{32} \Rightarrow v_0 \approx 17.80$ ft/sec. Then $y_{max} = \frac{(17.80 \sin 64°)^2}{2(32)} \approx 4.00$ ft and

   $R = \frac{v_0^2}{g} \sin 2\alpha \Rightarrow R = \frac{(17.80)^2}{32} \sin 128° \approx 7.80$ ft $\Rightarrow$ the engine traveled about 7.80 ft in 1 sec $\Rightarrow$ the engine
   velocity was about 7.80 ft/sec

25. (a) At the time t when the projectile hits the line OR we

     have $\tan \beta = \frac{y}{x}$; $x = [v_0 \cos (\alpha - \beta)]t$ and

     $y = [v_0 \sin (\alpha - \beta)]t - \frac{1}{2} gt^2 < 0$ since R is

     below level ground. Therefore let

     $|y| = \frac{1}{2} gt^2 - [v_0 \sin (\alpha - \beta)]t > 0$

     so that $\tan \beta = \frac{\left[\frac{1}{2} gt^2 - (v_0 \sin (\alpha - \beta))t\right]}{[v_0 \cos (\alpha - \beta)]t} = \frac{\left[\frac{1}{2} gt - v_0 \sin (\alpha - \beta)\right]}{v_0 \cos (\alpha - \beta)}$

     $\Rightarrow v_0 \cos (\alpha - \beta) \tan \beta = \frac{1}{2} gt - v_0 \sin (\alpha - \beta)$

     $\Rightarrow t = \frac{2v_0 \sin (\alpha - \beta) + 2v_0 \cos (\alpha - \beta) \tan \beta}{g}$, which is the time

     when the projectile hits the downhill slope. Therefore,

     $x = [v_0 \cos (\alpha - \beta)] \left[\frac{2v_0 \sin (\alpha - \beta) + 2v_0 \cos (\alpha - \beta) \tan \beta}{g}\right] = \frac{2v_0^2}{g} [\cos^2 (\alpha - \beta) \tan \beta + \sin (\alpha - \beta) \cos (\alpha - \beta)]$. If x is

     maximized, then OR is maximized: $\frac{dx}{d\alpha} = \frac{2v_0^2}{g} [-\sin 2(\alpha - \beta) \tan \beta + \cos 2(\alpha - \beta)] = 0$

     $\Rightarrow -\sin 2(\alpha - \beta) \tan \beta + \cos 2(\alpha - \beta) = 0 \Rightarrow \tan \beta = \cot 2(\alpha - \beta) \Rightarrow 2(\alpha - \beta) = 90° - \beta$

     $\Rightarrow \alpha - \beta = \frac{1}{2}(90° - \beta) \Rightarrow \alpha = \frac{1}{2}(90° + \beta) = \frac{1}{2}$ of $\angle AOR$.

(b) At the time t when the projectile hits OR we have

$\tan \beta = \frac{y}{x}$; $x = [v_0 \cos(\alpha + \beta)]t$ and

$y = [v_0 \sin(\alpha + \beta)]t - \frac{1}{2}gt^2$

$\Rightarrow \tan \beta = \frac{[v_0 \sin(\alpha+\beta)]t - \frac{1}{2}gt^2}{[v_0 \cos(\alpha+\beta)]t} = \frac{[v_0 \sin(\alpha+\beta) - \frac{1}{2}gt]}{v_0 \cos(\alpha+\beta)}$

$\Rightarrow v_0 \cos(\alpha + \beta) \tan \beta = v_0 \sin(\alpha + \beta) - \frac{1}{2}gt$

$\Rightarrow t = \frac{2v_0 \sin(\alpha+\beta) - 2v_0 \cos(\alpha+\beta)\tan\beta}{g}$, which is the time

when the projectile hits the uphill slope. Therefore,

$x = [v_0 \cos(\alpha+\beta)]\left[\frac{2v_0 \sin(\alpha+\beta) - 2v_0 \cos(\alpha+\beta)\tan\beta}{g}\right] = \frac{2v_0^2}{g}[\sin(\alpha+\beta)\cos(\alpha+\beta) - \cos^2(\alpha+\beta)\tan\beta]$. If x is

maximized, then OR is maximized: $\frac{dx}{d\alpha} = \frac{2v_0^2}{g}[\cos 2(\alpha+\beta) + \sin 2(\alpha+\beta)\tan\beta] = 0$

$\Rightarrow \cos 2(\alpha+\beta) + \sin 2(\alpha+\beta)\tan\beta = 0 \Rightarrow \cot 2(\alpha+\beta) + \tan\beta = 0 \Rightarrow \cot 2(\alpha+\beta) = -\tan\beta$

$= \tan(-\beta) \Rightarrow 2(\alpha+\beta) = 90° - (-\beta) = 90° + \beta \Rightarrow \alpha = \frac{1}{2}(90° - \beta) = \frac{1}{2}$ of $\angle AOR$. Therefore v_0 would bisect

$\angle AOR$ for maximum range uphill.

27. (a) (Assuming that "x" is zero at the point of impact:)

$\mathbf{r}(t) = (x(t))\mathbf{i} + (y(t))\mathbf{j}$; where $x(t) = (35 \cos 27°)t$ and $y(t) = 4 + (35 \sin 27°)t - 16t^2$.

(b) $y_{max} = \frac{(v_0 \sin \alpha)^2}{2g} + 4 = \frac{(35 \sin 27°)^2}{64} + 4 \approx 7.945$ feet, which is reached at t $= \frac{v_0 \sin \alpha}{g} = \frac{35 \sin 27°}{32} \approx 0.497$ seconds.

(c) For the time, solve $y = 4 + (35 \sin 27°)t - 16t^2 = 0$ for t, using the quadratic formula

$t = \frac{35 \sin 27° + \sqrt{(-35 \sin 27°)^2 + 256}}{32} \approx 1.201$ sec. Then the range is about $x(1.201) = (35 \cos 27°)(1.201)$

≈ 37.453 feet.

(d) For the time, solve $y = 4 + (35 \sin 27°)t - 16t^2 = 7$ for t, using the quadratic formula

$t = \frac{35 \sin 27° + \sqrt{(-35 \sin 27°)^2 - 192}}{32} \approx 0.254$ and 0.740 seconds. At those times the ball is about

$x(0.254) = (35 \cos 27°)(0.254) \approx 7.921$ feet and $x(0.740) = (35 \cos 27°)(0.740) \approx 23.077$ feet the impact point,

or about $37.453 - 7.921 \approx 29.532$ feet and $37.453 - 23.077 \approx 14.376$ feet from the landing spot.

(e) Yes. It changes things because the ball won't clear the net ($y_{max} \approx 7.945$).

29. $\frac{d^2\mathbf{r}}{dt^2} + k\frac{d\mathbf{r}}{dt} = -g\mathbf{j} \Rightarrow P(t) = k$ and $Q(t) = -g\mathbf{j} \Rightarrow \int P(t)\,dt = kt \Rightarrow v(t) = e^{\int P(t)\,dt} = e^{kt} \Rightarrow \frac{d\mathbf{r}}{dt} = \frac{1}{v(t)}\int v(t)\,Q(t)\,dt$

$= -ge^{-kt}\int e^{kt}\,\mathbf{j}\,dt = -ge^{-kt}\left[\frac{e^{kt}}{k}\mathbf{j} + C_1\right] = -\frac{g}{k}\mathbf{j} + Ce^{-kt}$, where $\mathbf{C} = -gC_1$; apply the initial condition:

$\frac{d\mathbf{r}}{dt}\Big|_{t=0} = (v_0 \cos \alpha)\mathbf{i} + (v_0 \sin \alpha)\mathbf{j} = -\frac{g}{k}\mathbf{j} + \mathbf{C} \Rightarrow \mathbf{C} = (v_0 \cos \alpha)\mathbf{i} + \left(\frac{g}{k} + v_0 \sin \alpha\right)\mathbf{j}$

$\Rightarrow \frac{d\mathbf{r}}{dt} = (v_0 e^{-kt}\cos \alpha)\mathbf{i} + \left(-\frac{g}{k} + e^{-kt}\left(\frac{g}{k} + v_0 \sin \alpha\right)\right)\mathbf{j}, \mathbf{r} = \int\left[(v_0 e^{-kt}\cos \alpha)\mathbf{i} + \left(-\frac{g}{k} + e^{-kt}\left(\frac{g}{k} + v_0 \sin \alpha\right)\right)\mathbf{j}\right]dt$

$= \left(-\frac{v_0}{k}e^{-kt}\cos \alpha\right)\mathbf{i} + \left(-\frac{gt}{k} - \frac{e^{-kt}}{k}\left(\frac{g}{k} + v_0 \sin \alpha\right)\right)\mathbf{j} + C_2$; apply the initial condition:

$\mathbf{r}(0) = \mathbf{0} = \left(-\frac{v_0}{k}\cos \alpha\right)\mathbf{i} + \left(-\frac{g}{k^2} - \frac{v_0 \sin \alpha}{k}\right)\mathbf{j} + C_2 \Rightarrow C_2 = \left(\frac{v_0}{k}\cos \alpha\right)\mathbf{i} + \left(\frac{g}{k^2} + \frac{v_0 \sin \alpha}{k}\right)\mathbf{j}$

$\Rightarrow \mathbf{r}(t) = \left(\frac{v_0}{k}(1 - e^{-kt})\cos \alpha\right)\mathbf{i} + \left(\frac{v_0}{k}(1 - e^{-kt})\sin \alpha + \frac{g}{k^2}(1 - kt - e^{-kt})\right)\mathbf{j}$

31. (a) $\mathbf{r}(t) = (x(t))\mathbf{i} + (y(t))\mathbf{j}$; where $x(t) = \left(\frac{1}{0.08}\right)(1 - e^{-0.08t})(152 \cos 20° - 17.6)$ and

$y(t) = 3 + \left(\frac{152}{0.08}\right)(1 - e^{-0.08t})(\sin 20°) + \left(\frac{32}{0.08^2}\right)(1 - 0.08t - e^{-0.08t})$

(b) Solve graphically using a calculator or CAS: At t ≈ 1.527 seconds the ball reaches a maximum height of about 41.893 feet.

(c) Use a graphing calculator or CAS to find that $y = 0$ when the ball has traveled for ≈ 3.181 seconds. The range is about $x(3.181) = \left(\frac{1}{0.08}\right)\left(1 - e^{-0.08(3.181)}\right)(152 \cos 20° - 17.6) \approx 351.734$ feet.

(d) Use a graphing calculator or CAS to find that $y = 35$ for t ≈ 0.877 and 2.190 seconds, at which times the ball is about $x(0.877) \approx 106.028$ feet and $x(2.190) \approx 251.530$ feet from home plate.

(e) No; the range is less than 380 feet. To find the wind needed for a home run, first use the method of part (d) to find that
y = 20 at t ≈ 0.376 and 2.716 seconds. Then define $x(w) = \left(\frac{1}{0.08}\right)\left(1 - e^{-0.08(2.716)}\right)(152\cos 20° + w)$, and solve
$x(w) = 380$ to find w ≈ 12.846 ft/sec.

13.3 ARC LENGTH AND THE UNIT TANGENT VECTOR T

1. $\mathbf{r} = (2\cos t)\mathbf{i} + (2\sin t)\mathbf{j} + \sqrt{5}t\mathbf{k} \Rightarrow \mathbf{v} = (-2\sin t)\mathbf{i} + (2\cos t)\mathbf{j} + \sqrt{5}\mathbf{k}$

$\Rightarrow |\mathbf{v}| = \sqrt{(-2\sin t)^2 + (2\cos t)^2 + \left(\sqrt{5}\right)^2} = \sqrt{4\sin^2 t + 4\cos^2 t + 5} = 3; \mathbf{T} = \frac{\mathbf{v}}{|\mathbf{v}|}$

$= \left(-\frac{2}{3}\sin t\right)\mathbf{i} + \left(\frac{2}{3}\cos t\right)\mathbf{j} + \frac{\sqrt{5}}{3}\mathbf{k}$ and Length $= \int_0^\pi |\mathbf{v}|\,dt = \int_0^\pi 3\,dt = [3t]_0^\pi = 3\pi$

3. $\mathbf{r} = t\mathbf{i} + \frac{2}{3}t^{3/2}\mathbf{k} \Rightarrow \mathbf{v} = \mathbf{i} + t^{1/2}\mathbf{k} \Rightarrow |\mathbf{v}| = \sqrt{1^2 + (t^{1/2})^2} = \sqrt{1+t}; \mathbf{T} = \frac{\mathbf{v}}{|\mathbf{v}|} = \frac{1}{\sqrt{1+t}}\mathbf{i} + \frac{\sqrt{t}}{\sqrt{1+t}}\mathbf{k}$

and Length $= \int_0^8 \sqrt{1+t}\,dt = \left[\frac{2}{3}(1+t)^{3/2}\right]_0^8 = \frac{52}{3}$

5. $\mathbf{r} = (\cos^3 t)\mathbf{j} + (\sin^3 t)\mathbf{k} \Rightarrow \mathbf{v} = (-3\cos^2 t\sin t)\mathbf{j} + (3\sin^2 t\cos t)\mathbf{k} \Rightarrow |\mathbf{v}|$

$= \sqrt{(-3\cos^2 t\sin t)^2 + (3\sin^2 t\cos t)^2} = \sqrt{(9\cos^2 t\sin^2 t)(\cos^2 t + \sin^2 t)} = 3|\cos t\sin t|$;

$\mathbf{T} = \frac{\mathbf{v}}{|\mathbf{v}|} = \frac{-3\cos^2 t\sin t}{3|\cos t\sin t|}\mathbf{j} + \frac{3\sin^2 t\cos t}{3|\cos t\sin t|}\mathbf{k} = (-\cos t)\mathbf{j} + (\sin t)\mathbf{k}$, if $0 \le t \le \frac{\pi}{2}$, and

Length $= \int_0^{\pi/2} 3|\cos t\sin t|\,dt = \int_0^{\pi/2} 3\cos t\sin t\,dt = \int_0^{\pi/2}\frac{3}{2}\sin 2t\,dt = \left[-\frac{3}{4}\cos 2t\right]_0^{\pi/2} = \frac{3}{2}$

7. $\mathbf{r} = (t\cos t)\mathbf{i} + (t\sin t)\mathbf{j} + \frac{2\sqrt{2}}{3}t^{3/2}\mathbf{k} \Rightarrow \mathbf{v} = (\cos t - t\sin t)\mathbf{i} + (\sin t + t\cos t)\mathbf{j} + \left(\sqrt{2}t^{1/2}\right)\mathbf{k}$

$\Rightarrow |\mathbf{v}| = \sqrt{(\cos t - t\sin t)^2 + (\sin t + t\cos t)^2 + \left(\sqrt{2}t\right)^2} = \sqrt{1 + t^2 + 2t} = \sqrt{(t+1)^2} = |t+1| = t+1$, if $t \ge 0$;

$\mathbf{T} = \frac{\mathbf{v}}{|\mathbf{v}|} = \left(\frac{\cos t - t\sin t}{t+1}\right)\mathbf{i} + \left(\frac{\sin t + t\cos t}{t+1}\right)\mathbf{j} + \left(\frac{\sqrt{2}t^{1/2}}{t+1}\right)\mathbf{k}$ and Length $= \int_0^\pi (t+1)\,dt = \left[\frac{t^2}{2} + t\right]_0^\pi = \frac{\pi^2}{2} + \pi$

9. Let $P(t_0)$ denote the point. Then $\mathbf{v} = (5\cos t)\mathbf{i} - (5\sin t)\mathbf{j} + 12\mathbf{k}$ and $26\pi = \int_0^{t_0}\sqrt{25\cos^2 t + 25\sin^2 t + 144}\,dt$

$= \int_0^{t_0} 13\,dt = 13t_0 \Rightarrow t_0 = 2\pi$, and the point is $P(2\pi) = (5\sin 2\pi, 5\cos 2\pi, 24\pi) = (0, 5, 24\pi)$

11. $\mathbf{r} = (4\cos t)\mathbf{i} + (4\sin t)\mathbf{j} + 3t\mathbf{k} \Rightarrow \mathbf{v} = (-4\sin t)\mathbf{i} + (4\cos t)\mathbf{j} + 3\mathbf{k} \Rightarrow |\mathbf{v}| = \sqrt{(-4\sin t)^2 + (4\cos t)^2 + 3^2}$

$= \sqrt{25} = 5 \Rightarrow s(t) = \int_0^t 5\,d\tau = 5t \Rightarrow$ Length $= s\left(\frac{\pi}{2}\right) = \frac{5\pi}{2}$

13. $\mathbf{r} = (e^t\cos t)\mathbf{i} + (e^t\sin t)\mathbf{j} + e^t\mathbf{k} \Rightarrow \mathbf{v} = (e^t\cos t - e^t\sin t)\mathbf{i} + (e^t\sin t + e^t\cos t)\mathbf{j} + e^t\mathbf{k}$

$\Rightarrow |\mathbf{v}| = \sqrt{(e^t\cos t - e^t\sin t)^2 + (e^t\sin t + e^t\cos t)^2 + (e^t)^2} = \sqrt{3e^{2t}} = \sqrt{3}e^t \Rightarrow s(t) = \int_0^t \sqrt{3}e^\tau\,d\tau$

$= \sqrt{3}e^t - \sqrt{3} \Rightarrow$ Length $= s(0) - s(-\ln 4) = 0 - \left(\sqrt{3}e^{-\ln 4} - \sqrt{3}\right) = \frac{3\sqrt{3}}{4}$

15. $\mathbf{r} = \left(\sqrt{2}t\right)\mathbf{i} + \left(\sqrt{2}t\right)\mathbf{j} + (1 - t^2)\mathbf{k} \Rightarrow \mathbf{v} = \sqrt{2}\mathbf{i} + \sqrt{2}\mathbf{j} - 2t\mathbf{k} \Rightarrow |\mathbf{v}| = \sqrt{\left(\sqrt{2}\right)^2 + \left(\sqrt{2}\right)^2 + (-2t)^2} = \sqrt{4 + 4t^2}$

$= 2\sqrt{1 + t^2} \Rightarrow$ Length $= \int_0^1 2\sqrt{1 + t^2}\,dt = \left[2\left(\frac{t}{2}\sqrt{1+t^2} + \frac{1}{2}\ln\left(t + \sqrt{1+t^2}\right)\right)\right]_0^1 = \sqrt{2} + \ln\left(1 + \sqrt{2}\right)$

17. (a) $\mathbf{r} = (\cos t)\mathbf{i} + (\sin t)\mathbf{j} + (1 - \cos t)\mathbf{k}, 0 \le t \le 2\pi \Rightarrow x = \cos t, y = \sin t, z = 1 - \cos t \Rightarrow x^2 + y^2$

$= \cos^2 t + \sin^2 t = 1$, a right circular cylinder with the z-axis as the axis and radius $= 1$. Therefore
$P(\cos t, \sin t, 1 - \cos t)$ lies on the cylinder $x^2 + y^2 = 1; t = 0 \Rightarrow P(1, 0, 0)$ is on the curve; $t = \frac{\pi}{2} \Rightarrow Q(0, 1, 1)$
is on the curve; $t = \pi \Rightarrow R(-1, 0, 2)$ is on the curve. Then $\overrightarrow{PQ} = -\mathbf{i} + \mathbf{j} + \mathbf{k}$ and $\overrightarrow{PR} = -2\mathbf{i} + 2\mathbf{k}$

$$\Rightarrow \vec{PQ} \times \vec{PR} = \begin{vmatrix} \mathbf{i} & \mathbf{j} & \mathbf{k} \\ -1 & 1 & 1 \\ -2 & 0 & 2 \end{vmatrix} = 2\mathbf{i} + 2\mathbf{k} \text{ is a vector normal to the plane of P, Q, and R. Then the}$$

plane containing P, Q, and R has an equation $2x + 2z = 2(1) + 2(0)$ or $x + z = 1$. Any point on the curve will satisfy this equation since $x + z = \cos t + (1 - \cos t) = 1$. Therefore, any point on the curve lies on the intersection of the cylinder $x^2 + y^2 = 1$ and the plane $x + z = 1 \Rightarrow$ the curve is an ellipse.

(b) $\mathbf{v} = (-\sin t)\mathbf{i} + (\cos t)\mathbf{j} + (\sin t)\mathbf{k} \Rightarrow |\mathbf{v}| = \sqrt{\sin^2 t + \cos^2 t + \sin^2 t} = \sqrt{1 + \sin^2 t} \Rightarrow \mathbf{T} = \frac{\mathbf{v}}{|\mathbf{v}|}$

$= \frac{(-\sin t)\mathbf{i} + (\cos t)\mathbf{j} + (\sin t)\mathbf{k}}{\sqrt{1 + \sin^2 t}} \Rightarrow \mathbf{T}(0) = \mathbf{j}, \mathbf{T}\left(\frac{\pi}{2}\right) = \frac{-\mathbf{i} + \mathbf{k}}{\sqrt{2}}, \mathbf{T}(\pi) = -\mathbf{j}, \mathbf{T}\left(\frac{3\pi}{2}\right) = \frac{\mathbf{i} - \mathbf{k}}{\sqrt{2}}$

(c) $\mathbf{a} = (-\cos t)\mathbf{i} - (\sin t)\mathbf{j} + (\cos t)\mathbf{k}; \mathbf{n} = \mathbf{i} + \mathbf{k}$ is

normal to the plane $x + z = 1 \Rightarrow \mathbf{n} \cdot \mathbf{a} = -\cos t + \cos t$

$= 0 \Rightarrow \mathbf{a}$ is orthogonal to $\mathbf{n} \Rightarrow \mathbf{a}$ is parallel to the

plane; $\mathbf{a}(0) = -\mathbf{i} + \mathbf{k}, \mathbf{a}\left(\frac{\pi}{2}\right) = -\mathbf{j}, \mathbf{a}(\pi) = \mathbf{i} - \mathbf{k},$

$\mathbf{a}\left(\frac{3\pi}{2}\right) = \mathbf{j}$

(d) $|\mathbf{v}| = \sqrt{1 + \sin^2 t}$ (See part (b) $\Rightarrow L = \int_0^{2\pi} \sqrt{1 + \sin^2 t}\, dt$

(e) $L \approx 7.64$ (by *Mathematica*)

19. $\angle PQB = \angle QOB = t$ and $PQ = \text{arc}(AQ) = t$ since
    $PQ = $ length of the unwound string = length of arc (AQ);
    thus $x = OB + BC = OB + DP = \cos t + t \sin t$, and
    $y = PC = QB - QD = \sin t - t \cos t$

13.4 CURVATURE AND THE UNIT NORMAL VECTOR N

1. $\mathbf{r} = t\mathbf{i} + \ln(\cos t)\mathbf{j} \Rightarrow \mathbf{v} = \mathbf{i} + \left(\frac{-\sin t}{\cos t}\right)\mathbf{j} = \mathbf{i} - (\tan t)\mathbf{j} \Rightarrow |\mathbf{v}| = \sqrt{1^2 + (-\tan t)^2} = \sqrt{\sec^2 t} = |\sec t| = \sec t$, since

$-\frac{\pi}{2} < t < \frac{\pi}{2} \Rightarrow \mathbf{T} = \frac{\mathbf{v}}{|\mathbf{v}|} = \left(\frac{1}{\sec t}\right)\mathbf{i} - \left(\frac{\tan t}{\sec t}\right)\mathbf{j} = (\cos t)\mathbf{i} - (\sin t)\mathbf{j}; \frac{d\mathbf{T}}{dt} = (-\sin t)\mathbf{i} - (\cos t)\mathbf{j}$

$\Rightarrow \left|\frac{d\mathbf{T}}{dt}\right| = \sqrt{(-\sin t)^2 + (-\cos t)^2} = 1 \Rightarrow \mathbf{N} = \frac{\left(\frac{d\mathbf{T}}{dt}\right)}{\left|\frac{d\mathbf{T}}{dt}\right|} = (-\sin t)\mathbf{i} - (\cos t)\mathbf{j};$

$\kappa = \frac{1}{|\mathbf{v}|} \cdot \left|\frac{d\mathbf{T}}{dt}\right| = \frac{1}{\sec t} \cdot 1 = \cos t.$

3. $\mathbf{r} = (2t + 3)\mathbf{i} + (5 - t^2)\mathbf{j} \Rightarrow \mathbf{v} = 2\mathbf{i} - 2t\mathbf{j} \Rightarrow |\mathbf{v}| = \sqrt{2^2 + (-2t)^2} = 2\sqrt{1 + t^2} \Rightarrow \mathbf{T} = \frac{\mathbf{v}}{|\mathbf{v}|} = \frac{2}{2\sqrt{1+t^2}}\mathbf{i} + \frac{-2t}{2\sqrt{1+t^2}}\mathbf{j}$

$= \frac{1}{\sqrt{1+t^2}}\mathbf{i} - \frac{t}{\sqrt{1+t^2}}\mathbf{j}; \frac{d\mathbf{T}}{dt} = \frac{-t}{\left(\sqrt{1+t^2}\right)^3}\mathbf{i} - \frac{1}{\left(\sqrt{1+t^2}\right)^3}\mathbf{j} \Rightarrow \left|\frac{d\mathbf{T}}{dt}\right| = \sqrt{\left(\frac{-t}{\left(\sqrt{1+t^2}\right)^3}\right)^2 + \left(-\frac{1}{\left(\sqrt{1+t^2}\right)^3}\right)^2}$

$= \sqrt{\frac{1}{(1+t^2)^2}} = \frac{1}{1+t^2} \Rightarrow \mathbf{N} = \frac{\left(\frac{d\mathbf{T}}{dt}\right)}{\left|\frac{d\mathbf{T}}{dt}\right|} = \frac{-t}{\sqrt{1+t^2}}\mathbf{i} - \frac{1}{\sqrt{1+t^2}}\mathbf{j};$

$\kappa = \frac{1}{|\mathbf{v}|} \cdot \left|\frac{d\mathbf{T}}{dt}\right| = \frac{1}{2\sqrt{1+t^2}} \cdot \frac{1}{1+t^2} = \frac{1}{2(1+t^2)^{3/2}}$

5. (a) $\kappa(x) = \frac{1}{|\mathbf{v}(x)|} \cdot \left|\frac{d\mathbf{T}(x)}{dt}\right|$. Now, $\mathbf{v} = \mathbf{i} + f'(x)\mathbf{j} \Rightarrow |\mathbf{v}(x)| = \sqrt{1 + [f'(x)]^2} \Rightarrow \mathbf{T} = \frac{\mathbf{v}}{|\mathbf{v}|}$

$$= \left(1 + [f'(x)]^2\right)^{-1/2}\mathbf{i} + f'(x)\left(1 + [f'(x)]^2\right)^{-1/2}\mathbf{j}. \text{ Thus } \frac{d\mathbf{T}}{dt}(x) = \frac{-f'(x)f''(x)}{\left(1 + [f'(x)]^2\right)^{3/2}}\mathbf{i} + \frac{f''(x)}{\left(1 + [f'(x)]^2\right)^{3/2}}\mathbf{j}$$

$$\Rightarrow \left|\frac{d\mathbf{T}(x)}{dt}\right| = \sqrt{\left[\frac{-f'(x)f''(x)}{\left(1 + [f'(x)]^2\right)^{3/2}}\right]^2 + \left(\frac{f''(x)}{\left(1 + [f'(x)]^2\right)^{3/2}}\right)^2} = \sqrt{\frac{[f''(x)]^2\left(1 + [f'(x)]^2\right)}{\left(1 + [f'(x)]^2\right)^3}} = \frac{|f''(x)|}{|1 + [f'(x)]^2|}$$

Thus $\kappa(x) = \frac{1}{(1 + [f'(x)]^2)^{1/2}} \cdot \frac{|f''(x)|}{|1 + [f'(x)]^2|} = \frac{|f''(x)|}{\left(1 + [f'(x)]^2\right)^{3/2}}$

(b) $y = \ln(\cos x) \Rightarrow \frac{dy}{dx} = \left(\frac{1}{\cos x}\right)(-\sin x) = -\tan x \Rightarrow \frac{d^2y}{dx^2} = -\sec^2 x \Rightarrow \kappa = \frac{|-\sec^2 x|}{[1 + (-\tan x)^2]^{3/2}} = \frac{\sec^2 x}{|\sec^3 x|}$

$= \frac{1}{\sec x} = \cos x$, since $-\frac{\pi}{2} < x < \frac{\pi}{2}$

(c) Note that $f''(x) = 0$ at an inflection point.

7. (a) $\mathbf{r}(t) = f(t)\mathbf{i} + g(t)\mathbf{j} \Rightarrow \mathbf{v} = f'(t)\mathbf{i} + g'(t)\mathbf{j}$ is tangent to the curve at the point $(f(t), g(t))$;

$\mathbf{n} \cdot \mathbf{v} = [-g'(t)\mathbf{i} + f'(t)\mathbf{j}] \cdot [f'(t)\mathbf{i} + g'(t)\mathbf{j}] = -g'(t)f'(t) + f'(t)g'(t) = 0; -\mathbf{n} \cdot \mathbf{v} = -(\mathbf{n} \cdot \mathbf{v}) = 0;$ thus,

$\mathbf{n}$ and $-\mathbf{n}$ are both normal to the curve at the point

(b) $\mathbf{r}(t) = t\mathbf{i} + e^{2t}\mathbf{j} \Rightarrow \mathbf{v} = \mathbf{i} + 2e^{2t}\mathbf{j} \Rightarrow \mathbf{n} = -2e^{2t}\mathbf{i} + \mathbf{j}$ points toward the concave side of the curve; $\mathbf{N} = \frac{\mathbf{n}}{|\mathbf{n}|}$ and

$|\mathbf{n}| = \sqrt{4e^{4t} + 1} \Rightarrow \mathbf{N} = \frac{-2e^{2t}}{\sqrt{1 + 4e^{4t}}}\mathbf{i} + \frac{1}{\sqrt{1 + 4e^{4t}}}\mathbf{j}$

(c) $\mathbf{r}(t) = \sqrt{4 - t^2}\,\mathbf{i} + t\mathbf{j} \Rightarrow \mathbf{v} = \frac{-t}{\sqrt{4 - t^2}}\mathbf{i} + \mathbf{j} \Rightarrow \mathbf{n} = -\mathbf{i} - \frac{t}{\sqrt{4 - t^2}}\mathbf{j}$ points toward the concave side of the curve;

$\mathbf{N} = \frac{\mathbf{n}}{|\mathbf{n}|}$ and $|\mathbf{n}| = \sqrt{1 + \frac{t^2}{4 - t^2}} = \frac{2}{\sqrt{4 - t^2}} \Rightarrow \mathbf{N} = -\frac{1}{2}\left(\sqrt{4 - t^2}\,\mathbf{i} + t\mathbf{j}\right)$

9. $\mathbf{r} = (3\sin t)\mathbf{i} + (3\cos t)\mathbf{j} + 4t\mathbf{k} \Rightarrow \mathbf{v} = (3\cos t)\mathbf{i} + (-3\sin t)\mathbf{j} + 4\mathbf{k} \Rightarrow |\mathbf{v}| = \sqrt{(3\cos t)^2 + (-3\sin t)^2 + 4^2}$

$= \sqrt{25} = 5 \Rightarrow \mathbf{T} = \frac{\mathbf{v}}{|\mathbf{v}|} = \left(\frac{3}{5}\cos t\right)\mathbf{i} - \left(\frac{3}{5}\sin t\right)\mathbf{j} + \frac{4}{5}\mathbf{k} \Rightarrow \frac{d\mathbf{T}}{dt} = \left(-\frac{3}{5}\sin t\right)\mathbf{i} - \left(\frac{3}{5}\cos t\right)\mathbf{j}$

$\Rightarrow \left|\frac{d\mathbf{T}}{dt}\right| = \sqrt{\left(-\frac{3}{5}\sin t\right)^2 + \left(-\frac{3}{5}\cos t\right)^2} = \frac{3}{5} \Rightarrow \mathbf{N} = \frac{\left(\frac{d\mathbf{T}}{dt}\right)}{\left|\frac{d\mathbf{T}}{dt}\right|} = (-\sin t)\mathbf{i} - (\cos t)\mathbf{j}; \kappa = \frac{1}{5} \cdot \frac{3}{5} = \frac{3}{25}$

11. $\mathbf{r} = (e^t \cos t)\mathbf{i} + (e^t \sin t)\mathbf{j} + 2\mathbf{k} \Rightarrow \mathbf{v} = (e^t \cos t - e^t \sin t)\mathbf{i} + (e^t \sin t + e^t \cos t)\mathbf{j} \Rightarrow$

$|\mathbf{v}| = \sqrt{(e^t \cos t - e^t \sin t)^2 + (e^t \sin t + e^t \cos t)^2} = \sqrt{2e^{2t}} = e^t\sqrt{2}$;

$\mathbf{T} = \frac{\mathbf{v}}{|\mathbf{v}|} = \left(\frac{\cos t - \sin t}{\sqrt{2}}\right)\mathbf{i} + \left(\frac{\sin t + \cos t}{\sqrt{2}}\right)\mathbf{j} \Rightarrow \frac{d\mathbf{T}}{dt} = \left(\frac{-\sin t - \cos t}{\sqrt{2}}\right)\mathbf{i} + \left(\frac{\cos t - \sin t}{\sqrt{2}}\right)\mathbf{j}$

$\Rightarrow \left|\frac{d\mathbf{T}}{dt}\right| = \sqrt{\left(\frac{-\sin t - \cos t}{\sqrt{2}}\right)^2 + \left(\frac{\cos t - \sin t}{\sqrt{2}}\right)^2} = 1 \Rightarrow \mathbf{N} = \frac{\left(\frac{d\mathbf{T}}{dt}\right)}{\left|\frac{d\mathbf{T}}{dt}\right|} = \left(\frac{-\cos t - \sin t}{\sqrt{2}}\right)\mathbf{i} + \left(\frac{-\sin t + \cos t}{\sqrt{2}}\right)\mathbf{j}$;

$\kappa = \frac{1}{|\mathbf{v}|} \cdot \left|\frac{d\mathbf{T}}{dt}\right| = \frac{1}{e^t\sqrt{2}} \cdot 1 = \frac{1}{e^t\sqrt{2}}$

13. $\mathbf{r} = \left(\frac{t^3}{3}\right)\mathbf{i} + \left(\frac{t^2}{2}\right)\mathbf{j}, t > 0 \Rightarrow \mathbf{v} = t^2\mathbf{i} + t\mathbf{j} \Rightarrow |\mathbf{v}| = \sqrt{t^4 + t^2} = t\sqrt{t^2 + 1}$, since $t > 0 \Rightarrow \mathbf{T} = \frac{\mathbf{v}}{|\mathbf{v}|}$

$= \frac{t}{\sqrt{t^2 + t}}\mathbf{i} + \frac{1}{\sqrt{t^2 + 1}}\mathbf{j} \Rightarrow \frac{d\mathbf{T}}{dt} = \frac{1}{(t^2 + 1)^{3/2}}\mathbf{i} - \frac{t}{(t^2 + 1)^{3/2}}\mathbf{j} \Rightarrow \left|\frac{d\mathbf{T}}{dt}\right| = \sqrt{\left(\frac{1}{(t^2 + 1)^{3/2}}\right)^2 + \left(\frac{-t}{(t^2 + 1)^{3/2}}\right)^2}$

$= \sqrt{\frac{1 + t^2}{(t^2 + 1)^3}} = \frac{1}{t^2 + 1} \Rightarrow \mathbf{N} = \frac{\left(\frac{d\mathbf{T}}{dt}\right)}{\left|\frac{d\mathbf{T}}{dt}\right|} = \frac{1}{\sqrt{t^2 + 1}}\mathbf{i} - \frac{t}{\sqrt{t^2 + 1}}\mathbf{j}; \kappa = \frac{1}{|\mathbf{v}|} \cdot \left|\frac{d\mathbf{T}}{dt}\right| = \frac{1}{t\sqrt{t^2 + 1}} \cdot \frac{1}{t^2 + 1} = \frac{1}{t(t^2 + 1)^{3/2}}$.

15. $\mathbf{r} = t\mathbf{i} + \left(a\cosh\frac{t}{a}\right)\mathbf{j}, a > 0 \Rightarrow \mathbf{v} = \mathbf{i} + \left(\sinh\frac{t}{a}\right)\mathbf{j} \Rightarrow |\mathbf{v}| = \sqrt{1 + \sinh^2\left(\frac{t}{a}\right)} = \sqrt{\cosh^2\left(\frac{t}{a}\right)} = \cosh\frac{t}{a}$

$\Rightarrow \mathbf{T} = \frac{\mathbf{v}}{|\mathbf{v}|} = \left(\text{sech}\,\frac{t}{a}\right)\mathbf{i} + \left(\tanh\frac{t}{a}\right)\mathbf{j} \Rightarrow \frac{d\mathbf{T}}{dt} = \left(-\frac{1}{a}\,\text{sech}\,\frac{t}{a}\tanh\frac{t}{a}\right)\mathbf{i} + \left(\frac{1}{a}\,\text{sech}^2\,\frac{t}{a}\right)\mathbf{j}$

$\Rightarrow \left|\frac{d\mathbf{T}}{dt}\right| = \sqrt{\frac{1}{a^2}\text{sech}^2\left(\frac{t}{a}\right)\tanh^2\left(\frac{t}{a}\right) + \frac{1}{a^2}\text{sech}^4\left(\frac{t}{a}\right)} = \frac{1}{a}\text{sech}\left(\frac{t}{a}\right) \Rightarrow \mathbf{N} = \frac{\left(\frac{d\mathbf{T}}{dt}\right)}{\left|\frac{d\mathbf{T}}{dt}\right|} = \left(-\tanh\frac{t}{a}\right)\mathbf{i} + \left(\text{sech}\,\frac{t}{a}\right)\mathbf{j}$;

$\kappa = \frac{1}{|\mathbf{v}|} \cdot \left|\frac{d\mathbf{T}}{dt}\right| = \frac{1}{\cosh\frac{t}{a}} \cdot \frac{1}{a}\text{sech}\left(\frac{t}{a}\right) = \frac{1}{a}\text{sech}^2\left(\frac{t}{a}\right)$.

17. $y = ax^2 \Rightarrow y' = 2ax \Rightarrow y'' = 2a$; from Exercise 5(a), $\kappa(x) = \frac{|2a|}{(1+4a^2x^2)^{3/2}} = |2a|\left(1 + 4a^2x^2\right)^{-3/2}$

$\Rightarrow \kappa'(x) = -\frac{3}{2}|2a|\left(1 + 4a^2x^2\right)^{-5/2}(8a^2x)$; thus, $\kappa'(x) = 0 \Rightarrow x = 0$. Now, $\kappa'(x) > 0$ for $x < 0$ and $\kappa'(x) < 0$ for $x > 0$ so that $\kappa(x)$ has an absolute maximum at $x = 0$ which is the vertex of the parabola. Since $x = 0$ is the only critical point for $\kappa(x)$, the curvature has no minimum value.

19. $\kappa = \frac{a}{a^2+b^2} \Rightarrow \frac{d\kappa}{da} = \frac{-a^2+b^2}{(a^2+b^2)^2}$; $\frac{d\kappa}{da} = 0 \Rightarrow -a^2 + b^2 = 0 \Rightarrow a = \pm b \Rightarrow a = b$ since $a, b \geq 0$. Now, $\frac{d\kappa}{da} > 0$ if $a < b$ and $\frac{d\kappa}{da} < 0$ if $a > b \Rightarrow \kappa$ is at a maximum for $a = b$ and $\kappa(b) = \frac{b}{b^2+b^2} = \frac{1}{2b}$ is the maximum value of κ.

21. $\mathbf{r} = t\mathbf{i} + (\sin t)\mathbf{j} \Rightarrow \mathbf{v} = \mathbf{i} + (\cos t)\mathbf{j} \Rightarrow |\mathbf{v}| = \sqrt{1^2 + (\cos t)^2} = \sqrt{1 + \cos^2 t} \Rightarrow \left|\mathbf{v}\left(\frac{\pi}{2}\right)\right| = \sqrt{1 + \cos^2\left(\frac{\pi}{2}\right)} = 1$; $\mathbf{T} = \frac{\mathbf{v}}{|\mathbf{v}|}$

$= \frac{\mathbf{i} + \cos t\,\mathbf{j}}{\sqrt{1+\cos^2 t}} \Rightarrow \frac{d\mathbf{T}}{dt} = \frac{\sin t\cos t}{(1+\cos^2 t)^{3/2}}\mathbf{i} + \frac{-\sin t}{(1+\cos^2 t)^{3/2}}\mathbf{j} \Rightarrow \left|\frac{d\mathbf{T}}{dt}\right| = \frac{|\sin t|}{1+\cos^2 t}$; $\left|\frac{d\mathbf{T}}{dt}\right|_{t=\frac{\pi}{2}} = \frac{|\sin\frac{\pi}{2}|}{1+\cos^2\left(\frac{\pi}{2}\right)} = \frac{1}{1} = 1$. Thus $\kappa\left(\frac{\pi}{2}\right) = \frac{1}{1}\cdot 1 = 1$

$\Rightarrow \rho = \frac{1}{1} = 1$ and the center is $\left(\frac{\pi}{2}, 0\right) \Rightarrow \left(x - \frac{\pi}{2}\right)^2 + y^2 = 1$

23. $y = x^2 \Rightarrow f'(x) = 2x$ and $f''(x) = 2$

$\Rightarrow \kappa = \frac{|2|}{(1+(2x)^2)^{3/2}} = \frac{2}{(1+4x^2)^{3/2}}$

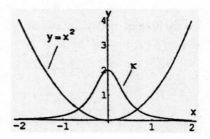

25. $y = \sin x \Rightarrow f'(x) = \cos x$ and $f''(x) = -\sin x$

$\Rightarrow \kappa = \frac{|-\sin x|}{(1+\cos^2 x)^{3/2}} = \frac{|\sin x|}{(1+\cos^2 x)^{3/2}}$

13.5 TORSION AND THE UNIT BINORMAL VECTOR B

1. By Exercise 9 in Section 13.4, $\mathbf{T} = \left(\frac{3}{5}\cos t\right)\mathbf{i} + \left(-\frac{3}{5}\sin t\right)\mathbf{j} + \frac{4}{5}\mathbf{k}$ and $\mathbf{N} = (-\sin t)\mathbf{i} - (\cos t)\mathbf{j}$ so that $\mathbf{B} = \mathbf{T} \times \mathbf{N}$

$= \begin{vmatrix} \mathbf{i} & \mathbf{j} & \mathbf{k} \\ \frac{3}{5}\cos t & -\frac{3}{5}\sin t & \frac{4}{5} \\ -\sin t & -\cos t & 0 \end{vmatrix} = \left(\frac{4}{5}\cos t\right)\mathbf{i} - \left(\frac{4}{5}\sin t\right)\mathbf{j} - \frac{3}{5}\mathbf{k}$. Also $\mathbf{v} = (3\cos t)\mathbf{i} + (-3\sin t)\mathbf{j} + 4\mathbf{k}$

$\Rightarrow \mathbf{a} = (-3\sin t)\mathbf{i} + (-3\cos t)\mathbf{j} \Rightarrow \frac{d\mathbf{a}}{dt} = (-3\cos t)\mathbf{i} + (3\sin t)\mathbf{j}$ and $\mathbf{v} \times \mathbf{a} = \begin{vmatrix} \mathbf{i} & \mathbf{j} & \mathbf{k} \\ 3\cos t & -3\sin t & 4 \\ -3\sin t & -3\cos t & 0 \end{vmatrix}$

$= (12\cos t)\mathbf{i} - (12\sin t)\mathbf{j} - 9\mathbf{k} \Rightarrow |\mathbf{v} \times \mathbf{a}|^2 = (12\cos t)^2 + (-12\sin t)^2 + (-9)^2 = 225$. Thus

$\tau = \frac{\begin{vmatrix} 3\cos t & -3\sin t & 4 \\ -3\sin t & -3\sin t & 0 \\ -3\cos t & 3\sin t & 0 \end{vmatrix}}{225} = \frac{4\cdot(-9\sin^2 t - 9\cos^2 t)}{225} = \frac{-36}{225} = -\frac{4}{25}$

3. By Exercise 11 in Section 13.4, $\mathbf{T} = \left(\frac{\cos t - \sin t}{\sqrt{2}}\right)\mathbf{i} + \left(\frac{\sin t + \cos t}{\sqrt{2}}\right)\mathbf{j}$ and $\mathbf{N} = \left(\frac{-\cos t - \sin t}{\sqrt{2}}\right)\mathbf{i} + \left(\frac{-\sin t + \cos t}{\sqrt{2}}\right)\mathbf{j}$; Thus

$\mathbf{B} = \mathbf{T} \times \mathbf{N} = \begin{vmatrix} \mathbf{i} & \mathbf{j} & \mathbf{k} \\ \frac{\cos t - \sin t}{\sqrt{2}} & \frac{\sin t + \cos t}{\sqrt{2}} & 0 \\ \frac{-\cos t - \sin t}{\sqrt{2}} & \frac{-\sin t + \cos t}{\sqrt{2}} & 0 \end{vmatrix} = \left[\left(\frac{\cos^2 t - 2\cos t\sin t + \sin^2 t}{2}\right) + \left(\frac{\sin^2 t + 2\sin t\cos t + \cos^2 t}{2}\right)\right]\mathbf{k}$

$= \left[\left(\frac{1-\sin(2t)}{2}\right) + \left(\frac{1+\sin(2t)}{2}\right)\right]\mathbf{k} = \mathbf{k}$. Also, $\mathbf{v} = (e^t\cos t - e^t\sin t)\mathbf{i} + (e^t\sin t + e^t\cos t)\mathbf{j}$

$\Rightarrow \mathbf{a} = [e^t(-\sin t - \cos t) + e^t(\cos t - \sin t)]\mathbf{i} + [e^t(\cos t - \sin t) + e^t(\sin t + \cos t)]\mathbf{j} = (-2e^t\sin t)\mathbf{i} + (2e^t\cos t)\mathbf{j}$

$\Rightarrow \frac{d\mathbf{a}}{dt} = -2e^t(\cos t + \sin t)\mathbf{i} + 2e^t(-\sin t + \cos t)\mathbf{j}$. Thus $\mathbf{v}\times\mathbf{a} = \begin{vmatrix} \mathbf{i} & \mathbf{j} & \mathbf{k} \\ e^t(\cos t - \sin t) & e^t(\sin t + \cos t) & 0 \\ -2e^t\sin t & 2e^t\cos t & 0 \end{vmatrix} = 2e^{2t}\mathbf{k}$

$\Rightarrow |\mathbf{v}\times\mathbf{a}|^2 = (2e^{2t})^2 = 4e^{4t}$. Thus $\tau = \dfrac{\begin{vmatrix} e^t(\cos t - \sin t) & e^t(\sin t + \cos t) & 0 \\ -2e^t\sin t & 2e^t\cos t & 0 \\ -2e^t(\cos t + \sin t) & 2e^t(-\sin t + \cos t) & 0 \end{vmatrix}}{4e^{4t}} = 0$

5. By Exercise 13 in Section 13.4, $\mathbf{T} = \frac{t}{(t^2+1)^{1/2}}\mathbf{i} + \frac{1}{(t^2+1)^{1/2}}\mathbf{j}$ and $\mathbf{N} = \frac{1}{\sqrt{t^2+1}}\mathbf{i} - \frac{t}{\sqrt{t^2+1}}\mathbf{j}$ so that $\mathbf{B} = \mathbf{T}\times\mathbf{N}$

$= \begin{vmatrix} \mathbf{i} & \mathbf{j} & \mathbf{k} \\ \frac{t}{\sqrt{t^2+1}} & \frac{1}{\sqrt{t^2+1}} & 0 \\ \frac{1}{\sqrt{t^2+1}} & \frac{-t}{\sqrt{t^2+1}} & 0 \end{vmatrix} = -\mathbf{k}$. Also, $\mathbf{v} = t^2\mathbf{i} + t\mathbf{j} \Rightarrow \mathbf{a} = 2t\mathbf{i} + \mathbf{j} \Rightarrow \frac{d\mathbf{a}}{dt} = 2\mathbf{i}$ so that $\begin{vmatrix} t^2 & t & 0 \\ 2t & 1 & 0 \\ 2 & 0 & 0 \end{vmatrix} = 0 \Rightarrow \tau = 0$

7. By Exercise 15 in Section 13.4, $\mathbf{T} = \frac{\mathbf{v}}{|\mathbf{v}|} = \left(\text{sech }\frac{t}{a}\right)\mathbf{i} + \left(\tanh \frac{t}{a}\right)\mathbf{j}$ and $\mathbf{N} = \left(-\tanh\frac{t}{a}\right)\mathbf{i} + \left(\text{sech }\frac{t}{a}\right)\mathbf{j}$ so that $\mathbf{B} = \mathbf{T}\times\mathbf{N}$

$= \begin{vmatrix} \mathbf{i} & \mathbf{j} & \mathbf{k} \\ \text{sech}\left(\frac{t}{a}\right) & \tanh\left(\frac{t}{a}\right) & 0 \\ -\tanh\left(\frac{t}{a}\right) & \text{sech}\left(\frac{t}{a}\right) & 0 \end{vmatrix} = \mathbf{k}$. Also, $\mathbf{v} = \mathbf{i} + \left(\sinh\frac{t}{a}\right)\mathbf{j} \Rightarrow \mathbf{a} = \left(\frac{1}{a}\cosh\frac{t}{a}\right)\mathbf{j} \Rightarrow \frac{d\mathbf{a}}{dt} = \frac{1}{a^2}\sinh\left(\frac{t}{a}\right)\mathbf{j}$ so that

$\begin{vmatrix} 1 & \sinh\left(\frac{t}{a}\right) & 0 \\ 0 & \frac{1}{a}\cosh\left(\frac{t}{a}\right) & 0 \\ 0 & \frac{1}{a^2}\sinh\left(\frac{t}{a}\right) & 0 \end{vmatrix} = 0 \Rightarrow \tau = 0$

9. $\mathbf{r} = (a\cos t)\mathbf{i} + (a\sin t)\mathbf{j} + bt\mathbf{k} \Rightarrow \mathbf{v} = (-a\sin t)\mathbf{i} + (a\cos t)\mathbf{j} + b\mathbf{k} \Rightarrow |\mathbf{v}| = \sqrt{(-a\sin t)^2 + (a\cos t)^2 + b^2}$

$= \sqrt{a^2+b^2} \Rightarrow a_T = \frac{d}{dt}|\mathbf{v}| = 0; \mathbf{a} = (-a\cos t)\mathbf{i} + (-a\sin t)\mathbf{j} \Rightarrow |\mathbf{a}| = \sqrt{(-a\cos t)^2 + (-a\sin t)^2} = \sqrt{a^2} = |a|$

$\Rightarrow a_N = \sqrt{|\mathbf{a}|^2 - a_T^2} = \sqrt{|\mathbf{a}|^2 - 0^2} = |\mathbf{a}| = |a| \Rightarrow \mathbf{a} = (0)\mathbf{T} + |a|\mathbf{N} = |a|\mathbf{N}$

11. $\mathbf{r} = (t+1)\mathbf{i} + 2t\mathbf{j} + t^2\mathbf{k} \Rightarrow \mathbf{v} = \mathbf{i} + 2\mathbf{j} + 2t\mathbf{k} \Rightarrow |\mathbf{v}| = \sqrt{1^2 + 2^2 + (2t)^2} = \sqrt{5+4t^2} \Rightarrow a_T = \frac{1}{2}(5+4t^2)^{-1/2}(8t)$

$= 4t(5+4t^2)^{-1/2} \Rightarrow a_T(1) = \frac{4}{\sqrt{9}} = \frac{4}{3}; \mathbf{a} = 2\mathbf{k} \Rightarrow \mathbf{a}(1) = 2\mathbf{k} \Rightarrow |\mathbf{a}(1)| = 2 \Rightarrow a_N = \sqrt{|\mathbf{a}|^2 - a_T^2} = \sqrt{2^2 - \left(\frac{4}{3}\right)^2}$

$= \sqrt{\frac{20}{9}} = \frac{2\sqrt{5}}{3} \Rightarrow \mathbf{a}(1) = \frac{4}{3}\mathbf{T} + \frac{2\sqrt{5}}{3}\mathbf{N}$

13. $\mathbf{r} = t^2\mathbf{i} + \left(t + \frac{1}{3}t^3\right)\mathbf{j} + \left(t - \frac{1}{3}t^3\right)\mathbf{k} \Rightarrow \mathbf{v} = 2t\mathbf{i} + (1+t^2)\mathbf{j} + (1-t^2)\mathbf{k} \Rightarrow |\mathbf{v}| = \sqrt{(2t)^2 + (1+t^2)^2 + (1-t^2)^2}$

$= \sqrt{2(t^4 + 2t^2 + 1)} = \sqrt{2}(1+t^2) \Rightarrow a_T = 2t\sqrt{2} \Rightarrow a_T(0) = 0; \mathbf{a} = 2\mathbf{i} + 2t\mathbf{j} - 2t\mathbf{k} \Rightarrow \mathbf{a}(0) = 2\mathbf{i} \Rightarrow |\mathbf{a}(0)| = 2$

$\Rightarrow a_N = \sqrt{|\mathbf{a}|^2 - a_T^2} = \sqrt{2^2 - 0^2} = 2 \Rightarrow \mathbf{a}(0) = (0)\mathbf{T} + 2\mathbf{N} = 2\mathbf{N}$

15. $\mathbf{r} = (\cos t)\mathbf{i} + (\sin t)\mathbf{j} - \mathbf{k} \Rightarrow \mathbf{v} = (-\sin t)\mathbf{i} + (\cos t)\mathbf{j} \Rightarrow |\mathbf{v}| = \sqrt{(-\sin t)^2 + (\cos t)^2} = 1 \Rightarrow \mathbf{T} = \frac{\mathbf{v}}{|\mathbf{v}|}$

$= (-\sin t)\mathbf{i} + (\cos t)\mathbf{j} \Rightarrow \mathbf{T}\left(\frac{\pi}{4}\right) = -\frac{\sqrt{2}}{2}\mathbf{i} + \frac{\sqrt{2}}{2}\mathbf{j}; \frac{d\mathbf{T}}{dt} = (-\cos t)\mathbf{i} - (\sin t)\mathbf{j} \Rightarrow \left|\frac{d\mathbf{T}}{dt}\right| = \sqrt{(-\cos t)^2 + (-\sin t)^2}$

$= 1 \Rightarrow \mathbf{N} = \frac{\left(\frac{d\mathbf{T}}{dt}\right)}{\left|\frac{d\mathbf{T}}{dt}\right|} = (-\cos t)\mathbf{i} - (\sin t)\mathbf{j} \Rightarrow \mathbf{N}\left(\frac{\pi}{4}\right) = -\frac{\sqrt{2}}{2}\mathbf{i} - \frac{\sqrt{2}}{2}\mathbf{j}; \mathbf{B} = \mathbf{T}\times\mathbf{N} = \begin{vmatrix} \mathbf{i} & \mathbf{j} & \mathbf{k} \\ -\sin t & \cos t & 0 \\ -\cos t & -\sin t & 0 \end{vmatrix} = \mathbf{k}$

$\Rightarrow \mathbf{B}\left(\frac{\pi}{4}\right) = \mathbf{k}$, the normal to the osculating plane; $\mathbf{r}\left(\frac{\pi}{4}\right) = \frac{\sqrt{2}}{2}\mathbf{i} + \frac{\sqrt{2}}{2}\mathbf{j} - \mathbf{k} \Rightarrow P = \left(\frac{\sqrt{2}}{2}, \frac{\sqrt{2}}{2}, -1\right)$ lies on the

osculating plane $\Rightarrow 0\left(x - \frac{\sqrt{2}}{2}\right) + 0\left(y - \frac{\sqrt{2}}{2}\right) + (z - (-1)) = 0 \Rightarrow z = -1$ is the osculating plane; $\mathbf{T}$ is normal

to the normal plane $\Rightarrow \left(-\frac{\sqrt{2}}{2}\right)\left(x - \frac{\sqrt{2}}{2}\right) + \left(\frac{\sqrt{2}}{2}\right)\left(y - \frac{\sqrt{2}}{2}\right) + 0(z - (-1)) = 0 \Rightarrow -\frac{\sqrt{2}}{2}x + \frac{\sqrt{2}}{2}y = 0$

$\Rightarrow \; -x + y = 0$ is the normal plane; $\mathbf{N}$ is normal to the rectifying plane

$\Rightarrow \left(-\frac{\sqrt{2}}{2}\right)\left(x - \frac{\sqrt{2}}{2}\right) + \left(-\frac{\sqrt{2}}{2}\right)\left(y - \frac{\sqrt{2}}{2}\right) + 0(z - (-1)) = 0 \;\Rightarrow\; -\frac{\sqrt{2}}{2}x - \frac{\sqrt{2}}{2}y = -1 \;\Rightarrow\; x + y = \sqrt{2}$ is the

rectifying plane

17. Yes. If the car is moving along a curved path, then $\kappa \neq 0$ and $a_N = \kappa\,|\mathbf{v}|^2 \neq 0 \;\Rightarrow\; \mathbf{a} = a_T\mathbf{T} + a_N\mathbf{N} \neq \mathbf{0}$.

19. $\mathbf{a} \perp \mathbf{v} \;\Rightarrow\; \mathbf{a} \perp \mathbf{T} \;\Rightarrow\; a_T = 0 \;\Rightarrow\; \frac{d}{dt}\,|\mathbf{v}| = 0 \;\Rightarrow\; |\mathbf{v}|$ is constant

21. $\mathbf{a} = a_T\mathbf{T} + a_N\mathbf{N}$, where $a_T = \frac{d}{dt}\,|\mathbf{v}| = \frac{d}{dt}\,(\text{constant}) = 0$ and $a_N = \kappa\,|\mathbf{v}|^2 \;\Rightarrow\; \mathbf{F} = m\mathbf{a} = m\kappa\,|\mathbf{v}|^2\mathbf{N} \;\Rightarrow\; |\mathbf{F}| = m\kappa\,|\mathbf{v}|^2$
$= \left(m\,|\mathbf{v}|^2\right)\kappa$, a constant multiple of the curvature κ of the trajectory

23. From Example 1, $|\mathbf{v}| = t$ and $a_N = t$ so that $a_N = \kappa\,|\mathbf{v}|^2 \;\Rightarrow\; \kappa = \frac{a_N}{|\mathbf{v}|^2} = \frac{t}{t^2} = \frac{1}{t}, t \neq 0 \;\Rightarrow\; \rho = \frac{1}{\kappa} = t$

25. If a plane curve is sufficiently differentiable the torsion is zero as the following argument shows:
$\mathbf{r} = f(t)\mathbf{i} + g(t)\mathbf{j} \;\Rightarrow\; \mathbf{v} = f'(t)\mathbf{i} + g'(t)\mathbf{j} \;\Rightarrow\; \mathbf{a} = f''(t)\mathbf{i} + g''(t)\mathbf{j} \;\Rightarrow\; \frac{d\mathbf{a}}{dt} = f'''(t)\mathbf{i} + g'''(t)\mathbf{j}$

$\Rightarrow \tau = \dfrac{\begin{vmatrix} f'(t) & g'(t) & 0 \\ f''(t) & g''(t) & 0 \\ f'''(t) & g'''(t) & 0 \end{vmatrix}}{|\mathbf{v} \times \mathbf{a}|^2} = 0$

27. $\mathbf{r}(t) = f(t)\mathbf{i} + g(t)\mathbf{j} + h(t)\mathbf{k} \;\Rightarrow\; \mathbf{v} = f'(t)\mathbf{i} + g'(t)\mathbf{j} + h'(t)\mathbf{k}; \mathbf{v} \cdot \mathbf{k} = 0 \;\Rightarrow\; h'(t) = 0 \;\Rightarrow\; h(t) = C$
$\Rightarrow \mathbf{r}(t) = f(t)\mathbf{i} + g(t)\mathbf{j} + C\mathbf{k}$ and $\mathbf{r}(a) = f(a)\mathbf{i} + g(a)\mathbf{j} + C\mathbf{k} = \mathbf{0} \;\Rightarrow\; f(a) = 0, g(a) = 0$ and $C = 0 \;\Rightarrow\; h(t) = 0$.

13.6 PLANETARY MOTION AND SATELLITES

1. $\frac{T^2}{a^3} = \frac{4\pi^2}{GM} \;\Rightarrow\; T^2 = \frac{4\pi^2}{GM}\,a^3 \;\Rightarrow\; T^2 = \frac{4\pi^2}{(6.6726\times10^{-11}\ \text{Nm}^2\text{kg}^{-2})\,(5.975\times10^{24}\ \text{kg})}\,(6,808,000\ \text{m})^3$
$\approx 3.125 \times 10^7\ \text{sec}^2 \;\Rightarrow\; T \approx \sqrt{3125 \times 10^4\ \text{sec}^2} \approx 55.90 \times 10^2\ \text{sec} \approx 93.2\ \text{min}$

3. $92.25\ \text{min} = 5535\ \text{sec}$ and $\frac{T^2}{a^3} = \frac{4\pi^2}{GM} \;\Rightarrow\; a^3 = \frac{GM}{4\pi^2}\,T^2$
$\Rightarrow a^3 = \frac{(6.6726\times10^{-11}\ \text{Nm}^2\text{kg}^{-2})\,(5.975\times10^{24}\ \text{kg})}{4\pi^2}\,(5535\ \text{sec})^2 = 3.094 \times 10^{20}\ \text{m}^3 \;\Rightarrow\; a \approx \sqrt[3]{3.094 \times 10^{20}\ \text{m}^3}$
$= 6.764 \times 10^6\ \text{m} \approx 6764\ \text{km}$. Note that $6764\ \text{km} \approx \frac{1}{2}(12{,}757\ \text{km} + 183\ \text{km} + 589\ \text{km})$.

5. $2a = \text{diameter of Mars} + \text{perigee height} + \text{apogee height} = D + 1499\ \text{km} + 35{,}800\ \text{km}$
$\Rightarrow 2(21{,}900)\ \text{km} = D + 37{,}299\ \text{km} \;\Rightarrow\; D = 6501\ \text{km}$

7. (a) Period of the satellite $=$ rotational period of the Earth $\;\Rightarrow\;$ period of the satellite $= 1436.1\ \text{min}$
$= 86{,}166\ \text{sec};\ a^3 = \frac{GMT^2}{4\pi^2} \;\Rightarrow\; a^3 = \frac{(6.6726\times10^{-11}\ \text{Nm}^2\text{kg}^{-2})\,(5.975\times10^{24}\ \text{kg})\,(86{,}166\ \text{sec})^2}{4\pi^2}$
$\approx 7.4980 \times 10^{22}\ \text{m}^3 \;\Rightarrow\; a \approx \sqrt[3]{74.980 \times 10^{21}\ \text{m}^3} \approx 4.2168 \times 10^7\ \text{m} = 42{,}168\ \text{km}$
   (b) The radius of the Earth is approximately $6379\ \text{km} \;\Rightarrow\;$ the height of the orbit is $42{,}168 - 6379 = 35{,}789\ \text{km}$
   (c) Symcom 3, GOES 4, and Intelsat 5

9. Period of the Moon $= 2.36055 \times 10^6\ \text{sec} \;\Rightarrow\; a^3 = \frac{GMT^2}{4\pi^2}$
$= \frac{(6.6726\times10^{-11}\ \text{Nm}^2\text{kg}^{-2})\,(5.975\times10^{24}\ \text{kg})\,(2.36055\times10^6\ \text{sec})^2}{4\pi^2} \approx 5.627 \times 10^{25}\ \text{m}^3 \;\Rightarrow\; a \approx \sqrt[3]{5.627 \times 10^{25}\ \text{m}^3}$
$\approx 3.832 \times 10^8\ \text{m} = 383{,}200\ \text{km}$ from the center of the Earth.

11. Solar System: $\dfrac{T^2}{a^3} = \dfrac{4\pi^2}{(6.6726\times10^{-11}\ \mathrm{Nm^2kg^{-2}})\,(1.99\times10^{30}\ \mathrm{kg})} \approx 2.97\times10^{-19}\ \mathrm{sec^2/m^3};$

Earth: $\dfrac{T^2}{a^3} = \dfrac{4\pi^2}{(6.6726\times10^{-11}\ \mathrm{Nm^2kg^{-2}})\,(5.975\times10^{24}\ \mathrm{kg})} \approx 9.902\times10^{-14}\ \mathrm{sec^2/m^3};$

Moon: $\dfrac{T^2}{a^3} = \dfrac{4\pi^2}{(6.6726\times10^{-11}\ \mathrm{Nm^2kg^{-2}})\,(7.354\times10^{22}\ \mathrm{kg})} \approx 8.045\times10^{-12}\ \mathrm{sec^2/m^3};$

13. $r = \dfrac{GM}{v^2} \Rightarrow v^2 = \dfrac{GM}{r} \Rightarrow v = \sqrt{\dfrac{GM}{r}}$ which is constant since G, M, and r (the radius of orbit) are constant

15. $T = \left(\dfrac{2\pi a^2}{r_0 v_0}\right)\sqrt{1-e^2} \Rightarrow T^2 = \left(\dfrac{4\pi^2 a^4}{r_0^2 v_0^2}\right)(1-e^2) = \left(\dfrac{4\pi^2 a^4}{r_0^2 v_0^2}\right)\left[1-\left(\dfrac{r_0 v_0^2}{GM}-1\right)^2\right]$ (from Equation 32)

$= \left(\dfrac{4\pi^2 a^4}{r_0^2 v_0^2}\right)\left[-\dfrac{r_0^2 v_0^4}{G^2 M^2}+2\left(\dfrac{r_0 v_0^2}{GM}\right)\right] = \left(\dfrac{4\pi^2 a^4}{r_0^2 v_0^2}\right)\left[\dfrac{2GMr_0 v_0^2 - r_0^2 v_0^4}{G^2 M^2}\right] = \dfrac{(4\pi^2 a^4)\,(2GM - r_0 v_0^2)}{r_0 G^2 M^2}$

$= (4\pi^2 a^4)\left(\dfrac{2GM - r_0 v_0^2}{2 r_0 GM}\right)\left(\dfrac{2}{GM}\right) = (4\pi^2 a^4)\left(\dfrac{1}{2a}\right)\left(\dfrac{2}{GM}\right)$ (from Equation 35) $\Rightarrow T^2 = \dfrac{4\pi^2 a^3}{GM} \Rightarrow \dfrac{T^2}{a^3} = \dfrac{4\pi^2}{GM}$

17. The graph of the path of planet B is the limaçon
at the right.

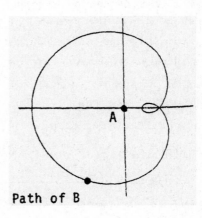

Path of B

CHAPTER 13 PRACTICE EXERCISES

1. $\mathbf{r}(t) = (4\cos t)\mathbf{i} + \left(\sqrt{2}\sin t\right)\mathbf{j} \Rightarrow x = 4\cos t$

and $y = \sqrt{2}\sin t \Rightarrow \dfrac{x^2}{16} + \dfrac{y^2}{2} = 1$;

$\mathbf{v} = (-4\sin t)\mathbf{i} + \left(\sqrt{2}\cos t\right)\mathbf{j}$ and

$\mathbf{a} = (-4\cos t)\mathbf{i} - \left(\sqrt{2}\sin t\right)\mathbf{j}$; $\mathbf{r}(0) = 4\mathbf{i}$, $\mathbf{v}(0) = \sqrt{2}\mathbf{j}$,

$\mathbf{a}(0) = -4\mathbf{i}$; $\mathbf{r}\left(\dfrac{\pi}{4}\right) = 2\sqrt{2}\mathbf{i} + \mathbf{j}$, $\mathbf{v}\left(\dfrac{\pi}{4}\right) = -2\sqrt{2}\mathbf{i} + \mathbf{j}$,

$\mathbf{a}\left(\dfrac{\pi}{4}\right) = -2\sqrt{2}\mathbf{i} - \mathbf{j}$; $|\mathbf{v}| = \sqrt{16\sin^2 t + 2\cos^2 t}$

$\Rightarrow a_T = \dfrac{d}{dt}|\mathbf{v}| = \dfrac{14\sin t\cos t}{\sqrt{16\sin^2 t + 2\cos^2 t}}$; at $t = 0$: $a_T = 0$, $a_N = \sqrt{|\mathbf{a}|^2 - 0} = 4$, $\kappa = \dfrac{a_N}{|\mathbf{v}|^2} = \dfrac{4}{2} = 2$;

at $t = \dfrac{\pi}{4}$: $a_T = \dfrac{7}{\sqrt{8+1}} = \dfrac{7}{3}$, $a_N = \sqrt{9 - \dfrac{49}{9}} = \dfrac{4\sqrt{2}}{3}$, $\kappa = \dfrac{a_N}{|\mathbf{v}|^2} = \dfrac{4\sqrt{2}}{27}$

3. $\mathbf{r} = \dfrac{1}{\sqrt{1+t^2}}\mathbf{i} + \dfrac{t}{\sqrt{1+t^2}}\mathbf{j} \Rightarrow \mathbf{v} = -t(1+t^2)^{-3/2}\mathbf{i} + (1+t^2)^{-3/2}\mathbf{j}$

$\Rightarrow |\mathbf{v}| = \sqrt{\left[-t(1+t^2)^{-3/2}\right]^2 + \left[(1+t^2)^{-3/2}\right]^2} = \dfrac{1}{1+t^2}$. We want to maximize $|\mathbf{v}|$: $\dfrac{d\,|\mathbf{v}|}{dt} = \dfrac{-2t}{(1+t^2)^2}$ and

$\dfrac{d\,|\mathbf{v}|}{dt} = 0 \Rightarrow \dfrac{-2t}{(1+t^2)^2} = 0 \Rightarrow t = 0$. For $t < 0$, $\dfrac{-2t}{(1+t^2)^2} > 0$; for $t > 0$, $\dfrac{-2t}{(1+t^2)^2} < 0 \Rightarrow |\mathbf{v}|_{max}$ occurs when

$t = 0 \Rightarrow |\mathbf{v}|_{max} = 1$

5. $\mathbf{v} = 3\mathbf{i} + 4\mathbf{j}$ and $\mathbf{a} = 5\mathbf{i} + 15\mathbf{j} \Rightarrow \mathbf{v}\times\mathbf{a} = \begin{vmatrix} \mathbf{i} & \mathbf{j} & \mathbf{k} \\ 3 & 4 & 0 \\ 5 & 15 & 0 \end{vmatrix} = 25\mathbf{k} \Rightarrow |\mathbf{v}\times\mathbf{a}| = 25;\ |\mathbf{v}| = \sqrt{3^2 + 4^2} = 5$

$\Rightarrow \kappa = \frac{|v \times a|}{|v|^3} = \frac{25}{5^3} = \frac{1}{5}$

7. $\mathbf{r} = x\mathbf{i} + y\mathbf{j} \Rightarrow \mathbf{v} = \frac{dx}{dt}\mathbf{i} + \frac{dy}{dt}\mathbf{j}$ and $\mathbf{v} \cdot \mathbf{i} = y \Rightarrow \frac{dx}{dt} = y$. Since the particle moves around the unit circle
$x^2 + y^2 = 1, 2x\frac{dx}{dt} + 2y\frac{dy}{dt} = 0 \Rightarrow \frac{dy}{dt} = -\frac{x}{y}\frac{dx}{dt} \Rightarrow \frac{dy}{dt} = -\frac{x}{y}(y) = -x$. Since $\frac{dx}{dt} = y$ and $\frac{dy}{dt} = -x$, we have
$\mathbf{v} = y\mathbf{i} - x\mathbf{j} \Rightarrow$ at $(1,0)$, $\mathbf{v} = -\mathbf{j}$ and the motion is clockwise.

9. $\frac{d\mathbf{r}}{dt}$ orthogonal to $\mathbf{r} \Rightarrow 0 = \frac{d\mathbf{r}}{dt} \cdot \mathbf{r} = \frac{1}{2}\frac{d\mathbf{r}}{dt} \cdot \mathbf{r} + \frac{1}{2}\mathbf{r} \cdot \frac{d\mathbf{r}}{dt} = \frac{1}{2}\frac{d}{dt}(\mathbf{r} \cdot \mathbf{r}) \Rightarrow \mathbf{r} \cdot \mathbf{r} = K$, a constant. If $\mathbf{r} = x\mathbf{i} + y\mathbf{j}$, where
x and y are differentiable functions of t, then $\mathbf{r} \cdot \mathbf{r} = x^2 + y^2 \Rightarrow x^2 + y^2 = K$, which is the equation of a circle
centered at the origin.

11. $y = y_0 + (v_0 \sin \alpha)t - \frac{1}{2}gt^2 \Rightarrow y = 6.5 + (44 \text{ ft/sec})(\sin 45°)(3 \text{ sec}) - \frac{1}{2}(32 \text{ ft/sec}^2)(3 \text{ sec})^2 = 6.5 + 66\sqrt{2} - 144$
$\approx -44.16 \text{ ft} \Rightarrow$ the shot put is on the ground. Now, $y = 0 \Rightarrow 6.5 + 22\sqrt{2}t - 16t^2 = 0 \Rightarrow t \approx 2.13$ sec (the
positive root) $\Rightarrow x \approx (44 \text{ ft/sec})(\cos 45°)(2.13 \text{ sec}) \approx 66.27$ ft or about 66 ft, 3 in. from the stopboard

13. $x = (v_0 \cos \alpha)t$ and $y = (v_0 \sin \alpha)t - \frac{1}{2}gt^2 \Rightarrow \tan \phi = \frac{y}{x} = \frac{(v_0 \sin \alpha)t - \frac{1}{2}gt^2}{(v_0 \cos \alpha)t} = \frac{(v_0 \sin \alpha) - \frac{1}{2}gt}{v_0 \cos \alpha}$
$\Rightarrow v_0 \cos \alpha \tan \phi = v_0 \sin \alpha - \frac{1}{2}gt \Rightarrow t = \frac{2v_0 \sin \alpha - 2v_0 \cos \alpha \tan \phi}{g}$, which is the time when the golf ball
hits the upward slope. At this time

$x = (v_0 \cos \alpha)\left(\frac{2v_0 \sin \alpha - 2v_0 \cos \alpha \tan \phi}{g}\right)$
$= \left(\frac{2}{g}\right)(v_0^2 \sin \alpha \cos \alpha - v_0^2 \cos^2 \alpha \tan \phi)$. Now
$OR = \frac{x}{\cos \phi} \Rightarrow OR = \left(\frac{2}{g}\right)\left(\frac{v_0^2 \sin \alpha \cos \alpha - v_0^2 \cos^2 \alpha \tan \phi}{\cos \phi}\right)$
$= \left(\frac{2v_0^2 \cos \alpha}{g}\right)\left(\frac{\sin \alpha}{\cos \phi} - \frac{\cos \alpha \tan \phi}{\cos \phi}\right)$
$= \left(\frac{2v_0^2 \cos \alpha}{g}\right)\left(\frac{\sin \alpha \cos \phi - \cos \alpha \sin \phi}{\cos^2 \phi}\right)$
$= \left(\frac{2v_0^2 \cos \alpha}{g \cos^2 \phi}\right)[\sin(\alpha - \phi)]$. The distance OR is maximized
when x is maximized: $\frac{dx}{d\alpha} = \left(\frac{2v_0^2}{g}\right)(\cos 2\alpha + \sin 2\alpha \tan \phi) = 0 \Rightarrow (\cos 2\alpha + \sin 2\alpha \tan \phi) = 0 \Rightarrow \cot 2\alpha + \tan \phi = 0$
$\Rightarrow \cot 2\alpha = \tan(-\phi) \Rightarrow 2\alpha = \frac{\pi}{2} + \phi \Rightarrow \alpha = \frac{\phi}{2} + \frac{\pi}{4}$

15. (a) $R = \frac{v_0^2}{g}\sin 2\alpha \Rightarrow 109.5 \text{ ft} = \left(\frac{v_0^2}{32 \text{ ft/sec}^2}\right)(\sin 90°) \Rightarrow v_0^2 = 3504 \text{ ft}^2/\text{sec}^2 \Rightarrow v_0 = \sqrt{3504 \text{ ft}^2/\text{sec}^2}$
≈ 59.19 ft/sec
(b) $x = (v_0 \cos \alpha)t$ and $y = 4 + (v_0 \sin \alpha)t - \frac{1}{2}gt^2$; when the cork hits the ground, $x = 177.75$ ft and $y = 0$
$\Rightarrow 177.75 = \left(v_0 \frac{1}{\sqrt{2}}\right)t$ and $0 = 4 + \left(v_0 \frac{1}{\sqrt{2}}\right)t - 16t^2 \Rightarrow 16t^2 = 4 + 177.75 \Rightarrow t = \frac{\sqrt{181.75}}{4}$
$\Rightarrow v_0 = \frac{(177.75)\sqrt{2}}{t} = \frac{4(177.75)\sqrt{2}}{\sqrt{181.75}} \approx 74.58$ ft/sec

17. $x^2 = (v_0^2 \cos^2 \alpha)t^2$ and $\left(y + \frac{1}{2}gt^2\right)^2 = (v_0^2 \sin^2 \alpha)t^2 \Rightarrow x^2 + \left(y + \frac{1}{2}gt^2\right)^2 = v_0^2 t^2$

19. $\mathbf{r}(t) = \left[\int_0^t \cos\left(\frac{1}{2}\pi\theta^2\right)d\theta\right]\mathbf{i} + \left[\int_0^t \sin\left(\frac{1}{2}\pi\theta^2\right)d\theta\right]\mathbf{j} \Rightarrow \mathbf{v}(t) = \cos\left(\frac{\pi t^2}{2}\right)\mathbf{i} + \sin\left(\frac{\pi t^2}{2}\right)\mathbf{j} \Rightarrow |\mathbf{v}| = 1;$

$\mathbf{a}(t) = -\pi t \sin\left(\frac{\pi t^2}{2}\right)\mathbf{i} + \pi t \cos\left(\frac{\pi t^2}{2}\right)\mathbf{j} \Rightarrow \mathbf{v} \times \mathbf{a} = \begin{vmatrix} \mathbf{i} & \mathbf{j} & \mathbf{k} \\ \cos\left(\frac{\pi t^2}{2}\right) & \sin\left(\frac{\pi t^2}{2}\right) & 0 \\ -\pi t \sin\left(\frac{\pi t^2}{2}\right) & \pi t \cos\left(\frac{\pi t^2}{2}\right) & 0 \end{vmatrix}$

$= \pi t\mathbf{k} \Rightarrow \kappa = \frac{|v \times a|}{|v|^3} = \pi t; |\mathbf{v}(t)| = \frac{ds}{dt} = 1 \Rightarrow s = t + C; \mathbf{r}(0) = \mathbf{0} \Rightarrow s(0) = 0 \Rightarrow C = 0 \Rightarrow \kappa = \pi s$

21. $\mathbf{r} = (2 \cos t)\mathbf{i} + (2 \sin t)\mathbf{j} + t^2\mathbf{k} \Rightarrow \mathbf{v} = (-2 \sin t)\mathbf{i} + (2 \cos t)\mathbf{j} + 2t\mathbf{k} \Rightarrow |\mathbf{v}| = \sqrt{(-2 \sin t)^2 + (2 \cos t)^2 + (2t)^2}$

$= 2\sqrt{1 + t^2} \Rightarrow \text{Length} = \int_0^{\pi/4} 2\sqrt{1 + t^2}\, dt = \left[t\sqrt{1 + t^2} + \ln\left|t + \sqrt{1 + t^2}\right|\right]_0^{\pi/4} = \frac{\pi}{4}\sqrt{1 + \frac{\pi^2}{16}} + \ln\left(\frac{\pi}{4} + \sqrt{1 + \frac{\pi^2}{16}}\right)$

23. $\mathbf{r} = \frac{4}{9}(1 + t)^{3/2}\mathbf{i} + \frac{4}{9}(1 - t)^{3/2}\mathbf{j} + \frac{1}{3}t\mathbf{k} \Rightarrow \mathbf{v} = \frac{2}{3}(1 + t)^{1/2}\mathbf{i} - \frac{2}{3}(1 - t)^{1/2}\mathbf{j} + \frac{1}{3}\mathbf{k}$

$\Rightarrow |\mathbf{v}| = \sqrt{\left[\frac{2}{3}(1 + t)^{1/2}\right]^2 + \left[-\frac{2}{3}(1 - t)^{1/2}\right]^2 + \left(\frac{1}{3}\right)^2} = 1 \Rightarrow \mathbf{T} = \frac{2}{3}(1 + t)^{1/2}\mathbf{i} - \frac{2}{3}(1 - t)^{1/2}\mathbf{j} + \frac{1}{3}\mathbf{k}$

$\Rightarrow \mathbf{T}(0) = \frac{2}{3}\mathbf{i} - \frac{2}{3}\mathbf{j} + \frac{1}{3}\mathbf{k}; \frac{d\mathbf{T}}{dt} = \frac{1}{3}(1 + t)^{-1/2}\mathbf{i} + \frac{1}{3}(1 - t)^{-1/2}\mathbf{j} \Rightarrow \frac{d\mathbf{T}}{dt}(0) = \frac{1}{3}\mathbf{i} + \frac{1}{3}\mathbf{j} \Rightarrow \left|\frac{d\mathbf{T}}{dt}(0)\right| = \frac{\sqrt{2}}{3}$

$\Rightarrow \mathbf{N}(0) = \frac{1}{\sqrt{2}}\mathbf{i} + \frac{1}{\sqrt{2}}\mathbf{j}; \mathbf{B}(0) = \mathbf{T}(0) \times \mathbf{N}(0) = \begin{vmatrix} \mathbf{i} & \mathbf{j} & \mathbf{k} \\ \frac{2}{3} & -\frac{2}{3} & \frac{1}{3} \\ \frac{1}{\sqrt{2}} & \frac{1}{\sqrt{2}} & 0 \end{vmatrix} = -\frac{1}{3\sqrt{2}}\mathbf{i} + \frac{1}{3\sqrt{2}}\mathbf{j} + \frac{4}{3\sqrt{2}}\mathbf{k};$

$\mathbf{a} = \frac{1}{3}(1 + t)^{-1/2}\mathbf{i} + \frac{1}{3}(1 - t)^{-1/2}\mathbf{j} \Rightarrow \mathbf{a}(0) = \frac{1}{3}\mathbf{i} + \frac{1}{3}\mathbf{j} \text{ and } \mathbf{v}(0) = \frac{2}{3}\mathbf{i} - \frac{2}{3}\mathbf{j} + \frac{1}{3}\mathbf{k} \Rightarrow \mathbf{v}(0) \times \mathbf{a}(0)$

$= \begin{vmatrix} \mathbf{i} & \mathbf{j} & \mathbf{k} \\ \frac{2}{3} & -\frac{2}{3} & \frac{1}{3} \\ \frac{1}{3} & \frac{1}{3} & 0 \end{vmatrix} = -\frac{1}{9}\mathbf{i} + \frac{1}{9}\mathbf{j} + \frac{4}{9}\mathbf{k} \Rightarrow |\mathbf{v} \times \mathbf{a}| = \frac{\sqrt{2}}{3} \Rightarrow \kappa(0) = \frac{|\mathbf{v} \times \mathbf{a}|}{|\mathbf{v}|^3} = \frac{\left(\frac{\sqrt{2}}{3}\right)}{1^3} = \frac{\sqrt{2}}{3};$

$\dot{\mathbf{a}} = -\frac{1}{6}(1 + t)^{-3/2}\mathbf{i} + \frac{1}{6}(1 - t)^{-3/2}\mathbf{j} \Rightarrow \dot{\mathbf{a}}(0) = -\frac{1}{6}\mathbf{i} + \frac{1}{6}\mathbf{j} \Rightarrow \tau(0) = \dfrac{\begin{vmatrix} \frac{2}{3} & -\frac{2}{3} & \frac{1}{3} \\ \frac{1}{3} & \frac{1}{3} & 0 \\ -\frac{1}{6} & \frac{1}{6} & 0 \end{vmatrix}}{|\mathbf{v} \times \mathbf{a}|^2} = \frac{\left(\frac{1}{3}\right)\left(\frac{2}{18}\right)}{\left(\frac{\sqrt{2}}{3}\right)^2} = \frac{1}{6};$

$t = 0 \Rightarrow \left(\frac{4}{9}, \frac{4}{9}, 0\right)$ is the point on the curve

25. $\mathbf{r} = t\mathbf{i} + \frac{1}{2}e^{2t}\mathbf{j} \Rightarrow \mathbf{v} = \mathbf{i} + e^{2t}\mathbf{j} \Rightarrow |\mathbf{v}| = \sqrt{1 + e^{4t}} \Rightarrow \mathbf{T} = \frac{1}{\sqrt{1 + e^{4t}}}\mathbf{i} + \frac{e^{2t}}{\sqrt{1 + e^{4t}}}\mathbf{j} \Rightarrow \mathbf{T}(\ln 2) = \frac{1}{\sqrt{17}}\mathbf{i} + \frac{4}{\sqrt{17}}\mathbf{j};$

$\frac{d\mathbf{T}}{dt} = \frac{-2e^{4t}}{(1 + e^{4t})^{3/2}}\mathbf{i} + \frac{2e^{2t}}{(1 + e^{4t})^{3/2}}\mathbf{j} \Rightarrow \frac{d\mathbf{T}}{dt}(\ln 2) = \frac{-32}{17\sqrt{17}}\mathbf{i} + \frac{8}{17\sqrt{17}}\mathbf{j} \Rightarrow \mathbf{N}(\ln 2) = -\frac{4}{\sqrt{17}}\mathbf{i} + \frac{1}{\sqrt{17}}\mathbf{j};$

$\mathbf{B}(\ln 2) = \mathbf{T}(\ln 2) \times \mathbf{N}(\ln 2) = \begin{vmatrix} \mathbf{i} & \mathbf{j} & \mathbf{k} \\ \frac{1}{\sqrt{17}} & \frac{4}{\sqrt{17}} & 0 \\ -\frac{4}{\sqrt{17}} & \frac{1}{\sqrt{17}} & 0 \end{vmatrix} = \mathbf{k}; \mathbf{a} = 2e^{2t}\mathbf{j} \Rightarrow \mathbf{a}(\ln 2) = 8\mathbf{j} \text{ and } \mathbf{v}(\ln 2) = \mathbf{i} + 4\mathbf{j}$

$\Rightarrow \mathbf{v}(\ln 2) \times \mathbf{a}(\ln 2) = \begin{vmatrix} \mathbf{i} & \mathbf{j} & \mathbf{k} \\ 1 & 4 & 0 \\ 0 & 8 & 0 \end{vmatrix} = 8\mathbf{k} \Rightarrow |\mathbf{v} \times \mathbf{a}| = 8 \text{ and } |\mathbf{v}(\ln 2)| = \sqrt{17} \Rightarrow \kappa(\ln 2) = \frac{8}{17\sqrt{17}}; \dot{\mathbf{a}} = 4e^{2t}\mathbf{j}$

$\Rightarrow \dot{\mathbf{a}}(\ln 2) = 16\mathbf{j} \Rightarrow \tau(\ln 2) = \dfrac{\begin{vmatrix} 1 & 4 & 0 \\ 0 & 8 & 0 \\ 0 & 16 & 0 \end{vmatrix}}{|\mathbf{v} \times \mathbf{a}|^2} = 0; t = \ln 2 \Rightarrow (\ln 2, 2, 0)$ is on the curve

27. $\mathbf{r} = (2 + 3t + 3t^2)\mathbf{i} + (4t + 4t^2)\mathbf{j} - (6 \cos t)\mathbf{k} \Rightarrow \mathbf{v} = (3 + 6t)\mathbf{i} + (4 + 8t)\mathbf{j} + (6 \sin t)\mathbf{k}$

$\Rightarrow |\mathbf{v}| = \sqrt{(3 + 6t)^2 + (4 + 8t)^2 + (6 \sin t)^2} = \sqrt{25 + 100t + 100t^2 + 36 \sin^2 t}$

$\Rightarrow \frac{d|\mathbf{v}|}{dt} = \frac{1}{2}(25 + 100t + 100t^2 + 36 \sin^2 t)^{-1/2}(100 + 200t + 72 \sin t \cos t) \Rightarrow a_T(0) = \frac{d|\mathbf{v}|}{dt}(0) = 10;$

$\mathbf{a} = 6\mathbf{i} + 8\mathbf{j} + (6 \cos t)\mathbf{k} \Rightarrow |\mathbf{a}| = \sqrt{6^2 + 8^2 + (6 \cos t)^2} = \sqrt{100 + 36 \cos^2 t} \Rightarrow |\mathbf{a}(0)| = \sqrt{136}$

$\Rightarrow a_N = \sqrt{|\mathbf{a}|^2 - a_T^2} = \sqrt{136 - 10^2} = \sqrt{36} = 6 \Rightarrow \mathbf{a}(0) = 10\mathbf{T} + 6\mathbf{N}$

29. $\mathbf{r} = (\sin t)\mathbf{i} + \left(\sqrt{2} \cos t\right)\mathbf{j} + (\sin t)\mathbf{k} \Rightarrow \mathbf{v} = (\cos t)\mathbf{i} - \left(\sqrt{2} \sin t\right)\mathbf{j} + (\cos t)\mathbf{k}$

$\Rightarrow |\mathbf{v}| = \sqrt{(\cos t)^2 + \left(-\sqrt{2} \sin t\right)^2 + (\cos t)^2} = \sqrt{2} \Rightarrow \mathbf{T} = \frac{\mathbf{v}}{|\mathbf{v}|} = \left(\frac{1}{\sqrt{2}} \cos t\right)\mathbf{i} - (\sin t)\mathbf{j} + \left(\frac{1}{\sqrt{2}} \cos t\right)\mathbf{k};$

$\frac{d\mathbf{T}}{dt} = \left(-\frac{1}{\sqrt{2}} \sin t\right)\mathbf{i} - (\cos t)\mathbf{j} - \left(\frac{1}{\sqrt{2}} \sin t\right)\mathbf{k} \Rightarrow \left|\frac{d\mathbf{T}}{dt}\right| = \sqrt{\left(-\frac{1}{\sqrt{2}} \sin t\right)^2 + (-\cos t)^2 + \left(-\frac{1}{\sqrt{2}} \sin t\right)^2} = 1$

$\Rightarrow \mathbf{N} = \frac{\left(\frac{d\mathbf{T}}{dt}\right)}{\left|\frac{d\mathbf{T}}{dt}\right|} = \left(-\frac{1}{\sqrt{2}}\sin t\right)\mathbf{i} - (\cos t)\mathbf{j} - \left(\frac{1}{\sqrt{2}}\sin t\right)\mathbf{k}$; $\mathbf{B} = \mathbf{T}\times\mathbf{N} = \begin{vmatrix} \mathbf{i} & \mathbf{j} & \mathbf{k} \\ \frac{1}{\sqrt{2}}\cos t & -\sin t & \frac{1}{\sqrt{2}}\cos t \\ -\frac{1}{\sqrt{2}}\sin t & -\cos t & -\frac{1}{\sqrt{2}}\sin t \end{vmatrix}$

$= \frac{1}{\sqrt{2}}\mathbf{i} - \frac{1}{\sqrt{2}}\mathbf{k}$; $\mathbf{a} = (-\sin t)\mathbf{i} - \left(\sqrt{2}\cos t\right)\mathbf{j} - (\sin t)\mathbf{k} \Rightarrow \mathbf{v}\times\mathbf{a} = \begin{vmatrix} \mathbf{i} & \mathbf{j} & \mathbf{k} \\ \cos t & -\sqrt{2}\sin t & \cos t \\ -\sin t & -\sqrt{2}\cos t & -\sin t \end{vmatrix}$

$= \sqrt{2}\mathbf{i} - \sqrt{2}\mathbf{k} \Rightarrow |\mathbf{v}\times\mathbf{a}| = \sqrt{4} = 2 \Rightarrow \kappa = \frac{|\mathbf{v}\times\mathbf{a}|}{|\mathbf{v}|^3} = \frac{2}{\left(\sqrt{2}\right)^3} = \frac{1}{\sqrt{2}}$; $\dot{\mathbf{a}} = (-\cos t)\mathbf{i} + \left(\sqrt{2}\sin t\right)\mathbf{j} - (\cos t)\mathbf{k}$

$\Rightarrow \tau = \frac{\begin{vmatrix} \cos t & -\sqrt{2}\sin t & \cos t \\ -\sin t & -\sqrt{2}\cos t & -\sin t \\ -\cos t & \sqrt{2}\sin t & -\cos t \end{vmatrix}}{|\mathbf{v}\times\mathbf{a}|^2} = \frac{(\cos t)\left(\sqrt{2}\right) - \left(\sqrt{2}\sin t\right)(0) + (\cos t)\left(-\sqrt{2}\right)}{4} = 0$

31. $\mathbf{r} = 2\mathbf{i} + \left(4\sin\frac{t}{2}\right)\mathbf{j} + \left(3 - \frac{t}{\pi}\right)\mathbf{k} \Rightarrow 0 = \mathbf{r}\cdot(\mathbf{i} - \mathbf{j}) = 2(1) + \left(4\sin\frac{t}{2}\right)(-1) \Rightarrow 0 = 2 - 4\sin\frac{t}{2} \Rightarrow \sin\frac{t}{2} = \frac{1}{2} \Rightarrow \frac{t}{2} = \frac{\pi}{6}$
$\Rightarrow t = \frac{\pi}{3}$ (for the first time)

33. $\mathbf{r} = e^t\mathbf{i} + (\sin t)\mathbf{j} + \ln(1 - t)\mathbf{k} \Rightarrow \mathbf{v} = e^t\mathbf{i} + (\cos t)\mathbf{j} - \left(\frac{1}{1-t}\right)\mathbf{k} \Rightarrow \mathbf{v}(0) = \mathbf{i} + \mathbf{j} - \mathbf{k}$; $\mathbf{r}(0) = \mathbf{i} \Rightarrow (1, 0, 0)$ is on the line
$\Rightarrow x = 1 + t$, $y = t$, and $z = -t$ are parametric equations of the line

35. (a) $\Delta SOT \approx \Delta TOD \Rightarrow \frac{DO}{OT} = \frac{OT}{SO} \Rightarrow \frac{y_0}{6380} = \frac{6380}{6380+437}$
$\Rightarrow y_0 = \frac{6380^2}{6817} \Rightarrow y_0 \approx 5971$ km;

(b) $VA = \int_{5971}^{6380} 2\pi x \sqrt{1 + \left(\frac{dx}{dy}\right)^2}\, dy$

$= 2\pi\int_{5971}^{6817} \sqrt{6380^2 - y^2}\left(\frac{6380}{\sqrt{6380^2 - y^2}}\right)dy$

$= 2\pi\int_{5971}^{6817} 6380\, dy = 2\pi\left[6380y\right]_{5971}^{6817}$
$= 16{,}395{,}469\text{ km}^2 \approx 1.639\times 10^7\text{ km}^2$;

(c) percentage visible $\approx \frac{16{,}395{,}469\text{ km}^2}{4\pi(6380\text{ km})^2} \approx 3.21\%$

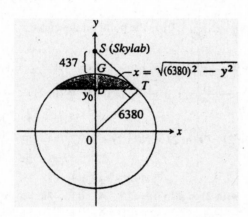

CHAPTER 13 ADDITIONAL AND ADVANCED EXERCISES

1. (a) The velocity of the boat at (x, y) relative to land is the sum of the velocity due to the rower and the
velocity of the river, or $\mathbf{v} = \left[-\frac{1}{250}(y - 50)^2 + 10\right]\mathbf{i} - 20\mathbf{j}$. Now, $\frac{dy}{dt} = -20 \Rightarrow y = -20t + c$; $y(0) = 100$
$\Rightarrow c = 100 \Rightarrow y = -20t + 100 \Rightarrow \mathbf{v} = \left[-\frac{1}{250}(-20t + 50)^2 + 10\right]\mathbf{i} - 20\mathbf{j} = \left(-\frac{8}{5}t^2 + 8t\right)\mathbf{i} - 20\mathbf{j}$
$\Rightarrow \mathbf{r}(t) = \left(-\frac{8}{15}t^3 + 4t^2\right)\mathbf{i} - 20t\mathbf{j} + \mathbf{C}_1$; $\mathbf{r}(0) = 0\mathbf{i} + 100\mathbf{j} \Rightarrow 100\mathbf{j} = \mathbf{C}_1 \Rightarrow \mathbf{r}(t)$
$= \left(-\frac{8}{15}t^3 + 4t^2\right)\mathbf{i} + (100 - 20t)\mathbf{j}$

(b) The boat reaches the shore when $y = 0 \Rightarrow 0 = -20t + 100$ from part (a) $\Rightarrow t = 5$
$\Rightarrow \mathbf{r}(5) = \left(-\frac{8}{15}\cdot 125 + 4\cdot 25\right)\mathbf{i} + (100 - 20\cdot 5)\mathbf{j} = \left(-\frac{200}{3} + 100\right)\mathbf{i} = \frac{100}{3}\mathbf{i}$; the distance downstream is
therefore $\frac{100}{3}$ m

3. (a) $\mathbf{r}(\theta) = (a\cos\theta)\mathbf{i} + (a\sin\theta)\mathbf{j} + b\theta\mathbf{k} \Rightarrow \frac{d\mathbf{r}}{dt} = [(-a\sin\theta)\mathbf{i} + (a\cos\theta)\mathbf{j} + b\mathbf{k}]\frac{d\theta}{dt}$; $|\mathbf{v}| = \sqrt{2gz} = \left|\frac{d\mathbf{r}}{dt}\right|$
$= \sqrt{a^2 + b^2}\,\frac{d\theta}{dt} \Rightarrow \frac{d\theta}{dt} = \sqrt{\frac{2gz}{a^2 + b^2}} = \sqrt{\frac{2gb\theta}{a^2 + b^2}} \Rightarrow \frac{d\theta}{dt}\Big|_{\theta=2\pi} = \sqrt{\frac{4\pi gb}{a^2 + b^2}} = 2\sqrt{\frac{\pi gb}{a^2 + b^2}}$

(b) $\frac{d\theta}{dt} = \sqrt{\frac{2gb\theta}{a^2 + b^2}} \Rightarrow \frac{d\theta}{\sqrt{\theta}} = \sqrt{\frac{2gb}{a^2 + b^2}}\, dt \Rightarrow 2\theta^{1/2} = \sqrt{\frac{2gb}{a^2 + b^2}}\, t + C$; $t = 0 \Rightarrow \theta = 0 \Rightarrow C = 0$
$\Rightarrow 2\theta^{1/2} = \sqrt{\frac{2gb}{a^2 + b^2}}\, t \Rightarrow \theta = \frac{gbt^2}{2(a^2 + b^2)}$; $z = b\theta \Rightarrow z = \frac{gb^2t^2}{2(a^2 + b^2)}$

(c) $\mathbf{v}(t) = \frac{d\mathbf{r}}{dt} = [(-a\sin\theta)\mathbf{i} + (a\cos\theta)\mathbf{j} + b\mathbf{k}]\frac{d\theta}{dt} = [(-a\sin\theta)\mathbf{i} + (a\cos\theta)\mathbf{j} + b\mathbf{k}]\left(\frac{gbt}{a^2+b^2}\right)$, from part (b)

$\Rightarrow \mathbf{v}(t) = \left[\frac{(-a\sin\theta)\mathbf{i}+(a\cos\theta)\mathbf{j}+b\mathbf{k}}{\sqrt{a^2+b^2}}\right]\left(\frac{gbt}{\sqrt{a^2+b^2}}\right) = \frac{gbt}{\sqrt{a^2+b^2}}\,\mathbf{T}$;

$\frac{d^2\mathbf{r}}{dt^2} = [(-a\cos\theta)\mathbf{i} - (a\sin\theta)\mathbf{j}]\left(\frac{d\theta}{dt}\right)^2 + [(-a\sin\theta)\mathbf{i} + (a\cos\theta)\mathbf{j} + b\mathbf{k}]\frac{d^2\theta}{dt^2}$

$= \left(\frac{gbt}{a^2+b^2}\right)^2[(-a\cos\theta)\mathbf{i} - (a\sin\theta)\mathbf{j}] + [(-a\sin\theta)\mathbf{i} + (a\cos\theta)\mathbf{j} + b\mathbf{k}]\left(\frac{gb}{a^2+b^2}\right)$

$= \left[\frac{(-a\sin\theta)\mathbf{i}+(a\cos\theta)\mathbf{j}+b\mathbf{k}}{\sqrt{a^2+b^2}}\right]\left(\frac{gb}{\sqrt{a^2+b^2}}\right) + a\left(\frac{gbt}{a^2+b^2}\right)^2[(-\cos\theta)\mathbf{i} - (\sin\theta)\mathbf{j}]$

$= \frac{gb}{\sqrt{a^2+b^2}}\,\mathbf{T} + a\left(\frac{gbt}{a^2+b^2}\right)^2\mathbf{N}$ (there is no component in the direction of $\mathbf{B}$).

5. $r = \frac{(1+e)r_0}{1+e\cos\theta} \Rightarrow \frac{dr}{d\theta} = \frac{(1+e)r_0(e\sin\theta)}{(1+e\cos\theta)^2}$; $\frac{dr}{d\theta} = 0 \Rightarrow \frac{(1+e)r_0(e\sin\theta)}{(1+e\cos\theta)^2} = 0 \Rightarrow (1+e)r_0(e\sin\theta) = 0$

$\Rightarrow \sin\theta = 0 \Rightarrow \theta = 0$ or π. Note that $\frac{dr}{d\theta} > 0$ when $\sin\theta > 0$ and $\frac{dr}{d\theta} < 0$ when $\sin\theta < 0$. Since $\sin\theta < 0$ on

$-\pi < \theta < 0$ and $\sin\theta > 0$ on $0 < \theta < \pi$, r is a minimum when $\theta = 0$ and $r(0) = \frac{(1+e)r_0}{1+e\cos 0} = r_0$

7. (a) $\mathbf{v} = \frac{dx}{dt}\mathbf{i} + \frac{dy}{dt}\mathbf{j}$ and $\mathbf{v} = \frac{dr}{dt}\mathbf{u}_r + r\frac{d\theta}{dt}\mathbf{u}_\theta = \left(\frac{dr}{dt}\right)[(\cos\theta)\mathbf{i} + (\sin\theta)\mathbf{j}] + \left(r\frac{d\theta}{dt}\right)[(-\sin\theta)\mathbf{i} + (\cos\theta)\mathbf{j}] \Rightarrow \mathbf{v}\cdot\mathbf{i} = \frac{dx}{dt}$ and

$\mathbf{v}\cdot\mathbf{i} = \frac{dr}{dt}\cos\theta - r\frac{d\theta}{dt}\sin\theta \Rightarrow \frac{dx}{dt} = \frac{dr}{dt}\cos\theta - r\frac{d\theta}{dt}\sin\theta$; $\mathbf{v}\cdot\mathbf{j} = \frac{dy}{dt}$ and $\mathbf{v}\cdot\mathbf{j} = \frac{dr}{dt}\sin\theta + r\frac{d\theta}{dt}\cos\theta$

$\Rightarrow \frac{dy}{dt} = \frac{dr}{dt}\sin\theta + r\frac{d\theta}{dt}\cos\theta$

(b) $\mathbf{u}_r = (\cos\theta)\mathbf{i} + (\sin\theta)\mathbf{j} \Rightarrow \mathbf{v}\cdot\mathbf{u}_r = \frac{dx}{dt}\cos\theta + \frac{dy}{dt}\sin\theta$

$= \left(\frac{dr}{dt}\cos\theta - r\frac{d\theta}{dt}\sin\theta\right)(\cos\theta) + \left(\frac{dr}{dt}\sin\theta + r\frac{d\theta}{dt}\cos\theta\right)(\sin\theta)$ by part (a),

$\Rightarrow \mathbf{v}\cdot\mathbf{u}_r = \frac{dr}{dt}$; therefore, $\frac{dr}{dt} = \frac{dx}{dt}\cos\theta + \frac{dy}{dt}\sin\theta$;

$\mathbf{u}_\theta = -(\sin\theta)\mathbf{i} + (\cos\theta)\mathbf{j} \Rightarrow \mathbf{v}\cdot\mathbf{u}_\theta = -\frac{dx}{dt}\sin\theta + \frac{dy}{dt}\cos\theta$

$= \left(\frac{dr}{dt}\cos\theta - r\frac{d\theta}{dt}\sin\theta\right)(-\sin\theta) + \left(\frac{dr}{dt}\sin\theta + r\frac{d\theta}{dt}\cos\theta\right)(\cos\theta)$ by part (a) $\Rightarrow \mathbf{v}\cdot\mathbf{u}_\theta = r\frac{d\theta}{dt}$;

therefore, $r\frac{d\theta}{dt} = -\frac{dx}{dt}\sin\theta + \frac{dy}{dt}\cos\theta$

9. (a) Let $r = 2 - t$ and $\theta = 3t \Rightarrow \frac{dr}{dt} = -1$ and $\frac{d\theta}{dt} = 3 \Rightarrow \frac{d^2r}{dt^2} = \frac{d^2\theta}{dt^2} = 0$. The halfway point is $(1,3) \Rightarrow t = 1$;

$\mathbf{v} = \frac{dr}{dt}\mathbf{u}_r + r\frac{d\theta}{dt}\mathbf{u}_\theta \Rightarrow \mathbf{v}(1) = -\mathbf{u}_r + 3\mathbf{u}_\theta$; $\mathbf{a} = \left[\frac{d^2r}{dt^2} - r\left(\frac{d\theta}{dt}\right)^2\right]\mathbf{u}_r + \left[r\frac{d^2\theta}{dt^2} + 2\frac{dr}{dt}\frac{d\theta}{dt}\right]\mathbf{u}_\theta \Rightarrow \mathbf{a}(1) = -9\mathbf{u}_r - 6\mathbf{u}_\theta$

(b) It takes the beetle 2 min to crawl to the origin $\Rightarrow$ the rod has revolved 6 radians

$\Rightarrow L = \int_0^6 \sqrt{[f(\theta)]^2 + [f'(\theta)]^2}\,d\theta = \int_0^6 \sqrt{\left(2 - \frac{\theta}{3}\right)^2 + \left(-\frac{1}{3}\right)^2}\,d\theta = \int_0^6 \sqrt{4 - \frac{4\theta}{3} + \frac{\theta^2}{9} + \frac{1}{9}}\,d\theta$

$= \int_0^6 \sqrt{\frac{37 - 12\theta + \theta^2}{9}}\,d\theta = \frac{1}{3}\int_0^6 \sqrt{(\theta - 6)^2 + 1}\,d\theta = \frac{1}{3}\left[\frac{(\theta-6)}{2}\sqrt{(\theta-6)^2+1} + \frac{1}{2}\ln\left|\theta - 6 + \sqrt{(\theta-6)^2+1}\right|\right]_0^6$

$= \sqrt{37} - \frac{1}{6}\ln\left(\sqrt{37} - 6\right) \approx 6.5$ in.

11. (a) $\mathbf{u}_r \times \mathbf{u}_\theta = \begin{vmatrix} \mathbf{i} & \mathbf{j} & \mathbf{k} \\ \cos\theta & \sin\theta & 0 \\ -\sin\theta & \cos\theta & 0 \end{vmatrix} = \mathbf{k} \Rightarrow$ a right-handed frame of unit vectors

(b) $\frac{d\mathbf{u}_r}{d\theta} = (-\sin\theta)\mathbf{i} + (\cos\theta)\mathbf{j} = \mathbf{u}_\theta$ and $\frac{d\mathbf{u}_\theta}{d\theta} = (-\cos\theta)\mathbf{i} - (\sin\theta)\mathbf{j} = -\mathbf{u}_r$

(c) From Eq. (7), $\mathbf{v} = \dot{r}\mathbf{u}_r + r\dot{\theta}\mathbf{u}_\theta + \dot{z}\mathbf{k} \Rightarrow \mathbf{a} = \dot{\mathbf{v}} = (\ddot{r}\mathbf{u}_r + \dot{r}\dot{\mathbf{u}}_r) + (\dot{r}\dot{\theta}\mathbf{u}_\theta + r\ddot{\theta}\mathbf{u}_\theta + r\dot{\theta}\dot{\mathbf{u}}_\theta) + \ddot{z}\mathbf{k}$

$= \left(\ddot{r} - r\dot{\theta}^2\right)\mathbf{u}_r + \left(r\ddot{\theta} + 2\dot{r}\dot{\theta}\right)\mathbf{u}_\theta + \ddot{z}\mathbf{k}$

NOTES:

CHAPTER 14 PARTIAL DERIVATIVES

14.1 FUNCTIONS OF SEVERAL VARIABLES

1. (a) Domain: all points in the xy-plane
   (b) Range: all real numbers
   (c) level curves are straight lines $y - x = c$ parallel to the line $y = x$
   (d) no boundary points
   (e) both open and closed
   (f) unbounded

3. (a) Domain: all points in the xy-plane
   (b) Range: $z \geq 0$
   (c) level curves: for $f(x, y) = 0$, the origin; for $f(x, y) = c > 0$, ellipses with center $(0, 0)$ and major and minor axes along the x- and y-axes, respectively
   (d) no boundary points
   (e) both open and closed
   (f) unbounded

5. (a) Domain: all points in the xy-plane
   (b) Range: all real numbers
   (c) level curves are hyperbolas with the x- and y-axes as asymptotes when $f(x, y) \neq 0$, and the x- and y-axes when $f(x, y) = 0$
   (d) no boundary points
   (e) both open and closed
   (f) unbounded

7. (a) Domain: all (x, y) satisfying $x^2 + y^2 < 16$
   (b) Range: $z \geq \frac{1}{4}$
   (c) level curves are circles centered at the origin with radii $r < 4$
   (d) boundary is the circle $x^2 + y^2 = 16$
   (e) open
   (f) bounded

9. (a) Domain: $(x, y) \neq (0, 0)$
   (b) Range: all real numbers
   (c) level curves are circles with center $(0, 0)$ and radii $r > 0$
   (d) boundary is the single point $(0, 0)$
   (e) open
   (f) unbounded

11. (a) Domain: all (x, y) satisfying $-1 \leq y - x \leq 1$
    (b) Range: $-\frac{\pi}{2} \leq z \leq \frac{\pi}{2}$
    (c) level curves are straight lines of the form $y - x = c$ where $-1 \leq c \leq 1$
    (d) boundary is the two straight lines $y = 1 + x$ and $y = -1 + x$
    (e) closed
    (f) unbounded

13. f   **15.** a   **17.** d

19. (a)

(b)

21. (a)

(b)

23. (a)

(b)

25. (a)

(b)

27. (a)

(b)

29. $f(x, y) = 16 - x^2 - y^2$ and $\left(2\sqrt{2}, \sqrt{2}\right) \Rightarrow z = 16 - \left(2\sqrt{2}\right)^2 - \left(\sqrt{2}\right)^2 = 6 \Rightarrow 6 = 16 - x^2 - y^2 \Rightarrow x^2 + y^2 = 10$

31. $f(x, y) = \int_x^y \frac{1}{1+t^2}\, dt$ at $\left(-\sqrt{2}, \sqrt{2}\right) \Rightarrow z = \tan^{-1} y - \tan^{-1} x$; at $\left(-\sqrt{2}, \sqrt{2}\right) \Rightarrow z = \tan^{-1}\sqrt{2} - \tan^{-1}\left(-\sqrt{2}\right)$
    $= 2\tan^{-1}\sqrt{2} \Rightarrow \tan^{-1} y - \tan^{-1} x = 2\tan^{-1}\sqrt{2}$

33.

$f(x, y, z) = x^2 + y^2 + z^2 = 1$

35.

$f(x, y, z) = x + z = 1$

37.

$f(x, y, z) = x^2 + y^2 = 1$

39.

$f(x, y, z) = z - x^2 - y^2 = 1$
or $z = x^2 + y^2 + 1$

41. $f(x, y, z) = \sqrt{x - y} - \ln z$ at $(3, -1, 1) \Rightarrow w = \sqrt{x - y} - \ln z$; at $(3, -1, 1) \Rightarrow w = \sqrt{3 - (-1)} - \ln 1 = 2$
    $\Rightarrow \sqrt{x - y} - \ln z = 2$

43. $g(x, y, z) = \sum_{n=0}^{\infty} \frac{(x+y)^n}{n!\, z^n}$ at $(\ln 2, \ln 4, 3) \Rightarrow w = \sum_{n=0}^{\infty} \frac{(x+y)^n}{n!\, z^n} = e^{(x+y)/z}$; at $(\ln 2, \ln 4, 3) \Rightarrow w = e^{(\ln 2 + \ln 4)/3}$
    $= e^{(\ln 8)/3} = e^{\ln 2} = 2 \Rightarrow 2 = e^{(x+y)/z} \Rightarrow \frac{x+y}{z} = \ln 2$

45. $f(x, y, z) = xyz$ and $x = 20 - t$, $y = t$, $z = 20 \Rightarrow w = (20 - t)(t)(20)$ along the line $\Rightarrow w = 400t - 20t^2$
    $\Rightarrow \frac{dw}{dt} = 400 - 40t$; $\frac{dw}{dt} = 0 \Rightarrow 400 - 40t = 0 \Rightarrow t = 10$ and $\frac{d^2w}{dt^2} = -40$ for all $t \Rightarrow$ yes, maximum at $t = 10$
    $\Rightarrow x = 20 - 10 = 10$, $y = 10$, $z = 20 \Rightarrow$ maximum of f along the line is $f(10, 10, 20) = (10)(10)(20) = 2000$

47. $w = 4\left(\frac{Th}{d}\right)^{1/2} = 4\left[\frac{(290\text{ K})(16.8\text{ km})}{5\text{ K/km}}\right]^{1/2} \approx 124.86\text{ km} \Rightarrow$ must be $\frac{1}{2}(124.86) \approx 63$ km south of Nantucket

14.2 LIMITS AND CONTINUITY

1. $\displaystyle\lim_{(x,y)\to(0,0)} \frac{3x^2-y^2+5}{x^2+y^2+2} = \frac{3(0)^2-0^2+5}{0^2+0^2+2} = \frac{5}{2}$

3. $\displaystyle\lim_{(x,y)\to(3,4)} \sqrt{x^2+y^2-1} = \sqrt{3^2+4^2-1} = \sqrt{24} = 2\sqrt{6}$

5. $\displaystyle\lim_{(x,y)\to\left(0,\frac{\pi}{4}\right)} \sec x \tan y = (\sec 0)\left(\tan \frac{\pi}{4}\right) = (1)(1) = 1$

7. $\displaystyle\lim_{(x,y)\to(0,\ln 2)} e^{x-y} = e^{0-\ln 2} = e^{\ln\left(\frac{1}{2}\right)} = \frac{1}{2}$

9. $\displaystyle\lim_{(x,y)\to(0,0)} \frac{e^y \sin x}{x} = \lim_{(x,y)\to(0,0)} (e^y)\left(\frac{\sin x}{x}\right) = e^0 \cdot \lim_{x\to 0}\left(\frac{\sin x}{x}\right) = 1\cdot 1 = 1$

11. $\displaystyle\lim_{(x,y)\to(1,0)} \frac{x\sin y}{x^2+1} = \frac{1\cdot\sin 0}{1^2+1} = \frac{0}{2} = 0$

13. $\displaystyle\lim_{\substack{(x,y)\to(1,1)\\ x\ne y}} \frac{x^2-2xy+y^2}{x-y} = \lim_{(x,y)\to(1,1)} \frac{(x-y)^2}{x-y} = \lim_{(x,y)\to(1,1)} (x-y) = (1-1) = 0$

15. $\displaystyle\lim_{\substack{(x,y)\to(1,1)\\ x\ne 1}} \frac{xy-y-2x+2}{x-1} = \lim_{\substack{(x,y)\to(1,1)\\ x\ne 1}} \frac{(x-1)(y-2)}{x-1} = \lim_{(x,y)\to(1,1)} (y-2) = (1-2) = -1$

17. $\displaystyle\lim_{\substack{(x,y)\to(0,0)\\ x\ne y}} \frac{x-y+2\sqrt{x}-2\sqrt{y}}{\sqrt{x}-\sqrt{y}} = \lim_{\substack{(x,y)\to(0,0)\\ x\ne y}} \frac{(\sqrt{x}-\sqrt{y})(\sqrt{x}+\sqrt{y}+2)}{\sqrt{x}-\sqrt{y}} = \lim_{(x,y)\to(0,0)} (\sqrt{x}+\sqrt{y}+2)$

$= \left(\sqrt{0}+\sqrt{0}+2\right) = 2$

Note: (x,y) must approach $(0,0)$ through the first quadrant only with $x\ne y$.

19. $\displaystyle\lim_{\substack{(x,y)\to(2,0)\\ 2x-y\ne 4}} \frac{\sqrt{2x-y}-2}{2x-y-4} = \lim_{\substack{(x,y)\to(2,0)\\ 2x-y\ne 4}} \frac{\sqrt{2x-y}-2}{(\sqrt{2x-y}+2)(\sqrt{2x-y}-2)} = \lim_{(x,y)\to(2,0)} \frac{1}{\sqrt{2x-y}+2}$

$= \frac{1}{\sqrt{(2)(2)-0}+2} = \frac{1}{2+2} = \frac{1}{4}$

21. $\displaystyle\lim_{P\to(1,3,4)} \left(\frac{1}{x}+\frac{1}{y}+\frac{1}{z}\right) = \frac{1}{1}+\frac{1}{3}+\frac{1}{4} = \frac{12+4+3}{12} = \frac{19}{12}$

23. $\displaystyle\lim_{P\to(3,3,0)} (\sin^2 x + \cos^2 y + \sec^2 z) = (\sin^2 3 + \cos^2 3) + \sec^2 0 = 1 + 1^2 = 2$

25. $\displaystyle\lim_{P\to(\pi,0,3)} ze^{-2y}\cos 2x = 3e^{-2(0)}\cos 2\pi = (3)(1)(1) = 3$

27. (a) All (x,y)

    (b) All (x,y) except $(0,0)$

29. (a) All (x,y) except where $x=0$ or $y=0$

    (b) All (x,y)

31. (a) All (x,y,z)

(b) All (x, y, z) except the interior of the cylinder $x^2 + y^2 = 1$

33. (a) All (x, y, z) with $z \neq 0$

(b) All (x, y, z) with $x^2 + z^2 \neq 1$

35. $\displaystyle\lim_{\substack{(x,y) \to (0,0) \\ \text{along } y = x \\ x > 0}} -\frac{x}{\sqrt{x^2+y^2}} = \lim_{x \to 0^+} -\frac{x}{\sqrt{x^2+x^2}} = \lim_{x \to 0^+} -\frac{x}{\sqrt{2}\,|x|} = \lim_{x \to 0^+} -\frac{x}{\sqrt{2}\,x} = \lim_{x \to 0^+} -\frac{1}{\sqrt{2}} = -\frac{1}{\sqrt{2}}$;

$\displaystyle\lim_{\substack{(x,y) \to (0,0) \\ \text{along } y = x \\ x < 0}} -\frac{x}{\sqrt{x^2+y^2}} = \lim_{x \to 0^-} -\frac{x}{\sqrt{2}\,|x|} = \lim_{x \to 0^-} -\frac{x}{\sqrt{2}(-x)} = \lim_{x \to 0^-} \frac{1}{\sqrt{2}} = \frac{1}{\sqrt{2}}$

37. $\displaystyle\lim_{\substack{(x,y) \to (0,0) \\ \text{along } y = kx^2}} \frac{x^4 - y^2}{x^4 + y^2} = \lim_{x \to 0} \frac{x^4 - (kx^2)^2}{x^4 + (kx^2)^2} = \lim_{x \to 0} \frac{x^4 - k^2x^4}{x^4 + k^2x^4} = \frac{1-k^2}{1+k^2} \Rightarrow$ different limits for different values of k

39. $\displaystyle\lim_{\substack{(x,y) \to (0,0) \\ \text{along } y = kx \\ k \neq -1}} \frac{x - y}{x + y} = \lim_{x \to 0} \frac{x - kx}{x + kx} = \frac{1-k}{1+k} \Rightarrow$ different limits for different values of k, $k \neq -1$

41. $\displaystyle\lim_{\substack{(x,y) \to (0,0) \\ \text{along } y = kx^2 \\ k \neq 0}} \frac{x^2 + y}{y} = \lim_{x \to 0} \frac{x^2 + kx^2}{kx^2} = \frac{1+k}{k} \Rightarrow$ different limits for different values of k, $k \neq 0$

43. No, the limit depends only on the values $f(x, y)$ has when $(x, y) \neq (x_0, y_0)$

45. $\displaystyle\lim_{(x,y) \to (0,0)} \left(1 - \frac{x^2y^2}{3}\right) = 1$ and $\displaystyle\lim_{(x,y) \to (0,0)} 1 = 1 \Rightarrow \displaystyle\lim_{(x,y) \to (0,0)} \frac{\tan^{-1} xy}{xy} = 1$, by the Sandwich Theorem

47. The limit is 0 since $\left|\sin\left(\frac{1}{x}\right)\right| \leq 1 \Rightarrow -1 \leq \sin\left(\frac{1}{x}\right) \leq 1 \Rightarrow -y \leq y\sin\left(\frac{1}{x}\right) \leq y$ for $y \geq 0$, and $-y \geq y\sin\left(\frac{1}{x}\right) \geq y$ for $y \leq 0$. Thus as $(x, y) \to (0, 0)$, both $-y$ and y approach $0 \Rightarrow y\sin\left(\frac{1}{x}\right) \to 0$, by the Sandwich Theorem.

49. (a) $f(x, y)\big|_{y=mx} = \frac{2m}{1+m^2} = \frac{2\tan\theta}{1+\tan^2\theta} = \sin 2\theta$. The value of $f(x, y) = \sin 2\theta$ varies with θ, which is the line's angle of inclination.

(b) Since $f(x, y)\big|_{y=mx} = \sin 2\theta$ and since $-1 \leq \sin 2\theta \leq 1$ for every θ, $\displaystyle\lim_{(x,y) \to (0,0)} f(x, y)$ varies from -1 to 1 along $y = mx$.

51. $\displaystyle\lim_{(x,y) \to (0,0)} \frac{x^3 - xy^2}{x^2 + y^2} = \lim_{r \to 0} \frac{r^3 \cos^3\theta - (r\cos\theta)(r^2\sin^2\theta)}{r^2\cos^2\theta + r^2\sin^2\theta} = \lim_{r \to 0} \frac{r(\cos^3\theta - \cos\theta\sin^2\theta)}{1} = 0$

53. $\displaystyle\lim_{(x,y) \to (0,0)} \frac{y^2}{x^2 + y^2} = \lim_{r \to 0} \frac{r^2\sin^2\theta}{r^2} = \lim_{r \to 0} (\sin^2\theta) = \sin^2\theta$; the limit does not exist since $\sin^2\theta$ is between 0 and 1 depending on θ

55. $\displaystyle\lim_{(x,y) \to (0,0)} \tan^{-1}\left[\frac{|x| + |y|}{x^2 + y^2}\right] = \lim_{r \to 0} \tan^{-1}\left[\frac{|r\cos\theta| + |r\sin\theta|}{r^2}\right] = \lim_{r \to 0} \tan^{-1}\left[\frac{|r|(|\cos\theta| + |\sin\theta|)}{r^2}\right]$;

if $r \to 0^+$, then $\displaystyle\lim_{r \to 0^+} \tan^{-1}\left[\frac{|r|(|\cos\theta| + |\sin\theta|)}{r^2}\right] = \lim_{r \to 0^+} \tan^{-1}\left[\frac{|\cos\theta| + |\sin\theta|}{r}\right] = \frac{\pi}{2}$; if $r \to 0^-$, then

$\displaystyle\lim_{r \to 0^-} \tan^{-1}\left[\frac{|r|(|\cos\theta| + |\sin\theta|)}{r^2}\right] = \lim_{r \to 0^-} \tan^{-1}\left(\frac{|\cos\theta| + |\sin\theta|}{-r}\right) = \frac{\pi}{2} \Rightarrow$ the limit is $\frac{\pi}{2}$

57. $\displaystyle\lim_{(x,y) \to (0,0)} \ln\left(\frac{3x^2 - x^2y^2 + 3y^2}{x^2 + y^2}\right) = \lim_{r \to 0} \ln\left(\frac{3r^2 \cos^2\theta - r^4 \cos^2\theta \sin^2\theta + 3r^2 \sin^2\theta}{r^2}\right)$

    $= \displaystyle\lim_{r \to 0} \ln\left(3 - r^2 \cos^2\theta \sin^2\theta\right) = \ln 3 \Rightarrow$ define $f(0,0) = \ln 3$

59. Let $\delta = 0.1$. Then $\sqrt{x^2 + y^2} < \delta \Rightarrow \sqrt{x^2 + y^2} < 0.1 \Rightarrow x^2 + y^2 < 0.01 \Rightarrow |x^2 + y^2 - 0| < 0.01 \Rightarrow |f(x,y) - f(0,0)|$

    $< 0.01 = \epsilon.$

61. Let $\delta = 0.005$. Then $|x| < \delta$ and $|y| < \delta \Rightarrow |f(x,y) - f(0,0)| = \left|\frac{x+y}{x^2+1} - 0\right| = \left|\frac{x+y}{x^2+1}\right| \le |x+y| < |x| + |y|$

    $< 0.005 + 0.005 = 0.01 = \epsilon.$

63. Let $\delta = \sqrt{0.015}$. Then $\sqrt{x^2 + y^2 + z^2} < \delta \Rightarrow |f(x,y,z) - f(0,0,0)| = |x^2 + y^2 + z^2 - 0| = |x^2 + y^2 + z^2|$

    $= \left(\sqrt{x^2 + t^2 + x^2}\right)^2 < \left(\sqrt{0.015}\right)^2 = 0.015 = \epsilon.$

65. Let $\delta = 0.005$. Then $|x| < \delta$, $|y| < \delta$, and $|z| < \delta \Rightarrow |f(x,y,z) - f(0,0,0)| = \left|\frac{x+y+z}{x^2+y^2+z^2+1} - 0\right|$

    $= \left|\frac{x+y+z}{x^2+y^2+z^2+1}\right| \le |x+y+z| \le |x| + |y| + |z| < 0.005 + 0.005 + 0.005 = 0.015 = \epsilon.$

67. $\displaystyle\lim_{(x,y,z) \to (x_0,y_0,z_0)} f(x,y,z) = \lim_{(x,y,z) \to (x_0,y_0,z_0)} (x+y+z) = x_0 + y_0 + z_0 = f(x_0,y_0,z_0) \Rightarrow$ f is continuous at

    every (x_0, y_0, z_0)

14.3 PARTIAL DERIVATIVES

1. $\frac{\partial f}{\partial x} = 4x$, $\frac{\partial f}{\partial y} = -3$

3. $\frac{\partial f}{\partial x} = 2x(y+2)$, $\frac{\partial f}{\partial y} = x^2 - 1$

5. $\frac{\partial f}{\partial x} = 2y(xy - 1)$, $\frac{\partial f}{\partial y} = 2x(xy - 1)$

7. $\frac{\partial f}{\partial x} = \frac{x}{\sqrt{x^2 + y^2}}$, $\frac{\partial f}{\partial y} = \frac{y}{\sqrt{x^2 + y^2}}$

9. $\frac{\partial f}{\partial x} = -\frac{1}{(x+y)^2} \cdot \frac{\partial}{\partial x}(x+y) = -\frac{1}{(x+y)^2}$, $\frac{\partial f}{\partial y} = -\frac{1}{(x+y)^2} \cdot \frac{\partial}{\partial y}(x+y) = -\frac{1}{(x+y)^2}$

11. $\frac{\partial f}{\partial x} = \frac{(xy-1)(1) - (x+y)(y)}{(xy-1)^2} = \frac{-y^2 - 1}{(xy-1)^2}$, $\frac{\partial f}{\partial y} = \frac{(xy-1)(1) - (x+y)(x)}{(xy-1)^2} = \frac{-x^2 - 1}{(xy-1)^2}$

13. $\frac{\partial f}{\partial x} = e^{(x+y+1)} \cdot \frac{\partial}{\partial x}(x+y+1) = e^{(x+y+1)}$, $\frac{\partial f}{\partial y} = e^{(x+y+1)} \cdot \frac{\partial}{\partial y}(x+y+1) = e^{(x+y+1)}$

15. $\frac{\partial f}{\partial x} = \frac{1}{x+y} \cdot \frac{\partial}{\partial x}(x+y) = \frac{1}{x+y}$, $\frac{\partial f}{\partial y} = \frac{1}{x+y} \cdot \frac{\partial}{\partial y}(x+y) = \frac{1}{x+y}$

17. $\frac{\partial f}{\partial x} = 2\sin(x-3y) \cdot \frac{\partial}{\partial x}\sin(x-3y) = 2\sin(x-3y)\cos(x-3y) \cdot \frac{\partial}{\partial x}(x-3y) = 2\sin(x-3y)\cos(x-3y)$,

    $\frac{\partial f}{\partial y} = 2\sin(x-3y) \cdot \frac{\partial}{\partial y}\sin(x-3y) = 2\sin(x-3y)\cos(x-3y) \cdot \frac{\partial}{\partial y}(x-3y) = -6\sin(x-3y)\cos(x-3y)$

19. $\frac{\partial f}{\partial x} = yx^{y-1}$, $\frac{\partial f}{\partial y} = x^y \ln x$

21. $\frac{\partial f}{\partial x} = -g(x)$, $\frac{\partial f}{\partial y} = g(y)$

23. $f_x = 1 + y^2, f_y = 2xy, f_z = -4z$

25. $f_x = 1, f_y = -\dfrac{y}{\sqrt{y^2 + z^2}}, f_z = -\dfrac{z}{\sqrt{y^2 + z^2}}$

27. $f_x = \dfrac{yz}{\sqrt{1 - x^2 y^2 z^2}}, f_y = \dfrac{xz}{\sqrt{1 - x^2 y^2 z^2}}, f_z = \dfrac{xy}{\sqrt{1 - x^2 y^2 z^2}}$

29. $f_x = \dfrac{1}{x + 2y + 3z}, f_y = \dfrac{2}{x + 2y + 3z}, f_z = \dfrac{3}{x + 2y + 3z}$

31. $f_x = -2x e^{-(x^2 + y^2 + z^2)}, f_y = -2y e^{-(x^2 + y^2 + z^2)}, f_z = -2z e^{-(x^2 + y^2 + z^2)}$

33. $f_x = \text{sech}^2 (x + 2y + 3z), f_y = 2 \, \text{sech}^2 (x + 2y + 3z), f_z = 3 \, \text{sech}^2 (x + 2y + 3z)$

35. $\dfrac{\partial f}{\partial t} = -2\pi \sin(2\pi t - \alpha), \dfrac{\partial f}{\partial \alpha} = \sin(2\pi t - \alpha)$

37. $\dfrac{\partial h}{\partial \rho} = \sin \phi \cos \theta, \dfrac{\partial h}{\partial \phi} = \rho \cos \phi \cos \theta, \dfrac{\partial h}{\partial \theta} = -\rho \sin \phi \sin \theta$

39. $W_p = V, W_v = P + \dfrac{\delta v^2}{2g}, W_\delta = \dfrac{V v^2}{2g}, W_v = \dfrac{2V \delta v}{2g} = \dfrac{V \delta v}{g}, W_g = -\dfrac{V \delta v^2}{2g^2}$

41. $\dfrac{\partial f}{\partial x} = 1 + y, \dfrac{\partial f}{\partial y} = 1 + x, \dfrac{\partial^2 f}{\partial x^2} = 0, \dfrac{\partial^2 f}{\partial y^2} = 0, \dfrac{\partial^2 f}{\partial y \partial x} = \dfrac{\partial^2 f}{\partial x \partial y} = 1$

43. $\dfrac{\partial g}{\partial x} = 2xy + y \cos x, \dfrac{\partial g}{\partial y} = x^2 - \sin y + \sin x, \dfrac{\partial^2 g}{\partial x^2} = 2y - y \sin x, \dfrac{\partial^2 g}{\partial y^2} = -\cos y, \dfrac{\partial^2 g}{\partial y \partial x} = \dfrac{\partial^2 g}{\partial x \partial y} = 2x + \cos x$

45. $\dfrac{\partial r}{\partial x} = \dfrac{1}{x+y}, \dfrac{\partial r}{\partial y} = \dfrac{1}{x+y}, \dfrac{\partial^2 r}{\partial x^2} = \dfrac{-1}{(x+y)^2}, \dfrac{\partial^2 r}{\partial y^2} = \dfrac{-1}{(x+y)^2}, \dfrac{\partial^2 r}{\partial y \partial x} = \dfrac{\partial^2 r}{\partial x \partial y} = \dfrac{-1}{(x+y)^2}$

47. $\dfrac{\partial w}{\partial x} = \dfrac{2}{2x+3y}, \dfrac{\partial w}{\partial y} = \dfrac{3}{2x+3y}, \dfrac{\partial^2 w}{\partial y \partial x} = \dfrac{-6}{(2x+3y)^2},$ and $\dfrac{\partial^2 w}{\partial x \partial y} = \dfrac{-6}{(2x+3y)^2}$

49. $\dfrac{\partial w}{\partial x} = y^2 + 2xy^3 + 3x^2 y^4, \dfrac{\partial w}{\partial y} = 2xy + 3x^2 y^2 + 4x^3 y^3, \dfrac{\partial^2 w}{\partial y \partial x} = 2y + 6xy^2 + 12x^2 y^3,$ and
$\dfrac{\partial^2 w}{\partial x \partial y} = 2y + 6xy^2 + 12x^2 y^3$

51. (a) x first   (b) y first   (c) x first   (d) x first   (e) y first   (f) y first

53. $f_x(1, 2) = \lim\limits_{h \to 0} \dfrac{f(1 + h, 2) - f(1, 2)}{h} = \lim\limits_{h \to 0} \dfrac{[1 - (1+h) + 2 - 6(1+h)^2] - (2 - 6)}{h} = \lim\limits_{h \to 0} \dfrac{-h - 6(1 + 2h + h^2) + 6}{h}$
$= \lim\limits_{h \to 0} \dfrac{-13h - 6h^2}{h} = \lim\limits_{h \to 0} (-13 - 6h) = -13,$
$f_y(1, 2) = \lim\limits_{h \to 0} \dfrac{f(1, 2 + h) - f(1, 2)}{h} = \lim\limits_{h \to 0} \dfrac{[1 - 1 + (2 + h) - 3(2 + h)] - (2 - 6)}{h} = \lim\limits_{h \to 0} \dfrac{(2 - 6 - 2h) - (2 - 6)}{h}$
$= \lim\limits_{h \to 0} (-2) = -2$

55. $f_z(x_0, y_0, z_0) = \lim\limits_{h \to 0} \dfrac{f(x_0, y_0, z_0 + h) - f(x_0, y_0, z_0)}{h};$
$f_z(1, 2, 3) = \lim\limits_{h \to 0} \dfrac{f(1, 2, 3 + h) - f(1, 2, 3)}{h} = \lim\limits_{h \to 0} \dfrac{2(3 + h)^2 - 2(9)}{h} = \lim\limits_{h \to 0} \dfrac{12h + 2h^2}{h} = \lim\limits_{h \to 0} (12 + 2h) = 12$

57. $y + \left(3z^2 \dfrac{\partial z}{\partial x}\right) x + z^3 - 2y \dfrac{\partial z}{\partial x} = 0 \Rightarrow (3xz^2 - 2y) \dfrac{\partial z}{\partial x} = -y - z^3 \Rightarrow$ at $(1, 1, 1)$ we have $(3 - 2) \dfrac{\partial z}{\partial x} = -1 - 1$ or
$\dfrac{\partial z}{\partial x} = -2$

59. $a^2 = b^2 + c^2 - 2bc \cos A \Rightarrow 2a = (2bc \sin A)\frac{\partial A}{\partial a} \Rightarrow \frac{\partial A}{\partial a} = \frac{a}{bc \sin A}$; also $0 = 2b - 2c \cos A + (2bc \sin A)\frac{\partial A}{\partial b}$

$\Rightarrow 2c \cos A - 2b = (2bc \sin A)\frac{\partial A}{\partial b} \Rightarrow \frac{\partial A}{\partial b} = \frac{c \cos A - b}{bc \sin A}$

61. Differentiating each equation implicitly gives $1 = v_x \ln u + \left(\frac{v}{u}\right)u_x$ and $0 = u_x \ln v + \left(\frac{u}{v}\right)v_x$ or

$$\left. \begin{matrix} (\ln u)\,v_x & + \left(\frac{v}{u}\right)u_x = 1 \\ \left(\frac{u}{v}\right)v_x + (\ln v)\,u_x = 0 \end{matrix} \right\} \Rightarrow v_x = \frac{\begin{vmatrix} 1 & \frac{v}{u} \\ 0 & \ln v \end{vmatrix}}{\begin{vmatrix} \ln u & \frac{v}{u} \\ \frac{u}{v} & \ln v \end{vmatrix}} = \frac{\ln v}{(\ln u)(\ln v) - 1}$$

63. $\frac{\partial f}{\partial x} = 2x,\ \frac{\partial f}{\partial y} = 2y,\ \frac{\partial f}{\partial z} = -4z \Rightarrow \frac{\partial^2 f}{\partial x^2} = 2,\ \frac{\partial^2 f}{\partial y^2} = 2,\ \frac{\partial^2 f}{\partial z^2} = -4 \Rightarrow \frac{\partial^2 f}{\partial x^2} + \frac{\partial^2 f}{\partial y^2} + \frac{\partial^2 f}{\partial z^2} = 2 + 2 + (-4) = 0$

65. $\frac{\partial f}{\partial x} = -2e^{-2y} \sin 2x,\ \frac{\partial f}{\partial y} = -2e^{-2y} \cos 2x,\ \frac{\partial^2 f}{\partial x^2} = -4e^{-2y} \cos 2x,\ \frac{\partial^2 f}{\partial y^2} = 4e^{-2y} \cos 2x \Rightarrow \frac{\partial^2 f}{\partial x^2} + \frac{\partial^2 f}{\partial y^2}$

$= -4e^{-2y} \cos 2x + 4e^{-2y} \cos 2x = 0$

67. $\frac{\partial f}{\partial x} = -\frac{1}{2}\left(x^2 + y^2 + z^2\right)^{-3/2}(2x) = -x\left(x^2 + y^2 + z^2\right)^{-3/2},\ \frac{\partial f}{\partial y} = -\frac{1}{2}\left(x^2 + y^2 + z^2\right)^{-3/2}(2y)$

$= -y\left(x^2 + y^2 + z^2\right)^{-3/2},\ \frac{\partial f}{\partial z} = -\frac{1}{2}\left(x^2 + y^2 + z^2\right)^{-3/2}(2z) = -z\left(x^2 + y^2 + z^2\right)^{-3/2};$

$\frac{\partial^2 f}{\partial x^2} = -\left(x^2 + y^2 + z^2\right)^{-3/2} + 3x^2\left(x^2 + y^2 + z^2\right)^{-5/2},\ \frac{\partial^2 f}{\partial y^2} = -\left(x^2 + y^2 + z^2\right)^{-3/2} + 3y^2\left(x^2 + y^2 + z^2\right)^{-5/2},$

$\frac{\partial^2 f}{\partial z^2} = -\left(x^2 + y^2 + z^2\right)^{-3/2} + 3z^2\left(x^2 + y^2 + z^2\right)^{-5/2} \Rightarrow \frac{\partial^2 f}{\partial x^2} + \frac{\partial^2 f}{\partial y^2} + \frac{\partial^2 f}{\partial z^2}$

$= \left[-\left(x^2 + y^2 + z^2\right)^{-3/2} + 3x^2\left(x^2 + y^2 + z^2\right)^{-5/2}\right] + \left[-\left(x^2 + y^2 + z^2\right)^{-3/2} + 3y^2\left(x^2 + y^2 + z^2\right)^{-5/2}\right]$

$+ \left[-\left(x^2 + y^2 + z^2\right)^{-3/2} + 3z^2\left(x^2 + y^2 + z^2\right)^{-5/2}\right] = -3\left(x^2 + y^2 + z^2\right)^{-3/2} + \left(3x^2 + 3y^2 + 3z^2\right)\left(x^2 + y^2 + z^2\right)^{-5/2}$

$= 0$

69. $\frac{\partial w}{\partial x} = \cos(x + ct),\ \frac{\partial w}{\partial t} = c \cos(x + ct);\ \frac{\partial^2 w}{\partial x^2} = -\sin(x + ct),\ \frac{\partial^2 w}{\partial t^2} = -c^2 \sin(x + ct) \Rightarrow \frac{\partial^2 w}{\partial t^2} = c^2\left[-\sin(x + ct)\right]$

$= c^2 \frac{\partial^2 w}{\partial x^2}$

71. $\frac{\partial w}{\partial x} = \cos(x + ct) - 2\sin(2x + 2ct),\ \frac{\partial w}{\partial t} = c \cos(x + ct) - 2c \sin(2x + 2ct);$

$\frac{\partial^2 w}{\partial x^2} = -\sin(x + ct) - 4\cos(2x + 2ct),\ \frac{\partial^2 w}{\partial t^2} = -c^2 \sin(x + ct) - 4c^2 \cos(2x + 2ct)$

$\Rightarrow \frac{\partial^2 w}{\partial t^2} = c^2\left[-\sin(x + ct) - 4\cos(2x + 2ct)\right] = c^2 \frac{\partial^2 w}{\partial x^2}$

73. $\frac{\partial w}{\partial x} = 2\sec^2(2x - 2ct),\ \frac{\partial w}{\partial t} = -2c\sec^2(2x - 2ct);\ \frac{\partial^2 w}{\partial x^2} = 8\sec^2(2x - 2ct)\tan(2x - 2ct),$

$\frac{\partial^2 w}{\partial t^2} = 8c^2\sec^2(2x - 2ct)\tan(2x - 2ct) \Rightarrow \frac{\partial^2 w}{\partial t^2} = c^2[8\sec^2(2x - 2ct)\tan(2x - 2ct)] = c^2 \frac{\partial^2 w}{\partial x^2}$

75. $\frac{\partial w}{\partial t} = \frac{\partial f}{\partial u}\frac{\partial u}{\partial t} = \frac{\partial f}{\partial u}(ac) \Rightarrow \frac{\partial^2 w}{\partial t^2} = (ac)\left(\frac{\partial^2 f}{\partial u^2}\right)(ac) = a^2 c^2 \frac{\partial^2 f}{\partial u^2};\ \frac{\partial w}{\partial x} = \frac{\partial f}{\partial u}\frac{\partial u}{\partial x} = \frac{\partial f}{\partial u}\cdot a \Rightarrow \frac{\partial^2 w}{\partial x^2} = \left(a\frac{\partial^2 f}{\partial u^2}\right)\cdot a$

$= a^2 \frac{\partial^2 f}{\partial u^2} \Rightarrow \frac{\partial^2 w}{\partial t^2} = a^2 c^2 \frac{\partial^2 f}{\partial u^2} = c^2\left(a^2 \frac{\partial^2 f}{\partial u^2}\right) = c^2 \frac{\partial^2 w}{\partial x^2}$

77. Yes, since $f_{xx}, f_{yy}, f_{xy},$ and f_{yx} are all continuous on R, use the same reasoning as in Exercise 76 with

$f_x(x, y) = f_x(x_0, y_0) + f_{xx}(x_0, y_0)\,\Delta x + f_{xy}(x_0, y_0)\,\Delta y + \epsilon_1 \Delta x + \epsilon_2 \Delta y$ and

$f_y(x, y) = f_y(x_0, y_0) + f_{yx}(x_0, y_0)\,\Delta x + f_{yy}(x_0, y_0)\,\Delta y + \hat{\epsilon}_1 \Delta x + \hat{\epsilon}_2 \Delta y.$ Then $\displaystyle \lim_{(x, y) \to (x_0, y_0)} f_x(x, y) = f_x(x_0, y_0)$

and $\displaystyle \lim_{(x, y) \to (x_0, y_0)} f_y(x, y) = f_y(x_0, y_0).$

14.4 THE CHAIN RULE

1. (a) $\frac{\partial w}{\partial x} = 2x,\ \frac{\partial w}{\partial y} = 2y,\ \frac{dx}{dt} = -\sin t,\ \frac{dy}{dt} = \cos t \Rightarrow \frac{dw}{dt} = -2x \sin t + 2y \cos t = -2 \cos t \sin t + 2 \sin t \cos t$
$= 0;\ w = x^2 + y^2 = \cos^2 t + \sin^2 t = 1 \Rightarrow \frac{dw}{dt} = 0$

(b) $\frac{dw}{dt}(\pi) = 0$

3. (a) $\frac{\partial w}{\partial x} = \frac{1}{z},\ \frac{\partial w}{\partial y} = \frac{1}{z},\ \frac{\partial w}{\partial z} = \frac{-(x+y)}{z^2},\ \frac{dx}{dt} = -2\cos t \sin t,\ \frac{dy}{dt} = 2 \sin t \cos t,\ \frac{dz}{dt} = -\frac{1}{t^2}$

$\Rightarrow \frac{dw}{dt} = -\frac{2}{z} \cos t \sin t + \frac{2}{z} \sin t \cos t + \frac{x+y}{z^2 t^2} = \frac{\cos^2 t + \sin^2 t}{\left(\frac{1}{t^2}\right)(t^2)} = 1;\ w = \frac{x}{z} + \frac{y}{z} = \frac{\cos^2 t}{\left(\frac{1}{t}\right)} + \frac{\sin^2 t}{\left(\frac{1}{t}\right)} = t \Rightarrow \frac{dw}{dt} = 1$

(b) $\frac{dw}{dt}(3) = 1$

5. (a) $\frac{\partial w}{\partial x} = 2ye^x,\ \frac{\partial w}{\partial y} = 2e^x,\ \frac{\partial w}{\partial z} = -\frac{1}{z},\ \frac{dx}{dt} = \frac{2t}{t^2+1},\ \frac{dy}{dt} = \frac{1}{t^2+1},\ \frac{dz}{dt} = e^t \Rightarrow \frac{dw}{dt} = \frac{4yte^x}{t^2+1} + \frac{2e^x}{t^2+1} - \frac{e^t}{z}$

$= \frac{(4t)(\tan^{-1} t)(t^2+1)}{t^2+1} + \frac{2(t^2+1)}{t^2+1} - \frac{e^t}{e^t} = 4t \tan^{-1} t + 1;\ w = 2ye^x - \ln z = (2 \tan^{-1} t)(t^2+1) - t$

$\Rightarrow \frac{dw}{dt} = \left(\frac{2}{t^2+1}\right)(t^2+1) + (2 \tan^{-1} t)(2t) - 1 = 4t \tan^{-1} t + 1$

(b) $\frac{dw}{dt}(1) = (4)(1)\left(\frac{\pi}{4}\right) + 1 = \pi + 1$

7. (a) $\frac{\partial z}{\partial u} = \frac{\partial z}{\partial x} \frac{\partial x}{\partial u} + \frac{\partial z}{\partial y} \frac{\partial y}{\partial u} = (4e^x \ln y)\left(\frac{\cos v}{u \cos v}\right) + \left(\frac{4e^x}{y}\right)(\sin v) = \frac{4e^x \ln y}{u} + \frac{4e^x \sin v}{y}$

$= \frac{4(u \cos v) \ln (u \sin v)}{u} + \frac{4(u \cos v)(\sin v)}{u \sin v} = (4 \cos v) \ln (u \sin v) + 4 \cos v;$

$\frac{\partial z}{\partial v} = \frac{\partial z}{\partial x} \frac{\partial x}{\partial v} + \frac{\partial z}{\partial y} \frac{\partial y}{\partial v} = (4e^x \ln y)\left(\frac{-u \sin v}{u \cos v}\right) + \left(\frac{4e^x}{y}\right)(u \cos v) = -(4e^x \ln y)(\tan v) + \frac{4e^x u \cos v}{y}$

$= [-4(u \cos v) \ln (u \sin v)](\tan v) + \frac{4(u \cos v)(u \cos v)}{u \sin v} = (-4u \sin v) \ln (u \sin v) + \frac{4u \cos^2 v}{\sin v};$

$z = 4e^x \ln y = 4(u \cos v) \ln (u \sin v) \Rightarrow \frac{\partial z}{\partial u} = (4 \cos v) \ln (u \sin v) + 4(u \cos v)\left(\frac{\sin v}{u \sin v}\right)$

$= (4 \cos v) \ln (u \sin v) + 4 \cos v;$ also $\frac{\partial z}{\partial v} = (-4u \sin v) \ln (u \sin v) + 4(u \cos v)\left(\frac{u \cos v}{u \sin v}\right)$

$= (-4u \sin v) \ln (u \sin v) + \frac{4u \cos^2 v}{\sin v}$

(b) At $\left(2, \frac{\pi}{4}\right)$: $\frac{\partial z}{\partial u} = 4 \cos \frac{\pi}{4} \ln \left(2 \sin \frac{\pi}{4}\right) + 4 \cos \frac{\pi}{4} = 2\sqrt{2} \ln \sqrt{2} + 2\sqrt{2} = \sqrt{2}(\ln 2 + 2);$

$\frac{\partial z}{\partial v} = (-4)(2) \sin \frac{\pi}{4} \ln \left(2 \sin \frac{\pi}{4}\right) + \frac{(4)(2)\left(\cos^2 \frac{\pi}{4}\right)}{\left(\sin \frac{\pi}{4}\right)} = -4\sqrt{2} \ln \sqrt{2} + 4\sqrt{2} = -2\sqrt{2} \ln 2 + 4\sqrt{2}$

9. (a) $\frac{\partial w}{\partial u} = \frac{\partial w}{\partial x} \frac{\partial x}{\partial u} + \frac{\partial w}{\partial y} \frac{\partial y}{\partial u} + \frac{\partial w}{\partial z} \frac{\partial z}{\partial u} = (y+z)(1) + (x+z)(1) + (y+x)(v) = x + y + 2z + v(y+x)$

$= (u+v) + (u-v) + 2uv + v(2u) = 2u + 4uv;\ \frac{\partial w}{\partial v} = \frac{\partial w}{\partial x} \frac{\partial x}{\partial v} + \frac{\partial w}{\partial y} \frac{\partial y}{\partial v} + \frac{\partial w}{\partial z} \frac{\partial z}{\partial v}$

$= (y+z)(1) + (x+z)(-1) + (y+x)(u) = y - x + (y+x)u = -2v + (2u)u = -2v + 2u^2;$

$w = xy + yz + xz = (u^2 - v^2) + (u^2 v - uv^2) + (u^2 v + uv^2) = u^2 - v^2 + 2u^2 v \Rightarrow \frac{\partial w}{\partial u} = 2u + 4uv$ and

$\frac{\partial w}{\partial v} = -2v + 2u^2$

(b) At $\left(\frac{1}{2}, 1\right)$: $\frac{\partial w}{\partial u} = 2\left(\frac{1}{2}\right) + 4\left(\frac{1}{2}\right)(1) = 3$ and $\frac{\partial w}{\partial v} = -2(1) + 2\left(\frac{1}{2}\right)^2 = -\frac{3}{2}$

11. (a) $\frac{\partial u}{\partial x} = \frac{\partial u}{\partial p} \frac{\partial p}{\partial x} + \frac{\partial u}{\partial q} \frac{\partial q}{\partial x} + \frac{\partial u}{\partial r} \frac{\partial r}{\partial x} = \frac{1}{q-r} + \frac{r-p}{(q-r)^2} + \frac{p-q}{(q-r)^2} = \frac{q-r+r-p+p-q}{(q-r)^2} = 0;$

$\frac{\partial u}{\partial y} = \frac{\partial u}{\partial p} \frac{\partial p}{\partial y} + \frac{\partial u}{\partial q} \frac{\partial q}{\partial y} + \frac{\partial u}{\partial r} \frac{\partial r}{\partial y} = \frac{1}{q-r} - \frac{r-p}{(q-r)^2} + \frac{p-q}{(q-r)^2} = \frac{q-r-r+p+p-q}{(q-r)^2} = \frac{2p-2r}{(q-r)^2}$

$= \frac{(2x+2y+2z)-(2x+2y-2z)}{(2z-2y)^2} = \frac{z}{(z-y)^2};\ \frac{\partial u}{\partial z} = \frac{\partial u}{\partial p} \frac{\partial p}{\partial z} + \frac{\partial u}{\partial q} \frac{\partial q}{\partial z} + \frac{\partial u}{\partial r} \frac{\partial r}{\partial z}$

$= \frac{1}{q-r} + \frac{r-p}{(q-r)^2} - \frac{p-q}{(q-r)^2} = \frac{q-r+r-p-p+q}{(q-r)^2} = \frac{2q-2p}{(q-r)^2} = \frac{-4y}{(2z-2y)^2} = -\frac{y}{(z-y)^2};$

$u = \frac{p-q}{q-r} = \frac{2y}{2z-2y} = \frac{y}{z-y} \Rightarrow \frac{\partial u}{\partial x} = 0,\ \frac{\partial u}{\partial y} = \frac{(z-y)-y(-1)}{(z-y)^2} = \frac{z}{(z-y)^2},$ and $\frac{\partial u}{\partial z} = \frac{(z-y)(0)-y(1)}{(z-y)^2}$

$= -\frac{y}{(z-y)^2}$

(b) At $\left(\sqrt{3}, 2, 1\right)$: $\frac{\partial u}{\partial x} = 0,\ \frac{\partial u}{\partial y} = \frac{1}{(1-2)^2} = 1,$ and $\frac{\partial u}{\partial z} = \frac{-2}{(1-2)^2} = -2$

13. $\dfrac{dz}{dt} = \dfrac{\partial z}{\partial x}\dfrac{dx}{dt} + \dfrac{\partial z}{\partial y}\dfrac{dy}{dt}$

15. $\dfrac{\partial w}{\partial u} = \dfrac{\partial w}{\partial x}\dfrac{\partial x}{\partial u} + \dfrac{\partial w}{\partial y}\dfrac{\partial y}{\partial u} + \dfrac{\partial w}{\partial z}\dfrac{\partial z}{\partial u}$ $\dfrac{\partial w}{\partial v} = \dfrac{\partial w}{\partial x}\dfrac{\partial x}{\partial v} + \dfrac{\partial w}{\partial y}\dfrac{\partial y}{\partial v} + \dfrac{\partial w}{\partial z}\dfrac{\partial z}{\partial v}$

17. $\dfrac{\partial w}{\partial u} = \dfrac{\partial w}{\partial x}\dfrac{\partial x}{\partial u} + \dfrac{\partial w}{\partial y}\dfrac{\partial y}{\partial u}$ $\dfrac{\partial w}{\partial v} = \dfrac{\partial w}{\partial x}\dfrac{\partial x}{\partial v} + \dfrac{\partial w}{\partial y}\dfrac{\partial y}{\partial v}$

19. $\dfrac{\partial z}{\partial t} = \dfrac{\partial z}{\partial x}\dfrac{\partial x}{\partial t} + \dfrac{\partial z}{\partial y}\dfrac{\partial y}{\partial t}$ $\dfrac{\partial z}{\partial s} = \dfrac{\partial z}{\partial x}\dfrac{\partial x}{\partial s} + \dfrac{\partial z}{\partial y}\dfrac{\partial y}{\partial s}$

21. $\dfrac{\partial w}{\partial s} = \dfrac{dw}{du}\dfrac{\partial u}{\partial s}$ $\dfrac{\partial w}{\partial t} = \dfrac{dw}{du}\dfrac{\partial u}{\partial t}$

23. $\frac{\partial w}{\partial r} = \frac{\partial w}{\partial x}\frac{dx}{dr} + \frac{\partial w}{\partial y}\frac{dy}{dr} = \frac{\partial w}{\partial x}\frac{dx}{dr}$ since $\frac{dy}{dr} = 0$ $\frac{\partial w}{\partial s} = \frac{\partial w}{\partial x}\frac{dx}{ds} + \frac{\partial w}{\partial y}\frac{dy}{ds} = \frac{\partial w}{\partial y}\frac{dy}{ds}$ since $\frac{dx}{ds} = 0$

25. Let $F(x, y) = x^3 - 2y^2 + xy = 0 \Rightarrow F_x(x, y) = 3x^2 + y$

and $F_y(x, y) = -4y + x \Rightarrow \frac{dy}{dx} = -\frac{F_x}{F_y} = -\frac{3x^2 + y}{(-4y + x)}$

$\Rightarrow \frac{dy}{dx}(1, 1) = \frac{4}{3}$

27. Let $F(x, y) = x^2 + xy + y^2 - 7 = 0 \Rightarrow F_x(x, y) = 2x + y$ and $F_y(x, y) = x + 2y \Rightarrow \frac{dy}{dx} = -\frac{F_x}{F_y} = -\frac{2x + y}{x + 2y}$

$\Rightarrow \frac{dy}{dx}(1, 2) = -\frac{4}{5}$

28. Let $F(x, y) = xe^y + \sin xy + y - \ln 2 = 0 \Rightarrow F_x(x, y) = e^y + y\cos xy$ and $F_y(x, y) = xe^y + x\sin xy + 1$

$\Rightarrow \frac{dy}{dx} = -\frac{F_x}{F_y} = -\frac{e^y + y\cos xy}{xe^y + x\sin xy + 1} \Rightarrow \frac{dy}{dx}(0, \ln 2) = -(2 + \ln 2)$

29. Let $F(x, y, z) = z^3 - xy + yz + y^3 - 2 = 0 \Rightarrow F_x(x, y, z) = -y$, $F_y(x, y, z) = -x + z + 3y^2$, $F_z(x, y, z) = 3z^2 + y$

$\Rightarrow \frac{\partial z}{\partial x} = -\frac{F_x}{F_z} = -\frac{-y}{3z^2 + y} = \frac{y}{3z^2 + y} \Rightarrow \frac{\partial z}{\partial x}(1, 1, 1) = \frac{1}{4}$; $\frac{\partial z}{\partial y} = -\frac{F_y}{F_z} = -\frac{-x + z + 3y^2}{3z^2 + y} = \frac{x - z - 3y^2}{3z^2 + y}$

$\Rightarrow \frac{\partial z}{\partial y}(1, 1, 1) = -\frac{3}{4}$

31. Let $F(x, y, z) = \sin(x + y) + \sin(y + z) + \sin(x + z) = 0 \Rightarrow F_x(x, y, z) = \cos(x + y) + \cos(x + z)$,

$F_y(x, y, z) = \cos(x + y) + \cos(y + z)$, $F_z(x, y, z) = \cos(y + z) + \cos(x + z) \Rightarrow \frac{\partial z}{\partial x} = -\frac{F_x}{F_z}$

$= -\frac{\cos(x + y) + \cos(x + z)}{\cos(y + z) + \cos(x + z)} \Rightarrow \frac{\partial z}{\partial x}(\pi, \pi, \pi) = -1$; $\frac{\partial z}{\partial y} = -\frac{F_y}{F_z} = -\frac{\cos(x + y) + \cos(y + z)}{\cos(y + z) + \cos(x + z)} \Rightarrow \frac{\partial z}{\partial y}(\pi, \pi, \pi) = -1$

33. $\frac{\partial w}{\partial r} = \frac{\partial w}{\partial x}\frac{\partial x}{\partial r} + \frac{\partial w}{\partial y}\frac{\partial y}{\partial r} + \frac{\partial w}{\partial z}\frac{\partial z}{\partial r} = 2(x + y + z)(1) + 2(x + y + z)[-\sin(r + s)] + 2(x + y + z)[\cos(r + s)]$

$= 2(x + y + z)[1 - \sin(r + s) + \cos(r + s)] = 2[r - s + \cos(r + s) + \sin(r + s)][1 - \sin(r + s) + \cos(r + s)]$

$\Rightarrow \frac{\partial w}{\partial r}\big|_{r=1, s=-1} = 2(3)(2) = 12$

35. $\frac{\partial w}{\partial v} = \frac{\partial w}{\partial x}\frac{\partial x}{\partial v} + \frac{\partial w}{\partial y}\frac{\partial y}{\partial v} = \left(2x - \frac{y}{x^2}\right)(-2) + \left(\frac{1}{x}\right)(1) = \left[2(u - 2v + 1) - \frac{2u + v - 2}{(u - 2v + 1)^2}\right](-2) + \frac{1}{u - 2v + 1}$

$\Rightarrow \frac{\partial w}{\partial v}\big|_{u=0, v=0} = -7$

37. $\frac{\partial z}{\partial u} = \frac{dz}{dx}\frac{\partial x}{\partial u} = \left(\frac{5}{1 + x^2}\right)e^u = \left[\frac{5}{1 + (e^u + \ln v)^2}\right]e^u \Rightarrow \frac{\partial z}{\partial u}\big|_{u=\ln 2, v=1} = \left[\frac{5}{1 + (2)^2}\right](2) = 2$;

$\frac{\partial z}{\partial v} = \frac{dz}{dx}\frac{\partial x}{\partial v} = \left(\frac{5}{1 + x^2}\right)\left(\frac{1}{v}\right) = \left[\frac{5}{1 + (e^u + \ln v)^2}\right]\left(\frac{1}{v}\right) \Rightarrow \frac{\partial z}{\partial v}\big|_{u=\ln 2, v=1} = \left[\frac{5}{1 + (2)^2}\right](1) = 1$

39. $V = IR \Rightarrow \frac{\partial V}{\partial I} = R$ and $\frac{\partial V}{\partial R} = I$; $\frac{dV}{dt} = \frac{\partial V}{\partial I}\frac{dI}{dt} + \frac{\partial V}{\partial R}\frac{dR}{dt} = R\frac{dI}{dt} + I\frac{dR}{dt} \Rightarrow -0.01$ volts/sec

$= (600 \text{ ohms})\frac{dI}{dt} + (0.04 \text{ amps})(0.5 \text{ ohms/sec}) \Rightarrow \frac{dI}{dt} = -0.00005$ amps/sec

41. $\frac{\partial f}{\partial x} = \frac{\partial f}{\partial u}\frac{\partial u}{\partial x} + \frac{\partial f}{\partial v}\frac{\partial v}{\partial x} + \frac{\partial f}{\partial w}\frac{\partial w}{\partial x} = \frac{\partial f}{\partial u}(1) + \frac{\partial f}{\partial v}(0) + \frac{\partial f}{\partial w}(-1) = \frac{\partial f}{\partial u} - \frac{\partial f}{\partial w}$,

$\frac{\partial f}{\partial y} = \frac{\partial f}{\partial u}\frac{\partial u}{\partial y} + \frac{\partial f}{\partial v}\frac{\partial v}{\partial y} + \frac{\partial f}{\partial w}\frac{\partial w}{\partial y} = \frac{\partial f}{\partial u}(-1) + \frac{\partial f}{\partial v}(1) + \frac{\partial f}{\partial w}(0) = -\frac{\partial f}{\partial u} + \frac{\partial f}{\partial v}$, and

$\frac{\partial f}{\partial z} = \frac{\partial f}{\partial u}\frac{\partial u}{\partial z} + \frac{\partial f}{\partial v}\frac{\partial v}{\partial z} + \frac{\partial f}{\partial w}\frac{\partial w}{\partial z} = \frac{\partial f}{\partial u}(0) + \frac{\partial f}{\partial v}(-1) + \frac{\partial f}{\partial w}(1) = -\frac{\partial f}{\partial v} + \frac{\partial f}{\partial w} \Rightarrow \frac{\partial f}{\partial x} + \frac{\partial f}{\partial y} + \frac{\partial f}{\partial z} = 0$

43. $w_x = \frac{\partial w}{\partial x} = \frac{\partial w}{\partial u}\frac{\partial u}{\partial x} + \frac{\partial w}{\partial v}\frac{\partial v}{\partial x} = x\frac{\partial w}{\partial u} + y\frac{\partial w}{\partial v} \Rightarrow w_{xx} = \frac{\partial w}{\partial u} + x\frac{\partial}{\partial x}\left(\frac{\partial w}{\partial u}\right) + y\frac{\partial}{\partial x}\left(\frac{\partial w}{\partial v}\right)$

$= \frac{\partial w}{\partial u} + x\left(\frac{\partial^2 w}{\partial u^2}\frac{\partial u}{\partial x} + \frac{\partial^2 w}{\partial v\partial u}\frac{\partial v}{\partial x}\right) + y\left(\frac{\partial^2 w}{\partial u\partial v}\frac{\partial u}{\partial x} + \frac{\partial^2 w}{\partial v^2}\frac{\partial v}{\partial x}\right) = \frac{\partial w}{\partial u} + x\left(x\frac{\partial^2 w}{\partial u^2} + y\frac{\partial^2 w}{\partial v\partial u}\right) + y\left(x\frac{\partial^2 w}{\partial u\partial v} + y\frac{\partial^2 w}{\partial v^2}\right)$

$= \frac{\partial w}{\partial u} + x^2\frac{\partial^2 w}{\partial u^2} + 2xy\frac{\partial^2 w}{\partial v\partial u} + y^2\frac{\partial^2 w}{\partial v^2}; w_y = \frac{\partial w}{\partial y} = \frac{\partial w}{\partial u}\frac{\partial u}{\partial y} + \frac{\partial w}{\partial v}\frac{\partial v}{\partial y} = -y\frac{\partial w}{\partial u} + x\frac{\partial w}{\partial v}$

$\Rightarrow w_{yy} = -\frac{\partial w}{\partial u} - y\left(\frac{\partial^2 w}{\partial u^2}\frac{\partial u}{\partial y} + \frac{\partial^2 w}{\partial v\partial u}\frac{\partial v}{\partial y}\right) + x\left(\frac{\partial^2 w}{\partial u\partial v}\frac{\partial u}{\partial y} + \frac{\partial^2 w}{\partial v^2}\frac{\partial v}{\partial y}\right)$

$= -\frac{\partial w}{\partial u} - y\left(-y\frac{\partial^2 w}{\partial u^2} + x\frac{\partial^2 w}{\partial v\partial u}\right) + x\left(-y\frac{\partial^2 w}{\partial u\partial v} + x\frac{\partial^2 w}{\partial v^2}\right) = -\frac{\partial w}{\partial u} + y^2\frac{\partial^2 w}{\partial u^2} - 2xy\frac{\partial^2 w}{\partial v\partial u} + x^2\frac{\partial^2 w}{\partial v^2}$; thus

$w_{xx} + w_{yy} = (x^2 + y^2)\frac{\partial^2 w}{\partial u^2} + (x^2 + y^2)\frac{\partial^2 w}{\partial v^2} = (x^2 + y^2)(w_{uu} + w_{vv}) = 0$, since $w_{uu} + w_{vv} = 0$

45. $f_x(x, y, z) = \cos t, f_y(x, y, z) = \sin t$, and $f_z(x, y, z) = t^2 + t - 2 \Rightarrow \frac{df}{dt} = \frac{\partial f}{\partial x}\frac{dx}{dt} + \frac{\partial f}{\partial y}\frac{dy}{dt} + \frac{\partial f}{\partial z}\frac{dz}{dt}$

$= (\cos t)(-\sin t) + (\sin t)(\cos t) + (t^2 + t - 2)(1) = t^2 + t - 2; \frac{df}{dt} = 0 \Rightarrow t^2 + t - 2 = 0 \Rightarrow t = -2$

or $t = 1; t = -2 \Rightarrow x = \cos(-2), y = \sin(-2), z = -2$ for the point $(\cos(-2), \sin(-2), -2); t = 1 \Rightarrow x = \cos 1$,
$y = \sin 1, z = 1$ for the point $(\cos 1, \sin 1, 1)$

47. (a) $\frac{\partial T}{\partial x} = 8x - 4y$ and $\frac{\partial T}{\partial y} = 8y - 4x \Rightarrow \frac{dT}{dt} = \frac{\partial T}{\partial x}\frac{dx}{dt} + \frac{\partial T}{\partial y}\frac{dy}{dt} = (8x - 4y)(-\sin t) + (8y - 4x)(\cos t)$

$= (8\cos t - 4\sin t)(-\sin t) + (8\sin t - 4\cos t)(\cos t) = 4\sin^2 t - 4\cos^2 t \Rightarrow \frac{d^2 T}{dt^2} = 16\sin t\cos t;$

$\frac{dT}{dt} = 0 \Rightarrow 4\sin^2 t - 4\cos^2 t = 0 \Rightarrow \sin^2 t = \cos^2 t \Rightarrow \sin t = \cos t$ or $\sin t = -\cos t \Rightarrow t = \frac{\pi}{4}, \frac{5\pi}{4}, \frac{3\pi}{4}, \frac{7\pi}{4}$ on
the interval $0 \le t \le 2\pi$;

$\left.\frac{d^2 T}{dt^2}\right|_{t=\frac{\pi}{4}} = 16\sin\frac{\pi}{4}\cos\frac{\pi}{4} > 0 \Rightarrow T$ has a minimum at $(x, y) = \left(\frac{\sqrt{2}}{2}, \frac{\sqrt{2}}{2}\right);$

$\left.\frac{d^2 T}{dt^2}\right|_{t=\frac{3\pi}{4}} = 16\sin\frac{3\pi}{4}\cos\frac{3\pi}{4} < 0 \Rightarrow T$ has a maximum at $(x, y) = \left(-\frac{\sqrt{2}}{2}, \frac{\sqrt{2}}{2}\right);$

$\left.\frac{d^2 T}{dt^2}\right|_{t=\frac{5\pi}{4}} = 16\sin\frac{5\pi}{4}\cos\frac{5\pi}{4} > 0 \Rightarrow T$ has a minimum at $(x, y) = \left(-\frac{\sqrt{2}}{2}, -\frac{\sqrt{2}}{2}\right);$

$\left.\frac{d^2 T}{dt^2}\right|_{t=\frac{7\pi}{4}} = 16\sin\frac{7\pi}{4}\cos\frac{7\pi}{4} < 0 \Rightarrow T$ has a maximum at $(x, y) = \left(\frac{\sqrt{2}}{2}, -\frac{\sqrt{2}}{2}\right)$

(b) $T = 4x^2 - 4xy + 4y^2 \Rightarrow \frac{\partial T}{\partial x} = 8x - 4y$, and $\frac{\partial T}{\partial y} = 8y - 4x$ so the extreme values occur at the four points

found in part (a): $T\left(-\frac{\sqrt{2}}{2}, \frac{\sqrt{2}}{2}\right) = T\left(\frac{\sqrt{2}}{2}, -\frac{\sqrt{2}}{2}\right) = 4\left(\frac{1}{2}\right) - 4\left(-\frac{1}{2}\right) + 4\left(\frac{1}{2}\right) = 6$, the maximum and

$T\left(\frac{\sqrt{2}}{2}, \frac{\sqrt{2}}{2}\right) = T\left(-\frac{\sqrt{2}}{2}, -\frac{\sqrt{2}}{2}\right) = 4\left(\frac{1}{2}\right) - 4\left(\frac{1}{2}\right) + 4\left(\frac{1}{2}\right) = 2$, the minimum

49. $G(u, x) = \int_a^u g(t, x) dt$ where $u = f(x) \Rightarrow \frac{dG}{dx} = \frac{\partial G}{\partial u}\frac{du}{dx} + \frac{\partial G}{\partial x}\frac{dx}{dx} = g(u, x)f'(x) + \int_a^u g_x(t, x) dt$; thus

$F(x) = \int_0^{x^2}\sqrt{t^4 + x^3}\, dt \Rightarrow F'(x) = \sqrt{(x^2)^4 + x^3}\,(2x) + \int_0^{x^2}\frac{\partial}{\partial x}\sqrt{t^4 + x^3}\, dt = 2x\sqrt{x^8 + x^3} + \int_0^{x^2}\frac{3x^2}{2\sqrt{t^4 + x^3}}\, dt$

14.5 DIRECTIONAL DERIVATIVES AND GRADIENT VECTORS

1. $\frac{\partial f}{\partial x} = -1, \frac{\partial f}{\partial y} = 1 \Rightarrow \nabla f = -\mathbf{i} + \mathbf{j}; f(2, 1) = -1$

$\Rightarrow -1 = y - x$ is the level curve

3. $\frac{\partial g}{\partial x} = -2x \Rightarrow \frac{\partial g}{\partial x}(-1,0) = 2; \frac{\partial g}{\partial y} = 1$

    $\Rightarrow \nabla g = 2\mathbf{i} + \mathbf{j}; g(-1,0) = -1$

    $\Rightarrow -1 = y - x^2$ is the level curve

5. $\frac{\partial f}{\partial x} = 2x + \frac{z}{x} \Rightarrow \frac{\partial f}{\partial x}(1,1,1) = 3; \frac{\partial f}{\partial y} = 2y \Rightarrow \frac{\partial f}{\partial y}(1,1,1) = 2; \frac{\partial f}{\partial z} = -4z + \ln x \Rightarrow \frac{\partial f}{\partial z}(1,1,1) = -4;$ ·

    thus $\nabla f = 3\mathbf{i} + 2\mathbf{j} - 4\mathbf{k}$

7. $\frac{\partial f}{\partial x} = -\frac{x}{(x^2+y^2+z^2)^{3/2}} + \frac{1}{x} \Rightarrow \frac{\partial f}{\partial x}(-1,2,-2) = -\frac{26}{27}; \frac{\partial f}{\partial y} = -\frac{y}{(x^2+y^2+z^2)^{3/2}} + \frac{1}{y} \Rightarrow \frac{\partial f}{\partial y}(-1,2,-2) = \frac{23}{54};$

    $\frac{\partial f}{\partial z} = -\frac{z}{(x^2+y^2+z^2)^{3/2}} + \frac{1}{z} \Rightarrow \frac{\partial f}{\partial z}(-1,2,-2) = -\frac{23}{54};$ thus $\nabla f = -\frac{26}{27}\mathbf{i} + \frac{23}{54}\mathbf{j} - \frac{23}{54}\mathbf{k}$

9. $\mathbf{u} = \frac{\mathbf{v}}{|\mathbf{v}|} = \frac{4\mathbf{i}+3\mathbf{j}}{\sqrt{4^2+3^2}} = \frac{4}{5}\mathbf{i} + \frac{3}{5}\mathbf{j}; f_x(x,y) = 2y \Rightarrow f_x(5,5) = 10; f_y(x,y) = 2x - 6y \Rightarrow f_y(5,5) = -20$

    $\Rightarrow \nabla f = 10\mathbf{i} - 20\mathbf{j} \Rightarrow (D_{\mathbf{u}}f)_{P_0} = \nabla f \cdot \mathbf{u} = 10\left(\frac{4}{5}\right) - 20\left(\frac{3}{5}\right) = -4$

11. $\mathbf{u} = \frac{\mathbf{v}}{|\mathbf{v}|} = \frac{12\mathbf{i}+5\mathbf{j}}{\sqrt{12^2+5^2}} = \frac{12}{13}\mathbf{i} + \frac{5}{13}\mathbf{j}; g_x(x,y) = 1 + \frac{y^2}{x^2} + \frac{2y\sqrt{3}}{2xy\sqrt{4x^2y^2-1}} \Rightarrow g_x(1,1) = 3; g_y(x,y)$

    $= -\frac{2y}{x} + \frac{2x\sqrt{3}}{2xy\sqrt{4x^2y^2-1}} \Rightarrow g_y(1,1) = -1 \Rightarrow \nabla g = 3\mathbf{i} - \mathbf{j} \Rightarrow (D_{\mathbf{u}}g)_{P_0} = \nabla g \cdot \mathbf{u} = \frac{36}{13} - \frac{5}{13} = \frac{31}{13}$

13. $\mathbf{u} = \frac{\mathbf{v}}{|\mathbf{v}|} = \frac{3\mathbf{i}+6\mathbf{j}-2\mathbf{k}}{\sqrt{3^2+6^2+(-2)^2}} = \frac{3}{7}\mathbf{i} + \frac{6}{7}\mathbf{j} - \frac{2}{7}\mathbf{k}; f_x(x,y,z) = y + z \Rightarrow f_x(1,-1,2) = 1; f_y(x,y,z) = x + z$

    $\Rightarrow f_y(1,-1,2) = 3; f_z(x,y,z) = y + x \Rightarrow f_z(1,-1,2) = 0 \Rightarrow \nabla f = \mathbf{i} + 3\mathbf{j} \Rightarrow (D_{\mathbf{u}}f)_{P_0} = \nabla f \cdot \mathbf{u} = \frac{3}{7} + \frac{18}{7} = 3$

15. $\mathbf{u} = \frac{\mathbf{v}}{|\mathbf{v}|} = \frac{2\mathbf{i}+\mathbf{j}-2\mathbf{k}}{\sqrt{2^2+1^2+(-2)^2}} = \frac{2}{3}\mathbf{i} + \frac{1}{3}\mathbf{j} - \frac{2}{3}\mathbf{k}; g_x(x,y,z) = 3e^x \cos yz \Rightarrow g_x(0,0,0) = 3; g_y(x,y,z) = -3ze^x \sin yz$

    $\Rightarrow g_y(0,0,0) = 0; g_z(x,y,z) = -3ye^x \sin yz \Rightarrow g_z(0,0,0) = 0 \Rightarrow \nabla g = 3\mathbf{i} \Rightarrow (D_{\mathbf{u}}g)_{P_0} = \nabla g \cdot \mathbf{u} = 2$

17. $\nabla f = (2x+y)\mathbf{i} + (x+2y)\mathbf{j} \Rightarrow \nabla f(-1,1) = -\mathbf{i} + \mathbf{j} \Rightarrow \mathbf{u} = \frac{\nabla f}{|\nabla f|} = \frac{-\mathbf{i}+\mathbf{j}}{\sqrt{(-1)^2+1^2}} = -\frac{1}{\sqrt{2}}\mathbf{i} + \frac{1}{\sqrt{2}}\mathbf{j};$ f increases

    most rapidly in the direction $\mathbf{u} = -\frac{1}{\sqrt{2}}\mathbf{i} + \frac{1}{\sqrt{2}}\mathbf{j}$ and decreases most rapidly in the direction $-\mathbf{u} = \frac{1}{\sqrt{2}}\mathbf{i} - \frac{1}{\sqrt{2}}\mathbf{j};$

    $(D_{\mathbf{u}}f)_{P_0} = \nabla f \cdot \mathbf{u} = |\nabla f| = \sqrt{2}$ and $(D_{-\mathbf{u}}f)_{P_0} = -\sqrt{2}$

19. $\nabla f = \frac{1}{y}\mathbf{i} - \left(\frac{x}{y^2} + z\right)\mathbf{j} - y\mathbf{k} \Rightarrow \nabla f(4,1,1) = \mathbf{i} - 5\mathbf{j} - \mathbf{k} \Rightarrow \mathbf{u} = \frac{\nabla f}{|\nabla f|} = \frac{\mathbf{i}-5\mathbf{j}-\mathbf{k}}{\sqrt{1^2+(-5)^2+(-1)^2}}$

    $= \frac{1}{3\sqrt{3}}\mathbf{i} - \frac{5}{3\sqrt{3}}\mathbf{j} - \frac{1}{3\sqrt{3}}\mathbf{k};$ f increases most rapidly in the direction of $\mathbf{u} = \frac{1}{3\sqrt{3}}\mathbf{i} - \frac{5}{3\sqrt{3}}\mathbf{j} - \frac{1}{3\sqrt{3}}\mathbf{k}$ and decreases

    most rapidly in the direction $-\mathbf{u} = -\frac{1}{3\sqrt{3}}\mathbf{i} + \frac{5}{3\sqrt{3}}\mathbf{j} + \frac{1}{3\sqrt{3}}\mathbf{k}; (D_{\mathbf{u}}f)_{P_0} = \nabla f \cdot \mathbf{u} = |\nabla f| = 3\sqrt{3}$ and

    $(D_{-\mathbf{u}}f)_{P_0} = -3\sqrt{3}$

21. $\nabla f = \left(\frac{1}{x} + \frac{1}{x}\right)\mathbf{i} + \left(\frac{1}{y} + \frac{1}{y}\right)\mathbf{j} + \left(\frac{1}{z} + \frac{1}{z}\right)\mathbf{k} \Rightarrow \nabla f(1,1,1) = 2\mathbf{i} + 2\mathbf{j} + 2\mathbf{k} \Rightarrow \mathbf{u} = \frac{\nabla f}{|\nabla f|} = \frac{1}{\sqrt{3}}\mathbf{i} + \frac{1}{\sqrt{3}}\mathbf{j} + \frac{1}{\sqrt{3}}\mathbf{k};$

    f increases most rapidly in the direction $\mathbf{u} = \frac{1}{\sqrt{3}}\mathbf{i} + \frac{1}{\sqrt{3}}\mathbf{j} + \frac{1}{\sqrt{3}}\mathbf{k}$ and decreases most rapidly in the direction

    $-\mathbf{u} = -\frac{1}{\sqrt{3}}\mathbf{i} - \frac{1}{\sqrt{3}}\mathbf{j} - \frac{1}{\sqrt{3}}\mathbf{k}; (D_{\mathbf{u}}f)_{P_0} = \nabla f \cdot \mathbf{u} = |\nabla f| = 2\sqrt{3}$ and $(D_{-\mathbf{u}}f)_{P_0} = -2\sqrt{3}$

23. $\nabla f = 2xi + 2yj \Rightarrow \nabla f\left(\sqrt{2}, \sqrt{2}\right) = 2\sqrt{2}i + 2\sqrt{2}j$

$\Rightarrow$ Tangent line: $2\sqrt{2}\left(x - \sqrt{2}\right) + 2\sqrt{2}\left(y - \sqrt{2}\right) = 0$

$\Rightarrow \sqrt{2}x + \sqrt{2}y = 4$

25. $\nabla f = yi + xj \Rightarrow \nabla f(2, -2) = -2i + 2j$

$\Rightarrow$ Tangent line: $-2(x - 2) + 2(y + 2) = 0$

$\Rightarrow y = x - 4$

27. $\nabla f = yi + (x + 2y)j \Rightarrow \nabla f(3, 2) = 2i + 7j$; a vector orthogonal to ∇f is $\mathbf{v} = 7i - 2j \Rightarrow \mathbf{u} = \frac{\mathbf{v}}{|\mathbf{v}|} = \frac{7i - 2j}{\sqrt{7^2 + (-2)^2}}$

$= \frac{7}{\sqrt{53}}i - \frac{2}{\sqrt{53}}j$ and $-\mathbf{u} = -\frac{7}{\sqrt{53}}i + \frac{2}{\sqrt{53}}j$ are the directions where the derivative is zero

29. $\nabla f = (2x - 3y)i + (-3x + 8y)j \Rightarrow \nabla f(1, 2) = -4i + 13j \Rightarrow |\nabla f(1, 2)| = \sqrt{(-4)^2 + (13)^2} = \sqrt{185}$; no, the maximum rate of change is $\sqrt{185} < 14$

31. $\nabla f = f_x(1, 2)i + f_y(1, 2)j$ and $\mathbf{u}_1 = \frac{i + j}{\sqrt{1^2 + 1^2}} = \frac{1}{\sqrt{2}}i + \frac{1}{\sqrt{2}}j \Rightarrow (D_{\mathbf{u}_1}f)(1, 2) = f_x(1, 2)\left(\frac{1}{\sqrt{2}}\right) + f_y(1, 2)\left(\frac{1}{\sqrt{2}}\right)$

$= 2\sqrt{2} \Rightarrow f_x(1, 2) + f_y(1, 2) = 4$; $\mathbf{u}_2 = -j \Rightarrow (D_{\mathbf{u}_2}f)(1, 2) = f_x(1, 2)(0) + f_y(1, 2)(-1) = -3 \Rightarrow -f_y(1, 2) = -3$

$\Rightarrow f_y(1, 2) = 3$; then $f_x(1, 2) + 3 = 4 \Rightarrow f_x(1, 2) = 1$; thus $\nabla f(1, 2) = i + 3j$ and $\mathbf{u} = \frac{\mathbf{v}}{|\mathbf{v}|} = \frac{-i - 2j}{\sqrt{(-1)^2 + (-2)^2}}$

$= -\frac{1}{\sqrt{5}}i - \frac{2}{\sqrt{5}}j \Rightarrow (D_{\mathbf{u}}f)_{P_0} = \nabla f \cdot \mathbf{u} = -\frac{1}{\sqrt{5}} - \frac{6}{\sqrt{5}} = -\frac{7}{\sqrt{5}}$

33. The directional derivative is the scalar component. With ∇f evaluated at P_0, the scalar component of ∇f in the direction of $\mathbf{u}$ is $\nabla f \cdot \mathbf{u} = (D_{\mathbf{u}}f)_{P_0}$.

35. If (x, y) is a point on the line, then $\mathbf{T}(x, y) = (x - x_0)i + (y - y_0)j$ is a vector parallel to the line $\Rightarrow \mathbf{T} \cdot \mathbf{N} = 0$ $\Rightarrow A(x - x_0) + B(y - y_0) = 0$, as claimed.

14.6 TANTGENT PLANES AND DIFFERENTIALS

1. (a) $\nabla f = 2xi + 2yj + 2zk \Rightarrow \nabla f(1, 1, 1) = 2i + 2j + 2k \Rightarrow$ Tangent plane: $2(x - 1) + 2(y - 1) + 2(z - 1) = 0$
$\Rightarrow x + y + z = 3$;
   (b) Normal line: $x = 1 + 2t, y = 1 + 2t, z = 1 + 2t$

3. (a) $\nabla f = -2xi + 2k \Rightarrow \nabla f(2, 0, 2) = -4i + 2k \Rightarrow$ Tangent plane: $-4(x - 2) + 2(z - 2) = 0$
$\Rightarrow -4x + 2z + 4 = 0 \Rightarrow -2x + z + 2 = 0$;
   (b) Normal line: $x = 2 - 4t, y = 0, z = 2 + 2t$

5. (a) $\nabla f = (-\pi \sin \pi x - 2xy + ze^{xz})i + (-x^2 + z)j + (xe^{xz} + y)k \Rightarrow \nabla f(0, 1, 2) = 2i + 2j + k \Rightarrow$ Tangent plane:
$2(x - 0) + 2(y - 1) + 1(z - 2) = 0 \Rightarrow 2x + 2y + z - 4 = 0$;
   (b) Normal line: $x = 2t, y = 1 + 2t, z = 2 + t$

7. (a) $\nabla f = i + j + k$ for all points $\Rightarrow \nabla f(0, 1, 0) = i + j + k \Rightarrow$ Tangent plane: $1(x - 0) + 1(y - 1) + 1(z - 0) = 0$
$\Rightarrow x + y + z - 1 = 0$;

(b) Normal line: $x = t, y = 1 + t, z = t$

9. $z = f(x, y) = \ln(x^2 + y^2) \Rightarrow f_x(x, y) = \frac{2x}{x^2 + y^2}$ and $f_y(x, y) = \frac{2y}{x^2 + y^2} \Rightarrow f_x(1, 0) = 2$ and $f_y(1, 0) = 0 \Rightarrow$ from

Eq. (4) the tangent plane at $(1, 0, 0)$ is $2(x - 1) - z = 0$ or $2x - z - 2 = 0$

11. $z = f(x, y) = \sqrt{y - x} \Rightarrow f_x(x, y) = -\frac{1}{2}(y - x)^{-1/2}$ and $f_y(x, y) = \frac{1}{2}(y - x)^{-1/2} \Rightarrow f_x(1, 2) = -\frac{1}{2}$ and $f_y(1, 2) = \frac{1}{2}$

$\Rightarrow$ from Eq. (4) the tangent plane at $(1, 2, 1)$ is $-\frac{1}{2}(x - 1) + \frac{1}{2}(y - 2) - (z - 1) = 0 \Rightarrow x - y + 2z - 1 = 0$

13. $\nabla f = \mathbf{i} + 2y\mathbf{j} + 2\mathbf{k} \Rightarrow \nabla f(1, 1, 1) = \mathbf{i} + 2\mathbf{j} + 2\mathbf{k}$ and $\nabla g = \mathbf{i}$ for all points; $\mathbf{v} = \nabla f \times \nabla g$

$\Rightarrow \mathbf{v} = \begin{vmatrix} \mathbf{i} & \mathbf{j} & \mathbf{k} \\ 1 & 2 & 2 \\ 1 & 0 & 0 \end{vmatrix} = 2\mathbf{j} - 2\mathbf{k} \Rightarrow$ Tangent line: $x = 1, y = 1 + 2t, z = 1 - 2t$

15. $\nabla f = 2x\mathbf{i} + 2\mathbf{j} + 2\mathbf{k} \Rightarrow \nabla f\left(1, 1, \frac{1}{2}\right) = 2\mathbf{i} + 2\mathbf{j} + 2\mathbf{k}$ and $\nabla g = \mathbf{j}$ for all points; $\mathbf{v} = \nabla f \times \nabla g$

$\Rightarrow \mathbf{v} = \begin{vmatrix} \mathbf{i} & \mathbf{j} & \mathbf{k} \\ 2 & 2 & 2 \\ 0 & 1 & 0 \end{vmatrix} = -2\mathbf{i} + 2\mathbf{k} \Rightarrow$ Tangent line: $x = 1 - 2t, y = 1, z = \frac{1}{2} + 2t$

17. $\nabla f = (3x^2 + 6xy^2 + 4y)\mathbf{i} + (6x^2y + 3y^2 + 4x)\mathbf{j} - 2z\mathbf{k} \Rightarrow \nabla f(1, 1, 3) = 13\mathbf{i} + 13\mathbf{j} - 6\mathbf{k}; \nabla g = 2x\mathbf{i} + 2y\mathbf{j} + 2z\mathbf{k}$

$\Rightarrow \nabla g(1, 1, 3) = 2\mathbf{i} + 2\mathbf{j} + 6\mathbf{k}; \mathbf{v} = \nabla f \times \nabla g \Rightarrow \mathbf{v} = \begin{vmatrix} \mathbf{i} & \mathbf{j} & \mathbf{k} \\ 13 & 13 & -6 \\ 2 & 2 & 6 \end{vmatrix} = 90\mathbf{i} - 90\mathbf{j} \Rightarrow$ Tangent line:

$x = 1 + 90t, y = 1 - 90t, z = 3$

19. $\nabla f = \left(\frac{x}{x^2 + y^2 + z^2}\right)\mathbf{i} + \left(\frac{y}{x^2 + y^2 + z^2}\right)\mathbf{j} + \left(\frac{z}{x^2 + y^2 + z^2}\right)\mathbf{k} \Rightarrow \nabla f(3, 4, 12) = \frac{3}{169}\mathbf{i} + \frac{4}{169}\mathbf{j} + \frac{12}{169}\mathbf{k};$

$\mathbf{u} = \frac{\mathbf{v}}{|\mathbf{v}|} = \frac{3\mathbf{i} + 6\mathbf{j} - 2\mathbf{k}}{\sqrt{3^2 + 6^2 + (-2)^2}} = \frac{3}{7}\mathbf{i} + \frac{6}{7}\mathbf{j} - \frac{2}{7}\mathbf{k} \Rightarrow \nabla f \cdot \mathbf{u} = \frac{9}{1183}$ and $df = (\nabla f \cdot \mathbf{u})\, ds = \left(\frac{9}{1183}\right)(0.1) \approx 0.0008$

21. $\nabla g = (1 + \cos z)\mathbf{i} + (1 - \sin z)\mathbf{j} + (-x\sin z - y\cos z)\mathbf{k} \Rightarrow \nabla g(2, -1, 0) = 2\mathbf{i} + \mathbf{j} + \mathbf{k}; \mathbf{A} = \overrightarrow{P_0 P_1} = -2\mathbf{i} + 2\mathbf{j} + 2\mathbf{k}$

$\Rightarrow \mathbf{u} = \frac{\mathbf{v}}{|\mathbf{v}|} = \frac{-2\mathbf{i} + 2\mathbf{j} + 2\mathbf{k}}{\sqrt{(-2)^2 + 2^2 + 2^2}} = -\frac{1}{\sqrt{3}}\mathbf{i} + \frac{1}{\sqrt{3}}\mathbf{j} + \frac{1}{\sqrt{3}}\mathbf{k} \Rightarrow \nabla g \cdot \mathbf{u} = 0$ and $dg = (\nabla g \cdot \mathbf{u})\, ds = (0)(0.2) = 0$

23. (a) The unit tangent vector at $\left(\frac{1}{2}, \frac{\sqrt{3}}{2}\right)$ in the direction of motion is $\mathbf{u} = \frac{\sqrt{3}}{2}\mathbf{i} - \frac{1}{2}\mathbf{j}$;

$\nabla T = (\sin 2y)\mathbf{i} + (2x\cos 2y)\mathbf{j} \Rightarrow \nabla T\left(\frac{1}{2}, \frac{\sqrt{3}}{2}\right) = \left(\sin\sqrt{3}\right)\mathbf{i} + \left(\cos\sqrt{3}\right)\mathbf{j} \Rightarrow D_\mathbf{u}T\left(\frac{1}{2}, \frac{\sqrt{3}}{2}\right) = \nabla T \cdot \mathbf{u}$

$= \frac{\sqrt{3}}{2}\sin\sqrt{3} - \frac{1}{2}\cos\sqrt{3} \approx 0.935°$ C/ft

(b) $\mathbf{r}(t) = (\sin 2t)\mathbf{i} + (\cos 2t)\mathbf{j} \Rightarrow \mathbf{v}(t) = (2\cos 2t)\mathbf{i} - (2\sin 2t)\mathbf{j}$ and $|\mathbf{v}| = 2; \frac{dT}{dt} = \frac{\partial T}{\partial x}\frac{dx}{dt} + \frac{\partial T}{\partial y}\frac{dy}{dt}$

$= \nabla T \cdot \mathbf{v} = \left(\nabla T \cdot \frac{\mathbf{v}}{|\mathbf{v}|}\right)|\mathbf{v}| = (D_\mathbf{u}T)\,|\mathbf{v}|$, where $\mathbf{u} = \frac{\mathbf{v}}{|\mathbf{v}|}$; at $\left(\frac{1}{2}, \frac{\sqrt{3}}{2}\right)$ we have $\mathbf{u} = \frac{\sqrt{3}}{2}\mathbf{i} - \frac{1}{2}\mathbf{j}$ from part (a)

$\Rightarrow \frac{dT}{dt} = \left(\frac{\sqrt{3}}{2}\sin\sqrt{3} - \frac{1}{2}\cos\sqrt{3}\right) \cdot 2 = \sqrt{3}\sin\sqrt{3} - \cos\sqrt{3} \approx 1.87°$ C/sec

25. (a) $f(0, 0) = 1, f_x(x, y) = 2x \Rightarrow f_x(0, 0) = 0, f_y(x, y) = 2y \Rightarrow f_y(0, 0) = 0 \Rightarrow L(x, y) = 1 + 0(x - 0) + 0(y - 0) = 1$

(b) $f(1, 1) = 3, f_x(1, 1) = 2, f_y(1, 1) = 2 \Rightarrow L(x, y) = 3 + 2(x - 1) + 2(y - 1) = 2x + 2y - 1$

27. (a) $f(0, 0) = 5, f_x(x, y) = 3$ for all $(x, y), f_y(x, y) = -4$ for all $(x, y) \Rightarrow L(x, y) = 5 + 3(x - 0) - 4(y - 0)$

$= 3x - 4y + 5$

(b) $f(1, 1) = 4, f_x(1, 1) = 3, f_y(1, 1) = -4 \Rightarrow L(x, y) = 4 + 3(x - 1) - 4(y - 1) = 3x - 4y + 5$

29. (a) $f(0,0) = 1, f_x(x,y) = e^x \cos y \Rightarrow f_x(0,0) = 1, f_y(x,y) = -e^x \sin y \Rightarrow f_y(0,0) = 0$
    $\Rightarrow L(x,y) = 1 + 1(x-0) + 0(y-0) = x + 1$

    (b) $f\left(0, \frac{\pi}{2}\right) = 0, f_x\left(0, \frac{\pi}{2}\right) = 0, f_y\left(0, \frac{\pi}{2}\right) = -1 \Rightarrow L(x,y) = 0 + 0(x-0) - 1\left(y - \frac{\pi}{2}\right) = -y + \frac{\pi}{2}$

31. $f(2,1) = 3, f_x(x,y) = 2x - 3y \Rightarrow f_x(2,1) = 1, f_y(x,y) = -3x \Rightarrow f_y(2,1) = -6 \Rightarrow L(x,y) = 3 + 1(x-2) - 6(y-1)$
    $= 7 + x - 6y; f_{xx}(x,y) = 2, f_{yy}(x,y) = 0, f_{xy}(x,y) = -3 \Rightarrow M = 3;$ thus $|E(x,y)| \leq \left(\frac{1}{2}\right)(3)\left(|x-2| + |y-1|\right)^2$
    $\leq \left(\frac{3}{2}\right)(0.1 + 0.1)^2 = 0.06$

33. $f(0,0) = 1, f_x(x,y) = \cos y \Rightarrow f_x(0,0) = 1, f_y(x,y) = 1 - x \sin y \Rightarrow f_y(0,0) = 1$
    $\Rightarrow L(x,y) = 1 + 1(x-0) + 1(y-0) = x + y + 1; f_{xx}(x,y) = 0, f_{yy}(x,y) = -x \cos y, f_{xy}(x,y) = -\sin y \Rightarrow M = 1;$
    thus $|E(x,y)| \leq \left(\frac{1}{2}\right)(1)\left(|x| + |y|\right)^2 \leq \left(\frac{1}{2}\right)(0.2 + 0.2)^2 = 0.08$

35. $f(0,0) = 1, f_x(x,y) = e^x \cos y \Rightarrow f_x(0,0) = 1, f_y(x,y) = -e^x \sin y \Rightarrow f_y(0,0) = 0$
    $\Rightarrow L(x,y) = 1 + 1(x-0) + 0(y-0) = 1 + x; f_{xx}(x,y) = e^x \cos y, f_{yy}(x,y) = -e^x \cos y, f_{xy}(x,y) = -e^x \sin y;$
    $|x| \leq 0.1 \Rightarrow -0.1 \leq x \leq 0.1$ and $|y| \leq 0.1 \Rightarrow -0.1 \leq y \leq 0.1;$ thus the max of $|f_{xx}(x,y)|$ on R is $e^{0.1} \cos(0.1)$
    $\leq 1.11,$ the max of $|f_{yy}(x,y)|$ on R is $e^{0.1} \cos(0.1) \leq 1.11,$ and the max of $|f_{xy}(x,y)|$ on R is $e^{0.1} \sin(0.1)$
    $\leq 0.12 \Rightarrow M = 1.11;$ thus $|E(x,y)| \leq \left(\frac{1}{2}\right)(1.11)\left(|x| + |y|\right)^2 \leq (0.555)(0.1 + 0.1)^2 = 0.0222$

37. (a) $f(1,1,1) = 3, f_x(1,1,1) = y + z|_{(1,1,1)} = 2, f_y(1,1,1) = x + z|_{(1,1,1)} = 2, f_z(1,1,1) = y + x|_{(1,1,1)} = 2$
    $\Rightarrow L(x,y,z) = 3 + 2(x-1) + 2(y-1) + 2(z-1) = 2x + 2y + 2z - 3$

    (b) $f(1,0,0) = 0, f_x(1,0,0) = 0, f_y(1,0,0) = 1, f_z(1,0,0) = 1 \Rightarrow L(x,y,z) = 0 + 0(x-1) + (y-0) + (z-0)$
    $= y + z$

    (c) $f(0,0,0) = 0, f_x(0,0,0) = 0, f_y(0,0,0) = 0, f_z(0,0,0) = 0 \Rightarrow L(x,y,z) = 0$

39. (a) $f(1,0,0) = 1, f_x(1,0,0) = \left.\frac{x}{\sqrt{x^2 + y^2 + z^2}}\right|_{(1,0,0)} = 1, f_y(1,0,0) = \left.\frac{y}{\sqrt{x^2 + y^2 + z^2}}\right|_{(1,0,0)} = 0,$
    $f_z(1,0,0) = \left.\frac{z}{\sqrt{x^2 + y^2 + z^2}}\right|_{(1,0,0)} = 0 \Rightarrow L(x,y,z) = 1 + 1(x-1) + 0(y-0) + 0(z-0) = x$

    (b) $f(1,1,0) = \sqrt{2}, f_x(1,1,0) = \frac{1}{\sqrt{2}}, f_y(1,1,0) = \frac{1}{\sqrt{2}}, f_z(1,1,0) = 0$
    $\Rightarrow L(x,y,z) = \sqrt{2} + \frac{1}{\sqrt{2}}(x-1) + \frac{1}{\sqrt{2}}(y-1) + 0(z-0) = \frac{1}{\sqrt{2}}x + \frac{1}{\sqrt{2}}y$

    (c) $f(1,2,2) = 3, f_x(1,2,2) = \frac{1}{3}, f_y(1,2,2) = \frac{2}{3}, f_z(1,2,2) = \frac{2}{3} \Rightarrow L(x,y,z) = 3 + \frac{1}{3}(x-1) + \frac{2}{3}(y-2) + \frac{2}{3}(z-2)$
    $= \frac{1}{3}x + \frac{2}{3}y + \frac{2}{3}z$

41. (a) $f(0,0,0) = 2, f_x(0,0,0) = e^x|_{(0,0,0)} = 1, f_y(0,0,0) = -\sin(y+z)|_{(0,0,0)} = 0,$
    $f_z(0,0,0) = -\sin(y+z)|_{(0,0,0)} = 0 \Rightarrow L(x,y,z) = 2 + 1(x-0) + 0(y-0) + 0(z-0) = 2 + x$

    (b) $f\left(0, \frac{\pi}{2}, 0\right) = 1, f_x\left(0, \frac{\pi}{2}, 0\right) = 1, f_y\left(0, \frac{\pi}{2}, 0\right) = -1, f_z\left(0, \frac{\pi}{2}, 0\right) = -1 \Rightarrow L(x,y,z)$
    $= 1 + 1(x-0) - 1\left(y - \frac{\pi}{2}\right) - 1(z-0) = x - y - z + \frac{\pi}{2} + 1$

    (c) $f\left(0, \frac{\pi}{4}, \frac{\pi}{4}\right) = 1, f_x\left(0, \frac{\pi}{4}, \frac{\pi}{4}\right) = 1, f_y\left(0, \frac{\pi}{4}, \frac{\pi}{4}\right) = -1, f_z\left(0, \frac{\pi}{4}, \frac{\pi}{4}\right) = -1 \Rightarrow L(x,y,z)$
    $= 1 + 1(x-0) - 1\left(y - \frac{\pi}{4}\right) - 1\left(z - \frac{\pi}{4}\right) = x - y - z + \frac{\pi}{2} + 1$

43. $f(x,y,z) = xz - 3yz + 2$ at $P_0(1,1,2) \Rightarrow f(1,1,2) = -2; f_x = z, f_y = -3z, f_z = x - 3y \Rightarrow L(x,y,z)$
    $= -2 + 2(x-1) - 6(y-1) - 2(z-2) = 2x - 6y - 2z + 6; f_{xx} = 0, f_{yy} = 0, f_{zz} = 0, f_{xy} = 0, f_{yz} = -3$
    $\Rightarrow M = 3;$ thus, $|E(x,y,z)| \leq \left(\frac{1}{2}\right)(3)(0.01 + 0.01 + 0.02)^2 = 0.0024$

45. $f(x,y,z) = xy + 2yz - 3xz$ at $P_0(1,1,0) \Rightarrow f(1,1,0) = 1; f_x = y - 3z, f_y = x + 2z, f_z = 2y - 3x$
    $\Rightarrow L(x,y,z) = 1 + (x-1) + (y-1) - (z-0) = x + y - z - 1; f_{xx} = 0, f_{yy} = 0, f_{zz} = 0, f_{xy} = 1, f_{xz} = -3,$

$f_{yz} = 2 \Rightarrow M = 3$; thus $|E(x, y, z)| \le \left(\frac{1}{2}\right)(3)(0.01 + 0.01 + 0.01)^2 = 0.00135$

47. $T_x(x, y) = e^y + e^{-y}$ and $T_y(x, y) = x\left(e^y - e^{-y}\right) \Rightarrow dT = T_x(x, y)\, dx + T_y(x, y)\, dy$

$= \left(e^y + e^{-y}\right) dx + x\left(e^y - e^{-y}\right) dy \Rightarrow dT\big|_{(2, \ln 2)} = 2.5\, dx + 3.0\, dy$. If $|dx| \le 0.1$ and $|dy| \le 0.02$, then the

maximum possible error in the computed value of T is $(2.5)(0.1) + (3.0)(0.02) = 0.31$ in magnitude.

49. $V_r = 2\pi rh$ and $V_h = \pi r^2 \Rightarrow dV = V_r\, dr + V_h\, dh \Rightarrow dV = 2\pi rh\, dr + \pi r^2\, dh \Rightarrow dV\big|_{(5,12)} = 120\pi\, dr + 25\pi\, dh$;

$|dr| \le 0.1$ cm and $|dh| \le 0.1$ cm $\Rightarrow dV \le (120\pi)(0.1) + (25\pi)(0.1) = 14.5\pi$ cm^3; $V(5, 12) = 300\pi$ cm^3

$\Rightarrow$ maximum percentage error is $\pm \frac{14.5\pi}{300\pi} \times 100 = \pm 4.83\%$

51. $A = xy \Rightarrow dA = x\, dy + y\, dx$; if $x > y$ then a 1-unit change in y gives a greater change in dA than a 1-unit

change in x. Thus, pay more attention to y which is the smaller of the two dimensions.

53. (a) $r^2 = x^2 + y^2 \Rightarrow 2r\, dr = 2x\, dx + 2y\, dy \Rightarrow dr = \frac{x}{r}\, dx + \frac{y}{r}\, dy \Rightarrow dr\big|_{(3,4)} = \left(\frac{3}{5}\right)(\pm 0.01) + \left(\frac{4}{5}\right)(\pm 0.01)$

$= \pm \frac{0.07}{5} = \pm 0.014 \Rightarrow \left|\frac{dr}{r} \times 100\right| = \left|\pm \frac{0.014}{5} \times 100\right| = 0.28\%$; $d\theta = \frac{\left(-\frac{y}{x^2}\right)}{\left(\frac{y}{x}\right)^2 + 1}\, dx + \frac{\left(\frac{1}{x}\right)}{\left(\frac{y}{x}\right)^2 + 1}\, dy$

$= \frac{-y}{y^2 + x^2}\, dx + \frac{x}{y^2 + x^2}\, dy \Rightarrow d\theta\big|_{(3,4)} = \left(\frac{-4}{25}\right)(\pm 0.01) + \left(\frac{3}{25}\right)(\pm 0.01) = \frac{\mp 0.04}{25} + \frac{\pm 0.03}{25}$

$\Rightarrow$ maximum change in $d\theta$ occurs when dx and dy have opposite signs (dx = 0.01 and dy = −0.01 or vice

versa) $\Rightarrow d\theta = \frac{\pm 0.07}{25} \approx \pm 0.0028$; $\theta = \tan^{-1}\left(\frac{4}{3}\right) \approx 0.927255218 \Rightarrow \left|\frac{d\theta}{\theta} \times 100\right| = \left|\frac{\pm 0.0028}{0.927255218} \times 100\right|$

$\approx 0.30\%$

(b) the radius r is more sensitive to changes in y, and the angle θ is more sensitive to changes in x

55. $f(a, b, c, d) = \begin{vmatrix} a & b \\ c & d \end{vmatrix} = ad - bc \Rightarrow f_a = d,\ f_b = -c,\ f_c = -b,\ f_d = a \Rightarrow df = d\, da - c\, db - b\, dc + a\, dd$; since

$|a|$ is much greater than $|b|$, $|c|$, and $|d|$, the function f is most sensitive to a change in d.

57. $Q_K = \frac{1}{2}\left(\frac{2KM}{h}\right)^{-1/2}\left(\frac{2M}{h}\right)$, $Q_M = \frac{1}{2}\left(\frac{2KM}{h}\right)^{-1/2}\left(\frac{2K}{h}\right)$, and $Q_h = \frac{1}{2}\left(\frac{2KM}{h}\right)^{-1/2}\left(\frac{-2KM}{h^2}\right)$

$\Rightarrow dQ = \frac{1}{2}\left(\frac{2KM}{h}\right)^{-1/2}\left(\frac{2M}{h}\right) dK + \frac{1}{2}\left(\frac{2KM}{h}\right)^{-1/2}\left(\frac{2K}{h}\right) dM + \frac{1}{2}\left(\frac{2KM}{h}\right)^{-1/2}\left(\frac{-2KM}{h^2}\right) dh$

$= \frac{1}{2}\left(\frac{2KM}{h}\right)^{-1/2}\left[\frac{2M}{h}\, dK + \frac{2K}{h}\, dM - \frac{2KM}{h^2}\, dh\right] \Rightarrow dQ\big|_{(2,20,0.0.05)}$

$= \frac{1}{2}\left[\frac{(2)(2)(20)}{0.05}\right]^{-1/2}\left[\frac{(2)(20)}{0.05}\, dK + \frac{(2)(2)}{0.05}\, dM - \frac{(2)(2)(20)}{(0.05)^2}\, dh\right] = (0.0125)(800\, dK + 80\, dM - 32{,}000\, dh)$

$\Rightarrow$ Q is most sensitive to changes in h

59. $z = f(x, y) \Rightarrow g(x, y, z) = f(x, y) - z = 0 \Rightarrow g_x(x, y, z) = f_x(x, y),\ g_y(x, y, z) = f_y(x, y)$ and $g_z(x, y, z) = -1$

$\Rightarrow g_x(x_0, y_0, f(x_0, y_0)) = f_x(x_0, y_0),\ g_y(x_0, y_0, f(x_0, y_0)) = f_y(x_0, y_0)$ and $g_z(x_0, y_0, f(x_0, y_0)) = -1 \Rightarrow$ the tangent

plane at the point P_0 is $f_x(x_0, y_0)(x - x_0) + f_y(x_0, y_0)(y - y_0) - [z - f(x_0, y_0)] = 0$ or

$z = f_x(x_0, y_0)(x - x_0) + f_y(x_0, y_0)(y - y_0) + f(x_0, y_0)$

61. $\nabla f = 2x\mathbf{i} + 2y\mathbf{j} + 2z\mathbf{k} = (2\cos t)\mathbf{i} + (2\sin t)\mathbf{j} + 2t\mathbf{k}$ and $\mathbf{v} = (-\sin t)\mathbf{i} + (\cos t)\mathbf{j} + \mathbf{k} \Rightarrow \mathbf{u} = \frac{\mathbf{v}}{|\mathbf{v}|}$

$= \frac{(-\sin t)\mathbf{i} + (\cos t)\mathbf{j} + \mathbf{k}}{\sqrt{(\sin t)^2 + (\cos t)^2 + 1^2}} = \left(\frac{-\sin t}{\sqrt{2}}\right)\mathbf{i} + \left(\frac{\cos t}{\sqrt{2}}\right)\mathbf{j} + \frac{1}{\sqrt{2}}\mathbf{k} \Rightarrow (D_\mathbf{u}f)_{P_0} = \nabla f \cdot \mathbf{u}$

$= (2\cos t)\left(\frac{-\sin t}{\sqrt{2}}\right) + (2\sin t)\left(\frac{\cos t}{\sqrt{2}}\right) + (2t)\left(\frac{1}{\sqrt{2}}\right) = \frac{2t}{\sqrt{2}} \Rightarrow (D_\mathbf{u}f)\left(\frac{-\pi}{4}\right) = \frac{-\pi}{2\sqrt{2}},\ (D_\mathbf{u}f)(0) = 0$ and

$(D_\mathbf{u}f)\left(\frac{\pi}{4}\right) = \frac{\pi}{2\sqrt{2}}$

63. $\mathbf{r} = \sqrt{t}\,\mathbf{i} + \sqrt{t}\,\mathbf{j} + (2t - 1)\mathbf{k} \Rightarrow \mathbf{v} = \frac{1}{2}t^{-1/2}\mathbf{i} + \frac{1}{2}t^{-1/2}\mathbf{j} + 2\mathbf{k}$; $t = 1 \Rightarrow x = 1, y = 1, z = 1 \Rightarrow P_0 = (1, 1, 1)$ and

$\mathbf{v}(1) = \frac{1}{2}\mathbf{i} + \frac{1}{2}\mathbf{j} + 2\mathbf{k}$; $f(x, y, z) = x^2 + y^2 - z - 1 = 0 \Rightarrow \nabla f = 2x\mathbf{i} + 2y\mathbf{j} - \mathbf{k} \Rightarrow \nabla f(1, 1, 1) = 2\mathbf{i} + 2\mathbf{j} - \mathbf{k}$;

now $\mathbf{v}(1) \cdot \nabla f(1, 1, 1) = 0$, thus the curve is tangent to the surface when $t = 1$

14.7 EXTREME VALUES AND SADDLE POINTS

1. $f_x(x, y) = 2x + y + 3 = 0$ and $f_y(x, y) = x + 2y - 3 = 0 \Rightarrow x = -3$ and $y = 3 \Rightarrow$ critical point is $(-3, 3)$; $f_{xx}(-3, 3) = 2, f_{yy}(-3, 3) = 2, f_{xy}(-3, 3) = 1 \Rightarrow f_{xx}f_{yy} - f_{xy}^2 = 3 > 0$ and $f_{xx} > 0 \Rightarrow$ local minimum of $f(-3, 3) = -5$

3. $f_x(x, y) = 2y - 10x + 4 = 0$ and $f_y(x, y) = 2x - 4y + 4 = 0 \Rightarrow x = \frac{2}{3}$ and $y = \frac{4}{3} \Rightarrow$ critical point is $\left(\frac{2}{3}, \frac{4}{3}\right)$; $f_{xx}\left(\frac{2}{3}, \frac{4}{3}\right) = -10, f_{yy}\left(\frac{2}{3}, \frac{4}{3}\right) = -4, f_{xy}\left(\frac{2}{3}, \frac{4}{3}\right) = 2 \Rightarrow f_{xx}f_{yy} - f_{xy}^2 = 36 > 0$ and $f_{xx} < 0 \Rightarrow$ local maximum of $f\left(\frac{2}{3}, \frac{4}{3}\right) = 0$

5. $f_x(x, y) = 2x + y + 3 = 0$ and $f_y(x, y) = x + 2 = 0 \Rightarrow x = -2$ and $y = 1 \Rightarrow$ critical point is $(-2, 1)$; $f_{xx}(-2, 1) = 2, f_{yy}(-2, 1) = 0, f_{xy}(-2, 1) = 1 \Rightarrow f_{xx}f_{yy} - f_{xy}^2 = -1 < 0 \Rightarrow$ saddle point

7. $f_x(x, y) = 5y - 14x + 3 = 0$ and $f_y(x, y) = 5x - 6 = 0 \Rightarrow x = \frac{6}{5}$ and $y = \frac{69}{25} \Rightarrow$ critical point is $\left(\frac{6}{5}, \frac{69}{25}\right)$; $f_{xx}\left(\frac{6}{5}, \frac{69}{25}\right) = -14, f_{yy}\left(\frac{6}{5}, \frac{69}{25}\right) = 0, f_{xy}\left(\frac{6}{5}, \frac{69}{25}\right) = 5 \Rightarrow f_{xx}f_{yy} - f_{xy}^2 = -25 < 0 \Rightarrow$ saddle point

9. $f_x(x, y) = 2x - 4y = 0$ and $f_y(x, y) = -4x + 2y + 6 = 0 \Rightarrow x = 2$ and $y = 1 \Rightarrow$ critical point is $(2, 1)$; $f_{xx}(2, 1) = 2, f_{yy}(2, 1) = 2, f_{xy}(2, 1) = -4 \Rightarrow f_{xx}f_{yy} - f_{xy}^2 = -12 < 0 \Rightarrow$ saddle point

11. $f_x(x, y) = 4x + 3y - 5 = 0$ and $f_y(x, y) = 3x + 8y + 2 = 0 \Rightarrow x = 2$ and $y = -1 \Rightarrow$ critical point is $(2, -1)$; $f_{xx}(2, -1) = 4, f_{yy}(2, -1) = 8, f_{xy}(2, -1) = 3 \Rightarrow f_{xx}f_{yy} - f_{xy}^2 = 23 > 0$ and $f_{xx} > 0 \Rightarrow$ local minimum of $f(2, -1) = -6$

13. $f_x(x, y) = 2x - 2 = 0$ and $f_y(x, y) = -2y + 4 = 0 \Rightarrow x = 1$ and $y = 2 \Rightarrow$ critical point is $(1, 2)$; $f_{xx}(1, 2) = 2, f_{yy}(1, 2) = -2, f_{xy}(1, 2) = 0 \Rightarrow f_{xx}f_{yy} - f_{xy}^2 = -4 < 0 \Rightarrow$ saddle point

15. $f_x(x, y) = 2x + 2y = 0$ and $f_y(x, y) = 2x = 0 \Rightarrow x = 0$ and $y = 0 \Rightarrow$ critical point is $(0, 0)$; $f_{xx}(0, 0) = 2, f_{yy}(0, 0) = 0, f_{xy}(0, 0) = 2 \Rightarrow f_{xx}f_{yy} - f_{xy}^2 = -4 < 0 \Rightarrow$ saddle point

17. $f_x(x, y) = 3x^2 - 2y = 0$ and $f_y(x, y) = -3y^2 - 2x = 0 \Rightarrow x = 0$ and $y = 0$, or $x = -\frac{2}{3}$ and $y = \frac{2}{3} \Rightarrow$ critical points are $(0, 0)$ and $\left(-\frac{2}{3}, \frac{2}{3}\right)$; for $(0, 0)$: $f_{xx}(0, 0) = 6x|_{(0,0)} = 0, f_{yy}(0, 0) = -6y|_{(0,0)} = 0, f_{xy}(0, 0) = -2$ $\Rightarrow f_{xx}f_{yy} - f_{xy}^2 = -4 < 0 \Rightarrow$ saddle point; for $\left(-\frac{2}{3}, \frac{2}{3}\right)$: $f_{xx}\left(-\frac{2}{3}, \frac{2}{3}\right) = -4, f_{yy}\left(-\frac{2}{3}, \frac{2}{3}\right) = -4, f_{xy}\left(-\frac{2}{3}, \frac{2}{3}\right) = -2$ $\Rightarrow f_{xx}f_{yy} - f_{xy}^2 = 12 > 0$ and $f_{xx} < 0 \Rightarrow$ local maximum of $f\left(-\frac{2}{3}, \frac{2}{3}\right) = \frac{170}{27}$

19. $f_x(x, y) = 12x - 6x^2 + 6y = 0$ and $f_y(x, y) = 6y + 6x = 0 \Rightarrow x = 0$ and $y = 0$, or $x = 1$ and $y = -1 \Rightarrow$ critical points are $(0, 0)$ and $(1, -1)$; for $(0, 0)$: $f_{xx}(0, 0) = 12 - 12x|_{(0,0)} = 12, f_{yy}(0, 0) = 6, f_{xy}(0, 0) = 6 \Rightarrow f_{xx}f_{yy} - f_{xy}^2$ $= 36 > 0$ and $f_{xx} > 0 \Rightarrow$ local minimum of $f(0, 0) = 0$; for $(1, -1)$: $f_{xx}(1, -1) = 0, f_{yy}(1, -1) = 6,$ $f_{xy}(1, -1) = 6 \Rightarrow f_{xx}f_{yy} - f_{xy}^2 = -36 < 0 \Rightarrow$ saddle point

21. $f_x(x, y) = 27x^2 - 4y = 0$ and $f_y(x, y) = y^2 - 4x = 0 \Rightarrow x = 0$ and $y = 0$, or $x = \frac{4}{9}$ and $y = \frac{4}{3} \Rightarrow$ critical points are $(0, 0)$ and $\left(\frac{4}{9}, \frac{4}{3}\right)$; for $(0, 0)$: $f_{xx}(0, 0) = 54x|_{(0,0)} = 0, f_{yy}(0, 0) = 2y|_{(0,0)} = 0, f_{xy}(0, 0) = -4 \Rightarrow f_{xx}f_{yy} - f_{xy}^2$ $= -16 < 0 \Rightarrow$ saddle point; for $\left(\frac{4}{9}, \frac{4}{3}\right)$: $f_{xx}\left(\frac{4}{9}, \frac{4}{3}\right) = 24, f_{yy}\left(\frac{4}{9}, \frac{4}{3}\right) = \frac{8}{3}, f_{xy}\left(\frac{4}{9}, \frac{4}{3}\right) = -4 \Rightarrow f_{xx}f_{yy} - f_{xy}^2 = 48 > 0$ and $f_{xx} > 0 \Rightarrow$ local minimum of $f\left(\frac{4}{9}, \frac{4}{3}\right) = -\frac{64}{81}$

23. $f_x(x, y) = 3x^2 + 6x = 0 \Rightarrow x = 0$ or $x = -2$; $f_y(x, y) = 3y^2 - 6y = 0 \Rightarrow y = 0$ or $y = 2 \Rightarrow$ the critical points are $(0, 0), (0, 2), (-2, 0),$ and $(-2, 2)$; for $(0, 0)$: $f_{xx}(0, 0) = 6x + 6|_{(0,0)} = 6, f_{yy}(0, 0) = 6y - 6|_{(0,0)} = -6,$

$f_{xy}(0,0) = 0 \Rightarrow f_{xx}f_{yy} - f_{xy}^2 = -36 < 0 \Rightarrow$ saddle point; for $(0,2)$: $f_{xx}(0,2) = 6, f_{yy}(0,2) = 6, f_{xy}(0,2) = 0$

$\Rightarrow f_{xx}f_{yy} - f_{xy}^2 = 36 > 0$ and $f_{xx} > 0 \Rightarrow$ local minimum of $f(0,2) = -12$; for $(-2,0)$: $f_{xx}(-2,0) = -6$,

$f_{yy}(-2,0) = -6, f_{xy}(-2,0) = 0 \Rightarrow f_{xx}f_{yy} - f_{xy}^2 = 36 > 0$ and $f_{xx} < 0 \Rightarrow$ local maximum of $f(-2,0) = -4$;

for $(-2,2)$: $f_{xx}(-2,2) = -6, f_{yy}(-2,2) = 6, f_{xy}(-2,2) = 0 \Rightarrow f_{xx}f_{yy} - f_{xy}^2 = -36 < 0 \Rightarrow$ saddle point

25. $f_x(x,y) = 4y - 4x^3 = 0$ and $f_y(x,y) = 4x - 4y^3 = 0 \Rightarrow x = y \Rightarrow x(1 - x^2) = 0 \Rightarrow x = 0, 1, -1 \Rightarrow$ the critical

points are $(0,0), (1,1)$, and $(-1,-1)$; for $(0,0)$: $f_{xx}(0,0) = -12x^2|_{(0,0)} = 0, f_{yy}(0,0) = -12y^2|_{(0,0)} = 0$,

$f_{xy}(0,0) = 4 \Rightarrow f_{xx}f_{yy} - f_{xy}^2 = -16 < 0 \Rightarrow$ saddle point; for $(1,1)$: $f_{xx}(1,1) = -12, f_{yy}(1,1) = -12, f_{xy}(1,1) = 4$

$\Rightarrow f_{xx}f_{yy} - f_{xy}^2 = 128 > 0$ and $f_{xx} < 0 \Rightarrow$ local maximum of $f(1,1) = 2$; for $(-1,-1)$: $f_{xx}(-1,-1) = -12$,

$f_{yy}(-1,-1) = -12, f_{xy}(-1,-1) = 4 \Rightarrow f_{xx}f_{yy} - f_{xy}^2 = 128 > 0$ and $f_{xx} < 0 \Rightarrow$ local maximum of $f(-1,-1) = 2$

27. $f_x(x,y) = \dfrac{-2x}{(x^2+y^2-1)^2} = 0$ and $f_y(x,y) = \dfrac{-2y}{(x^2+y^2-1)^2} = 0 \Rightarrow x = 0$ and $y = 0 \Rightarrow$ the critical point is $(0,0)$;

$f_{xx} = \dfrac{4x^2 - 2y^2 + 2}{(x^2+y^2-1)^3}, f_{yy} = \dfrac{-2x^2 + 4y^2 + 2}{(x^2+y^2-1)^3}, f_{xy} = \dfrac{8xy}{(x^2+y^2-1)^3}$; $f_{xx}(0,0) = -2, f_{yy}(0,0) = -2, f_{xy}(0,0) = 0$

$\Rightarrow f_{xx}f_{yy} - f_{xy}^2 = 4 > 0$ and $f_{xx} < 0 \Rightarrow$ local maximum of $f(0,0) = -1$

29. $f_x(x,y) = y \cos x = 0$ and $f_y(x,y) = \sin x = 0 \Rightarrow x = n\pi$, n an integer, and $y = 0 \Rightarrow$ the critical points are

$(n\pi, 0)$, n an integer (Note: $\cos x$ and $\sin x$ cannot both be 0 for the same x, so $\sin x$ must be 0 and $y = 0$);

$f_{xx} = -y \sin x, f_{yy} = 0, f_{xy} = \cos x$; $f_{xx}(n\pi, 0) = 0, f_{yy}(n\pi, 0) = 0, f_{xy}(n\pi, 0) = 1$ if n is even and $f_{xy}(n\pi, 0) = -1$

if n is odd $\Rightarrow f_{xx}f_{yy} - f_{xy}^2 = -1 < 0 \Rightarrow$ saddle point.

31. (i)   On OA, $f(x,y) = f(0,y) = y^2 - 4y + 1$ on $0 \le y \le 2$;

       $f'(0,y) = 2y - 4 = 0 \Rightarrow y = 2$;

       $f(0,0) = 1$ and $f(0,2) = -3$

   (ii)   On AB, $f(x,y) = f(x,2) = 2x^2 - 4x - 3$ on $0 \le x \le 1$;

       $f'(x,2) = 4x - 4 = 0 \Rightarrow x = 1$;

       $f(0,2) = -3$ and $f(1,2) = -5$

  (iii)   On OB, $f(x,y) = f(x,2x) = 6x^2 - 12x + 1$ on

       $0 \le x \le 1$; endpoint values have been found above;

       $f'(x,2x) = 12x - 12 = 0 \Rightarrow x = 1$ and $y = 2$, but $(1,2)$ is not an interior point of OB

  (iv)   For interior points of the triangular region, $f_x(x,y) = 4x - 4 = 0$ and $f_y(x,y) = 2y - 4 = 0$

       $\Rightarrow x = 1$ and $y = 2$, but $(1,2)$ is not an interior point of the region. Therefore, the absolute maximum is

       1 at $(0,0)$ and the absolute minimum is -5 at $(1,2)$.

33. (i)   On OA, $f(x,y) = f(0,y) = y^2$ on $0 \le y \le 2$;

       $f'(0,y) = 2y = 0 \Rightarrow y = 0$ and $x = 0$; $f(0,0) = 0$ and

       $f(0,2) = 4$

   (ii)   On OB, $f(x,y) = f(x,0) = x^2$ on $0 \le x \le 1$;

       $f'(x,0) = 2x = 0 \Rightarrow x = 0$ and $y = 0$; $f(0,0) = 0$ and

       $f(1,0) = 1$

  (iii)   On AB, $f(x,y) = f(x,-2x+2) = 5x^2 - 8x + 4$ on

       $0 \le x \le 1$; $f'(x, -2x+2) = 10x - 8 = 0 \Rightarrow x = \frac{4}{5}$

       and $y = \frac{2}{5}$; $f\left(\frac{4}{5}, \frac{2}{5}\right) = \frac{4}{5}$; endpoint values have been found above.

  (iv)   For interior points of the triangular region, $f_x(x,y) = 2x = 0$ and $f_y(x,y) = 2y = 0 \Rightarrow x = 0$ and $y = 0$, but $(0,0)$ is

       not an interior point of the region. Therefore the absolute maximum is 4 at $(0,2)$ and the absolute minimum is 0 at

       $(0,0)$.

35. (i) On OC, $T(x, y) = T(x, 0) = x^2 - 6x + 2$ on

$0 \le x \le 5$; $T'(x, 0) = 2x - 6 = 0 \Rightarrow x = 3$ and

$y = 0$; $T(3, 0) = -7$, $T(0, 0) = 2$, and $T(5, 0) = -3$

(ii) On CB, $T(x, y) = T(5, y) = y^2 + 5y - 3$ on

$-3 \le y \le 0$; $T'(5, y) = 2y + 5 = 0 \Rightarrow y = -\frac{5}{2}$ and

$x = 5$; $T\left(5, -\frac{5}{2}\right) = -\frac{37}{4}$ and $T(5, -3) = -9$

(iii) On AB, $T(x, y) = T(x, -3) = x^2 - 9x + 11$ on

$0 \le x \le 5$; $T'(x, -3) = 2x - 9 = 0 \Rightarrow x = \frac{9}{2}$ and

$y = -3$; $T\left(\frac{9}{2}, -3\right) = -\frac{37}{4}$ and $T(0, -3) = 11$

(iv) On AO, $T(x, y) = T(0, y) = y^2 + 2$ on $-3 \le y \le 0$; $T'(0, y) = 2y = 0 \Rightarrow y = 0$ and $x = 0$, but $(0, 0)$ is

not an interior point of AO

(v) For interior points of the rectangular region, $T_x(x, y) = 2x + y - 6 = 0$ and $T_y(x, y) = x + 2y = 0 \Rightarrow x = 4$

and $y = -2$, an interior critical point with $T(4, -2) = -10$. Therefore the absolute maximum is 11 at

$(0, -3)$ and the absolute minimum is -10 at $(4, -2)$.

37. (i) On AB, $f(x, y) = f(1, y) = 3 \cos y$ on $-\frac{\pi}{4} \le y \le \frac{\pi}{4}$;

$f'(1, y) = -3 \sin y = 0 \Rightarrow y = 0$ and $x = 1$;

$f(1, 0) = 3$, $f\left(1, -\frac{\pi}{4}\right) = \frac{3\sqrt{2}}{2}$, and $f\left(1, \frac{\pi}{4}\right) = \frac{3\sqrt{2}}{2}$

(ii) On CD, $f(x, y) = f(3, y) = 3 \cos y$ on $-\frac{\pi}{4} \le y \le \frac{\pi}{4}$;

$f'(3, y) = -3 \sin y = 0 \Rightarrow y = 0$ and $x = 3$;

$f(3, 0) = 3$, $f\left(3, -\frac{\pi}{4}\right) = \frac{3\sqrt{2}}{2}$ and $f\left(3, \frac{\pi}{4}\right) = \frac{3\sqrt{2}}{2}$

(iii) On BC, $f(x, y) = f\left(x, \frac{\pi}{4}\right) = \frac{\sqrt{2}}{2}(4x - x^2)$ on

$1 \le x \le 3$; $f'\left(x, \frac{\pi}{4}\right) = \sqrt{2}(2 - x) = 0 \Rightarrow x = 2$ and $y = \frac{\pi}{4}$; $f\left(2, \frac{\pi}{4}\right) = 2\sqrt{2}$, $f\left(1, \frac{\pi}{4}\right) = \frac{3\sqrt{2}}{2}$, and

$f\left(3, \frac{\pi}{4}\right) = \frac{3\sqrt{2}}{2}$

(iv) On AD, $f(x, y) = f\left(x, -\frac{\pi}{4}\right) = \frac{\sqrt{2}}{2}(4x - x^2)$ on $1 \le x \le 3$; $f'\left(x, -\frac{\pi}{4}\right) = \sqrt{2}(2 - x) = 0 \Rightarrow x = 2$ and $y = -\frac{\pi}{4}$;

$f\left(2, -\frac{\pi}{4}\right) = 2\sqrt{2}$, $f\left(1, -\frac{\pi}{4}\right) = \frac{3\sqrt{2}}{2}$, and $f\left(3, -\frac{\pi}{4}\right) = \frac{3\sqrt{2}}{2}$

(v) For interior points of the region, $f_x(x, y) = (4 - 2x) \cos y = 0$ and $f_y(x, y) = -(4x - x^2) \sin y = 0 \Rightarrow x = 2$

and $y = 0$, which is an interior critical point with $f(2, 0) = 4$. Therefore the absolute maximum is 4 at

$(2, 0)$ and the absolute minimum is $\frac{3\sqrt{2}}{2}$ at $\left(3, -\frac{\pi}{4}\right)$, $\left(3, \frac{\pi}{4}\right)$, $\left(1, -\frac{\pi}{4}\right)$, and $\left(1, \frac{\pi}{4}\right)$.

39. Let $F(a, b) = \int_a^b (6 - x - x^2) \, dx$ where $a \le b$. The boundary of the domain of F is the line $a = b$ in the

ab-plane, and $F(a, a) = 0$, so F is identically 0 on the boundary of its domain. For interior critical points we

have: $\frac{\partial F}{\partial a} = -(6 - a - a^2) = 0 \Rightarrow a = -3, 2$ and $\frac{\partial F}{\partial b} = (6 - b - b^2) = 0 \Rightarrow b = -3, 2$. Since $a \le b$, there is only

one interior critical point $(-3, 2)$ and $F(-3, 2) = \int_{-3}^2 (6 - x - x^2) \, dx$ gives the area under the parabola

$y = 6 - x - x^2$ that is above the x-axis. Therefore, $a = -3$ and $b = 2$.

41. $T_x(x, y) = 2x - 1 = 0$ and $T_y(x, y) = 4y = 0 \Rightarrow x = \frac{1}{2}$ and $y = 0$ with $T\left(\frac{1}{2}, 0\right) = -\frac{1}{4}$; on the boundary

$x^2 + y^2 = 1$: $T(x, y) = -x^2 - x + 2$ for $-1 \le x \le 1 \Rightarrow T'(x, y) = -2x - 1 = 0 \Rightarrow x = -\frac{1}{2}$ and $y = \pm \frac{\sqrt{3}}{2}$;

$T\left(-\frac{1}{2}, \frac{\sqrt{3}}{2}\right) = \frac{9}{4}$, $T\left(-\frac{1}{2}, -\frac{\sqrt{3}}{2}\right) = \frac{9}{4}$, $T(-1, 0) = 2$, and $T(1, 0) = 0 \Rightarrow$ the hottest is $2\frac{1}{4}°$ at $\left(-\frac{1}{2}, \frac{\sqrt{3}}{2}\right)$ and

$\left(-\frac{1}{2}, -\frac{\sqrt{3}}{2}\right)$; the coldest is $-\frac{1}{4}°$ at $\left(\frac{1}{2}, 0\right)$.

43. (a) $f_x(x, y) = 2x - 4y = 0$ and $f_y(x, y) = 2y - 4x = 0 \Rightarrow x = 0$ and $y = 0$; $f_{xx}(0, 0) = 2$, $f_{yy}(0, 0) = 2$,

$f_{xy}(0, 0) = -4 \Rightarrow f_{xx}f_{yy} - f_{xy}^2 = -12 < 0 \Rightarrow$ saddle point at $(0, 0)$

(b) $f_x(x, y) = 2x - 2 = 0$ and $f_y(x, y) = 2y - 4 = 0 \Rightarrow x = 1$ and $y = 2$; $f_{xx}(1, 2) = 2$, $f_{yy}(1, 2) = 2$,

$f_{xy}(1, 2) = 0 \Rightarrow f_{xx}f_{yy} - f_{xy}^2 = 4 > 0$ and $f_{xx} > 0 \Rightarrow$ local minimum at $(1, 2)$

(c) $f_x(x, y) = 9x^2 - 9 = 0$ and $f_y(x, y) = 2y + 4 = 0 \Rightarrow x = \pm 1$ and $y = -2$; $f_{xx}(1, -2) = 18x|_{(1, -2)} = 18$,

$f_{yy}(1, -2) = 2$, $f_{xy}(1, -2) = 0 \Rightarrow f_{xx}f_{yy} - f_{xy}^2 = 36 > 0$ and $f_{xx} > 0 \Rightarrow$ local minimum at $(1, -2)$;

$f_{xx}(-1, -2) = -18$, $f_{yy}(-1, -2) = 2$, $f_{xy}(-1, -2) = 0 \Rightarrow f_{xx}f_{yy} - f_{xy}^2 = -36 < 0 \Rightarrow$ saddle point at $(-1, -2)$

45. If $k = 0$, then $f(x, y) = x^2 + y^2 \Rightarrow f_x(x, y) = 2x = 0$ and $f_y(x, y) = 2y = 0 \Rightarrow x = 0$ and $y = 0 \Rightarrow (0, 0)$ is the only critical point. If $k \neq 0$, $f_x(x, y) = 2x + ky = 0 \Rightarrow y = -\frac{2}{k}x$; $f_y(x, y) = kx + 2y = 0 \Rightarrow kx + 2\left(-\frac{2}{k}x\right) = 0$ $\Rightarrow kx - \frac{4x}{k} = 0 \Rightarrow \left(k - \frac{4}{k}\right)x = 0 \Rightarrow x = 0$ or $k = \pm 2 \Rightarrow y = \left(-\frac{2}{k}\right)(0) = 0$ or $y = \pm x$; in any case $(0, 0)$ is a critical point.

47. No; for example $f(x, y) = xy$ has a saddle point at $(a, b) = (0, 0)$ where $f_x = f_y = 0$.

49. We want the point on $z = 10 - x^2 - y^2$ where the tangent plane is parallel to the plane $x + 2y + 3z = 0$. To find a normal vector to $z = 10 - x^2 - y^2$ let $w = z + x^2 + y^2 - 10$. Then $\nabla w = 2x\mathbf{i} + 2y\mathbf{j} + \mathbf{k}$ is normal to $z = 10 - x^2 - y^2$ at (x, y). The vector ∇w is parallel to $\mathbf{i} + 2\mathbf{j} + 3\mathbf{k}$ which is normal to the plane $x + 2y + 3z = 0$ if $6x\mathbf{i} + 6y\mathbf{j} + 3\mathbf{k} = \mathbf{i} + 2\mathbf{j} + 3\mathbf{k}$ or $x = \frac{1}{6}$ and $y = \frac{1}{3}$. Thus the point is $\left(\frac{1}{6}, \frac{1}{3}, 10 - \frac{1}{36} - \frac{1}{9}\right)$ or $\left(\frac{1}{6}, \frac{1}{3}, \frac{355}{36}\right)$.

51. No, because the domain $x \geq 0$ and $y \geq 0$ is unbounded since x and y can be as large as we please. Absolute extrema are guaranteed for continuous functions defined over closed <u>and bounded</u> domains in the plane. Since the domain is unbounded, the continuous function $f(x, y) = x + y$ need not have an absolute maximum (although, in this case, it does have an absolute minimum value of $f(0, 0) = 0$).

53. (a) $\frac{df}{dt} = \frac{\partial f}{\partial x}\frac{dx}{dt} + \frac{\partial f}{\partial y}\frac{dy}{dt} = \frac{dx}{dt} + \frac{dy}{dt} = -2\sin t + 2\cos t = 0 \Rightarrow \cos t = \sin t \Rightarrow x = y$

   (i) On the semicircle $x^2 + y^2 = 4$, $y \geq 0$, we have $t = \frac{\pi}{4}$ and $x = y = \sqrt{2} \Rightarrow f\left(\sqrt{2}, \sqrt{2}\right) = 2\sqrt{2}$. At the endpoints, $f(-2, 0) = -2$ and $f(2, 0) = 2$. Therefore the absolute minimum is $f(-2, 0) = -2$ when $t = \pi$; the absolute maximum is $f\left(\sqrt{2}, \sqrt{2}\right) = 2\sqrt{2}$ when $t = \frac{\pi}{4}$.

   (ii) On the quartercircle $x^2 + y^2 = 4$, $x \geq 0$ and $y \geq 0$, the endpoints give $f(0, 2) = 2$ and $f(2, 0) = 2$. Therefore the absolute minimum is $f(2, 0) = 2$ and $f(0, 2) = 2$ when $t = 0, \frac{\pi}{2}$ respectively; the absolute maximum is $f\left(\sqrt{2}, \sqrt{2}\right) = 2\sqrt{2}$ when $t = \frac{\pi}{4}$.

(b) $\frac{dg}{dt} = \frac{\partial g}{\partial x}\frac{dx}{dt} + \frac{\partial g}{\partial y}\frac{dy}{dt} = y\frac{dx}{dt} + x\frac{dy}{dt} = -4\sin^2 t + 4\cos^2 t = 0 \Rightarrow \cos t = \pm\sin t \Rightarrow x = \pm y$.

   (i) On the semicircle $x^2 + y^2 = 4$, $y \geq 0$, we obtain $x = y = \sqrt{2}$ at $t = \frac{\pi}{4}$ and $x = -\sqrt{2}$, $y = \sqrt{2}$ at $t = \frac{3\pi}{4}$. Then $g\left(\sqrt{2}, \sqrt{2}\right) = 2$ and $g\left(-\sqrt{2}, \sqrt{2}\right) = -2$. At the endpoints, $g(-2, 0) = g(2, 0) = 0$. Therefore the absolute minimum is $g\left(-\sqrt{2}, \sqrt{2}\right) = -2$ when $t = \frac{3\pi}{4}$; the absolute maximum is $g\left(\sqrt{2}, \sqrt{2}\right) = 2$ when $t = \frac{\pi}{4}$.

   (ii) On the quartercircle $x^2 + y^2 = 4$, $x \geq 0$ and $y \geq 0$, the endpoints give $g(0, 2) = 0$ and $g(2, 0) = 0$. Therefore the absolute minimum is $g(2, 0) = 0$ and $g(0, 2) = 0$ when $t = 0, \frac{\pi}{2}$ respectively; the absolute maximum is $g\left(\sqrt{2}, \sqrt{2}\right) = 2$ when $t = \frac{\pi}{4}$.

(c) $\frac{dh}{dt} = \frac{\partial h}{\partial x}\frac{dx}{dt} + \frac{\partial h}{\partial y}\frac{dy}{dt} = 4x\frac{dx}{dt} + 2y\frac{dy}{dt} = (8\cos t)(-2\sin t) + (4\sin t)(2\cos t) = -8\cos t\sin t = 0$ $\Rightarrow t = 0, \frac{\pi}{2}, \pi$ yielding the points $(2, 0), (0, 2)$ for $0 \leq t \leq \pi$.

   (i) On the semicircle $x^2 + y^2 = 4$, $y \geq 0$ we have $h(2, 0) = 8$, $h(0, 2) = 4$, and $h(-2, 0) = 8$. Therefore, the absolute minimum is $h(0, 2) = 4$ when $t = \frac{\pi}{2}$; the absolute maximum is $h(2, 0) = 8$ and $h(-2, 0) = 8$ when $t = 0, \pi$ respectively.

(ii)   On the quartercircle $x^2 + y^2 = 4$, $x \geq 0$ and $y \geq 0$ the absolute minimum is $h(0, 2) = 4$ when $t = \frac{\pi}{2}$; the absolute maximum is $h(2, 0) = 8$ when $t = 0$.

55. $\frac{df}{dt} = \frac{\partial f}{\partial x}\frac{dx}{dt} + \frac{\partial f}{\partial y}\frac{dy}{dt} = y\frac{dx}{dt} + x\frac{dy}{dt}$

   (i) $x = 2t$ and $y = t + 1 \Rightarrow \frac{df}{dt} = (t+1)(2) + (2t)(1) = 4t + 2 = 0 \Rightarrow t = -\frac{1}{2} \Rightarrow x = -1$ and $y = \frac{1}{2}$ with $f\left(-1, \frac{1}{2}\right) = -\frac{1}{2}$. The absolute minimum is $f\left(-1, \frac{1}{2}\right) = -\frac{1}{2}$ when $t = -\frac{1}{2}$; there is no absolute maximum.

   (ii)   For the endpoints: $t = -1 \Rightarrow x = -2$ and $y = 0$ with $f(-2, 0) = 0$; $t = 0 \Rightarrow x = 0$ and $y = 1$ with $f(0, 1) = 0$. The absolute minimum is $f\left(-1, \frac{1}{2}\right) = -\frac{1}{2}$ when $t = -\frac{1}{2}$; the absolute maximum is $f(0, 1) = 0$ and $f(-2, 0) = 0$ when $t = -1, 0$ respectively.

   (iii) There are no interior critical points. For the endpoints: $t = 0 \Rightarrow x = 0$ and $y = 1$ with $f(0, 1) = 0$; $t = 1 \Rightarrow x = 2$ and $y = 2$ with $f(2, 2) = 4$. The absolute minimum is $f(0, 1) = 0$ when $t = 0$; the absolute maximum is $f(2, 2) = 4$ when $t = 1$.

57. $m = \frac{(2)(-1) - 3(-14)}{(2)^2 - 3(10)} = -\frac{20}{13}$ and

   $b = \frac{1}{3}\left[-1 - \left(-\frac{20}{13}\right)(2)\right] = \frac{9}{13}$

   $\Rightarrow y = -\frac{20}{13}x + \frac{9}{13}$; $y\big|_{x=4} = -\frac{71}{13}$

k	x_k	y_k	x_k^2	$x_k y_k$
1	−1	2	1	−2
2	0	1	0	0
3	3	−4	9	−12
Σ	2	−1	10	−14

59. $m = \frac{(3)(5) - 3(8)}{(3)^2 - 3(5)} = \frac{3}{2}$ and

   $b = \frac{1}{3}\left[5 - \frac{3}{2}(3)\right] = \frac{1}{6}$

   $\Rightarrow y = \frac{3}{2}x + \frac{1}{6}$; $y\big|_{x=4} = \frac{37}{6}$

k	x_k	y_k	x_k^2	$x_k y_k$
1	0	0	0	0
2	1	2	1	2
3	2	3	4	6
Σ	3	5	5	8

61. $m = \frac{(162)(41.32) - 6(1192.8)}{(162)^2 - 6(5004)} \approx 0.122$ and

   $b = \frac{1}{6}[41.32 - (0.122)(162)] \approx 3.59$

   $\Rightarrow y = 0.122x + 3.59$

k	x_k	y_k	x_k^2	$x_k y_k$
1	12	5.27	144	63.24
2	18	5.68	324	102.24
3	24	6.25	576	150
4	30	7.21	900	216.3
5	36	8.20	1296	295.2
6	42	8.71	1764	365.82
Σ	162	41.32	5004	1192.8

63. (b) $m = \frac{(3201)(17,785) - 10(5,710,292)}{(3201)^2 - 10(1,430,389)}$

≈ 0.0427 and $b = \frac{1}{10}[17,785 - (0.0427)(3201)]$

$\approx 1764.8 \Rightarrow y = 0.0427K + 1764.8$

Köchel numbers

k	K_k	y_k	K^2	$K_k y_k$
1	1	1761	1	1761
2	75	1771	5625	132,825
3	155	1772	24,025	274,660
4	219	1775	47,961	388,725
5	271	1777	73,441	481,567
6	351	1780	123,201	624,780
7	425	1783	180,625	757,775
8	503	1786	253,009	898,358
9	575	1789	330,625	1,028,675
10	626	1791	391,876	1,121,166
Σ	3201	17,785	1,430,389	5,710,292

(c) $K = 364 \Rightarrow y = (0.0427)(364)$

$\Rightarrow y = (0.0427)(364) + 1764.8 \approx 1780$

14.8 LAGRANGE MULTIPLIERS

1. $\nabla f = y\mathbf{i} + x\mathbf{j}$ and $\nabla g = 2x\mathbf{i} + 4y\mathbf{j}$ so that $\nabla f = \lambda \nabla g \Rightarrow y\mathbf{i} + x\mathbf{j} = \lambda(2x\mathbf{i} + 4y\mathbf{j}) \Rightarrow y = 2x\lambda$ and $x = 4y\lambda$

$\Rightarrow x = 8x\lambda^2 \Rightarrow \lambda = \pm\frac{\sqrt{2}}{4}$ or $x = 0$.

CASE 1: If $x = 0$, then $y = 0$. But $(0,0)$ is not on the ellipse so $x \neq 0$.

CASE 2: $x \neq 0 \Rightarrow \lambda = \pm\frac{\sqrt{2}}{4} \Rightarrow x = \pm\sqrt{2}y \Rightarrow \left(\pm\sqrt{2}y\right)^2 + 2y^2 = 1 \Rightarrow y = \pm\frac{1}{2}$.

Therefore f takes on its extreme values at $\left(\pm\frac{\sqrt{2}}{2}, \frac{1}{2}\right)$ and $\left(\pm\frac{\sqrt{2}}{2}, -\frac{1}{2}\right)$. The extreme values of f on the ellipse

are $\pm\frac{\sqrt{2}}{2}$.

3. $\nabla f = -2x\mathbf{i} - 2y\mathbf{j}$ and $\nabla g = \mathbf{i} + 3\mathbf{j}$ so that $\nabla f = \lambda \nabla g \Rightarrow -2x\mathbf{i} - 2y\mathbf{j} = \lambda(\mathbf{i} + 3\mathbf{j}) \Rightarrow x = -\frac{\lambda}{2}$ and $y = -\frac{3\lambda}{2}$

$\Rightarrow \left(-\frac{\lambda}{2}\right) + 3\left(-\frac{3\lambda}{2}\right) = 10 \Rightarrow \lambda = -2 \Rightarrow x = 1$ and $y = 3 \Rightarrow$ f takes on its extreme value at $(1, 3)$ on the line.

The extreme value is $f(1, 3) = 49 - 1 - 9 = 39$.

5. We optimize $f(x, y) = x^2 + y^2$, the square of the distance to the origin, subject to the constraint

$g(x, y) = xy^2 - 54 = 0$. Thus $\nabla f = 2x\mathbf{i} + 2y\mathbf{j}$ and $\nabla g = y^2\mathbf{i} + 2xy\mathbf{j}$ so that $\nabla f = \lambda \nabla g \Rightarrow 2x\mathbf{i} + 2y\mathbf{j}$

$= \lambda\left(y^2\mathbf{i} + 2xy\mathbf{j}\right) \Rightarrow 2x = \lambda y^2$ and $2y = 2\lambda xy$.

CASE 1: If $y = 0$, then $x = 0$. But $(0, 0)$ does not satisfy the constraint $xy^2 = 54$ so $y \neq 0$.

CASE 2: If $y \neq 0$, then $2 = 2\lambda x \Rightarrow x = \frac{1}{\lambda} \Rightarrow 2\left(\frac{1}{\lambda}\right) = \lambda y^2 \Rightarrow y^2 = \frac{2}{\lambda^2}$. Then $xy^2 = 54 \Rightarrow \left(\frac{1}{\lambda}\right)\left(\frac{2}{\lambda^2}\right) = 54$

$\Rightarrow \lambda^3 = \frac{1}{27} \Rightarrow \lambda = \frac{1}{3} \Rightarrow x = 3$ and $y^2 = 18 \Rightarrow x = 3$ and $y = \pm3\sqrt{2}$.

Therefore $\left(3, \pm3\sqrt{2}\right)$ are the points on the curve $xy^2 = 54$ nearest the origin (since $xy^2 = 54$ has points

increasingly far away as y gets close to 0, no points are farthest away).

7. (a) $\nabla f = \mathbf{i} + \mathbf{j}$ and $\nabla g = y\mathbf{i} + x\mathbf{j}$ so that $\nabla f = \lambda \nabla g \Rightarrow \mathbf{i} + \mathbf{j} = \lambda(y\mathbf{i} + x\mathbf{j}) \Rightarrow 1 = \lambda y$ and $1 = \lambda x \Rightarrow y = \frac{1}{\lambda}$ and

$x = \frac{1}{\lambda} \Rightarrow \frac{1}{\lambda^2} = 16 \Rightarrow \lambda = \pm\frac{1}{4}$. Use $\lambda = \frac{1}{4}$ since $x > 0$ and $y > 0$. Then $x = 4$ and $y = 4 \Rightarrow$ the minimum

value is 8 at the point $(4, 4)$. Now, $xy = 16$, $x > 0$, $y > 0$ is a branch of a hyperbola in the first quadrant

with the x- and y-axes as asymptotes. The equations $x + y = c$ give a family of parallel lines with $m = -1$.

As these lines move away from the origin, the number c increases. Thus the minimum value of c occurs

where $x + y = c$ is tangent to the hyperbola's branch.

(b) $\nabla f = y\mathbf{i} + x\mathbf{j}$ and $\nabla g = \mathbf{i} + \mathbf{j}$ so that $\nabla f = \lambda \nabla g \Rightarrow y\mathbf{i} + x\mathbf{j} = \lambda(\mathbf{i} + \mathbf{j}) \Rightarrow y = \lambda = x$ y + y = 16 \Rightarrow y = 8

$\Rightarrow x = 8 \Rightarrow f(8, 8) = 64$ is the maximum value. The equations $xy = c$ ($x > 0$ and $y > 0$ or $x < 0$ and $y < 0$

to get a maximum value) give a family of hyperbolas in the first and third quadrants with the x- and y-axes as asymptotes. The maximum value of c occurs where the hyperbola $xy = c$ is tangent to the line $x + y = 16$.

9. $V = \pi r^2 h \Rightarrow 16\pi = \pi r^2 h \Rightarrow 16 = r^2 h \Rightarrow g(r, h) = r^2 h - 16$; $S = 2\pi rh + 2\pi r^2 \Rightarrow \nabla S = (2\pi h + 4\pi r)\mathbf{i} + 2\pi r\mathbf{j}$ and $\nabla g = 2rh\mathbf{i} + r^2\mathbf{j}$ so that $\nabla S = \lambda \nabla g \Rightarrow (2\pi rh + 4\pi r)\mathbf{i} + 2\pi r\mathbf{j} = \lambda(2rh\mathbf{i} + r^2\mathbf{j}) \Rightarrow 2\pi rh + 4\pi r = 2rh\lambda$ and $2\pi r = \lambda r^2 \Rightarrow r = 0$ or $\lambda = \frac{2\pi}{r}$. But $r = 0$ gives no physical can, so $r \neq 0 \Rightarrow \lambda = \frac{2\pi}{r} \Rightarrow 2\pi h + 4\pi r$ $= 2rh\left(\frac{2\pi}{r}\right) \Rightarrow 2r = h \Rightarrow 16 = r^2(2r) \Rightarrow r = 2 \Rightarrow h = 4$; thus $r = 2$ cm and $h = 4$ cm give the only extreme surface area of 24π cm^2. Since $r = 4$ cm and $h = 1$ cm $\Rightarrow V = 16\pi$ cm^3 and $S = 40\pi$ cm^2, which is a larger surface area, then 24π cm^2 must be the minimum surface area.

11. $A = (2x)(2y) = 4xy$ subject to $g(x, y) = \frac{x^2}{16} + \frac{y^2}{9} - 1 = 0$; $\nabla A = 4y\mathbf{i} + 4x\mathbf{j}$ and $\nabla g = \frac{x}{8}\mathbf{i} + \frac{2y}{9}\mathbf{j}$ so that ∇A $= \lambda \nabla g \Rightarrow 4y\mathbf{i} + 4x\mathbf{j} = \lambda\left(\frac{x}{8}\mathbf{i} + \frac{2y}{9}\mathbf{j}\right) \Rightarrow 4y = \left(\frac{x}{8}\right)\lambda$ and $4x = \left(\frac{2y}{9}\right)\lambda \Rightarrow \lambda = \frac{32y}{x}$ and $4x = \left(\frac{2y}{9}\right)\left(\frac{32y}{x}\right)$ $\Rightarrow y = \pm\frac{3}{4}x \Rightarrow \frac{x^2}{16} + \frac{\left(\pm\frac{3}{4}x\right)^2}{9} = 1 \Rightarrow x^2 = 8 \Rightarrow x = \pm 2\sqrt{2}$. We use $x = 2\sqrt{2}$ since x represents distance. Then $y = \frac{3}{4}\left(2\sqrt{2}\right) = \frac{3\sqrt{2}}{2}$, so the length is $2x = 4\sqrt{2}$ and the width is $2y = 3\sqrt{2}$.

13. $\nabla f = 2x\mathbf{i} + 2y\mathbf{j}$ and $\nabla g = (2x - 2)\mathbf{i} + (2y - 4)\mathbf{j}$ so that $\nabla f = \lambda \nabla g = 2x\mathbf{i} + 2y\mathbf{j} = \lambda[(2x - 2)\mathbf{i} + (2y - 4)\mathbf{j}]$ $\Rightarrow 2x = \lambda(2x - 2)$ and $2y = \lambda(2y - 4) \Rightarrow x = \frac{\lambda}{\lambda - 1}$ and $y = \frac{2\lambda}{\lambda - 1}, \lambda \neq 1 \Rightarrow y = 2x \Rightarrow x^2 - 2x + (2x)^2 - 4(2x)$ $= 0 \Rightarrow x = 0$ and $y = 0$, or $x = 2$ and $y = 4$. Therefore $f(0, 0) = 0$ is the minimum value and $f(2, 4) = 20$ is the maximum value. (Note that $\lambda = 1$ gives $2x = 2x - 2$ or $0 = -2$, which is impossible.)

15. $\nabla T = (8x - 4y)\mathbf{i} + (-4x + 2y)\mathbf{j}$ and $g(x, y) = x^2 + y^2 - 25 = 0 \Rightarrow \nabla g = 2x\mathbf{i} + 2y\mathbf{j}$ so that $\nabla T = \lambda \nabla g$ $\Rightarrow (8x - 4y)\mathbf{i} + (-4x + 2y)\mathbf{j} = \lambda(2x\mathbf{i} + 2y\mathbf{j}) \Rightarrow 8x - 4y = 2\lambda x$ and $-4x + 2y = 2\lambda y \Rightarrow y = \frac{-2x}{\lambda - 1}, \lambda \neq 1$ $\Rightarrow 8x - 4\left(\frac{-2x}{\lambda - 1}\right) = 2\lambda x \Rightarrow x = 0$, or $\lambda = 0$, or $\lambda = 5$.
    CASE 1: $x = 0 \Rightarrow y = 0$; but $(0, 0)$ is not on $x^2 + y^2 = 25$ so $x \neq 0$.
    CASE 2: $\lambda = 0 \Rightarrow y = 2x \Rightarrow x^2 + (2x)^2 = 25 \Rightarrow x = \pm\sqrt{5}$ and $y = 2x$.
    CASE 3: $\lambda = 5 \Rightarrow y = \frac{-2x}{4} = -\frac{x}{2} \Rightarrow x^2 + \left(-\frac{x}{2}\right)^2 = 25 \Rightarrow x = \pm 2\sqrt{5} \Rightarrow x = 2\sqrt{5}$ and $y = -\sqrt{5}$, or $x = -2\sqrt{5}$ and $y = \sqrt{5}$.
    Therefore $T\left(\sqrt{5}, 2\sqrt{5}\right) = 0° = T\left(-\sqrt{5}, -2\sqrt{5}\right)$ is the minimum value and $T\left(2\sqrt{5}, -\sqrt{5}\right) = 125°$ $= T\left(-2\sqrt{5}, \sqrt{5}\right)$ is the maximum value. (Note: $\lambda = 1 \Rightarrow x = 0$ from the equation $-4x + 2y = 2\lambda y$; but we found $x \neq 0$ in CASE 1.)

17. Let $f(x, y, z) = (x - 1)^2 + (y - 1)^2 + (z - 1)^2$ be the square of the distance from $(1, 1, 1)$. Then $\nabla f = 2(x - 1)\mathbf{i} + 2(y - 1)\mathbf{j} + 2(z - 1)\mathbf{k}$ and $\nabla g = \mathbf{i} + 2\mathbf{j} + 3\mathbf{k}$ so that $\nabla f = \lambda \nabla g$ $\Rightarrow 2(x - 1)\mathbf{i} + 2(y - 1)\mathbf{j} + 2(z - 1)\mathbf{k} = \lambda(\mathbf{i} + 2\mathbf{j} + 3\mathbf{k}) \Rightarrow 2(x - 1) = \lambda, 2(y - 1) = 2\lambda, 2(z - 1) = 3\lambda$ $\Rightarrow 2(y - 1) = 2[2(x - 1)]$ and $2(z - 1) = 3[2(x - 1)] \Rightarrow x = \frac{y + 1}{2} \Rightarrow z + 2 = 3\left(\frac{y+1}{2}\right)$ or $z = \frac{3y - 1}{2}$; thus $\frac{y+1}{2} + 2y + 3\left(\frac{3y-1}{2}\right) - 13 = 0 \Rightarrow y = 2 \Rightarrow x = \frac{3}{2}$ and $z = \frac{5}{2}$. Therefore the point $\left(\frac{3}{2}, 2, \frac{5}{2}\right)$ is closest (since no point on the plane is farthest from the point $(1, 1, 1)$).

19. Let $f(x, y, z) = x^2 + y^2 + z^2$ be the square of the distance from the origin. Then $\nabla f = 2x\mathbf{i} + 2y\mathbf{j} + 2z\mathbf{k}$ and $\nabla g = 2x\mathbf{i} - 2y\mathbf{j} - 2z\mathbf{k}$ so that $\nabla f = \lambda \nabla g \Rightarrow 2x\mathbf{i} + 2y\mathbf{j} + 2z\mathbf{k} = \lambda(2x\mathbf{i} - 2y\mathbf{j} - 2z\mathbf{k}) \Rightarrow 2x = 2x\lambda, 2y = -2y\lambda$, and $2z = -2z\lambda \Rightarrow x = 0$ or $\lambda = 1$.
    CASE 1: $\lambda = 1 \Rightarrow 2y = -2y \Rightarrow y = 0$; $2z = -2z \Rightarrow z = 0 \Rightarrow x^2 - 1 = 0 \Rightarrow x = \pm 1$ and $y = z = 0$.
    CASE 2: $x = 0 \Rightarrow -y^2 - z^2 = 1$, which has no solution.
    Therefore the points on the unit circle $x^2 + y^2 = 1$, are the points on the surface $x^2 + y^2 - z^2 = 1$ closest to the origin. The minimum distance is 1.

21. Let $f(x, y, z) = x^2 + y^2 + z^2$ be the square of the distance to the origin. Then $\nabla f = 2x\mathbf{i} + 2y\mathbf{j} + 2z\mathbf{k}$ and
$\nabla g = -y\mathbf{i} - x\mathbf{j} + 2z\mathbf{k}$ so that $\nabla f = \lambda \nabla g \Rightarrow 2x\mathbf{i} + 2y\mathbf{j} + 2z\mathbf{k} = \lambda(-y\mathbf{i} - x\mathbf{j} + 2z\mathbf{k}) \Rightarrow 2x = -y\lambda, 2y = -x\lambda$, and
$2z = 2z\lambda \Rightarrow \lambda = 1$ or $z = 0$.
CASE 1: $\lambda = 1 \Rightarrow 2x = -y$ and $2y = -x \Rightarrow y = 0$ and $x = 0 \Rightarrow z^2 - 4 = 0 \Rightarrow z = \pm 2$ and $x = y = 0$.
CASE 2: $z = 0 \Rightarrow -xy - 4 = 0 \Rightarrow y = -\frac{4}{x}$. Then $2x = \frac{4}{x}\lambda \Rightarrow \lambda = \frac{x^2}{2}$, and $-\frac{8}{x} = -x\lambda \Rightarrow -\frac{8}{x} = -x\left(\frac{x^2}{2}\right)$
$\Rightarrow x^4 = 16 \Rightarrow x = \pm 2$. Thus, $x = 2$ and $y = -2$, or $x = -2$ and $y = 2$.
Therefore we get four points: $(2, -2, 0), (-2, 2, 0), (0, 0, 2)$ and $(0, 0, -2)$. But the points $(0, 0, 2)$ and $(0, 0, -2)$
are closest to the origin since they are 2 units away and the others are $2\sqrt{2}$ units away.

23. $\nabla f = \mathbf{i} - 2\mathbf{j} + 5\mathbf{k}$ and $\nabla g = 2x\mathbf{i} + 2y\mathbf{j} + 2z\mathbf{k}$ so that $\nabla f = \lambda \nabla g \Rightarrow \mathbf{i} - 2\mathbf{j} + 5\mathbf{k} = \lambda(2x\mathbf{i} + 2y\mathbf{j} + 2z\mathbf{k}) \Rightarrow 1 = 2x\lambda,$
$-2 = 2y\lambda$, and $5 = 2z\lambda \Rightarrow x = \frac{1}{2\lambda}, y = -\frac{1}{\lambda} = -2x$, and $z = \frac{5}{2\lambda} = 5x \Rightarrow x^2 + (-2x)^2 + (5x)^2 = 30 \Rightarrow x = \pm 1$.
Thus, $x = 1, y = -2, z = 5$ or $x = -1, y = 2, z = -5$. Therefore $f(1, -2, 5) = 30$ is the maximum value and
$f(-1, 2, -5) = -30$ is the minimum value.

25. $f(x, y, z) = x^2 + y^2 + z^2$ and $g(x, y, z) = x + y + z - 9 = 0 \Rightarrow \nabla f = 2x\mathbf{i} + 2y\mathbf{j} + 2z\mathbf{k}$ and $\nabla g = \mathbf{i} + \mathbf{j} + \mathbf{k}$ so that
$\nabla f = \lambda \nabla g \Rightarrow 2x\mathbf{i} + 2y\mathbf{j} + 2z\mathbf{k} = \lambda(\mathbf{i} + \mathbf{j} + \mathbf{k}) \Rightarrow 2x = \lambda, 2y = \lambda$, and $2z = \lambda \Rightarrow x = y = z \Rightarrow x + x + x - 9 = 0$
$\Rightarrow x = 3, y = 3$, and $z = 3$.

27. $V = 6xyz$ and $g(x, y, z) = x^2 + y^2 + z^2 - 1 = 0 \Rightarrow \nabla V = 6yz\mathbf{i} + 6xz\mathbf{j} + 6xy\mathbf{k}$ and $\nabla g = 2x\mathbf{i} + 2y\mathbf{j} + 2z\mathbf{k}$ so that
$\nabla V = \lambda \nabla g \Rightarrow 3yz = \lambda x, 3xz = \lambda y$, and $3xy = \lambda z \Rightarrow 3xyz = \lambda x^2$ and $3xyz = \lambda y^2 \Rightarrow y = \pm x \Rightarrow z = \pm x$
$\Rightarrow x^2 + x^2 + x^2 = 1 \Rightarrow x = \frac{1}{\sqrt{3}}$ since $x > 0 \Rightarrow$ the dimensions of the box are $\frac{2}{\sqrt{3}}$ by $\frac{2}{\sqrt{3}}$ by $\frac{2}{\sqrt{3}}$ for maximum
volume. (Note that there is no minimum volume since the box could be made arbitrarily thin.)

29. $\nabla T = 16x\mathbf{i} + 4z\mathbf{j} + (4y - 16)\mathbf{k}$ and $\nabla g = 8x\mathbf{i} + 2y\mathbf{j} + 8z\mathbf{k}$ so that $\nabla T = \lambda \nabla g \Rightarrow 16x\mathbf{i} + 4z\mathbf{j} + (4y - 16)\mathbf{k}$
$= \lambda(8x\mathbf{i} + 2y\mathbf{j} + 8z\mathbf{k}) \Rightarrow 16x = 8x\lambda, 4z = 2y\lambda$, and $4y - 16 = 8z\lambda \Rightarrow \lambda = 2$ or $x = 0$.
CASE 1: $\lambda = 2 \Rightarrow 4z = 2y(2) \Rightarrow z = y$. Then $4z - 16 = 16z \Rightarrow z = -\frac{4}{3} \Rightarrow y = -\frac{4}{3}$. Then
$\qquad 4x^2 + \left(-\frac{4}{3}\right)^2 + 4\left(-\frac{4}{3}\right)^2 = 16 \Rightarrow x = \pm \frac{4}{3}$.
CASE 2: $x = 0 \Rightarrow \lambda = \frac{2z}{y} \Rightarrow 4y - 16 = 8z\left(\frac{2z}{y}\right) \Rightarrow y^2 - 4y = 4z^2 \Rightarrow 4(0)^2 + y^2 + (y^2 - 4y) - 16 = 0$
$\qquad \Rightarrow y^2 - 2y - 8 = 0 \Rightarrow (y - 4)(y + 2) = 0 \Rightarrow y = 4$ or $y = -2$. Now $y = 4 \Rightarrow 4z^2 = 4^2 - 4(4)$
$\qquad \Rightarrow z = 0$ and $y = -2 \Rightarrow 4z^2 = (-2)^2 - 4(-2) \Rightarrow z = \pm \sqrt{3}$.
The temperatures are $T\left(\pm \frac{4}{3}, -\frac{4}{3}, -\frac{4}{3}\right) = 642\frac{2}{3}°$, $T(0, 4, 0) = 600°$, $T\left(0, -2, \sqrt{3}\right) = \left(600 - 24\sqrt{3}\right)°$, and
$T\left(0, -2, -\sqrt{3}\right) = \left(600 + 24\sqrt{3}\right)° \approx 641.6°$. Therefore $\left(\pm \frac{4}{3}, -\frac{4}{3}, -\frac{4}{3}\right)$ are the hottest points on the space probe.

31. $\nabla U = (y + 2)\mathbf{i} + x\mathbf{j}$ and $\nabla g = 2\mathbf{i} + \mathbf{j}$ so that $\nabla U = \lambda \nabla g \Rightarrow (y + 2)\mathbf{i} + x\mathbf{j} = \lambda(2\mathbf{i} + \mathbf{j}) \Rightarrow y + 2 = 2\lambda$ and
$x = \lambda \Rightarrow y + 2 = 2x \Rightarrow y = 2x - 2 \Rightarrow 2x + (2x - 2) = 30 \Rightarrow x = 8$ and $y = 14$. Therefore $U(8, 14) = \$128$
is the maximum value of U under the constraint.

33. Let $g_1(x, y, z) = 2x - y = 0$ and $g_2(x, y, z) = y + z = 0 \Rightarrow \nabla g_1 = 2\mathbf{i} - \mathbf{j}, \nabla g_2 = \mathbf{j} + \mathbf{k}$, and $\nabla f = 2x\mathbf{i} + 2\mathbf{j} - 2z\mathbf{k}$
so that $\nabla f = \lambda \nabla g_1 + \mu \nabla g_2 \Rightarrow 2x\mathbf{i} + 2\mathbf{j} - 2z\mathbf{k} = \lambda(2\mathbf{i} - \mathbf{j}) + \mu(\mathbf{j} + \mathbf{k}) \Rightarrow 2x\mathbf{i} + 2\mathbf{j} - 2z\mathbf{k} = 2\lambda\mathbf{i} + (\mu - \lambda)\mathbf{j} + \mu\mathbf{k}$
$\Rightarrow 2x = 2\lambda, 2 = \mu - \lambda$, and $-2z = \mu \Rightarrow x = \lambda$. Then $2 = -2z - x \Rightarrow x = -2z - 2$ so that $2x - y = 0$
$\Rightarrow 2(-2z - 2) - y = 0 \Rightarrow -4z - 4 - y = 0$. This equation coupled with $y + z = 0$ implies $z = -\frac{4}{3}$ and $y = \frac{4}{3}$.
Then $x = \frac{2}{3}$ so that $\left(\frac{2}{3}, \frac{4}{3}, -\frac{4}{3}\right)$ is the point that gives the maximum value $f\left(\frac{2}{3}, \frac{4}{3}, -\frac{4}{3}\right) = \left(\frac{2}{3}\right)^2 + 2\left(\frac{4}{3}\right) - \left(-\frac{4}{3}\right)^2$
$= \frac{4}{3}$.

35. Let $f(x, y, z) = x^2 + y^2 + z^2$ be the square of the distance from the origin. We want to minimize $f(x, y, z)$ subject to the constraints $g_1(x, y, z) = y + 2z - 12 = 0$ and $g_2(x, y, z) = x + y - 6 = 0$. Thus $\nabla f = 2x\mathbf{i} + 2y\mathbf{j} + 2z\mathbf{k}$, $\nabla g_1 = \mathbf{j} + 2\mathbf{k}$, and $\nabla g_2 = \mathbf{i} + \mathbf{j}$ so that $\nabla f = \lambda \nabla g_1 + \mu \nabla g_2 \Rightarrow 2x = \mu, 2y = \lambda + \mu$, and $2z = 2\lambda$. Then $0 = y + 2z - 12 = \left(\frac{\lambda}{2} + \frac{\mu}{2}\right) + 2\lambda - 12 \Rightarrow \frac{5}{2}\lambda + \frac{1}{2}\mu = 12 \Rightarrow 5\lambda + \mu = 24; 0 = x + y - 6 = \frac{\mu}{2} + \left(\frac{\lambda}{2} + \frac{\mu}{2}\right) - 6$ $\Rightarrow \frac{1}{2}\lambda + \mu = 6 \Rightarrow \lambda + 2\mu = 12$. Solving these two equations for λ and μ gives $\lambda = 4$ and $\mu = 4 \Rightarrow x = \frac{\mu}{2} = 2$, $y = \frac{\lambda + \mu}{2} = 4$, and $z = \lambda = 4$. The point $(2, 4, 4)$ on the line of intersection is closest to the origin. (There is no maximum distance from the origin since points on the line can be arbitrarily far away.)

37. Let $g_1(x, y, z) = z - 1 = 0$ and $g_2(x, y, z) = x^2 + y^2 + z^2 - 10 = 0 \Rightarrow \nabla g_1 = \mathbf{k}, \nabla g_2 = 2x\mathbf{i} + 2y\mathbf{j} + 2z\mathbf{k}$, and $\nabla f = 2xyz\mathbf{i} + x^2z\mathbf{j} + x^2y\mathbf{k}$ so that $\nabla f = \lambda \nabla g_1 + \mu \nabla g_2 \Rightarrow 2xyz\mathbf{i} + x^2z\mathbf{j} + x^2y\mathbf{k} = \lambda(\mathbf{k}) + \mu(2x\mathbf{i} + 2y\mathbf{j} + 2z\mathbf{k})$ $\Rightarrow 2xyz = 2x\mu, x^2z = 2y\mu$, and $x^2y = 2z\mu + \lambda \Rightarrow xyz = x\mu \Rightarrow x = 0$ or $yz = \mu \Rightarrow \mu = y$ since $z = 1$.
CASE 1: $x = 0$ and $z = 1 \Rightarrow y^2 - 9 = 0$ (from g_2) $\Rightarrow y = \pm 3$ yielding the points $(0, \pm 3, 1)$.
CASE 2: $\mu = y \Rightarrow x^2z = 2y^2 \Rightarrow x^2 = 2y^2$ (since $z = 1$) $\Rightarrow 2y^2 + y^2 + 1 - 10 = 0$ (from g_2) $\Rightarrow 3y^2 - 9 = 0$
$$\Rightarrow y = \pm\sqrt{3} \Rightarrow x^2 = 2\left(\pm\sqrt{3}\right)^2 \Rightarrow x = \pm\sqrt{6} \text{ yielding the points } \left(\pm\sqrt{6}, \pm\sqrt{3}, 1\right).$$
Now $f(0, \pm 3, 1) = 1$ and $f\left(\pm\sqrt{6}, \pm\sqrt{3}, 1\right) = 6\left(\pm\sqrt{3}\right) + 1 = 1 \pm 6\sqrt{3}$. Therefore the maximum of f is $1 + 6\sqrt{3}$ at $\left(\pm\sqrt{6}, \sqrt{3}, 1\right)$, and the minimum of f is $1 - 6\sqrt{3}$ at $\left(\pm\sqrt{6}, -\sqrt{3}, 1\right)$.

39. Let $g_1(x, y, z) = y - x = 0$ and $g_2(x, y, z) = x^2 + y^2 + z^2 - 4 = 0$. Then $\nabla f = y\mathbf{i} + x\mathbf{j} + 2z\mathbf{k}, \nabla g_1 = -\mathbf{i} + \mathbf{j}$, and $\nabla g_2 = 2x\mathbf{i} + 2y\mathbf{j} + 2z\mathbf{k}$ so that $\nabla f = \lambda \nabla g_1 + \mu \nabla g_2 \Rightarrow y\mathbf{i} + x\mathbf{j} + 2z\mathbf{k} = \lambda(-\mathbf{i} + \mathbf{j}) + \mu(2x\mathbf{i} + 2y\mathbf{j} + 2z\mathbf{k})$ $\Rightarrow y = -\lambda + 2x\mu, x = \lambda + 2y\mu$, and $2z = 2z\mu \Rightarrow z = 0$ or $\mu = 1$.
CASE 1: $z = 0 \Rightarrow x^2 + y^2 - 4 = 0 \Rightarrow 2x^2 - 4 = 0$ (since $x = y$) $\Rightarrow x = \pm\sqrt{2}$ and $y = \pm\sqrt{2}$ yielding the points $\left(\pm\sqrt{2}, \pm\sqrt{2}, 0\right)$.
CASE 2: $\mu = 1 \Rightarrow y = -\lambda + 2x$ and $x = \lambda + 2y \Rightarrow x + y = 2(x + y) \Rightarrow 2x = 2(2x)$ since $x = y \Rightarrow x = 0 \Rightarrow y = 0$ $\Rightarrow z^2 - 4 = 0 \Rightarrow z = \pm 2$ yielding the points $(0, 0, \pm 2)$.
Now, $f(0, 0, \pm 2) = 4$ and $f\left(\pm\sqrt{2}, \pm\sqrt{2}, 0\right) = 2$. Therefore the maximum value of f is 4 at $(0, 0, \pm 2)$ and the minimum value of f is 2 at $\left(\pm\sqrt{2}, \pm\sqrt{2}, 0\right)$.

41. $\nabla f = \mathbf{i} + \mathbf{j}$ and $\nabla g = y\mathbf{i} + x\mathbf{j}$ so that $\nabla f = \lambda \nabla g \Rightarrow \mathbf{i} + \mathbf{j} = \lambda(y\mathbf{i} + x\mathbf{j}) \Rightarrow 1 = y\lambda$ and $1 = x\lambda \Rightarrow y = x$ $\Rightarrow y^2 = 16 \Rightarrow y = \pm 4 \Rightarrow (4, 4)$ and $(-4, -4)$ are candidates for the location of extreme values. But as $x \to \infty$, $y \to \infty$ and $f(x, y) \to \infty$; as $x \to -\infty, y \to 0$ and $f(x, y) \to -\infty$. Therefore no maximum or minimum value exists subject to the constraint.

43. (a) Maximize $f(a, b, c) = a^2b^2c^2$ subject to $a^2 + b^2 + c^2 = r^2$. Thus $\nabla f = 2ab^2c^2\mathbf{i} + 2a^2bc^2\mathbf{j} + 2a^2b^2c\mathbf{k}$ and $\nabla g = 2a\mathbf{i} + 2b\mathbf{j} + 2c\mathbf{k}$ so that $\nabla f = \lambda \nabla g \Rightarrow 2ab^2c^2 = 2a\lambda, 2a^2bc^2 = 2b\lambda$, and $2a^2b^2c = 2c\lambda$ $\Rightarrow 2a^2b^2c^2 = 2a^2\lambda = 2b^2\lambda = 2c^2\lambda \Rightarrow \lambda = 0$ or $a^2 = b^2 = c^2$.
CASE 1: $\lambda = 0 \Rightarrow a^2b^2c^2 = 0$.
CASE 2: $a^2 = b^2 = c^2 \Rightarrow f(a, b, c) = a^2a^2a^2$ and $3a^2 = r^2 \Rightarrow f(a, b, c) = \left(\frac{r^2}{3}\right)^3$ is the maximum value.

(b) The point $\left(\sqrt{a}, \sqrt{b}, \sqrt{c}\right)$ is on the sphere if $a + b + c = r^2$. Moreover, by part (a), $abc = f\left(\sqrt{a}, \sqrt{b}, \sqrt{c}\right)$
$\leq \left(\frac{r^2}{3}\right)^3 \Rightarrow (abc)^{1/3} \leq \frac{r^2}{3} = \frac{a + b + c}{3}$, as claimed.

14.9 PARTIAL DERIVATIVES WITH CONSTRAINED VARIABLES

1. $w = x^2 + y^2 + z^2$ and $z = x^2 + y^2$:

(a) $\begin{pmatrix} y \\ z \end{pmatrix} \rightarrow \begin{pmatrix} x = x(y,z) \\ y = y \\ z = z \end{pmatrix} \rightarrow w \Rightarrow \left(\frac{\partial w}{\partial y}\right)_z = \frac{\partial w}{\partial x}\frac{\partial x}{\partial y} + \frac{\partial w}{\partial y}\frac{\partial y}{\partial y} + \frac{\partial w}{\partial z}\frac{\partial z}{\partial y}; \frac{\partial z}{\partial y} = 0$ and $\frac{\partial z}{\partial y} = 2x\frac{\partial x}{\partial y} + 2y\frac{\partial y}{\partial y}$

$= 2x\frac{\partial x}{\partial y} + 2y \Rightarrow 0 = 2x\frac{\partial x}{\partial y} + 2y \Rightarrow \frac{\partial x}{\partial y} = -\frac{y}{x} \Rightarrow \left(\frac{\partial w}{\partial y}\right)_z = (2x)\left(-\frac{y}{x}\right) + (2y)(1) + (2z)(0) = -2y + 2y = 0$

(b) $\begin{pmatrix} x \\ z \end{pmatrix} \rightarrow \begin{pmatrix} x = x \\ y = y(x,z) \\ z = z \end{pmatrix} \rightarrow w \Rightarrow \left(\frac{\partial w}{\partial z}\right)_x = \frac{\partial w}{\partial x}\frac{\partial x}{\partial z} + \frac{\partial w}{\partial y}\frac{\partial y}{\partial z} + \frac{\partial w}{\partial z}\frac{\partial z}{\partial z}; \frac{\partial x}{\partial z} = 0$ and $\frac{\partial z}{\partial z} = 2x\frac{\partial x}{\partial z} + 2y\frac{\partial y}{\partial z}$

$\Rightarrow 1 = 2y\frac{\partial y}{\partial z} \Rightarrow \frac{\partial y}{\partial z} = \frac{1}{2y} \Rightarrow \left(\frac{\partial w}{\partial z}\right)_x = (2x)(0) + (2y)\left(\frac{1}{2y}\right) + (2z)(1) = 1 + 2z$

(c) $\begin{pmatrix} y \\ z \end{pmatrix} \rightarrow \begin{pmatrix} x = x(y,z) \\ y = y \\ z = z \end{pmatrix} \rightarrow w \Rightarrow \left(\frac{\partial w}{\partial z}\right)_y = \frac{\partial w}{\partial x}\frac{\partial x}{\partial z} + \frac{\partial w}{\partial y}\frac{\partial y}{\partial z} + \frac{\partial w}{\partial z}\frac{\partial z}{\partial z}; \frac{\partial y}{\partial z} = 0$ and $\frac{\partial z}{\partial z} = 2x\frac{\partial x}{\partial z} + 2y\frac{\partial y}{\partial z}$

$\Rightarrow 1 = 2x\frac{\partial x}{\partial z} \Rightarrow \frac{\partial x}{\partial z} = \frac{1}{2x} \Rightarrow \left(\frac{\partial w}{\partial z}\right)_y = (2x)\left(\frac{1}{2x}\right) + (2y)(0) + (2z)(1) = 1 + 2z$

3. $U = f(P, V, T)$ and $PV = nRT$

(a) $\begin{pmatrix} P \\ V \end{pmatrix} \rightarrow \begin{pmatrix} P = P \\ V = V \\ T = \frac{PV}{nR} \end{pmatrix} \rightarrow U \Rightarrow \left(\frac{\partial U}{\partial P}\right)_V = \frac{\partial U}{\partial P}\frac{\partial P}{\partial P} + \frac{\partial U}{\partial V}\frac{\partial V}{\partial P} + \frac{\partial U}{\partial T}\frac{\partial T}{\partial P} = \frac{\partial U}{\partial P} + \left(\frac{\partial U}{\partial V}\right)(0) + \left(\frac{\partial U}{\partial T}\right)\left(\frac{V}{nR}\right)$

$= \frac{\partial U}{\partial P} + \left(\frac{\partial U}{\partial T}\right)\left(\frac{V}{nR}\right)$

(b) $\begin{pmatrix} V \\ T \end{pmatrix} \rightarrow \begin{pmatrix} P = \frac{nRT}{V} \\ V = V \\ T = T \end{pmatrix} \rightarrow U \Rightarrow \left(\frac{\partial U}{\partial T}\right)_V = \frac{\partial U}{\partial P}\frac{\partial P}{\partial T} + \frac{\partial U}{\partial V}\frac{\partial V}{\partial T} + \frac{\partial U}{\partial T}\frac{\partial T}{\partial T} = \left(\frac{\partial U}{\partial P}\right)\left(\frac{nR}{V}\right) + \left(\frac{\partial U}{\partial V}\right)(0) + \frac{\partial U}{\partial T}$

$= \left(\frac{\partial U}{\partial P}\right)\left(\frac{nR}{V}\right) + \frac{\partial U}{\partial T}$

5. $w = x^2y^2 + yz - z^3$ and $x^2 + y^2 + z^2 = 6$

(a) $\begin{pmatrix} x \\ y \end{pmatrix} \rightarrow \begin{pmatrix} x = x \\ y = y \\ z = z(x,y) \end{pmatrix} \rightarrow w \Rightarrow \left(\frac{\partial w}{\partial y}\right)_x = \frac{\partial w}{\partial x}\frac{\partial x}{\partial y} + \frac{\partial w}{\partial y}\frac{\partial y}{\partial y} + \frac{\partial w}{\partial z}\frac{\partial z}{\partial y}$

$= (2xy^2)(0) + (2x^2y + z)(1) + (y - 3z^2)\frac{\partial z}{\partial y} = 2x^2y + z + (y - 3z^2)\frac{\partial z}{\partial y}$. Now $(2x)\frac{\partial x}{\partial y} + 2y + (2z)\frac{\partial z}{\partial y} = 0$ and

$\frac{\partial x}{\partial y} = 0 \Rightarrow 2y + (2z)\frac{\partial z}{\partial y} = 0 \Rightarrow \frac{\partial z}{\partial y} = -\frac{y}{z}$. At $(w,x,y,z) = (4,2,1,-1)$, $\frac{\partial z}{\partial y} = -\frac{1}{-1} = 1 \Rightarrow \left(\frac{\partial w}{\partial y}\right)_x\Big|_{(4,2,1,-1)}$

$= [(2)(2)^2(1) + (-1)] + [1 - 3(-1)^2](1) = 5$

(b) $\begin{pmatrix} y \\ z \end{pmatrix} \rightarrow \begin{pmatrix} x = x(y,z) \\ y = y \\ z = z \end{pmatrix} \rightarrow w \Rightarrow \left(\frac{\partial w}{\partial y}\right)_z = \frac{\partial w}{\partial x}\frac{\partial x}{\partial y} + \frac{\partial w}{\partial y}\frac{\partial y}{\partial y} + \frac{\partial w}{\partial z}\frac{\partial z}{\partial y}$

$= (2xy^2)\frac{\partial x}{\partial y} + (2x^2y + z)(1) + (y - 3z^2)(0) = (2xy^2)\frac{\partial x}{\partial y} + 2x^2y + z$. Now $(2x)\frac{\partial x}{\partial y} + 2y + (2z)\frac{\partial z}{\partial y} = 0$ and

$\frac{\partial z}{\partial y} = 0 \Rightarrow (2x)\frac{\partial x}{\partial y} + 2y = 0 \Rightarrow \frac{\partial x}{\partial y} = -\frac{y}{x}$. At $(w,x,y,z) = (4,2,1,-1)$, $\frac{\partial x}{\partial y} = -\frac{1}{2} \Rightarrow \left(\frac{\partial w}{\partial y}\right)_z\Big|_{(4,2,1,-1)}$

$= (2)(2)(1)^2\left(-\frac{1}{2}\right) + (2)(2)^2(1) + (-1) = 5$

7. $\begin{pmatrix} r \\ \theta \end{pmatrix} \rightarrow \begin{pmatrix} x = r\cos\theta \\ y = r\sin\theta \end{pmatrix} \Rightarrow \left(\frac{\partial x}{\partial r}\right)_\theta = \cos\theta$; $x^2 + y^2 = r^2 \Rightarrow 2x + 2y\frac{\partial y}{\partial x} = 2r\frac{\partial r}{\partial x}$ and $\frac{\partial y}{\partial x} = 0 \Rightarrow 2x = 2r\frac{\partial r}{\partial x}$

$\Rightarrow \frac{\partial r}{\partial x} = \frac{x}{r} \Rightarrow \left(\frac{\partial r}{\partial x}\right)_y = \frac{x}{\sqrt{x^2 + y^2}}$

9. If x is a differentiable function of y and z, then $f(x,y,z) = 0 \Rightarrow \frac{\partial f}{\partial x}\frac{\partial x}{\partial x} + \frac{\partial f}{\partial y}\frac{\partial y}{\partial x} + \frac{\partial f}{\partial z}\frac{\partial z}{\partial x} = 0 \Rightarrow \frac{\partial f}{\partial x} + \frac{\partial f}{\partial y}\frac{\partial y}{\partial x} = 0$

$\Rightarrow \left(\frac{\partial x}{\partial y}\right)_z = -\frac{\partial f/\partial y}{\partial f/\partial z}$. Similarly, if y is a differentiable function of x and z, $\left(\frac{\partial y}{\partial z}\right)_x = -\frac{\partial f/\partial z}{\partial f/\partial x}$ and if z is a differentiable function of x and y, $\left(\frac{\partial z}{\partial x}\right)_y = -\frac{\partial f/\partial x}{\partial f/\partial y}$. Then $\left(\frac{\partial x}{\partial y}\right)_z \left(\frac{\partial y}{\partial z}\right)_x \left(\frac{\partial z}{\partial x}\right)_y$

$= \left(-\frac{\partial f/\partial y}{\partial f/\partial z}\right)\left(-\frac{\partial f/\partial z}{\partial f/\partial x}\right)\left(-\frac{\partial f/\partial x}{\partial f/\partial y}\right) = -1.$

11. If x and y are independent, then $g(x, y, z) = 0 \Rightarrow \frac{\partial g}{\partial x}\frac{\partial x}{\partial y} + \frac{\partial g}{\partial y}\frac{\partial y}{\partial y} + \frac{\partial g}{\partial z}\frac{\partial z}{\partial y} = 0$ and $\frac{\partial x}{\partial y} = 0 \Rightarrow \frac{\partial g}{\partial y} + \frac{\partial g}{\partial z}\frac{\partial z}{\partial y} = 0$

$\Rightarrow \left(\frac{\partial z}{\partial y}\right)_x = -\frac{\partial g/\partial y}{\partial g/\partial z}$, as claimed.

14.10 TAYLOR'S FORMULA FOR TWO VARIABLES

1. $f(x, y) = xe^y \Rightarrow f_x = e^y, f_y = xe^y, f_{xx} = 0, f_{xy} = e^y, f_{yy} = xe^y$

$\Rightarrow f(x, y) \approx f(0, 0) + xf_x(0, 0) + yf_y(0, 0) + \frac{1}{2}[x^2 f_{xx}(0, 0) + 2xyf_{xy}(0, 0) + y^2 f_{yy}(0, 0)]$

$= 0 + x \cdot 1 + y \cdot 0 + \frac{1}{2}(x^2 \cdot 0 + 2xy \cdot 1 + y^2 \cdot 0) = x + xy$ quadratic approximation;

$f_{xxx} = 0, f_{xxy} = 0, f_{xyy} = e^y, f_{yyy} = xe^y$

$\Rightarrow f(x, y) \approx \text{quadratic} + \frac{1}{6}[x^3 f_{xxx}(0, 0) + 3x^2 y f_{xxy}(0, 0) + 3xy^2 f_{xyy}(0, 0) + y^3 f_{yyy}(0, 0)]$

$= x + xy + \frac{1}{6}(x^3 \cdot 0 + 3x^2 y \cdot 0 + 3xy^2 \cdot 1 + y^3 \cdot 0) = x + xy + \frac{1}{2}xy^2$, cubic approximation

3. $f(x, y) = y \sin x \Rightarrow f_x = y \cos x, f_y = \sin x, f_{xx} = -y \sin x, f_{xy} = \cos x, f_{yy} = 0$

$\Rightarrow f(x, y) \approx f(0, 0) + xf_x(0, 0) + yf_y(0, 0) + \frac{1}{2}[x^2 f_{xx}(0, 0) + 2xyf_{xy}(0, 0) + y^2 f_{yy}(0, 0)]$

$= 0 + x \cdot 0 + y \cdot 0 + \frac{1}{2}(x^2 \cdot 0 + 2xy \cdot 1 + y^2 \cdot 0) = xy$, quadratic approximation;

$f_{xxx} = -y \cos x, f_{xxy} = -\sin x, f_{xyy} = 0, f_{yyy} = 0$

$\Rightarrow f(x, y) \approx \text{quadratic} + \frac{1}{6}[x^3 f_{xxx}(0, 0) + 3x^2 y f_{xxy}(0, 0) + 3xy^2 f_{xyy}(0, 0) + y^3 f_{yyy}(0, 0)]$

$= xy + \frac{1}{6}(x^3 \cdot 0 + 3x^2 y \cdot 0 + 3xy^2 \cdot 0 + y^3 \cdot 0) = xy$, cubic approximation

5. $f(x, y) = e^x \ln(1 + y) \Rightarrow f_x = e^x \ln(1 + y), f_y = \frac{e^x}{1+y}, f_{xx} = e^x \ln(1 + y), f_{xy} = \frac{e^x}{1+y}, f_{yy} = -\frac{e^x}{(1+y)^2}$

$\Rightarrow f(x, y) \approx f(0, 0) + xf_x(0, 0) + yf_y(0, 0) + \frac{1}{2}[x^2 f_{xx}(0, 0) + 2xyf_{xy}(0, 0) + y^2 f_{yy}(0, 0)]$

$= 0 + x \cdot 0 + y \cdot 1 + \frac{1}{2}[x^2 \cdot 0 + 2xy \cdot 1 + y^2 \cdot (-1)] = y + \frac{1}{2}(2xy - y^2)$, quadratic approximation;

$f_{xxx} = e^x \ln(1 + y), f_{xxy} = \frac{e^x}{1+y}, f_{xyy} = -\frac{e^x}{(1+y)^2}, f_{yyy} = \frac{2e^x}{(1+y)^3}$

$\Rightarrow f(x, y) \approx \text{quadratic} + \frac{1}{6}[x^3 f_{xxx}(0, 0) + 3x^2 y f_{xxy}(0, 0) + 3xy^2 f_{xyy}(0, 0) + y^3 f_{yyy}(0, 0)]$

$= y + \frac{1}{2}(2xy - y^2) + \frac{1}{6}[x^3 \cdot 0 + 3x^2 y \cdot 1 + 3xy^2 \cdot (-1) + y^3 \cdot 2]$

$= y + \frac{1}{2}(2xy - y^2) + \frac{1}{6}(3x^2 y - 3xy^2 + 2y^3)$, cubic approximation

7. $f(x, y) = \sin(x^2 + y^2) \Rightarrow f_x = 2x \cos(x^2 + y^2), f_y = 2y \cos(x^2 + y^2), f_{xx} = 2 \cos(x^2 + y^2) - 4x^2 \sin(x^2 + y^2),$

$f_{xy} = -4xy \sin(x^2 + y^2), f_{yy} = 2 \cos(x^2 + y^2) - 4y^2 \sin(x^2 + y^2)$

$\Rightarrow f(x, y) \approx f(0, 0) + xf_x(0, 0) + yf_y(0, 0) + \frac{1}{2}[x^2 f_{xx}(0, 0) + 2xyf_{xy}(0, 0) + y^2 f_{yy}(0, 0)]$

$= 0 + x \cdot 0 + y \cdot 0 + \frac{1}{2}(x^2 \cdot 2 + 2xy \cdot 0 + y^2 \cdot 2) = x^2 + y^2$, quadratic approximation;

$f_{xxx} = -12x \sin(x^2 + y^2) - 8x^3 \cos(x^2 + y^2), f_{xxy} = -4y \sin(x^2 + y^2) - 8x^2 y \cos(x^2 + y^2),$

$f_{xyy} = -4x \sin(x^2 + y^2) - 8xy^2 \cos(x^2 + y^2), f_{yyy} = -12y \sin(x^2 + y^2) - 8y^3 \cos(x^2 + y^2)$

$\Rightarrow f(x, y) \approx \text{quadratic} + \frac{1}{6}[x^3 f_{xxx}(0, 0) + 3x^2 y f_{xxy}(0, 0) + 3xy^2 f_{xyy}(0, 0) + y^3 f_{yyy}(0, 0)]$

$= x^2 + y^2 + \frac{1}{6}(x^3 \cdot 0 + 3x^2 y \cdot 0 + 3xy^2 \cdot 0 + y^3 \cdot 0) = x^2 + y^2$, cubic approximation

9. $f(x, y) = \frac{1}{1-x-y} \Rightarrow f_x = \frac{1}{(1-x-y)^2} = f_y, f_{xx} = \frac{2}{(1-x-y)^3} = f_{xy} = f_{yy}$

$\Rightarrow f(x, y) \approx f(0, 0) + xf_x(0, 0) + yf_y(0, 0) + \frac{1}{2}[x^2 f_{xx}(0, 0) + 2xyf_{xy}(0, 0) + y^2 f_{yy}(0, 0)]$

$= 1 + x \cdot 1 + y \cdot 1 + \frac{1}{2}(x^2 \cdot 2 + 2xy \cdot 2 + y^2 \cdot 2) = 1 + (x + y) + (x^2 + 2xy + y^2)$

$= 1 + (x + y) + (x + y)^2$, quadratic approximation; $f_{xxx} = \frac{6}{(1-x-y)^4} = f_{xxy} = f_{xyy} = f_{yyy}$

$\Rightarrow f(x, y) \approx \text{quadratic} + \frac{1}{6}[x^3 f_{xxx}(0, 0) + 3x^2 y f_{xxy}(0, 0) + 3xy^2 f_{xyy}(0, 0) + y^3 f_{yyy}(0, 0)]$

$$= 1 + (x+y) + (x+y)^2 + \tfrac{1}{6}(x^3 \cdot 6 + 3x^2y \cdot 6 + 3xy^2 \cdot 6 + y^3 \cdot 6)$$

$$= 1 + (x+y) + (x+y)^2 + (x^3 + 3x^2y + 3xy^2 + y^3) = 1 + (x+y) + (x+y)^2 + (x+y)^3, \text{ cubic approximation}$$

11. $f(x, y) = \cos x \cos y \Rightarrow f_x = -\sin x \cos y, \ f_y = -\cos x \sin y, \ f_{xx} = -\cos x \cos y, \ f_{xy} = \sin x \sin y,$

$f_{yy} = -\cos x \cos y \Rightarrow f(x, y) \approx f(0, 0) + xf_x(0, 0) + yf_y(0, 0) + \tfrac{1}{2}[x^2f_{xx}(0, 0) + 2xyf_{xy}(0, 0) + y^2f_{yy}(0, 0)]$

$= 1 + x \cdot 0 + y \cdot 0 + \tfrac{1}{2}[x^2 \cdot (-1) + 2xy \cdot 0 + y^2 \cdot (-1)] = 1 - \tfrac{x^2}{2} - \tfrac{y^2}{2},$ quadratic approximation. Since all partial

derivatives of f are products of sines and cosines, the absolute value of these derivatives is less than or equal

to $1 \Rightarrow E(x, y) \leq \tfrac{1}{6}[(0.1)^3 + 3(0.1)^3 + 3(0.1)^3 + 0.1)^3] \leq 0.00134.$

CHAPTER 14 PRACTICE EXERCISES

1. Domain: All points in the xy-plane
   Range: $z \geq 0$

   Level curves are ellipses with major axis along the y-axis
   and minor axis along the x-axis.

3. Domain: All (x, y) such that $x \neq 0$ and $y \neq 0$
   Range: $z \neq 0$

   Level curves are hyperbolas with the x- and y-axes
   as asymptotes.

5. Domain: All points (x, y, z) in space
   Range: All real numbers

   Level surfaces are paraboloids of revolution with
   the z-axis as axis.

7. Domain: All (x, y, z) such that $(x, y, z) \neq (0, 0, 0)$
   Range: Positive real numbers

   Level surfaces are spheres with center $(0, 0, 0)$ and
   radius $r > 0$.

9. $\displaystyle\lim_{(x, y) \to (\pi, \ln 2)} e^y \cos x = e^{\ln 2} \cos \pi = (2)(-1) = -2$

11. $\displaystyle\lim_{\substack{(x, y) \to (1, 1) \\ x \neq \pm y}} \frac{x - y}{x^2 - y^2} = \lim_{\substack{(x, y) \to (1, 1) \\ x \neq \pm y}} \frac{x - y}{(x - y)(x + y)} = \lim_{(x, y) \to (1, 1)} \frac{1}{x + y} = \frac{1}{1 + 1} = \frac{1}{2}$

13. $\displaystyle\lim_{P \to (1, -1, e)} \ln |x + y + z| = \ln |1 + (-1) + e| = \ln e = 1$

15. Let $y = kx^2$, $k \neq 1$. Then $\lim\limits_{\substack{(x,y) \to (0,0) \\ y \neq x^2}} \frac{y}{x^2 - y} = \lim\limits_{(x, kx^2) \to (0,0)} \frac{kx^2}{x^2 - kx^2} = \frac{k}{1-k^2}$ which gives different limits for

different values of k $\Rightarrow$ the limit does not exist.

17. Let $y = kx$. Then $\lim\limits_{(x,y) \to (0,0)} \frac{x^2 - y^2}{x^2 + y^2} = \frac{x^2 - k^2x^2}{x^2 + k^2x^2} = \frac{1-k^2}{1+k^2}$ which gives different limits for different values

of k $\Rightarrow$ the limit does not exist so $f(0,0)$ cannot be defined in a way that makes f continuous at the origin.

19. $\frac{\partial g}{\partial r} = \cos\theta + \sin\theta$, $\frac{\partial g}{\partial \theta} = -r\sin\theta + r\cos\theta$

21. $\frac{\partial f}{\partial R_1} = -\frac{1}{R_1^2}$, $\frac{\partial f}{\partial R_2} = -\frac{1}{R_2^2}$, $\frac{\partial f}{\partial R_3} = -\frac{1}{R_3^2}$

23. $\frac{\partial P}{\partial n} = \frac{RT}{V}$, $\frac{\partial P}{\partial R} = \frac{nT}{V}$, $\frac{\partial P}{\partial T} = \frac{nR}{V}$, $\frac{\partial P}{\partial V} = -\frac{nRT}{V^2}$

25. $\frac{\partial g}{\partial x} = \frac{1}{y}$, $\frac{\partial g}{\partial y} = 1 - \frac{x}{y^2}$ $\Rightarrow$ $\frac{\partial^2 g}{\partial x^2} = 0$, $\frac{\partial^2 g}{\partial y^2} = \frac{2x}{y^3}$, $\frac{\partial^2 g}{\partial y \partial x} = \frac{\partial^2 g}{\partial x \partial y} = -\frac{1}{y^2}$

27. $\frac{\partial f}{\partial x} = 1 + y - 15x^2 + \frac{2x}{x^2+1}$, $\frac{\partial f}{\partial y} = x$ $\Rightarrow$ $\frac{\partial^2 f}{\partial x^2} = -30x + \frac{2-2x^2}{(x^2+1)^2}$, $\frac{\partial^2 f}{\partial y^2} = 0$, $\frac{\partial^2 f}{\partial y \partial x} = \frac{\partial^2 f}{\partial x \partial y} = 1$

29. $\frac{\partial w}{\partial x} = y\cos(xy + \pi)$, $\frac{\partial w}{\partial y} = x\cos(xy + \pi)$, $\frac{dx}{dt} = e^t$, $\frac{dy}{dt} = \frac{1}{t+1}$

$\Rightarrow$ $\frac{dw}{dt} = [y\cos(xy + \pi)]e^t + [x\cos(xy + \pi)]\left(\frac{1}{t+1}\right)$; $t = 0$ $\Rightarrow$ $x = 1$ and $y = 0$

$\Rightarrow$ $\frac{dw}{dt}\Big|_{t=0} = 0 \cdot 1 + [1 \cdot (-1)]\left(\frac{1}{0+1}\right) = -1$

31. $\frac{\partial w}{\partial x} = 2\cos(2x - y)$, $\frac{\partial w}{\partial y} = -\cos(2x - y)$, $\frac{\partial x}{\partial r} = 1$, $\frac{\partial x}{\partial s} = \cos s$, $\frac{\partial y}{\partial s} = s$, $\frac{\partial y}{\partial s} = r$

$\Rightarrow$ $\frac{\partial w}{\partial r} = [2\cos(2x - y)](1) + [-\cos(2x - y)](s)$; $r = \pi$ and $s = 0$ $\Rightarrow$ $x = \pi$ and $y = 0$

$\Rightarrow$ $\frac{\partial w}{\partial r}\Big|_{(\pi,0)} = (2\cos 2\pi) - (\cos 2\pi)(0) = 2$; $\frac{\partial w}{\partial s} = [2\cos(2x - y)](\cos s) + [-\cos(2x - y)](r)$

$\Rightarrow$ $\frac{\partial w}{\partial s}\Big|_{(\pi,0)} = (2\cos 2\pi)(\cos 0) - (\cos 2\pi)(\pi) = 2 - \pi$

33. $\frac{\partial f}{\partial x} = y + z$, $\frac{\partial f}{\partial y} = x + z$, $\frac{\partial f}{\partial z} = y + x$, $\frac{dx}{dt} = -\sin t$, $\frac{dy}{dt} = \cos t$, $\frac{dz}{dt} = -2\sin 2t$

$\Rightarrow$ $\frac{df}{dt} = -(y + z)(\sin t) + (x + z)(\cos t) - 2(y + x)(\sin 2t)$; $t = 1$ $\Rightarrow$ $x = \cos 1$, $y = \sin 1$, and $z = \cos 2$

$\Rightarrow$ $\frac{df}{dt}\Big|_{t=1} = -(\sin 1 + \cos 2)(\sin 1) + (\cos 1 + \cos 2)(\cos 1) - 2(\sin 1 + \cos 1)(\sin 2)$

35. $F(x,y) = 1 - x - y^2 - \sin xy$ $\Rightarrow$ $F_x = -1 - y\cos xy$ and $F_y = -2y - x\cos xy$ $\Rightarrow$ $\frac{dy}{dx} = -\frac{F_x}{F_y} = -\frac{-1 - y\cos xy}{-2y - x\cos xy}$

$= \frac{1 + y\cos xy}{-2y - x\cos xy}$ $\Rightarrow$ at $(x,y) = (0,1)$ we have $\frac{dy}{dx}\Big|_{(0,1)} = \frac{1+1}{-2} = -1$

37. $\nabla f = (-\sin x\cos y)\mathbf{i} - (\cos x\sin y)\mathbf{j}$ $\Rightarrow$ $\nabla f|_{(\frac{\pi}{4},\frac{\pi}{4})} = -\frac{1}{2}\mathbf{i} - \frac{1}{2}\mathbf{j}$ $\Rightarrow$ $|\nabla f| = \sqrt{\left(-\frac{1}{2}\right)^2 + \left(-\frac{1}{2}\right)^2} = \frac{1}{\sqrt{2}} = \frac{\sqrt{2}}{2}$;

$\mathbf{u} = \frac{\nabla f}{|\nabla f|} = -\frac{\sqrt{2}}{2}\mathbf{i} - \frac{\sqrt{2}}{2}\mathbf{j}$ $\Rightarrow$ f increases most rapidly in the direction $\mathbf{u} = -\frac{\sqrt{2}}{2}\mathbf{i} - \frac{\sqrt{2}}{2}\mathbf{j}$ and decreases most

rapidly in the direction $-\mathbf{u} = \frac{\sqrt{2}}{2}\mathbf{i} + \frac{\sqrt{2}}{2}\mathbf{j}$; $(D_{\mathbf{u}}f)_{P_0} = |\nabla f| = \frac{\sqrt{2}}{2}$ and $(D_{-\mathbf{u}}f)_{P_0} = -\frac{\sqrt{2}}{2}$;

$\mathbf{u}_1 = \frac{\mathbf{v}}{|\mathbf{v}|} = \frac{3\mathbf{i} + 4\mathbf{j}}{\sqrt{3^2 + 4^2}} = \frac{3}{5}\mathbf{i} + \frac{4}{5}\mathbf{j}$ $\Rightarrow$ $(D_{\mathbf{u}_1}f)_{P_0} = \nabla f \cdot \mathbf{u}_1 = \left(-\frac{1}{2}\right)\left(\frac{3}{5}\right) + \left(-\frac{1}{2}\right)\left(\frac{4}{5}\right) = -\frac{7}{10}$

39. $\nabla f = \left(\frac{2}{2x + 3y + 6z}\right)\mathbf{i} + \left(\frac{3}{2x + 3y + 6z}\right)\mathbf{j} + \left(\frac{6}{2x + 3y + 6z}\right)\mathbf{k}$ $\Rightarrow$ $\nabla f|_{(-1,-1,1)} = 2\mathbf{i} + 3\mathbf{j} + 6\mathbf{k}$;

$\mathbf{u} = \frac{\nabla f}{|\nabla f|} = \frac{2\mathbf{i} + 3\mathbf{j} + 6\mathbf{k}}{\sqrt{2^2 + 3^2 + 6^2}} = \frac{2}{7}\mathbf{i} + \frac{3}{7}\mathbf{j} + \frac{6}{7}\mathbf{k}$ $\Rightarrow$ f increases most rapidly in the direction $\mathbf{u} = \frac{2}{7}\mathbf{i} + \frac{3}{7}\mathbf{j} + \frac{6}{7}\mathbf{k}$ and

decreases most rapidly in the direction $-\mathbf{u} = -\frac{2}{7}\mathbf{i} - \frac{3}{7}\mathbf{j} - \frac{6}{7}\mathbf{k}$; $(D_{\mathbf{u}}f)_{P_0} = |\nabla f| = 7$, $(D_{-\mathbf{u}}f)_{P_0} = -7$;

$\mathbf{u_1} = \frac{\mathbf{v}}{|\mathbf{v}|} = \frac{2}{7}\mathbf{i} + \frac{3}{7}\mathbf{j} + \frac{6}{7}\mathbf{k} \Rightarrow (D_{\mathbf{u_1}}f)_{P_0} = (D_{\mathbf{u}}f)_{P_0} = 7$

41. $\mathbf{r} = (\cos 3t)\mathbf{i} + (\sin 3t)\mathbf{j} + 3t\mathbf{k} \Rightarrow \mathbf{v}(t) = (-3\sin 3t)\mathbf{i} + (3\cos 3t)\mathbf{j} + 3\mathbf{k} \Rightarrow \mathbf{v}\left(\frac{\pi}{3}\right) = -3\mathbf{j} + 3\mathbf{k}$

$\Rightarrow \mathbf{u} = -\frac{1}{\sqrt{2}}\mathbf{j} + \frac{1}{\sqrt{2}}\mathbf{k}; f(x,y,z) = xyz \Rightarrow \nabla f = yz\mathbf{i} + xz\mathbf{j} + xy\mathbf{k}; t = \frac{\pi}{3}$ yields the point on the helix $(-1, 0, \pi)$

$\Rightarrow \nabla f|_{(-1,0,\pi)} = -\pi\mathbf{j} \Rightarrow \nabla f \cdot \mathbf{u} = (-\pi\mathbf{j}) \cdot \left(-\frac{1}{\sqrt{2}}\mathbf{j} + \frac{1}{\sqrt{2}}\mathbf{k}\right) = \frac{\pi}{\sqrt{2}}$

43. (a) Let $\nabla f = a\mathbf{i} + b\mathbf{j}$ at $(1,2)$. The direction toward $(2,2)$ is determined by $\mathbf{v_1} = (2-1)\mathbf{i} + (2-2)\mathbf{j} = \mathbf{i} = \mathbf{u}$
so that $\nabla f \cdot \mathbf{u} = 2 \Rightarrow a = 2$. The direction toward $(1,1)$ is determined by $\mathbf{v_2} = (1-1)\mathbf{i} + (1-2)\mathbf{j} = -\mathbf{j} = \mathbf{u}$
so that $\nabla f \cdot \mathbf{u} = -2 \Rightarrow -b = -2 \Rightarrow b = 2$. Therefore $\nabla f = 2\mathbf{i} + 2\mathbf{j}; f_x(1,2) = f_y(1,2) = 2$.

(b) The direction toward $(4,6)$ is determined by $\mathbf{v_3} = (4-1)\mathbf{i} + (6-2)\mathbf{j} = 3\mathbf{i} + 4\mathbf{j} \Rightarrow \mathbf{u} = \frac{3}{5}\mathbf{i} + \frac{4}{5}\mathbf{j}$
$\Rightarrow \nabla f \cdot \mathbf{u} = \frac{14}{5}$.

45. $\nabla f = 2x\mathbf{i} + \mathbf{j} + 2z\mathbf{k} \Rightarrow$
$\nabla f|_{(0,-1,-1)} = \mathbf{j} - 2\mathbf{k}$,
$\nabla f|_{(0,0,0)} = \mathbf{j}$,
$\nabla f|_{(0,-1,1)} = \mathbf{j} + 2\mathbf{k}$

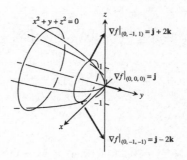

47. $\nabla f = 2x\mathbf{i} - \mathbf{j} - 5\mathbf{k} \Rightarrow \nabla f|_{(2,-1,1)} = 4\mathbf{i} - \mathbf{j} - 5\mathbf{k} \Rightarrow$ Tangent Plane: $4(x-2) - (y+1) - 5(z-1) = 0$
$\Rightarrow 4x - y - 5z = 4$; Normal Line: $x = 2 + 4t, y = -1 - t, z = 1 - 5t$

49. $\frac{\partial z}{\partial x} = \frac{2x}{x^2+y^2} \Rightarrow \frac{\partial z}{\partial x}\Big|_{(0,1,0)} = 0$ and $\frac{\partial z}{\partial y} = \frac{2y}{x^2+y^2} \Rightarrow \frac{\partial z}{\partial y}\Big|_{(0,1,0)} = 2$; thus the tangent plane is
$2(y-1) - (z-0) = 0$ or $2y - z - 2 = 0$

51. $\nabla f = (-\cos x)\mathbf{i} + \mathbf{j} \Rightarrow \nabla f|_{(\pi,1)} = \mathbf{i} + \mathbf{j} \Rightarrow$ the tangent
line is $(x - \pi) + (y - 1) = 0 \Rightarrow x + y = \pi + 1$; the
normal line is $y - 1 = 1(x - \pi) \Rightarrow y = x - \pi + 1$

53. Let $f(x,y,z) = x^2 + 2y + 2z - 4$ and $g(x,y,z) = y - 1$. Then $\nabla f = 2x\mathbf{i} + 2\mathbf{j} + 2\mathbf{k}|_{(1,1,\frac{1}{2})} = 2\mathbf{i} + 2\mathbf{j} + 2\mathbf{k}$

and $\nabla g = \mathbf{j} \Rightarrow \nabla f \times \nabla g = \begin{vmatrix} \mathbf{i} & \mathbf{j} & \mathbf{k} \\ 2 & 2 & 2 \\ 0 & 1 & 0 \end{vmatrix} = -2\mathbf{i} + 2\mathbf{k} \Rightarrow$ the line is $x = 1 - 2t, y = 1, z = \frac{1}{2} + 2t$

55. $f\left(\frac{\pi}{4}, \frac{\pi}{4}\right) = \frac{1}{2}, f_x\left(\frac{\pi}{4}, \frac{\pi}{4}\right) = \cos x \cos y|_{(\pi/4,\pi/4)} = \frac{1}{2}, f_y\left(\frac{\pi}{4}, \frac{\pi}{4}\right) = -\sin x \sin y|_{(\pi/4,\pi/4)} = -\frac{1}{2}$
$\Rightarrow L(x,y) = \frac{1}{2} + \frac{1}{2}\left(x - \frac{\pi}{4}\right) - \frac{1}{2}\left(y - \frac{\pi}{4}\right) = \frac{1}{2} + \frac{1}{2}x - \frac{1}{2}y; f_{xx}(x,y) = -\sin x \cos y, f_{yy}(x,y) = -\sin x \cos y,$ and
$f_{xy}(x,y) = -\cos x \sin y.$ Thus an upper bound for E depends on the bound M used for $|f_{xx}|, |f_{xy}|,$ and $|f_{yy}|.$
With $M = \frac{\sqrt{2}}{2}$ we have $|E(x,y)| \le \frac{1}{2}\left(\frac{\sqrt{2}}{2}\right)\left(|x - \frac{\pi}{4}| + |y - \frac{\pi}{4}|\right)^2 \le \frac{\sqrt{2}}{4}(0.2)^2 \le 0.0142;$
with $M = 1, |E(x,y)| \le \frac{1}{2}(1)\left(|x - \frac{\pi}{4}| + |y - \frac{\pi}{4}|\right)^2 = \frac{1}{2}(0.2)^2 = 0.02.$

57. $f(1,0,0) = 0, f_x(1,0,0) = y - 3z|_{(1,0,0)} = 0, f_y(1,0,0) = x + 2z|_{(1,0,0)} = 1, f_z(1,0,0) = 2y - 3x|_{(1,0,0)} = -3$
$\Rightarrow L(x,y,z) = 0(x-1) + (y-0) - 3(z-0) = y - 3z; f(1,1,0) = 1, f_x(1,1,0) = 1, f_y(1,1,0) = 1, f_z(1,1,0) = -1$

$\Rightarrow \ L(x, y, z) = 1 + (x - 1) + (y - 1) - 1(z - 0) = x + y - z - 1$

59. $V = \pi r^2 h \Rightarrow dV = 2\pi rh \ dr + \pi r^2 \ dh \Rightarrow dV\big|_{(1.5, 5280)} = 2\pi(1.5)(5280) \ dr + \pi(1.5)^2 \ dh = 15{,}840\pi \ dr + 2.25\pi \ dh.$
You should be more careful with the diameter since it has a greater effect on dV.

61. $dI = \frac{1}{R} dV - \frac{V}{R^2} dR \Rightarrow dI\big|_{(24, 100)} = \frac{1}{100} dV - \frac{24}{100^2} dR \Rightarrow dI\big|_{dV=-1, dR=-20} = -0.01 + (480)(.0001) = 0.038,$
or increases by 0.038 amps; % change in $V = (100)\left(-\frac{1}{24}\right) \approx -4.17\%$; % change in $R = \left(-\frac{20}{100}\right)(100) = -20\%$;
$I = \frac{24}{100} = 0.24 \Rightarrow$ estimated % change in $I = \frac{dI}{I} \times 100 = \frac{0.038}{0.24} \times 100 \approx 15.83\% \Rightarrow$ more sensitive to voltage change.

63. (a) $y = uv \Rightarrow dy = v \ du + u \ dv$; percentage change in $u \le 2\% \Rightarrow |du| \le 0.02$, and percentage change in $v \le 3\%$
$\Rightarrow |dv| \le 0.03; \ \frac{dy}{y} = \frac{v \ du + u \ dv}{uv} = \frac{du}{u} + \frac{dv}{v} \Rightarrow \left|\frac{dy}{y} \times 100\right| = \left|\frac{du}{u} \times 100 + \frac{dv}{v} \times 100\right| \le \left|\frac{du}{u} \times 100\right| + \left|\frac{dv}{v} \times 100\right|$
$\le 2\% + 3\% = 5\%$

(b) $z = u + v \Rightarrow \frac{dz}{z} = \frac{du + dv}{u + v} = \frac{du}{u + v} + \frac{dv}{u + v} \le \frac{du}{u} + \frac{dv}{v}$ (since $u > 0$, $v > 0$)
$\Rightarrow \left|\frac{dz}{z} \times 100\right| \le \left|\frac{du}{u} \times 100 + \frac{dv}{v} \times 100\right| = \left|\frac{dy}{y} \times 100\right|$

65. $f_x(x, y) = 2x - y + 2 = 0$ and $f_y(x, y) = -x + 2y + 2 = 0 \Rightarrow x = -2$ and $y = -2 \Rightarrow (-2, -2)$ is the critical point;
$f_{xx}(-2, -2) = 2, \ f_{yy}(-2, -2) = 2, \ f_{xy}(-2, -2) = -1 \Rightarrow f_{xx}f_{yy} - f_{xy}^2 = 3 > 0$ and $f_{xx} > 0 \Rightarrow$ local minimum value
of $f(-2, -2) = -8$

67. $f_x(x, y) = 6x^2 + 3y = 0$ and $f_y(x, y) = 3x + 6y^2 = 0 \Rightarrow y = -2x^2$ and $3x + 6(4x^4) = 0 \Rightarrow x(1 + 8x^3) = 0$
$\Rightarrow x = 0$ and $y = 0$, or $x = -\frac{1}{2}$ and $y = -\frac{1}{2} \Rightarrow$ the critical points are $(0, 0)$ and $\left(-\frac{1}{2}, -\frac{1}{2}\right)$. For $(0, 0)$:
$f_{xx}(0, 0) = 12x\big|_{(0,0)} = 0, \ f_{yy}(0, 0) = 12y\big|_{(0,0)} = 0, \ f_{xy}(0, 0) = 3 \Rightarrow f_{xx}f_{yy} - f_{xy}^2 = -9 < 0 \Rightarrow$ saddle point with
$f(0, 0) = 0$. For $\left(-\frac{1}{2}, -\frac{1}{2}\right)$: $f_{xx} = -6, \ f_{yy} = -6, \ f_{xy} = 3 \Rightarrow f_{xx}f_{yy} - f_{xy}^2 = 27 > 0$ and $f_{xx} < 0 \Rightarrow$ local maximum
value of $f\left(-\frac{1}{2}, -\frac{1}{2}\right) = \frac{1}{4}$

69. $f_x(x, y) = 3x^2 + 6x = 0$ and $f_y(x, y) = 3y^2 - 6y = 0 \Rightarrow x(x + 2) = 0$ and $y(y - 2) = 0 \Rightarrow x = 0$ or $x = -2$ and
$y = 0$ or $y = 2 \Rightarrow$ the critical points are $(0, 0), (0, 2), (-2, 0)$, and $(-2, 2)$. For $(0, 0)$: $f_{xx}(0, 0) = 6x + 6\big|_{(0,0)}$
$= 6, \ f_{yy}(0, 0) = 6y - 6\big|_{(0,0)} = -6, \ f_{xy}(0, 0) = 0 \Rightarrow f_{xx}f_{yy} - f_{xy}^2 = -36 < 0 \Rightarrow$ saddle point with $f(0, 0) = 0$. For
$(0, 2)$: $f_{xx}(0, 2) = 6, \ f_{yy}(0, 2) = 6, \ f_{xy}(0, 2) = 0 \Rightarrow f_{xx}f_{yy} - f_{xy}^2 = 36 > 0$ and $f_{xx} > 0 \Rightarrow$ local minimum value of
$f(0, 2) = -4$. For $(-2, 0)$: $f_{xx}(-2, 0) = -6, \ f_{yy}(-2, 0) = -6, \ f_{xy}(-2, 0) = 0 \Rightarrow f_{xx}f_{yy} - f_{xy}^2 = 36 > 0$ and $f_{xx} < 0$
$\Rightarrow$ local maximum value of $f(-2, 0) = 4$. For $(-2, 2)$: $f_{xx}(-2, 2) = -6, \ f_{yy}(-2, 2) = 6, \ f_{xy}(-2, 2) = 0$
$\Rightarrow f_{xx}f_{yy} - f_{xy}^2 = -36 < 0 \Rightarrow$ saddle point with $f(-2, 2) = 0$.

71. (i)   On OA, $f(x, y) = f(0, y) = y^2 + 3y$ for $0 \le y \le 4$
$\Rightarrow f'(0, y) = 2y + 3 = 0 \Rightarrow y = -\frac{3}{2}$. But $\left(0, -\frac{3}{2}\right)$
is not in the region.
Endpoints: $f(0, 0) = 0$ and $f(0, 4) = 28$.

(ii) On AB, $f(x, y) = f(x, -x + 4) = x^2 - 10x + 28$
for $0 \le x \le 4 \Rightarrow f'(x, -x + 4) = 2x - 10 = 0$
$\Rightarrow x = 5, \ y = -1$. But $(5, -1)$ is not in the region.
Endpoints: $f(4, 0) = 4$ and $f(0, 4) = 28$.

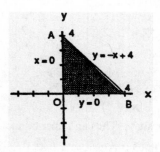

(iii) On OB, $f(x, y) = f(x, 0) = x^2 - 3x$ for $0 \le x \le 4 \Rightarrow f'(x, 0) = 2x - 3 \Rightarrow x = \frac{3}{2}$ and $y = 0 \Rightarrow \left(\frac{3}{2}, 0\right)$ is a
critical point with $f\left(\frac{3}{2}, 0\right) = -\frac{9}{4}$.
Endpoints: $f(0, 0) = 0$ and $f(4, 0) = 4$.

(iv) For the interior of the triangular region, $f_x(x, y) = 2x + y - 3 = 0$ and $f_y(x, y) = x + 2y + 3 = 0 \Rightarrow x = 3$

and $y = -3$. But $(3, -3)$ is not in the region. Therefore the absolute maximum is 28 at $(0, 4)$ and the absolute minimum is $-\frac{9}{4}$ at $\left(\frac{3}{2}, 0\right)$.

73. (i)   On AB, $f(x, y) = f(-2, y) = y^2 - y - 4$ for
$-2 \leq y \leq 2 \Rightarrow f'(-2, y) = 2y - 1 \Rightarrow y = \frac{1}{2}$ and
$x = -2 \Rightarrow \left(-2, \frac{1}{2}\right)$ is an interior critical point in AB
with $f\left(-2, \frac{1}{2}\right) = -\frac{17}{4}$. Endpoints: $f(-2, -2) = 2$ and
$f(2, 2) = -2$.

(ii)   On BC, $f(x, y) = f(x, 2) = -2$ for $-2 \leq x \leq 2$
$\Rightarrow f'(x, 2) = 0 \Rightarrow$ no critical points in the interior of
BC. Endpoints: $f(-2, 2) = -2$ and $f(2, 2) = -2$.

(iii) On CD, $f(x, y) = f(2, y) = y^2 - 5y + 4$ for
$-2 \leq y \leq 2 \Rightarrow f'(2, y) = 2y - 5 = 0 \Rightarrow y = \frac{5}{2}$ and $x = 2$. But $\left(2, \frac{5}{2}\right)$ is not in the region.
Endpoints: $f(2, -2) = 18$ and $f(2, 2) = -2$.

(iv)   On AD, $f(x, y) = f(x, -2) = 4x + 10$ for $-2 \leq x \leq 2 \Rightarrow f'(x, -2) = 4 \Rightarrow$ no critical points in the interior
of AD. Endpoints: $f(-2, -2) = 2$ and $f(2, -2) = 18$.

(v)   For the interior of the square, $f_x(x, y) = -y + 2 = 0$ and $f_y(x, y) = 2y - x - 3 = 0 \Rightarrow y = 2$ and $x = 1$
$\Rightarrow (1, 2)$ is an interior critical point of the square with $f(1, 2) = -2$. Therefore the absolute maximum
is 18 at $(2, -2)$ and the absolute minimum is $-\frac{17}{4}$ at $\left(-2, \frac{1}{2}\right)$.

75. (i)   On AB, $f(x, y) = f(x, x + 2) = -2x + 4$ for
$-2 \leq x \leq 2 \Rightarrow f'(x, x + 2) = -2 = 0 \Rightarrow$ no critical
points in the interior of AB. Endpoints: $f(-2, 0) = 8$
and $f(2, 4) = 0$.

(ii)   On BC, $f(x, y) = f(2, y) = -y^2 + 4y$ for $0 \leq y \leq 4$
$\Rightarrow f'(2, y) = -2y + 4 = 0 \Rightarrow y = 2$ and $x = 2$
$\Rightarrow (2, 2)$ is an interior critical point of BC with
$f(2, 2) = 4$. Endpoints: $f(2, 0) = 0$ and $f(2, 4) = 0$.

(iii) On AC, $f(x, y) = f(x, 0) = x^2 - 2x$ for $-2 \leq x \leq 2$
$\Rightarrow f'(x, 0) = 2x - 2 \Rightarrow x = 1$ and $y = 0 \Rightarrow (1, 0)$ is an interior critical point of AC with $f(1, 0) = -1$.
Endpoints: $f(-2, 0) = 8$ and $f(2, 0) = 0$.

(iv)   For the interior of the triangular region, $f_x(x, y) = 2x - 2 = 0$ and $f_y(x, y) = -2y + 4 = 0 \Rightarrow x = 1$ and
$y = 2 \Rightarrow (1, 2)$ is an interior critical point of the region with $f(1, 2) = 3$. Therefore the absolute maximum
is 8 at $(-2, 0)$ and the absolute minimum is -1 at $(1, 0)$.

77. (i)   On AB, $f(x, y) = f(-1, y) = y^3 - 3y^2 + 2$ for
$-1 \leq y \leq 1 \Rightarrow f'(-1, y) = 3y^2 - 6y = 0 \Rightarrow y = 0$
and $x = -1$, or $y = 2$ and $x = -1 \Rightarrow (-1, 0)$ is an
interior critical point of AB with $f(-1, 0) = 2$; $(-1, 2)$
is outside the boundary. Endpoints: $f(-1, -1) = -2$
and $f(-1, 1) = 0$.

(ii)   On BC, $f(x, y) = f(x, 1) = x^3 + 3x^2 - 2$ for
$-1 \leq x \leq 1 \Rightarrow f'(x, 1) = 3x^2 + 6x = 0 \Rightarrow x = 0$
and $y = 1$, or $x = -2$ and $y = 1 \Rightarrow (0, 1)$ is an
interior critical point of BC with $f(0, 1) = -2$; $(-2, 1)$ is outside the boundary. Endpoints: $f(-1, 1) = 0$ and
$f(1, 1) = 2$.

(iii) On CD, $f(x, y) = f(1, y) = y^3 - 3y^2 + 4$ for $-1 \leq y \leq 1 \Rightarrow f'(1, y) = 3y^2 - 6y = 0 \Rightarrow y = 0$ and $x = 1$, or
$y = 2$ and $x = 1 \Rightarrow (1, 0)$ is an interior critical point of CD with $f(1, 0) = 4$; $(1, 2)$ is outside the boundary.
Endpoints: $f(1, 1) = 2$ and $f(1, -1) = 0$.

(iv)   On AD, $f(x, y) = f(x, -1) = x^3 + 3x^2 - 4$ for $-1 \le x \le 1$ $\Rightarrow$ $f'(x, -1) = 3x^2 + 6x = 0$ $\Rightarrow$ $x = 0$ and $y = -1$, or $x = -2$ and $y = -1$ $\Rightarrow$ $(0, -1)$ is an interior point of AD with $f(0, -1) = -4$; $(-2, -1)$ is outside the boundary. Endpoints: $f(-1, -1) = -2$ and $f(1, -1) = 0$.

(v)   For the interior of the square, $f_x(x, y) = 3x^2 + 6x = 0$ and $f_y(x, y) = 3y^2 - 6y = 0$ $\Rightarrow$ $x = 0$ or $x = -2$, and $y = 0$ or $y = 2$ $\Rightarrow$ $(0, 0)$ is an interior critical point of the square region with $f(0, 0) = 0$; the points $(0, 2)$, $(-2, 0)$, and $(-2, 2)$ are outside the region. Therefore the absolute maximum is 4 at $(1, 0)$ and the absolute minimum is -4 at $(0, -1)$.

79. $\nabla f = 3x^2 \mathbf{i} + 2y\mathbf{j}$ and $\nabla g = 2x\mathbf{i} + 2y\mathbf{j}$ so that $\nabla f = \lambda \nabla g$ $\Rightarrow$ $3x^2 \mathbf{i} + 2y\mathbf{j} = \lambda(2x\mathbf{i} + 2y\mathbf{j})$ $\Rightarrow$ $3x^2 = 2x\lambda$ and $2y = 2y\lambda$ $\Rightarrow$ $\lambda = 1$ or $y = 0$.

CASE 1: $\lambda = 1$ $\Rightarrow$ $3x^2 = 2x$ $\Rightarrow$ $x = 0$ or $x = \frac{2}{3}$; $x = 0$ $\Rightarrow$ $y = \pm 1$ yielding the points $(0, 1)$ and $(0, -1)$; $x = \frac{2}{3}$ $\Rightarrow$ $y = \pm \frac{\sqrt{5}}{3}$ yielding the points $\left(\frac{2}{3}, \frac{\sqrt{5}}{3}\right)$ and $\left(\frac{2}{3}, -\frac{\sqrt{5}}{3}\right)$.

CASE 2: $y = 0$ $\Rightarrow$ $x^2 - 1 = 0$ $\Rightarrow$ $x = \pm 1$ yielding the points $(1, 0)$ and $(-1, 0)$.

Evaluations give $f(0, \pm 1) = 1$, $f\left(\frac{2}{3}, \pm \frac{\sqrt{5}}{3}\right) = \frac{23}{27}$, $f(1, 0) = 1$, and $f(-1, 0) = -1$. Therefore the absolute maximum is 1 at $(0, \pm 1)$ and $(1, 0)$, and the absolute minimum is -1 at $(-1, 0)$.

81. (i) $f(x, y) = x^2 + 3y^2 + 2y$ on $x^2 + y^2 = 1$ $\Rightarrow$ $\nabla f = 2x\mathbf{i} + (6y + 2)\mathbf{j}$ and $\nabla g = 2x\mathbf{i} + 2y\mathbf{j}$ so that $\nabla f = \lambda \nabla g$ $\Rightarrow$ $2x\mathbf{i} + (6y + 2)\mathbf{j} = \lambda(2x\mathbf{i} + 2y\mathbf{j})$ $\Rightarrow$ $2x = 2x\lambda$ and $6y + 2 = 2y\lambda$ $\Rightarrow$ $\lambda = 1$ or $x = 0$.

CASE 1: $\lambda = 1$ $\Rightarrow$ $6y + 2 = 2y$ $\Rightarrow$ $y = -\frac{1}{2}$ and $x = \pm \frac{\sqrt{3}}{2}$ yielding the points $\left(\pm \frac{\sqrt{3}}{2}, -\frac{1}{2}\right)$.

CASE 2: $x = 0$ $\Rightarrow$ $y^2 = 1$ $\Rightarrow$ $y = \pm 1$ yielding the points $(0, \pm 1)$.

Evaluations give $f\left(\pm \frac{\sqrt{3}}{2}, -\frac{1}{2}\right) = \frac{1}{2}$, $f(0, 1) = 5$, and $f(0, -1) = 1$. Therefore $\frac{1}{2}$ and 5 are the extreme values on the boundary of the disk.

(ii) For the interior of the disk, $f_x(x, y) = 2x = 0$ and $f_y(x, y) = 6y + 2 = 0$ $\Rightarrow$ $x = 0$ and $y = -\frac{1}{3}$ $\Rightarrow$ $\left(0, -\frac{1}{3}\right)$ is an interior critical point with $f\left(0, -\frac{1}{3}\right) = -\frac{1}{3}$. Therefore the absolute maximum of f on the disk is 5 at $(0, 1)$ and the absolute minimum of f on the disk is $-\frac{1}{3}$ at $\left(0, -\frac{1}{3}\right)$.

83. $\nabla f = \mathbf{i} - \mathbf{j} + \mathbf{k}$ and $\nabla g = 2x\mathbf{i} + 2y\mathbf{j} + 2z\mathbf{k}$ so that $\nabla f = \lambda \nabla g$ $\Rightarrow$ $\mathbf{i} - \mathbf{j} + \mathbf{k} = \lambda(2x\mathbf{i} + 2y\mathbf{j} + 2z\mathbf{k})$ $\Rightarrow$ $1 = 2x\lambda$, $-1 = 2y\lambda$, $1 = 2z\lambda$ $\Rightarrow$ $x = -y = z = \frac{1}{\lambda}$. Thus $x^2 + y^2 + z^2 = 1$ $\Rightarrow$ $3x^2 = 1$ $\Rightarrow$ $x = \pm \frac{1}{\sqrt{3}}$ yielding the points $\left(\frac{1}{\sqrt{3}}, -\frac{1}{\sqrt{3}}, \frac{1}{\sqrt{3}}\right)$ and $\left(-\frac{1}{\sqrt{3}}, \frac{1}{\sqrt{3}}, -\frac{1}{\sqrt{3}}\right)$. Evaluations give the absolute maximum value of $f\left(\frac{1}{\sqrt{3}}, -\frac{1}{\sqrt{3}}, \frac{1}{\sqrt{3}}\right) = \frac{3}{\sqrt{3}} = \sqrt{3}$ and the absolute minimum value of $f\left(-\frac{1}{\sqrt{3}}, \frac{1}{\sqrt{3}}, -\frac{1}{\sqrt{3}}\right) = -\sqrt{3}$.

85. The cost is $f(x, y, z) = 2axy + 2bxz + 2cyz$ subject to the constraint $xyz = V$. Then $\nabla f = \lambda \nabla g$ $\Rightarrow$ $2ay + 2bz = \lambda yz$, $2ax + 2cz = \lambda xz$, and $2bx + 2cy = \lambda xy$ $\Rightarrow$ $2axy + 2bxz = \lambda xyz$, $2axy + 2cyz = \lambda xyz$, and $2bxz + 2cyz = \lambda xyz$ $\Rightarrow$ $2axy + 2bxz = 2axy + 2cyz$ $\Rightarrow$ $y = \left(\frac{b}{c}\right)x$. Also $2axy + 2bxz = 2bxz + 2cyz$ $\Rightarrow$ $z = \left(\frac{a}{c}\right)x$. Then $x \left(\frac{b}{c}x\right)\left(\frac{a}{c}x\right) = V$ $\Rightarrow$ $x^3 = \frac{c^2 V}{ab}$ $\Rightarrow$ width $= x = \left(\frac{c^2 V}{ab}\right)^{1/3}$, Depth $= y = \left(\frac{b}{c}\right)\left(\frac{c^2 V}{ab}\right)^{1/3} = \left(\frac{b^2 V}{ac}\right)^{1/3}$, and Height $= z = \left(\frac{a}{c}\right)\left(\frac{c^2 V}{ab}\right)^{1/3} = \left(\frac{a^2 V}{bc}\right)^{1/3}$.

87. $\nabla f = (y + z)\mathbf{i} + x\mathbf{j} + x\mathbf{k}$, $\nabla g = 2x\mathbf{i} + 2y\mathbf{j}$, and $\nabla h = z\mathbf{i} + x\mathbf{k}$ so that $\nabla f = \lambda \nabla g + \mu \nabla h$ $\Rightarrow$ $(y + z)\mathbf{i} + x\mathbf{j} + x\mathbf{k} = \lambda(2x\mathbf{i} + 2y\mathbf{j}) + \mu(z\mathbf{i} + x\mathbf{k})$ $\Rightarrow$ $y + z = 2\lambda x + \mu z$, $x = 2\lambda y$, $x = \mu x$ $\Rightarrow$ $x = 0$ or $\mu = 1$.

CASE 1: $x = 0$ which is impossible since $xz = 1$.

CASE 2: $\mu = 1$ $\Rightarrow$ $y + z = 2\lambda x + z$ $\Rightarrow$ $y = 2\lambda x$ and $x = 2\lambda y$ $\Rightarrow$ $y = (2\lambda)(2\lambda y)$ $\Rightarrow$ $y = 0$ or $4\lambda^2 = 1$. If $y = 0$, then $x^2 = 1$ $\Rightarrow$ $x = \pm 1$ so with $xz = 1$ we obtain the points $(1, 0, 1)$

and $(-1, 0, -1)$. If $4\lambda^2 = 1$, then $\lambda = \pm\frac{1}{2}$. For $\lambda = -\frac{1}{2}$, $y = -x$ so $x^2 + y^2 = 1 \Rightarrow x^2 = \frac{1}{2}$

$\Rightarrow x = \pm\frac{1}{\sqrt{2}}$ with $xz = 1 \Rightarrow z = \pm\sqrt{2}$, and we obtain the points $\left(\frac{1}{\sqrt{2}}, -\frac{1}{\sqrt{2}}, \sqrt{2}\right)$ and

$\left(-\frac{1}{\sqrt{2}}, \frac{1}{\sqrt{2}}, -\sqrt{2}\right)$. For $\lambda = \frac{1}{2}$, $y = x \Rightarrow x^2 = \frac{1}{2} \Rightarrow x = \pm\frac{1}{\sqrt{2}}$ with $xz = 1 \Rightarrow z = \pm\sqrt{2}$,

and we obtain the points $\left(\frac{1}{\sqrt{2}}, \frac{1}{\sqrt{2}}, \sqrt{2}\right)$ and $\left(-\frac{1}{\sqrt{2}}, -\frac{1}{\sqrt{2}}, -\sqrt{2}\right)$.

Evaluations give $f(1, 0, 1) = 1$, $f(-1, 0, -1) = 1$, $f\left(\frac{1}{\sqrt{2}}, -\frac{1}{\sqrt{2}}, \sqrt{2}\right) = \frac{1}{2}$, $f\left(-\frac{1}{\sqrt{2}}, \frac{1}{\sqrt{2}}, -\sqrt{2}\right) = \frac{1}{2}$,

$f\left(\frac{1}{\sqrt{2}}, \frac{1}{\sqrt{2}}, \sqrt{2}\right) = \frac{3}{2}$, and $f\left(-\frac{1}{\sqrt{2}}, -\frac{1}{\sqrt{2}}, -\sqrt{2}\right) = \frac{3}{2}$. Therefore the absolute maximum is $\frac{3}{2}$ at

$\left(\frac{1}{\sqrt{2}}, \frac{1}{\sqrt{2}}, \sqrt{2}\right)$ and $\left(-\frac{1}{\sqrt{2}}, -\frac{1}{\sqrt{2}}, -\sqrt{2}\right)$, and the absolute minimum is $\frac{1}{2}$ at $\left(-\frac{1}{\sqrt{2}}, \frac{1}{\sqrt{2}}, -\sqrt{2}\right)$ and

$\left(\frac{1}{\sqrt{2}}, -\frac{1}{\sqrt{2}}, \sqrt{2}\right)$.

89. (a) y, z are independent with $w = x^2 e^{yz}$ and $z = x^2 - y^2 \Rightarrow \frac{\partial w}{\partial y} = \frac{\partial w}{\partial x}\frac{\partial x}{\partial y} + \frac{\partial w}{\partial y}\frac{\partial y}{\partial y} + \frac{\partial w}{\partial z}\frac{\partial z}{\partial y}$

$= (2xe^{yz})\frac{\partial x}{\partial y} + (zx^2 e^{yz})(1) + (yx^2 e^{yz})(0); z = x^2 - y^2 \Rightarrow 0 = 2x\frac{\partial x}{\partial y} - 2y \Rightarrow \frac{\partial x}{\partial y} = \frac{y}{x}$; therefore,

$\left(\frac{\partial w}{\partial y}\right)_z = (2xe^{yz})\left(\frac{y}{x}\right) + zx^2 e^{yz} = (2y + zx^2)e^{yz}$

(b) z, x are independent with $w = x^2 e^{yz}$ and $z = x^2 - y^2 \Rightarrow \frac{\partial w}{\partial z} = \frac{\partial w}{\partial x}\frac{\partial x}{\partial z} + \frac{\partial w}{\partial y}\frac{\partial y}{\partial z} + \frac{\partial w}{\partial z}\frac{\partial z}{\partial z}$

$= (2xe^{yz})(0) + (zx^2 e^{yz})\frac{\partial y}{\partial z} + (yx^2 e^{yz})(1); z = x^2 - y^2 \Rightarrow 1 = 0 - 2y\frac{\partial y}{\partial z} \Rightarrow \frac{\partial y}{\partial z} = -\frac{1}{2y}$; therefore,

$\left(\frac{\partial w}{\partial z}\right)_x = (zx^2 e^{yz})\left(-\frac{1}{2y}\right) + yx^2 e^{yz} = x^2 e^{yz}\left(y - \frac{z}{2y}\right)$

(c) z, y are independent with $w = x^2 e^{yz}$ and $z = x^2 - y^2 \Rightarrow \frac{\partial w}{\partial z} = \frac{\partial w}{\partial x}\frac{\partial x}{\partial z} + \frac{\partial w}{\partial y}\frac{\partial y}{\partial z} + \frac{\partial w}{\partial z}\frac{\partial z}{\partial z}$

$= (2xe^{yz})\frac{\partial x}{\partial z} + (zx^2 e^{yz})(0) + (yx^2 e^{yz})(1); z = x^2 - y^2 \Rightarrow 1 = 2x\frac{\partial x}{\partial z} - 0 \Rightarrow \frac{\partial x}{\partial z} = \frac{1}{2x}$; therefore,

$\left(\frac{\partial w}{\partial z}\right)_y = (2xe^{yz})\left(\frac{1}{2x}\right) + yx^2 e^{yz} = (1 + x^2 y)e^{yz}$

91. Note that $x = r\cos\theta$ and $y = r\sin\theta \Rightarrow r = \sqrt{x^2 + y^2}$ and $\theta = \tan^{-1}\left(\frac{y}{x}\right)$. Thus,

$\frac{\partial w}{\partial x} = \frac{\partial w}{\partial r}\frac{\partial r}{\partial x} + \frac{\partial w}{\partial\theta}\frac{\partial\theta}{\partial x} = \left(\frac{\partial w}{\partial r}\right)\left(\frac{x}{\sqrt{x^2+y^2}}\right) + \left(\frac{\partial w}{\partial\theta}\right)\left(\frac{-y}{x^2+y^2}\right) = (\cos\theta)\frac{\partial w}{\partial r} - \left(\frac{\sin\theta}{r}\right)\frac{\partial w}{\partial\theta}$;

$\frac{\partial w}{\partial y} = \frac{\partial w}{\partial r}\frac{\partial r}{\partial y} + \frac{\partial w}{\partial\theta}\frac{\partial\theta}{\partial y} = \left(\frac{\partial w}{\partial r}\right)\left(\frac{y}{\sqrt{x^2+y^2}}\right) + \left(\frac{\partial w}{\partial\theta}\right)\left(\frac{x}{x^2+y^2}\right) = (\sin\theta)\frac{\partial w}{\partial r} + \left(\frac{\cos\theta}{r}\right)\frac{\partial w}{\partial\theta}$

93. $\frac{\partial u}{\partial y} = b$ and $\frac{\partial u}{\partial x} = a \Rightarrow \frac{\partial w}{\partial x} = \frac{dw}{du}\frac{\partial u}{\partial x} = a\frac{dw}{du}$ and $\frac{\partial w}{\partial y} = \frac{dw}{du}\frac{\partial u}{\partial y} = b\frac{dw}{du} \Rightarrow \frac{1}{a}\frac{\partial w}{\partial x} = \frac{dw}{du}$ and $\frac{1}{b}\frac{\partial w}{\partial y} = \frac{dw}{du}$

$\Rightarrow \frac{1}{a}\frac{\partial w}{\partial x} = \frac{1}{b}\frac{\partial w}{\partial y} \Rightarrow b\frac{\partial w}{\partial x} = a\frac{\partial w}{\partial y}$

95. $e^u\cos v - x = 0 \Rightarrow (e^u\cos v)\frac{\partial u}{\partial x} - (e^u\sin v)\frac{\partial v}{\partial x} = 1; e^u\sin v - y = 0 \Rightarrow (e^u\sin v)\frac{\partial u}{\partial x} + (e^u\cos v)\frac{\partial v}{\partial x} = 0$.

Solving this system yields $\frac{\partial u}{\partial x} = e^{-u}\cos v$ and $\frac{\partial v}{\partial x} = -e^{-u}\sin v$. Similarly, $e^u\cos v - x = 0$

$\Rightarrow (e^u\cos v)\frac{\partial u}{\partial y} - (e^u\sin v)\frac{\partial v}{\partial y} = 0$ and $e^u\sin v - y = 0 \Rightarrow (e^u\sin v)\frac{\partial u}{\partial y} + (e^u\cos v)\frac{\partial v}{\partial y} = 1$. Solving this

second system yields $\frac{\partial u}{\partial y} = e^{-u}\sin v$ and $\frac{\partial v}{\partial y} = e^{-u}\cos v$. Therefore $\left(\frac{\partial u}{\partial x}\mathbf{i} + \frac{\partial u}{\partial y}\mathbf{j}\right) \cdot \left(\frac{\partial v}{\partial x}\mathbf{i} + \frac{\partial v}{\partial y}\mathbf{j}\right)$

$= [(e^{-u}\cos v)\mathbf{i} + (e^{-u}\sin v)\mathbf{j}] \cdot [(-e^{-u}\sin v)\mathbf{i} + (e^{-u}\cos v)\mathbf{j}] = 0 \Rightarrow$ the vectors are orthogonal $\Rightarrow$ the angle

between the vectors is the constant $\frac{\pi}{2}$.

97. $(y + z)^2 + (z - x)^2 = 16 \Rightarrow \nabla f = -2(z - x)\mathbf{i} + 2(y + z)\mathbf{j} + 2(y + 2z - x)\mathbf{k}$; if the normal line is parallel to the

yz-plane, then x is constant $\Rightarrow \frac{\partial f}{\partial x} = 0 \Rightarrow -2(z - x) = 0 \Rightarrow z = x \Rightarrow (y + z)^2 + (z - z)^2 = 16 \Rightarrow y + z = \pm 4$.

Let $x = t \Rightarrow z = t \Rightarrow y = -t \pm 4$. Therefore the points are $(t, -t \pm 4, t)$, t a real number.

99. $\nabla f = \lambda(x\mathbf{i} + y\mathbf{j} + z\mathbf{k}) \Rightarrow \frac{\partial f}{\partial x} = \lambda x \Rightarrow f(x, y, z) = \frac{1}{2}\lambda x^2 + g(y, z)$ for some function $g \Rightarrow \lambda y = \frac{\partial f}{\partial y} = \frac{\partial g}{\partial y}$

$\Rightarrow g(y, z) = \frac{1}{2}\lambda y^2 + h(z)$ for some function $h \Rightarrow \lambda z = \frac{\partial f}{\partial z} = \frac{\partial g}{\partial z} = h'(z) \Rightarrow h(z) = \frac{1}{2}\lambda z^2 + C$ for some arbitrary constant C $\Rightarrow$ g(y, z) $= \frac{1}{2} \lambda y^2 + \left(\frac{1}{2} \lambda z^2 + C \right) \Rightarrow$ f(x, y, z) $= \frac{1}{2} \lambda x^2 + \frac{1}{2} \lambda y^2 + \frac{1}{2} \lambda z^2 + C \Rightarrow$ f(0, 0, a) $= \frac{1}{2} \lambda a^2 + C \Rightarrow$ and f(0, 0, -a) $= \frac{1}{2} \lambda (-a)^2 + C \Rightarrow$ f(0, 0, a) = f(0, 0, -a) for any constant a, as claimed.

101. Let f(x, y, z) = xy + z - 2 $\Rightarrow \nabla$f = y$\mathbf{i}$ + x$\mathbf{j}$ + $\mathbf{k}$. At (1, 1, 1), we have ∇f = $\mathbf{i}$ + $\mathbf{j}$ + $\mathbf{k}$ $\Rightarrow$ the normal line is x = 1 + t, y = 1 + t, z = 1 + t, so at t = -1 $\Rightarrow$ x = 0, y = 0, z = 0 and the normal line passes through the origin.

CHAPTER 14 ADDITIONAL AND ADVANCED EXERCISES

1. By definition, $f_{xy}(0, 0) = \lim\limits_{h \to 0} \frac{f_x(0, h) - f_x(0, 0)}{h}$ so we need to calculate the first partial derivatives in the numerator. For (x, y) $\neq$ (0, 0) we calculate $f_x(x, y)$ by applying the differentiation rules to the formula for f(x, y): $f_x(x, y) = \frac{x^2 y - y^3}{x^2 + y^2} + (xy) \frac{(x^2 + y^2)(2x) - (x^2 - y^2)(2x)}{(x^2 + y^2)^2} = \frac{x^2 y - y^3}{x^2 + y^2} + \frac{4x^2 y^3}{(x^2 + y^2)^2} \Rightarrow f_x(0, h) = -\frac{h^3}{h^2} = -h$.

For (x, y) = (0, 0) we apply the definition: $f_x(0, 0) = \lim\limits_{h \to 0} \frac{f(h, 0) - f(0, 0)}{h} = \lim\limits_{h \to 0} \frac{0 - 0}{h} = 0$. Then by definition $f_{xy}(0, 0) = \lim\limits_{h \to 0} \frac{-h - 0}{h} = -1$. Similarly, $f_{yx}(0, 0) = \lim\limits_{h \to 0} \frac{f_y(h, 0) - f_y(0, 0)}{h}$, so for (x, y) $\neq$ (0, 0) we have $f_y(x, y) = \frac{x^3 - xy^2}{x^2 + y^2} - \frac{4x^3 y^2}{(x^2 + y^2)^2} \Rightarrow f_y(h, 0) = \frac{h^3}{h^2} = h$; for (x, y) = (0, 0) we obtain $f_y(0, 0) = \lim\limits_{h \to 0} \frac{f(0, h) - f(0, 0)}{h} = \lim\limits_{h \to 0} \frac{0 - 0}{h} = 0$. Then by definition $f_{yx}(0, 0) = \lim\limits_{h \to 0} \frac{h - 0}{h} = 1$. Note that $f_{xy}(0, 0) \neq f_{yx}(0, 0)$ in this case.

3. Substitution of u + u(x) and v = v(x) in g(u, v) gives g(u(x), v(x)) which is a function of the independent variable x. Then, $g(u, v) = \int_u^v f(t) \, dt \Rightarrow \frac{dg}{dx} = \frac{\partial g}{\partial u} \frac{du}{dx} + \frac{\partial g}{\partial v} \frac{dv}{dx} = \left(\frac{\partial}{\partial u} \int_u^v f(t) \, dt \right) \frac{du}{dx} + \left(\frac{\partial}{\partial v} \int_u^v f(t) \, dt \right) \frac{dv}{dx}$

$= \left(-\frac{\partial}{\partial u} \int_v^u f(t) \, dt \right) \frac{du}{dx} + \left(\frac{\partial}{\partial v} \int_u^v f(t) \, dt \right) \frac{dv}{dx} = -f(u(x)) \frac{du}{dx} + f(v(x)) \frac{dv}{dx} = f(v(x)) \frac{dv}{dx} - f(u(x)) \frac{du}{dx}$

5. (a) Let u = tx, v = ty, and w = f(u, v) = f(u(t, x), v(t, y)) = f(tx, ty) = t^{n}f(x, y), where t, x, and y are independent variables. Then $nt^{n-1} f(x, y) = \frac{\partial w}{\partial t} = \frac{\partial w}{\partial u} \frac{\partial u}{\partial t} + \frac{\partial w}{\partial v} \frac{\partial v}{\partial t} = x \frac{\partial w}{\partial u} + y \frac{\partial w}{\partial v}$. Now,

$\frac{\partial w}{\partial x} = \frac{\partial w}{\partial u} \frac{\partial u}{\partial x} + \frac{\partial w}{\partial v} \frac{\partial v}{\partial x} = \left(\frac{\partial w}{\partial u} \right)(t) + \left(\frac{\partial w}{\partial v} \right)(0) = t \frac{\partial w}{\partial u} \Rightarrow \frac{\partial w}{\partial u} = \left(\frac{1}{t} \right) \left(\frac{\partial w}{\partial x} \right)$. Likewise,

$\frac{\partial w}{\partial y} = \frac{\partial w}{\partial u} \frac{\partial u}{\partial y} + \frac{\partial w}{\partial v} \frac{\partial v}{\partial y} = \left(\frac{\partial w}{\partial u} \right)(0) + \left(\frac{\partial w}{\partial v} \right)(t) \Rightarrow \frac{\partial w}{\partial v} = \left(\frac{1}{t} \right) \left(\frac{\partial w}{\partial y} \right)$. Therefore,

$nt^{n-1} f(x, y) = x \frac{\partial w}{\partial u} + y \frac{\partial w}{\partial v} = \left(\frac{x}{t} \right) \left(\frac{\partial w}{\partial x} \right) + \left(\frac{y}{t} \right) \left(\frac{\partial w}{\partial y} \right)$. When t = 1, u = x, v = y, and w = f(x, y)

$\Rightarrow \frac{\partial w}{\partial x} = \frac{\partial f}{\partial x}$ and $\frac{\partial w}{\partial y} = \frac{\partial f}{\partial y} \Rightarrow$ nf(x, y) = x $\frac{\partial f}{\partial x}$ + y $\frac{\partial f}{\partial y}$, as claimed.

(b) From part (a), $nt^{n-1} f(x, y) = x \frac{\partial w}{\partial u} + y \frac{\partial w}{\partial v}$. Differentiating with respect to t again we obtain

$n(n - 1)t^{n-2} f(x, y) = x \frac{\partial^2 w}{\partial u^2} \frac{\partial u}{\partial t} + x \frac{\partial^2 w}{\partial v \partial u} \frac{\partial v}{\partial t} + y \frac{\partial^2 w}{\partial u \partial v} \frac{\partial u}{\partial t} + y \frac{\partial^2 w}{\partial v^2} \frac{\partial v}{\partial t} = x^2 \frac{\partial^2 w}{\partial u^2} + 2xy \frac{\partial^2 w}{\partial u \partial v} + y^2 \frac{\partial^2 w}{\partial v^2}$.

Also from part (a), $\frac{\partial^2 w}{\partial x^2} = \frac{\partial}{\partial x} \left(\frac{\partial w}{\partial x} \right) = \frac{\partial}{\partial x} \left(t \frac{\partial w}{\partial u} \right) = t \frac{\partial^2 w}{\partial u^2} \frac{\partial u}{\partial x} + t \frac{\partial^2 w}{\partial v \partial u} \frac{\partial v}{\partial x} = t^2 \frac{\partial^2 w}{\partial u^2}$, $\frac{\partial^2 w}{\partial y^2} = \frac{\partial}{\partial y} \left(\frac{\partial w}{\partial y} \right)$

$= \frac{\partial}{\partial y} \left(t \frac{\partial w}{\partial v} \right) = t \frac{\partial^2 w}{\partial u \partial v} \frac{\partial u}{\partial y} + t \frac{\partial^2 w}{\partial v^2} \frac{\partial v}{\partial y} = t^2 \frac{\partial^2 w}{\partial v^2}$, and $\frac{\partial^2 w}{\partial y \partial x} = \frac{\partial}{\partial y} \left(\frac{\partial w}{\partial x} \right) = \frac{\partial}{\partial y} \left(t \frac{\partial w}{\partial u} \right) = t \frac{\partial^2 w}{\partial u^2} \frac{\partial u}{\partial y} + t \frac{\partial^2 w}{\partial v \partial u} \frac{\partial v}{\partial y}$

$= t^2 \frac{\partial^2 w}{\partial v \partial u} \Rightarrow \left(\frac{1}{t^2} \right) \frac{\partial^2 w}{\partial x^2} = \frac{\partial^2 w}{\partial u^2}$, $\left(\frac{1}{t^2} \right) \frac{\partial^2 w}{\partial y^2} = \frac{\partial^2 w}{\partial v^2}$, and $\left(\frac{1}{t^2} \right) \frac{\partial^2 w}{\partial y \partial x} = \frac{\partial^2 w}{\partial v \partial u}$

$\Rightarrow n(n - 1)t^{n-2} f(x, y) = \left(\frac{x^2}{t^2} \right) \left(\frac{\partial^2 w}{\partial x^2} \right) + \left(\frac{2xy}{t^2} \right) \left(\frac{\partial^2 w}{\partial y \partial x} \right) + \left(\frac{y^2}{t^2} \right) \left(\frac{\partial^2 w}{\partial y^2} \right)$ for t $\neq$ 0. When t = 1, w = f(x, y) and

we have $n(n - 1)f(x, y) = x^2 \left(\frac{\partial^2 f}{\partial x^2} \right) + 2xy \left(\frac{\partial^2 f}{\partial x \partial y} \right) + y^2 \left(\frac{\partial^2 f}{\partial y^2} \right)$ as claimed.

7. (a) $\mathbf{r} = x\mathbf{i} + y\mathbf{j} + z\mathbf{k} \Rightarrow r = |\mathbf{r}| = \sqrt{x^2 + y^2 + z^2}$ and $\nabla r = \frac{x}{\sqrt{x^2 + y^2 + z^2}} \mathbf{i} + \frac{y}{\sqrt{x^2 + y^2 + z^2}} \mathbf{j} + \frac{z}{\sqrt{x^2 + y^2 + z^2}} \mathbf{k}$

$= \frac{\mathbf{r}}{r}$

(b) $r^n = \left(\sqrt{x^2 + y^2 + z^2} \right)^n$

$\Rightarrow \nabla (r^n) = nx \left(x^2 + y^2 + z^2 \right)^{(n/2)-1} \mathbf{i} + ny \left(x^2 + y^2 + z^2 \right)^{(n/2)-1} \mathbf{j} + nz \left(x^2 + y^2 + z^2 \right)^{(n/2)-1} \mathbf{k}$

$= nr^{n-2} \mathbf{r}$

(c) Let n = 2 in part (b). Then $\frac{1}{2}\nabla(r^2) = \mathbf{r} \Rightarrow \nabla\left(\frac{1}{2}r^2\right) = \mathbf{r} \Rightarrow \frac{r^2}{2} = \frac{1}{2}(x^2 + y^2 + z^2)$ is the function.

(d) $\mathbf{dr} = dx\mathbf{i} + dy\mathbf{j} + dz\mathbf{k} \Rightarrow \mathbf{r} \cdot \mathbf{dr} = x\,dx + y\,dy + z\,dz$, and $dr = r_x\,dx + r_y\,dy + r_z\,dz = \frac{x}{r}\,dx + \frac{y}{r}\,dy + \frac{z}{r}\,dz$

$\Rightarrow r\,dr = x\,dx + y\,dy + z\,dz = \mathbf{r} \cdot \mathbf{dr}$

(e) $\mathbf{A} = a\mathbf{i} + b\mathbf{j} + c\mathbf{k} \Rightarrow \mathbf{A} \cdot \mathbf{r} = ax + by + cz \Rightarrow \nabla(\mathbf{A} \cdot \mathbf{r}) = a\mathbf{i} + b\mathbf{j} + c\mathbf{k} = \mathbf{A}$

9. $f(x, y, z) = xz^2 - yz + \cos xy - 1 \Rightarrow \nabla f = (z^2 - y\sin xy)\mathbf{i} + (-z - x\sin xy)\mathbf{j} + (2xz - y)\mathbf{k} \Rightarrow \nabla f(0,0,1) = \mathbf{i} - \mathbf{j}$

$\Rightarrow$ the tangent plane is $x - y = 0$; $\mathbf{r} = (\ln t)\mathbf{i} + (t\ln t)\mathbf{j} + t\mathbf{k} \Rightarrow \mathbf{r}' = \left(\frac{1}{t}\right)\mathbf{i} + (\ln t + 1)\mathbf{j} + \mathbf{k}$; $x = y = 0$, $z = 1$

$\Rightarrow t = 1 \Rightarrow \mathbf{r}'(1) = \mathbf{i} + \mathbf{j} + \mathbf{k}$. Since $(\mathbf{i} + \mathbf{j} + \mathbf{k}) \cdot (\mathbf{i} - \mathbf{j}) = \mathbf{r}'(1) \cdot \nabla f = 0$, $\mathbf{r}$ is parallel to the plane, and

$\mathbf{r}(1) = 0\mathbf{i} + 0\mathbf{j} + \mathbf{k} \Rightarrow \mathbf{r}$ is contained in the plane.

11. $\frac{\partial z}{\partial x} = 3x^2 - 9y = 0$ and $\frac{\partial z}{\partial y} = 3y^2 - 9x = 0 \Rightarrow y = \frac{1}{3}x^2$ and $3\left(\frac{1}{3}x^2\right)^2 - 9x = 0 \Rightarrow \frac{1}{3}x^4 - 9x = 0$

$\Rightarrow x(x^3 - 27) = 0 \Rightarrow x = 0$ or $x = 3$. Now $x = 0 \Rightarrow y = 0$ or $(0, 0)$ and $x = 3 \Rightarrow y = 3$ or $(3, 3)$. Next

$\frac{\partial^2 z}{\partial x^2} = 6x$, $\frac{\partial^2 z}{\partial y^2} = 6y$, and $\frac{\partial^2 z}{\partial x \partial y} = -9$. For $(0, 0)$, $\frac{\partial^2 z}{\partial x^2}\frac{\partial^2 z}{\partial y^2} - \left(\frac{\partial^2 z}{\partial x \partial y}\right)^2 = -81 \Rightarrow$ no extremum (a saddle point),

and for $(3, 3)$, $\frac{\partial^2 z}{\partial x^2}\frac{\partial^2 z}{\partial y^2} - \left(\frac{\partial^2 z}{\partial x \partial y}\right)^2 = 243 > 0$ and $\frac{\partial^2 z}{\partial x^2} = 18 > 0 \Rightarrow$ a local minimum.

13. Let $f(x, y, z) = \frac{x^2}{a^2} + \frac{y^2}{b^2} + \frac{z^2}{c^2} - 1 \Rightarrow \nabla f = \frac{2x}{a^2}\mathbf{i} + \frac{2y}{b^2}\mathbf{j} + \frac{2z}{c^2}\mathbf{k} \Rightarrow$ an equation of the plane tangent at the point

$P_0(x_0, y_0, y_0)$ is $\left(\frac{2x_0}{a^2}\right)x + \left(\frac{2y_0}{b^2}\right)y + \left(\frac{2z_0}{c^2}\right)z = \frac{2x_0^2}{a^2} + \frac{2y_0^2}{b^2} + \frac{2z_0^2}{c^2} = 2$ or $\left(\frac{x_0}{a^2}\right)x + \left(\frac{y_0}{b^2}\right)y + \left(\frac{z_0}{c^2}\right)z = 1$.

The intercepts of the plane are $\left(\frac{a^2}{x_0}, 0, 0\right)$, $\left(0, \frac{b^2}{y_0}, 0\right)$ and $\left(0, 0, \frac{c^2}{z_0}\right)$. The volume of the tetrahedron formed

by the plane and the coordinate planes is $V = \left(\frac{1}{3}\right)\left(\frac{1}{2}\right)\left(\frac{a^2}{x_0}\right)\left(\frac{b^2}{y_0}\right)\left(\frac{c^2}{z_0}\right) \Rightarrow$ we need to maximize

$V(x, y, z) = \frac{(abc)^2}{6}(xyz)^{-1}$ subject to the constraint $f(x, y, z) = \frac{x^2}{a^2} + \frac{y^2}{b^2} + \frac{z^2}{c^2} = 1$. Thus,

$\left[-\frac{(abc)^2}{6}\right]\left(\frac{1}{x^2yz}\right) = \frac{2x}{a^2}\lambda$, $\left[-\frac{(abc)^2}{6}\right]\left(\frac{1}{xy^2z}\right) = \frac{2y}{b^2}\lambda$, and $\left[-\frac{(abc)^2}{6}\right]\left(\frac{1}{xyz^2}\right) = \frac{2z}{c^2}\lambda$. Multiply the first equation

by a^2yz, the second by b^2xz, and the third by c^2xy. Then equate the first and second $\Rightarrow a^2y^2 = b^2x^2$

$\Rightarrow y = \frac{b}{a}x$, $x > 0$; equate the first and third $\Rightarrow a^2z^2 = c^2x^2 \Rightarrow z = \frac{c}{a}x$, $x > 0$; substitute into $f(x, y, z) = 0$

$\Rightarrow x = \frac{a}{\sqrt{3}} \Rightarrow y = \frac{b}{\sqrt{3}} \Rightarrow z = \frac{c}{\sqrt{3}} \Rightarrow V = \frac{\sqrt{3}}{2}abc$.

15. Let (x_0, y_0) be any point in R. We must show $\lim_{(x, y) \to (x_0, y_0)} f(x, y) = f(x_0, y_0)$ or, equivalently that

$\lim_{(h, k) \to (0, 0)} |f(x_0 + h, y_0 + k) - f(x_0, y_0)| = 0$. Consider $f(x_0 + h, y_0 + k) - f(x_0, y_0)$

$= [f(x_0 + h, y_0 + k) - f(x_0, y_0 + k)] + [f(x_0, y_0 + k) - f(x_0, y_0)]$. Let $F(x) = f(x, y_0 + k)$ and apply the Mean Value

Theorem: there exists ξ with $x_0 < \xi < x_0 + h$ such that $F'(\xi)h = F(x_0 + h) - F(x_0) \Rightarrow hf_x(\xi, y_0 + k)$

$= f(x_0 + h, y_0 + k) - f(x_0, y_0 + k)$. Similarly, $kf_y(x_0, \eta) = f(x_0, y_0 + k) - f(x_0, y_0)$ for some η with

$y_0 < \eta < y_0 + k$. Then $|f(x_0 + h, y_0 + k) - f(x_0, y_0)| \le |hf_x(\xi, y_0 + k)| + |kf_y(x_0, \eta)|$. If M, N are positive real

numbers such that $|f_x| \le M$ and $|f_y| \le N$ for all (x, y) in the xy-plane, then $|f(x_0 + h, y_0 + k) - f(x_0, y_0)|$

$\le M|h| + N|k|$. As $(h, k) \to 0$, $|f(x_0 + h, y_0 + k) - f(x_0, y_0)| \to 0 \Rightarrow \lim_{(h, k) \to (0, 0)} |f(x_0 + h, y_0 + k) - f(x_0, y_0)|$

$= 0 \Rightarrow f$ is continuous at (x_0, y_0).

17. $\frac{\partial f}{\partial x} = 0 \Rightarrow f(x, y) = h(y)$ is a function of y only. Also, $\frac{\partial g}{\partial y} = \frac{\partial f}{\partial x} = 0 \Rightarrow g(x, y) = k(x)$ is a function of x only.

Moreover, $\frac{\partial f}{\partial y} = \frac{\partial g}{\partial x} \Rightarrow h'(y) = k'(x)$ for all x and y. This can happen only if $h'(y) = k'(x) = c$ is a constant.

Integration gives $h(y) = cy + c_1$ and $k(x) = cx + c_2$, where c_1 and c_2 are constants. Therefore $f(x, y) = cy + c_1$

and $g(x, y) = cx + c_2$. Then $f(1, 2) = g(1, 2) = 5 \Rightarrow 5 = 2c + c_1 = c + c_2$, and $f(0, 0) = 4 \Rightarrow c_1 = 4 \Rightarrow c = \frac{1}{2}$

$\Rightarrow c_2 = \frac{9}{2}$. Thus, $f(x, y) = \frac{1}{2}y + 4$ and $g(x, y) = \frac{1}{2}x + \frac{9}{2}$.

19. Since the particle is heat-seeking, at each point (x, y) it moves in the direction of maximal temperature increase, that is in the direction of $\nabla T(x, y) = (e^{-2y} \sin x) \mathbf{i} + (2e^{-2y} \cos x) \mathbf{j}$. Since $\nabla T(x, y)$ is parallel to the particle's velocity vector, it is tangent to the path $y = f(x)$ of the particle $\Rightarrow f'(x) = \frac{2e^{-2y} \cos x}{e^{-2y} \sin x} = 2 \cot x$.

Integration gives $f(x) = 2 \ln |\sin x| + C$ and $f\left(\frac{\pi}{4}\right) = 0 \Rightarrow 0 = 2 \ln \left|\sin \frac{\pi}{4}\right| + C \Rightarrow C = -2 \ln \frac{\sqrt{2}}{2} = \ln \left(\frac{2}{\sqrt{2}}\right)^2$

$= \ln 2$. Therefore, the path of the particle is the graph of $y = 2 \ln |\sin x| + \ln 2$.

21. (a) $\mathbf{k}$ is a vector normal to $z = 10 - x^2 - y^2$ at the point $(0, 0, 10)$. So directions tangential to S at $(0, 0, 10)$ will be unit vectors $\mathbf{u} = a\mathbf{i} + b\mathbf{j}$. Also, $\nabla T(x, y, z) = (2xy + 4) \mathbf{i} + (x^2 + 2yz + 14) \mathbf{j} + (y^2 + 1) \mathbf{k}$
$\Rightarrow \nabla T(0, 0, 10) = 4\mathbf{i} + 14\mathbf{j} + \mathbf{k}$. We seek the unit vector $\mathbf{u} = a\mathbf{i} + b\mathbf{j}$ such that $D_\mathbf{u}T(0, 0, 10)$
$= (4\mathbf{i} + 14\mathbf{j} + \mathbf{k}) \cdot (a\mathbf{i} + b\mathbf{j}) = (4\mathbf{i} + 14\mathbf{j}) \cdot (a\mathbf{i} + b\mathbf{j})$ is a maximum. The maximum will occur when $a\mathbf{i} + b\mathbf{j}$
has the same direction as $4\mathbf{i} + 14\mathbf{j}$, or $\mathbf{u} = \frac{1}{\sqrt{53}} (2\mathbf{i} + 7\mathbf{j})$.

(b) A vector normal to S at $(1, 1, 8)$ is $\mathbf{n} = 2\mathbf{i} + 2\mathbf{j} + \mathbf{k}$. Now, $\nabla T(1, 1, 8) = 6\mathbf{i} + 31\mathbf{j} + 2\mathbf{k}$ and we seek the unit vector $\mathbf{u}$ such that $D_\mathbf{u}T(1, 1, 8) = \nabla T \cdot \mathbf{u}$ has its largest value. Now write $\nabla T = \mathbf{v} + \mathbf{w}$, where $\mathbf{v}$ is parallel to ∇T and $\mathbf{w}$ is orthogonal to ∇T. Then $D_\mathbf{u}T = \nabla T \cdot \mathbf{u} = (\mathbf{v} + \mathbf{w}) \cdot \mathbf{u} = \mathbf{v} \cdot \mathbf{u} + \mathbf{w} \cdot \mathbf{u} = \mathbf{w} \cdot \mathbf{u}$. Thus
$D_\mathbf{u}T(1, 1, 8)$ is a maximum when $\mathbf{u}$ has the same direction as $\mathbf{w}$. Now, $\mathbf{w} = \nabla T - \left(\frac{\nabla T \cdot \mathbf{n}}{|\mathbf{n}|^2}\right) \mathbf{n}$
$= (6\mathbf{i} + 31\mathbf{j} + 2\mathbf{k}) - \left(\frac{12 + 62 + 2}{4 + 4 + 1}\right)(2\mathbf{i} + 2\mathbf{j} + \mathbf{k}) = \left(6 - \frac{152}{9}\right)\mathbf{i} + \left(31 - \frac{152}{9}\right)\mathbf{j} + \left(2 - \frac{76}{9}\right)\mathbf{k}$
$= -\frac{98}{9}\mathbf{i} + \frac{127}{9}\mathbf{j} - \frac{58}{9}\mathbf{k} \Rightarrow \mathbf{u} = \frac{\mathbf{w}}{|\mathbf{w}|} = -\frac{1}{\sqrt{29,097}}(98\mathbf{i} - 127\mathbf{j} + 58\mathbf{k})$.

23. $w = e^{rt} \sin \pi x \Rightarrow w_t = re^{rt} \sin \pi x$ and $w_x = \pi e^{rt} \cos \pi x \Rightarrow w_{xx} = -\pi^2 e^{rt} \sin \pi x$; $w_{xx} = \frac{1}{c^2} w_t$, where c^2 is the positive constant determined by the material of the rod $\Rightarrow -\pi^2 e^{rt} \sin \pi x = \frac{1}{c^2}(re^{rt} \sin \pi x)$
$\Rightarrow (r + c^2\pi^2) e^{rt} \sin \pi x = 0 \Rightarrow r = -c^2\pi^2 \Rightarrow w = e^{-c^2\pi^2 t} \sin \pi x$

CHAPTER 15 MULTIPLE INTEGRALS

15.1 DOUBLE INTEGRALS

1. $\int_0^3 \int_0^2 (4 - y^2) \, dy \, dx = \int_0^3 \left[4y - \frac{y^3}{3} \right]_0^2 dx = \frac{16}{3} \int_0^3 dx = 16$

3. $\int_{-1}^0 \int_{-1}^1 (x + y + 1) \, dx \, dy = \int_{-1}^0 \left[\frac{x^2}{2} + yx + x \right]_{-1}^1 dy$

   $= \int_{-1}^0 (2y + 2) \, dy = [y^2 + 2y]_{-1}^0 = 1$

5. $\int_0^\pi \int_0^x (x \sin y) \, dy \, dx = \int_0^\pi [-x \cos y]_0^x \, dx$

   $= \int_0^\pi (x - x \cos x) \, dx = \left[\frac{x^2}{2} - (\cos x + x \sin x) \right]_0^\pi$

   $= \frac{\pi^2}{2} + 2$

7. $\int_1^{\ln 8} \int_0^{\ln y} e^{x+y} \, dx \, dy = \int_1^{\ln 8} [e^{x+y}]_0^{\ln y} \, dy = \int_1^{\ln 8} (ye^y - e^y) \, dy$

   $= [(y - 1)e^y - e^y]_1^{\ln 8} = 8(\ln 8 - 1) - 8 + e$

   $= 8 \ln 8 - 16 + e$

9. $\int_0^1 \int_0^{y^2} 3y^3 e^{xy} \, dx \, dy = \int_0^1 [3y^2 e^{xy}]_0^{y^2} \, dy$

   $= \int_0^1 \left(3y^2 e^{y^3} - 3y^2 \right) dy = \left[e^{y^3} - y^3 \right]_0^1 = e - 2$

11. $\int_1^2 \int_x^{2x} \frac{x}{y} \, dy \, dx = \int_1^2 [x \ln y]_x^{2x} \, dx = (\ln 2) \int_1^2 x \, dx = \frac{3}{2} \ln 2$

13. $\int_0^1 \int_0^{1-x} (x^2 + y^2) \, dy \, dx = \int_0^1 \left[x^2 y + \frac{y^3}{3} \right]_0^{1-x} dx = \int_0^1 \left[x^2(1 - x) + \frac{(1-x)^3}{3} \right] dx = \int_0^1 \left[x^2 - x^3 + \frac{(1-x)^3}{3} \right] dx$

   $= \left[\frac{x^3}{3} - \frac{x^4}{4} - \frac{(1-x)^4}{12} \right]_0^1 = \left(\frac{1}{3} - \frac{1}{4} - 0 \right) - \left(0 - 0 - \frac{1}{12} \right) = \frac{1}{6}$

15. $\int_0^1 \int_0^{1-u} (v - \sqrt{u}) \, dv \, du = \int_0^1 \left[\frac{v^2}{2} - v\sqrt{u} \right]_0^{1-u} du = \int_0^1 \left[\frac{1 - 2u + u^2}{2} - \sqrt{u}(1 - u) \right] du$

   $= \int_0^1 \left(\frac{1}{2} - u + \frac{u^2}{2} - u^{1/2} + u^{3/2} \right) du = \left[\frac{u}{2} - \frac{u^2}{2} + \frac{u^3}{6} - \frac{2}{3} u^{3/2} + \frac{2}{5} u^{5/2} \right]_0^1 = \frac{1}{2} - \frac{1}{2} + \frac{1}{6} - \frac{2}{3} + \frac{2}{5} = -\frac{1}{2} + \frac{2}{5} = -\frac{1}{10}$

17. $\int_{-2}^{0}\int_{v}^{-v} 2\,dp\,dv = 2\int_{-2}^{0}[p]_{v}^{-v}\,dv = 2\int_{-2}^{0} -2v\,dv$

$\quad = -2\,[v^2]_{-2}^{0} = 8$

19. $\int_{-\pi/3}^{\pi/3}\int_{0}^{\sec t} 3\cos t\,du\,dt = \int_{-\pi/3}^{\pi/3}[(3\cos t)u]_{0}^{\sec t}$

$\quad = \int_{-\pi/3}^{\pi/3} 3\,dt = 2\pi$

21. $\int_{2}^{4}\int_{0}^{(4-y)/2} dx\,dy$

23. $\int_{0}^{1}\int_{x^2}^{x} dy\,dx$

25. $\int_{1}^{e}\int_{\ln y}^{1} dx\,dy$

27. $\int_{0}^{9}\int_{0}^{\frac{1}{2}\sqrt{9-y}} 16x\,dx\,dy$

29. $\int_{-1}^{1}\int_{0}^{\sqrt{1-x^2}} 3y\,dy\,dx$

31. $\int_0^\pi \int_x^\pi \frac{\sin y}{y}\, dy\, dx = \int_0^\pi \int_0^y \frac{\sin y}{y}\, dx\, dy = \int_0^\pi \sin y\, dy = 2$

33. $\int_0^1 \int_y^1 x^2 e^{xy}\, dx\, dy = \int_0^1 \int_0^x x^2 e^{xy}\, dy\, dx = \int_0^1 [xe^{xy}]_0^x\, dx$

$= \int_0^1 (xe^{x^2} - x)\, dx = \left[\frac{1}{2} e^{x^2} - \frac{x^2}{2}\right]_0^1 = \frac{e-2}{2}$

35. $\int_0^{2\sqrt{\ln 3}} \int_{y/2}^{\sqrt{\ln 3}} e^{x^2}\, dx\, dy = \int_0^{\sqrt{\ln 3}} \int_0^{2x} e^{x^2}\, dy\, dx$

$= \int_0^{\sqrt{\ln 3}} 2xe^{x^2}\, dx = [e^{x^2}]_0^{\sqrt{\ln 3}} = e^{\ln 3} - 1 = 2$

37. $\int_0^{1/16} \int_{y^{1/4}}^{1/2} \cos(16\pi x^5)\, dx\, dy = \int_0^{1/2} \int_0^{x^4} \cos(16\pi x^5)\, dy\, dx$

$= \int_0^{1/2} x^4 \cos(16\pi x^5)\, dx = \left[\frac{\sin(16\pi x^5)}{80\pi}\right]_0^{1/2} = \frac{1}{80\pi}$

39. $\iint\limits_R (y - 2x^2)\, dA$

$= \int_{-1}^0 \int_{-x-1}^{x+1} (y - 2x^2)\, dy\, dx + \int_0^1 \int_{x-1}^{1-x} (y - 2x^2)\, dy\, dx$

$= \int_{-1}^0 \left[\frac{1}{2} y^2 - 2x^2 y\right]_{-x-1}^{x+1}\, dx + \int_0^1 \left[\frac{1}{2} y^2 - 2x^2 y\right]_{x-1}^{1-x}\, dx$

$= \int_{-1}^0 \left[\frac{1}{2}(x+1)^2 - 2x^2(x+1) - \frac{1}{2}(-x-1)^2 + 2x^2(-x-1)\right] dx$

$\quad + \int_0^1 \left[\frac{1}{2}(1-x)^2 - 2x^2(1-x) - \frac{1}{2}(x-1)^2 + 2x^2(x-1)\right] dx$

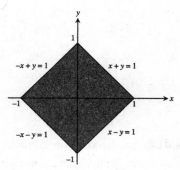

$= -4\int_{-1}^0 (x^3 + x^2)\, dx + 4\int_0^1 (x^3 - x^2)\, dx = -4\left[\frac{x^4}{4} + \frac{x^3}{3}\right]_{-1}^0 + 4\left[\frac{x^4}{4} - \frac{x^3}{3}\right]_0^1$

$= 4\left[\frac{(-1)^4}{4} + \frac{(-1)^3}{3}\right] + 4\left(\frac{1}{4} - \frac{1}{3}\right) = 8\left(\frac{3}{12} - \frac{4}{12}\right) = -\frac{8}{12} = -\frac{2}{3}$

41. $V = \int_0^1 \int_x^{2-x} (x^2 + y^2)\, dy\, dx = \int_0^1 \left[x^2 y + \frac{y^3}{3}\right]_x^{2-x}\, dx = \int_0^1 \left[2x^2 - \frac{7x^3}{3} + \frac{(2-x)^3}{3}\right] dx = \left[\frac{2x^3}{3} - \frac{7x^4}{12} - \frac{(2-x)^4}{12}\right]_0^1$

$= \left(\frac{2}{3} - \frac{7}{12} - \frac{1}{12}\right) - \left(0 - 0 - \frac{16}{12}\right) = \frac{4}{3}$

43. $V = \int_{-4}^1 \int_{3x}^{4-x^2} (x + 4)\, dy\, dx = \int_{-4}^1 [xy + 4y]_{3x}^{4-x^2}\, dx = \int_{-4}^1 [x(4 - x^2) + 4(4 - x^2) - 3x^2 - 12x]\, dx$

$= \int_{-4}^1 (-x^3 - 7x^2 - 8x + 16)\, dx = \left[-\frac{1}{4} x^4 - \frac{7}{3} x^3 - 4x^2 + 16x\right]_{-4}^1 = \left(-\frac{1}{4} - \frac{7}{3} + 12\right) - \left(\frac{64}{3} - 64\right)$

$= \frac{157}{3} - \frac{1}{4} = \frac{625}{12}$

45. $V = \int_0^2 \int_0^3 (4 - y^2)\, dx\, dy = \int_0^2 [4x - y^2 x]_0^3\, dy = \int_0^2 (12 - 3y^2)\, dy = [12y - y^3]_0^2 = 24 - 8 = 16$

47. $V = \int_0^2 \int_0^{2-x} (12 - 3y^2)\, dy\, dx = \int_0^2 [12y - y^3]_0^{2-x}\, dx = \int_0^2 [24 - 12x - (2-x)^3]\, dx$

$= \left[24x - 6x^2 + \frac{(2-x)^4}{4}\right]_0^2 = 20$

49. $V = \int_1^2 \int_{-1/x}^{1/x} (x + 1)\, dy\, dx = \int_1^2 [xy + y]_{-1/x}^{1/x}\, dx = \int_1^2 \left[1 + \frac{1}{x} - \left(-1 - \frac{1}{x}\right)\right] = 2\int_1^2 \left(1 + \frac{1}{x}\right)\, dx$

$= 2[x + \ln x]_1^2 = 2(1 + \ln 2)$

51. $\int_1^\infty \int_{e^{-x}}^1 \frac{1}{x^3 y}\, dy\, dx = \int_1^\infty \left[\frac{\ln y}{x^3}\right]_{e^{-x}}^1\, dx = \int_1^\infty -\left(\frac{-x}{x^3}\right)\, dx = -\lim_{b \to \infty} \left[\frac{1}{x}\right]_1^b = -\lim_{b \to \infty} \left(\frac{1}{b} - 1\right) = 1$

53. $\int_{-\infty}^\infty \int_{-\infty}^\infty \frac{1}{(x^2+1)(y^2+1)}\, dx\, dy = 2\int_0^\infty \left(\frac{2}{y^2+1}\right)\left(\lim_{b \to \infty} \tan^{-1} b - \tan^{-1} 0\right)\, dy = 2\pi \lim_{b \to \infty} \int_0^b \frac{1}{y^2+1}\, dy$

$= 2\pi \left(\lim_{b \to \infty} \tan^{-1} b - \tan^{-1} 0\right) = (2\pi)\left(\frac{\pi}{2}\right) = \pi^2$

55. $\iint\limits_R f(x,y)\, dA \approx \frac{1}{4} f\left(-\frac{1}{2}, 0\right) + \frac{1}{8} f(0,0) + \frac{1}{8} f\left(\frac{1}{4}, 0\right) + \frac{1}{4} f\left(\frac{1}{2}, 0\right) + \frac{1}{4} f\left(-\frac{1}{2}, \frac{1}{2}\right) + \frac{1}{8} f\left(0, \frac{1}{2}\right) + \frac{1}{8} f\left(\frac{1}{4}, \frac{1}{2}\right)$

$= \frac{1}{4}\left(-\frac{1}{2} + \frac{1}{2} + 0\right) + \frac{1}{8}\left(0 + \frac{1}{4} + \frac{1}{2} + \frac{3}{4}\right) = \frac{3}{16}$

57. The ray $\theta = \frac{\pi}{6}$ meets the circle $x^2 + y^2 = 4$ at the point $\left(\sqrt{3}, 1\right) \Rightarrow$ the ray is represented by the line $y = \frac{x}{\sqrt{3}}$.

Thus, $\iint\limits_R f(x,y)\, dA = \int_0^{\sqrt{3}} \int_{x/\sqrt{3}}^{\sqrt{4-x^2}} \sqrt{4-x^2}\, dy\, dx = \int_0^{\sqrt{3}} \left[(4 - x^2) - \frac{x}{\sqrt{3}}\sqrt{4-x^2}\right]\, dx = \left[4x - \frac{x^3}{3} + \frac{(4-x^2)^{3/2}}{3\sqrt{3}}\right]_0^{\sqrt{3}}$

$= \frac{20\sqrt{3}}{9}$

59. $V = \int_0^1 \int_x^{2-x} (x^2 + y^2)\, dy\, dx = \int_0^1 \left[x^2 y + \frac{y^3}{3}\right]_x^{2-x}\, dx$

$= \int_0^1 \left[2x^2 - \frac{7x^3}{3} + \frac{(2-x)^3}{3}\right]\, dx = \left[\frac{2x^3}{3} - \frac{7x^4}{12} - \frac{(2-x)^4}{12}\right]_0^1$

$= \left(\frac{2}{3} - \frac{7}{12} - \frac{1}{12}\right) - \left(0 - 0 - \frac{16}{12}\right) = \frac{4}{3}$

61. To maximize the integral, we want the domain to include all points where the integrand is positive and to exclude all points where the integrand is negative. These criteria are met by the points (x, y) such that $4 - x^2 - 2y^2 \geq 0$ or $x^2 + 2y^2 \leq 4$, which is the ellipse $x^2 + 2y^2 = 4$ together with its interior.

63. No, it is not possible By Fubini's theorem, the two orders of integration must give the same result.

65. $\int_{-b}^b \int_{-b}^b e^{-x^2-y^2}\, dx\, dy = \int_{-b}^b \int_{-b}^b e^{-y^2} e^{-x^2}\, dx\, dy = \int_{-b}^b e^{-y^2}\left(\int_{-b}^b e^{-x^2}\, dx\right)\, dy = \left(\int_{-b}^b e^{-x^2}\, dx\right)\left(\int_{-b}^b e^{-y^2}\, dy\right)$

$= \left(\int_{-b}^b e^{-x^2}\, dx\right)^2 = \left(2\int_0^b e^{-x^2}\, dx\right)^2 = 4\left(\int_0^b e^{-x^2}\, dx\right)^2$; taking limits as $b \to \infty$ gives the stated result.

15.2 AREAS, MOMENTS, AND CENTERS OF MASS

1. $\int_0^2 \int_0^{2-x} dy\, dx = \int_0^2 (2 - x)\, dx = \left[2x - \frac{x^2}{2}\right]_0^2 = 2,$

or $\int_0^2 \int_0^{2-y} dx\, dy = \int_0^2 (2 - y)\, dy = 2$

3. $\int_{-2}^1 \int_{y-2}^{-y^2} dx\, dy = \int_{-2}^1 (-y^2 - y + 2)\, dy$

$= \left[-\frac{y^3}{3} - \frac{y^2}{2} + 2y\right]_{-2}^1$

$= \left(-\frac{1}{3} - \frac{1}{2} + 2\right) - \left(\frac{8}{3} - 2 - 4\right) = \frac{9}{2}$

5. $\int_0^{\ln 2} \int_0^{e^x} dy\, dx = \int_0^{\ln 2} e^x\, dx = \left[e^x\right]_0^{\ln 2} = 2 - 1 = 1$

7. $\int_0^1 \int_{y^2}^{2y-y^2} dx\, dy = \int_0^1 (2y - 2y^2)\, dy = \left[y^2 - \frac{2}{3}y^3\right]_0^1$

$= \frac{1}{3}$

9. $\int_0^6 \int_{y^2/3}^{2y} dx\, dy = \int_0^6 \left(2y - \frac{y^2}{3}\right) dy = \left[y^2 - \frac{y^3}{9}\right]_0^6$

$= 36 - \frac{216}{9} = 12$

11. $\int_0^{\pi/4} \int_{\sin x}^{\cos x} dy\, dx$

$= \int_0^{\pi/4} (\cos x - \sin x)\, dx = \left[\sin x + \cos x\right]_0^{\pi/4}$

$= \left(\frac{\sqrt{2}}{2} + \frac{\sqrt{2}}{2}\right) - (0 + 1) = \sqrt{2} - 1$

13. $\int_{-1}^0 \int_{-2x}^{1-x} dy\, dx + \int_0^2 \int_{-x/2}^{1-x} dy\, dx$

$= \int_{-1}^0 (1 + x)\, dx + \int_0^2 \left(1 - \frac{x}{2}\right) dx$

$= \left[x + \frac{x^2}{2}\right]_{-1}^0 + \left[x - \frac{x^2}{4}\right]_0^2 = -\left(-1 + \frac{1}{2}\right) + (2 - 1) = \frac{3}{2}$

15. (a) average $= \frac{1}{\pi^2} \int_0^\pi \int_0^\pi \sin(x+y) \, dy \, dx = \frac{1}{\pi^2} \int_0^\pi [-\cos(x+y)]_0^\pi \, dx = \frac{1}{\pi^2} \int_0^\pi [-\cos(x+\pi) + \cos x] \, dx$

$= \frac{1}{\pi^2} [-\sin(x+\pi) + \sin x]_0^\pi = \frac{1}{\pi^2} [(-\sin 2\pi + \sin \pi) - (-\sin \pi + \sin 0)] = 0$

   (b) average $= \frac{1}{\left(\frac{\pi^2}{2}\right)} \int_0^\pi \int_0^{\pi/2} \sin(x+y) \, dy \, dx = \frac{2}{\pi^2} \int_0^\pi [-\cos(x+y)]_0^{\pi/2} \, dx = \frac{2}{\pi^2} \int_0^\pi [-\cos(x+\frac{\pi}{2}) + \cos x] \, dx$

$= \frac{2}{\pi^2} [-\sin(x+\frac{\pi}{2}) + \sin x]_0^\pi = \frac{2}{\pi^2} [(-\sin\frac{3\pi}{2} + \sin\pi) - (-\sin\frac{\pi}{2} + \sin 0)] = \frac{4}{\pi^2}$

17. average height $= \frac{1}{4} \int_0^2 \int_0^2 (x^2 + y^2) \, dy \, dx = \frac{1}{4} \int_0^2 \left[x^2 y + \frac{y^3}{3}\right]_0^2 dx = \frac{1}{4} \int_0^2 (2x^2 + \frac{8}{3}) \, dx = \frac{1}{2}\left[\frac{x^3}{3} + \frac{4x}{3}\right]_0^2 = \frac{8}{3}$

19. $M = \int_0^1 \int_x^{2-x^2} 3 \, dy \, dx = 3\int_0^1 (2 - x^2 - x) \, dx = \frac{7}{2}$; $M_y = \int_0^1 \int_x^{2-x^2} 3x \, dy \, dx = 3\int_0^1 [xy]_x^{2-x^2} dx$

$= 3\int_0^1 (2x - x^3 - x^2) \, dx = \frac{5}{4}$; $M_x = \int_0^1 \int_x^{2-x^2} 3y \, dy \, dx = \frac{3}{2}\int_0^1 [y^2]_x^{2-x^2} dx = \frac{3}{2}\int_0^1 (4 - 5x^2 + x^4) \, dx = \frac{19}{5}$

$\Rightarrow \bar{x} = \frac{5}{14}$ and $\bar{y} = \frac{38}{35}$

21. $M = \int_0^2 \int_{y^2/2}^{4-y} dx \, dy = \int_0^2 \left(4 - y - \frac{y^2}{2}\right) dy = \frac{14}{3}$; $M_y = \int_0^2 \int_{y^2/2}^{4-y} x \, dx \, dy = \frac{1}{2}\int_0^2 [x^2]_{y^2/2}^{4-y} dy$

$= \frac{1}{2}\int_0^2 \left(16 - 8y + y^2 - \frac{y^4}{4}\right) dy = \frac{128}{15}$; $M_x = \int_0^2 \int_{y^2/2}^{4-y} y \, dx \, dy = \int_0^2 \left(4y - y^2 - \frac{y^3}{2}\right) dy = \frac{10}{3}$

$\Rightarrow \bar{x} = \frac{64}{35}$ and $\bar{y} = \frac{5}{7}$

23. $M = 2\int_0^1 \int_0^{\sqrt{1-x^2}} dy \, dx = 2\int_0^1 \sqrt{1-x^2} \, dx = 2\left(\frac{\pi}{4}\right) = \frac{\pi}{2}$; $M_x = 2\int_0^1 \int_0^{\sqrt{1-x^2}} y \, dy \, dx = \int_0^1 [y^2]_0^{\sqrt{1-x^2}} dx$

$= \int_0^1 (1 - x^2) \, dx = \left[x - \frac{x^3}{3}\right]_0^1 = \frac{2}{3}$ $\Rightarrow \bar{y} = \frac{4}{3\pi}$ and $\bar{x} = 0$, by symmetry

25. $M = \int_0^a \int_0^{\sqrt{a^2-x^2}} dy \, dx = \frac{\pi a^2}{4}$; $M_y = \int_0^a \int_0^{\sqrt{a^2-x^2}} x \, dy \, dx = \int_0^a [xy]_0^{\sqrt{a^2-x^2}} dx = \int_0^a x\sqrt{a^2 - x^2} \, dx = \frac{a^3}{3}$

$\Rightarrow \bar{x} = \bar{y} = \frac{4a}{3\pi}$, by symmetry

27. $I_x = \int_{-2}^2 \int_{-\sqrt{4-x^2}}^{\sqrt{4-x^2}} y^2 \, dy \, dx = \int_{-2}^2 \left[\frac{y^3}{3}\right]_{-\sqrt{4-x^2}}^{\sqrt{4-x^2}} dx = \frac{2}{3}\int_{-2}^2 (4 - x^2)^{3/2} \, dx = 4\pi$; $I_y = 4\pi$, by symmetry;

$I_0 = I_x + I_y = 8\pi$

29. $M = \int_{-\infty}^0 \int_0^{e^x} dy \, dx = \int_{-\infty}^0 e^x \, dx = \lim_{b \to -\infty} \int_b^0 e^x \, dx = 1 - \lim_{b \to -\infty} e^b = 1$; $M_y = \int_{-\infty}^0 \int_0^{e^x} x \, dy \, dx = \int_{-\infty}^0 xe^x \, dx$

$= \lim_{b \to -\infty} \int_b^0 xe^x \, dx = \lim_{b \to -\infty} [xe^x - e^x]_b^0 = -1 - \lim_{b \to -\infty} (be^b - e^b) = -1$; $M_x = \int_{-\infty}^0 \int_0^{e^x} y \, dy \, dx$

$= \frac{1}{2}\int_{-\infty}^0 e^{2x} \, dx = \frac{1}{2} \lim_{b \to -\infty} \int_b^0 e^{2x} \, dx = \frac{1}{4} \Rightarrow \bar{x} = -1$ and $\bar{y} = \frac{1}{4}$

31. $M = \int_0^2 \int_{-y}^{y-y^2} (x+y) \, dx \, dy = \int_0^2 \left[\frac{x^2}{2} + xy\right]_{-y}^{y-y^2} dy = \int_0^2 \left(\frac{y^4}{2} - 2y^3 + 2y^2\right) dy = \left[\frac{y^5}{10} - \frac{y^4}{2} + \frac{2y^3}{3}\right]_0^2 = \frac{8}{15}$;

$I_x = \int_0^2 \int_{-y}^{y-y^2} y^2(x+y) \, dx \, dy = \int_0^2 \left[\frac{x^2 y^2}{2} + xy^3\right]_{-y}^{y-y^2} dy = \int_0^2 \left(\frac{y^6}{2} - 2y^5 + 2y^4\right) dy = \frac{64}{105}$;

$R_x = \sqrt{\frac{I_x}{M}} = \sqrt{\frac{8}{7}} = 2\sqrt{\frac{2}{7}}$

33. $M = \int_0^1 \int_x^{2-x} (6x + 3y + 3) \, dy \, dx = \int_0^1 \left[6xy + \frac{3}{2}y^2 + 3y\right]_x^{2-x} dx = \int_0^1 (12 - 12x^2) \, dx = 8$;

$M_y = \int_0^1 \int_x^{2-x} x(6x + 3y + 3) \, dy \, dx = \int_0^1 (12x - 12x^3) \, dx = 3$; $M_x = \int_0^1 \int_x^{2-x} y(6x + 3y + 3) \, dy \, dx$

$= \int_0^1 (14 - 6x - 6x^2 - 2x^3) \, dx = \frac{17}{2} \Rightarrow \bar{x} = \frac{3}{8}$ and $\bar{y} = \frac{17}{16}$

35. $M = \int_0^1 \int_0^6 (x + y + 1)\, dx\, dy = \int_0^1 (6y + 24)\, dy = 27$; $M_x = \int_0^1 \int_0^6 y(x + y + 1)\, dx\, dy = \int_0^1 y(6y + 24)\, dy = 14$;

$M_y = \int_0^1 \int_0^6 x(x + y + 1)\, dx\, dy = \int_0^1 (18y + 90)\, dy = 99 \Rightarrow \bar{x} = \frac{11}{3}$ and $\bar{y} = \frac{14}{27}$; $I_y = \int_0^1 \int_0^6 x^2(x + y + 1)\, dx\, dy$

$= 216 \int_0^1 \left(\frac{y}{3} + \frac{11}{6}\right)\, dy = 432$; $R_y = \sqrt{\frac{I_y}{M}} = 4$

37. $M = \int_{-1}^1 \int_0^{x^2} (7y + 1)\, dy\, dx = \int_{-1}^1 \left(\frac{7x^4}{2} + x^2\right)\, dx = \frac{31}{15}$; $M_x = \int_{-1}^1 \int_0^{x^2} y(7y + 1)\, dy\, dx = \int_{-1}^1 \left(\frac{7x^6}{3} + \frac{x^4}{2}\right)\, dx = \frac{13}{15}$;

$M_y = \int_{-1}^1 \int_0^{x^2} x(7y + 1)\, dy\, dx = \int_{-1}^1 \left(\frac{7x^5}{2} + x^3\right)\, dx = 0 \Rightarrow \bar{x} = 0$ and $\bar{y} = \frac{13}{31}$; $I_y = \int_{-1}^1 \int_0^{x^2} x^2(7y + 1)\, dy\, dx$

$= \int_{-1}^1 \left(\frac{7x^6}{2} + x^4\right)\, dx = \frac{7}{5}$; $R_y = \sqrt{\frac{I_y}{M}} = \sqrt{\frac{21}{31}}$

39. $M = \int_0^1 \int_{-y}^y (y + 1)\, dx\, dy = \int_0^1 (2y^2 + 2y)\, dy = \frac{5}{3}$; $M_x = \int_0^1 \int_{-y}^y y(y + 1)\, dx\, dy = 2\int_0^1 (y^3 + y^2)\, dy = \frac{7}{6}$;

$M_y = \int_0^1 \int_{-y}^y x(y + 1)\, dx\, dy = \int_0^1 0\, dy = 0 \Rightarrow \bar{x} = 0$ and $\bar{y} = \frac{7}{10}$; $I_x = \int_0^1 \int_{-y}^y y^2(y + 1)\, dx\, dy = \int_0^1 (2y^4 + 2y^3)\, dy$

$= \frac{9}{10} \Rightarrow R_x = \sqrt{\frac{I_x}{M}} = \frac{3\sqrt{6}}{10}$; $I_y = \int_0^1 \int_{-y}^y x^2(y + 1)\, dx\, dy = \frac{1}{3}\int_0^1 (2y^4 + 2y^3)\, dy = \frac{3}{10} \Rightarrow R_y = \sqrt{\frac{I_y}{M}} = \frac{3\sqrt{2}}{10}$;

$I_o = I_x + I_y = \frac{6}{5} \Rightarrow R_0 = \sqrt{\frac{I_o}{M}} = \frac{3\sqrt{2}}{5}$

41. $\int_{-5}^5 \int_{-2}^0 \frac{10{,}000 e^y}{1 + \frac{|x|}{2}}\, dy\, dx = 10{,}000\,(1 - e^{-2}) \int_{-5}^5 \frac{dx}{1 + \frac{|x|}{2}} = 10{,}000\,(1 - e^{-2}) \left[\int_{-5}^0 \frac{dx}{1 - \frac{x}{2}} + \int_0^5 \frac{dx}{1 + \frac{x}{2}}\right]$

$= 10{,}000\,(1 - e^{-2}) \left[-2 \ln\left(1 - \frac{x}{2}\right)\right]_{-5}^0 + 10{,}000\,(1 - e^{-2}) \left[2 \ln\left(1 + \frac{x}{2}\right)\right]_0^5$

$= 10{,}000\,(1 - e^{-2}) \left[2 \ln\left(1 + \frac{5}{2}\right)\right] + 10{,}000\,(1 - e^{-2}) \left[2 \ln\left(1 + \frac{5}{2}\right)\right] = 40{,}000\,(1 - e^{-2}) \ln\left(\frac{7}{2}\right) \approx 43{,}329$

43. $M = \int_{-1}^1 \int_0^{a(1-x^2)} dy\, dx = 2a \int_0^1 (1 - x^2)\, dx = 2a \left[x - \frac{x^3}{3}\right]_0^1 = \frac{4a}{3}$; $M_x = \int_{-1}^1 \int_0^{a(1-x^2)} y\, dy\, dx$

$= \frac{2a^2}{2} \int_0^1 (1 - 2x^2 + x^4)\, dx = a^2 \left[x - \frac{2x^3}{3} + \frac{x^5}{5}\right]_0^1 = \frac{8a^2}{15} \Rightarrow \bar{y} = \frac{M_x}{M} = \frac{\left(\frac{8a^2}{15}\right)}{\left(\frac{4a}{3}\right)} = \frac{2a}{5}$. The angle θ between the

x-axis and the line segment from the fulcrum to the center of mass on the y-axis plus $45°$ must be no more than

$90°$ if the center of mass is to lie on the left side of the line $x = 1 \Rightarrow \theta + \frac{\pi}{4} \leq \frac{\pi}{2} \Rightarrow \tan^{-1}\left(\frac{2a}{5}\right) \leq \frac{\pi}{4} \Rightarrow a \leq \frac{5}{2}$.

Thus, if $0 < a \leq \frac{5}{2}$, then the appliance will have to be tipped more than $45°$ to fall over.

45. $M = \int_0^1 \int_{-1/\sqrt{1-x^2}}^{1/\sqrt{1-x^2}} dy\, dx = \int_0^1 \frac{2}{\sqrt{1-x^2}}\, dx = \left[2 \sin^{-1} x\right]_0^1 = 2\left(\frac{\pi}{2} - 0\right) = \pi$; $M_y = \int_0^1 \int_{-1/\sqrt{1-x^2}}^{1/\sqrt{1-x^2}} x\, dy\, dx$

$= \int_0^1 \frac{2x}{\sqrt{1-x^2}}\, dx = \left[-2\,(1 - x^2)^{1/2}\right]_0^1 = 2 \Rightarrow \bar{x} = \frac{2}{\pi}$ and $\bar{y} = 0$ by symmetry

47. (a) $\frac{1}{2} = M = \int_0^1 \int_{y^2}^{2y-y^2} \delta\, dx\, dy = 2\delta \int_0^1 (y - y^2)\, dy = 2\delta \left[\frac{y^2}{2} - \frac{y^3}{3}\right]_0^1 = 2\delta\left(\frac{1}{6}\right) = \frac{\delta}{3} \Rightarrow \delta = \frac{3}{2}$

(b) average value $= \dfrac{\int_0^1 \int_{y^2}^{2y-y^2} (y+1)\, dx\, dy}{\int_0^1 \int_{y^2}^{2y-y^2} dx\, dy} = \dfrac{\left(\frac{1}{2}\right)}{\left(\frac{1}{3}\right)} = \frac{3}{2} = \delta$, so the values are the same

49. (a) $\bar{x} = \frac{M_y}{M} = 0 \Rightarrow M_y = \iint_R x\delta(x, y)\, dy\, dx = 0$

(b) $I_L = \iint_R (x - h)^2\, \delta(x, y)\, dA = \iint_R x^2\, \delta(x, y)\, dA - \iint_R 2hx\, \delta(x, y)\, dA + \iint_R h^2\, \delta(x, y)\, dA$

$= I_y - 0 + h^2 \iint_R \delta(x, y)\, dA = I_{c.m.} + mh^2$

51. $M_{x_{p_1 \cup p_2}} = \iint\limits_{R_1} y \, dA_1 + \iint\limits_{R_2} y \, dA_2 = M_{x_1} + M_{x_2} \Rightarrow \overline{x} = \frac{M_{x_1} + M_{x_2}}{m_1 + m_2}$; likewise, $\overline{y} = \frac{M_{y_1} + M_{y_2}}{m_1 + m_2}$;

thus $\mathbf{c} = \overline{x}\mathbf{i} + \overline{y}\mathbf{j} = \frac{1}{m_1 + m_2} [(M_{x_1} + M_{x_2})\mathbf{i} + (M_{y_1} + M_{y_2})\mathbf{j}] = \frac{1}{m_1 + m_2} [(m_1\overline{x}_1 + m_2\overline{x}_2)\mathbf{i} + (m_1\overline{y}_1 + m_2\overline{y}_2)\mathbf{j}]$

$= \frac{1}{m_1 + m_2} [m_1(\overline{x}_1\mathbf{i} + \overline{y}_1\mathbf{j}) + m_2(\overline{x}_2\mathbf{i} + \overline{y}_2\mathbf{j})] = \frac{m_1\mathbf{c}_1 + m_2\mathbf{c}_2}{m_1 + m_2}$

53. (a) $\mathbf{c} = \frac{8(\mathbf{i} + 3\mathbf{j}) + 2(3\mathbf{i} + 3.5\mathbf{j})}{8 + 2} = \frac{14\mathbf{i} + 31\mathbf{j}}{10} \Rightarrow \overline{x} = \frac{7}{5}$ and $\overline{y} = \frac{31}{10}$

(b) $\mathbf{c} = \frac{8(\mathbf{i} + 3\mathbf{j}) + 6(5\mathbf{i} + 2\mathbf{j})}{14} = \frac{38\mathbf{i} + 36\mathbf{j}}{14} \Rightarrow \overline{x} = \frac{19}{7}$ and $\overline{y} = \frac{18}{7}$

(c) $\mathbf{c} = \frac{2(3\mathbf{i} + 3.5\mathbf{j}) + 6(5\mathbf{i} + 2\mathbf{j})}{8} = \frac{36\mathbf{i} + 19\mathbf{j}}{8} \Rightarrow \overline{x} = \frac{9}{2}$ and $\overline{y} = \frac{19}{8}$

(d) $\mathbf{c} = \frac{8(\mathbf{i} + 3\mathbf{j}) + 2(3\mathbf{i} + 3.5\mathbf{j}) + 6(5\mathbf{i} + 2\mathbf{j})}{16} = \frac{44\mathbf{i} + 43\mathbf{j}}{16} \Rightarrow \overline{x} = \frac{11}{4}$ and $\overline{y} = \frac{43}{16}$

55. Place the midpoint of the triangle's base at the origin and above the semicircle. Then the center of mass of the triangle is $\left(0, \frac{h}{3}\right)$, and the center of mass of the disk is $\left(0, -\frac{4a}{3\pi}\right)$ from Exercise 25. From

Pappus's formula, $\mathbf{c} = \frac{(ah)\left(\frac{h}{3}\mathbf{j}\right) + \left(\frac{\pi a^2}{2}\right)\left(-\frac{4a}{3\pi}\mathbf{j}\right)}{\left(ah + \frac{\pi a^2}{2}\right)} = \frac{\left(\frac{ah^2 - 2a^3}{3}\right)\mathbf{j}}{\left(ah + \frac{\pi a^2}{2}\right)}$, so the centroid is on the boundary

if $ah^2 - 2a^3 = 0 \Rightarrow h^2 = 2a^2 \Rightarrow h = a\sqrt{2}$. In order for the center of mass to be inside T we must have $ah^2 - 2a^3 > 0$ or $h > a\sqrt{2}$.

15.3 DOUBLE INTEGRALS IN POLAR FORM

1. $\int_{-1}^{1} \int_{0}^{\sqrt{1-x^2}} dy \, dx = \int_{0}^{\pi} \int_{0}^{1} r \, dr \, d\theta = \frac{1}{2} \int_{0}^{\pi} d\theta = \frac{\pi}{2}$

3. $\int_{0}^{1} \int_{0}^{\sqrt{1-y^2}} (x^2 + y^2) \, dx \, dy = \int_{0}^{\pi/2} \int_{0}^{1} r^3 \, dr \, d\theta = \frac{1}{4} \int_{0}^{\pi/2} d\theta = \frac{\pi}{8}$

5. $\int_{-a}^{a} \int_{-\sqrt{a^2-x^2}}^{\sqrt{a^2-x^2}} dy \, dx = \int_{0}^{2\pi} \int_{0}^{a} r \, dr \, d\theta = \frac{a^2}{2} \int_{0}^{2\pi} d\theta = \pi a^2$

7. $\int_{0}^{6} \int_{0}^{y} x \, dx \, dy = \int_{\pi/4}^{\pi/2} \int_{0}^{6 \csc \theta} r^2 \cos \theta \, dr \, d\theta = 72 \int_{\pi/4}^{\pi/2} \cot \theta \csc^2 \theta \, d\theta = -36 \left[\cot^2 \theta\right]_{\pi/4}^{\pi/2} = 36$

9. $\int_{-1}^{0} \int_{-\sqrt{1-x^2}}^{0} \frac{2}{1 + \sqrt{x^2+y^2}} \, dy \, dx = \int_{\pi}^{3\pi/2} \int_{0}^{1} \frac{2r}{1+r} \, dr \, d\theta = 2 \int_{\pi}^{3\pi/2} \int_{0}^{1} \left(1 - \frac{1}{1+r}\right) dr \, d\theta = 2 \int_{\pi}^{3\pi/2} (1 - \ln 2) \, d\theta$

$= (1 - \ln 2)\pi$

11. $\int_{0}^{\ln 2} \int_{0}^{\sqrt{(\ln 2)^2 - y^2}} e^{\sqrt{x^2+y^2}} \, dx \, dy = \int_{0}^{\pi/2} \int_{0}^{\ln 2} re^r \, dr \, d\theta = \int_{0}^{\pi/2} (2 \ln 2 - 1) \, d\theta = \frac{\pi}{2}(2 \ln 2 - 1)$

13. $\int_{0}^{2} \int_{0}^{\sqrt{1 - (x-1)^2}} \frac{x+y}{x^2+y^2} \, dy \, dx = \int_{0}^{\pi/2} \int_{0}^{2 \cos \theta} \frac{r(\cos \theta + \sin \theta)}{r^2} r \, dr \, d\theta = \int_{0}^{\pi/2} (2 \cos^2 \theta + 2 \sin \theta \cos \theta) \, d\theta$

$= \left[\theta + \frac{\sin 2\theta}{2} + \sin^2 \theta\right]_{0}^{\pi/2} = \frac{\pi+2}{2} = \frac{\pi}{2} + 1$

15. $\int_{-1}^{1} \int_{-\sqrt{1-y^2}}^{\sqrt{1-y^2}} \ln(x^2 + y^2 + 1) \, dx \, dy = 4 \int_{0}^{\pi/2} \int_{0}^{1} \ln(r^2 + 1) r \, dr \, d\theta = 2 \int_{0}^{\pi/2} (\ln 4 - 1) \, d\theta = \pi(\ln 4 - 1)$

17. $\int_{0}^{\pi/2} \int_{0}^{2\sqrt{2 - \sin 2\theta}} r \, dr \, d\theta = 2 \int_{0}^{\pi/2} (2 - \sin 2\theta) \, d\theta = 2(\pi - 1)$

19. $A = 2 \int_{0}^{\pi/6} \int_{0}^{12 \cos 3\theta} r \, dr \, d\theta = 144 \int_{0}^{\pi/6} \cos^2 3\theta \, d\theta = 12\pi$

21. $A = \int_0^{\pi/2} \int_0^{1+\sin\theta} r \, dr \, d\theta = \frac{1}{2} \int_0^{\pi/2} \left(\frac{3}{2} + 2\sin\theta - \frac{\cos 2\theta}{2} \right) d\theta = \frac{3\pi}{8} + 1$

23. $M_x = \int_0^{\pi} \int_0^{1-\cos\theta} 3r^2 \sin\theta \, dr \, d\theta = \int_0^{\pi} (1-\cos\theta)^3 \sin\theta \, d\theta = 4$

25. $M = 2 \int_{\pi/6}^{\pi/2} \int_3^{6\sin\theta} dr \, d\theta = 2 \int_{\pi/6}^{\pi/2} (6\sin\theta - 3) \, d\theta = 6 \left[-2\cos\theta - \theta \right]_{\pi/6}^{\pi/2} = 6\sqrt{3} - 2\pi$

27. $M = 2 \int_0^{\pi} \int_0^{1+\cos\theta} r \, dr \, d\theta = \int_0^{\pi} (1+\cos\theta)^2 \, d\theta = \frac{3\pi}{2}$; $M_y = 2 \int_0^{\pi} \int_0^{1+\cos\theta} r^2 \cos\theta \, dr \, d\theta$

$= 2 \int_0^{\pi} \left(\frac{4\cos\theta}{3} + \frac{15}{24} + \cos 2\theta - \sin^2\theta \cos\theta + \frac{\cos 4\theta}{4} \right) d\theta = \frac{5\pi}{4} \Rightarrow \bar{x} = \frac{5}{6}$ and $\bar{y} = 0$, by symmetry

29. average $= \frac{4}{\pi a^2} \int_0^{\pi/2} \int_0^a r\sqrt{a^2 - r^2} \, dr \, d\theta = \frac{4}{3\pi a^2} \int_0^{\pi/2} a^3 \, d\theta = \frac{2a}{3}$

31. average $= \frac{1}{\pi a^2} \int_{-a}^a \int_{-\sqrt{a^2-x^2}}^{\sqrt{a^2-x^2}} \sqrt{x^2 + y^2} \, dy \, dx = \frac{1}{\pi a^2} \int_0^{2\pi} \int_0^a r^2 \, dr \, d\theta = \frac{a}{3\pi} \int_0^{2\pi} d\theta = \frac{2a}{3}$

33. $\int_0^{2\pi} \int_1^{\sqrt{e}} \left(\frac{\ln r^2}{r} \right) r \, dr \, d\theta = \int_0^{2\pi} \int_1^{\sqrt{e}} 2\ln r \, dr \, d\theta = 2 \int_0^{2\pi} [r \ln r - r]_1^{e^{1/2}} \, d\theta = 2 \int_0^{2\pi} \sqrt{e} \left[\left(\frac{1}{2} - 1 \right) + 1 \right] d\theta = 2\pi (2 - \sqrt{e})$

35. $V = 2 \int_0^{\pi/2} \int_1^{1+\cos\theta} r^2 \cos\theta \, dr \, d\theta = \frac{2}{3} \int_0^{\pi/2} (3\cos^2\theta + 3\cos^3\theta + \cos^4\theta) \, d\theta$

$= \frac{2}{3} \left[\frac{15\theta}{8} + \sin 2\theta + 3\sin\theta - \sin^3\theta + \frac{\sin 4\theta}{32} \right]_0^{\pi/2} = \frac{4}{3} + \frac{5\pi}{8}$

37. (a) $I^2 = \int_0^{\infty} \int_0^{\infty} e^{-(x^2+y^2)} \, dx \, dy = \int_0^{\pi/2} \int_0^{\infty} \left(e^{-r^2} \right) r \, dr \, d\theta = \int_0^{\pi/2} \left[\lim_{b \to \infty} \int_0^b re^{-r^2} \, dr \right] d\theta$

$= -\frac{1}{2} \int_0^{\pi/2} \lim_{b \to \infty} \left(e^{-b^2} - 1 \right) d\theta = \frac{1}{2} \int_0^{\pi/2} d\theta = \frac{\pi}{4} \Rightarrow I = \frac{\sqrt{\pi}}{2}$

(b) $\lim_{x \to \infty} \int_0^x \frac{2e^{-t^2}}{\sqrt{\pi}} \, dt = \frac{2}{\sqrt{\pi}} \int_0^{\infty} e^{-t^2} \, dt = \left(\frac{2}{\sqrt{\pi}} \right) \left(\frac{\sqrt{\pi}}{2} \right) = 1$, from part (a)

39. Over the disk $x^2 + y^2 \le \frac{3}{4}$: $\iint_R \frac{1}{1-x^2-y^2} \, dA = \int_0^{2\pi} \int_0^{\sqrt{3}/2} \frac{r}{1-r^2} \, dr \, d\theta = \int_0^{2\pi} \left[-\frac{1}{2} \ln (1-r^2) \right]_0^{\sqrt{3}/2} d\theta$

$= \int_0^{2\pi} \left(-\frac{1}{2} \ln \frac{1}{4} \right) d\theta = (\ln 2) \int_0^{2\pi} d\theta = \pi \ln 4$

Over the disk $x^2 + y^2 \le 1$: $\iint_R \frac{1}{1-x^2-y^2} \, dA = \int_0^{2\pi} \int_0^1 \frac{r}{1-r^2} \, dr \, d\theta = \int_0^{2\pi} \left[\lim_{a \to 1^-} \int_0^a \frac{r}{1-r^2} \, dr \right] d\theta$

$= \int_0^{2\pi} \lim_{a \to 1^-} \left[-\frac{1}{2} \ln (1-a^2) \right] d\theta = 2\pi \cdot \lim_{a \to 1^-} \left[-\frac{1}{2} \ln (1-a^2) \right] = 2\pi \cdot \infty$, so the integral does not exist over $x^2 + y^2 \le 1$

41. average $= \frac{1}{\pi a^2} \int_0^{2\pi} \int_0^a [(r\cos\theta - h)^2 + r^2 \sin^2\theta] r \, dr \, d\theta = \frac{1}{\pi a^2} \int_0^{2\pi} \int_0^a (r^3 - 2r^2 h \cos\theta + rh^2) \, dr \, d\theta$

$= \frac{1}{\pi a^2} \int_0^{2\pi} \left(\frac{a^4}{4} - \frac{2a^3 h \cos\theta}{3} + \frac{a^2 h^2}{2} \right) d\theta = \frac{1}{\pi} \int_0^{2\pi} \left(\frac{a^2}{4} - \frac{2ah\cos\theta}{3} + \frac{h^2}{2} \right) d\theta = \frac{1}{\pi} \left[\frac{a^2\theta}{4} - \frac{2ah\sin\theta}{3} + \frac{h^2\theta}{2} \right]_0^{2\pi}$

$= \frac{1}{2} (a^2 + 2h^2)$

15.4 TRIPLE INTEGRALS IN RECTANGULAR COORDINATES

1. $\int_0^1 \int_0^{1-x} \int_{x+z}^1 F(x, y, z) \, dy \, dz \, dx = \int_0^1 \int_0^{1-x} \int_{x+z}^1 dy \, dz \, dx = \int_0^1 \int_0^{1-x} (1-x-z) \, dz \, dx$

$= \int_0^1 \left[(1-x) - x(1-x) - \frac{(1-x)^2}{2} \right] dx = \int_0^1 \frac{(1-x)^2}{2} \, dx = \left[-\frac{(1-x)^3}{6} \right]_0^1 = \frac{1}{6}$

3. $\int_0^1 \int_0^{2-2x} \int_0^{3-3x-3y/2} dz\,dy\,dx$

$= \int_0^1 \int_0^{2-2x} \left(3 - 3x - \frac{3}{2}y\right) dy\,dx$

$= \int_0^1 \left[3(1-x) \cdot 2(1-x) - \frac{3}{4} \cdot 4(1-x)^2\right] dx$

$= 3 \int_0^1 (1-x)^2 dx = \left[-(1-x)^3\right]_0^1 = 1,$

$\int_0^2 \int_0^{1-y/2} \int_0^{3-3x-3y/2} dz\,dx\,dy, \quad \int_0^1 \int_0^{3-3x} \int_0^{2-2x-2z/3} dy\,dz\,dx,$

$\int_0^3 \int_0^{1-z/3} \int_0^{2-2x-2z/3} dy\,dx\,dz, \quad \int_0^2 \int_0^{3-3y/2} \int_0^{1-y/2-z/3} dx\,dz\,dy,$

$\int_0^3 \int_0^{2-2z/3} \int_0^{1-y/2-z/3} dx\,dy\,dz$

5. $\int_{-2}^2 \int_{-\sqrt{4-x^2}}^{\sqrt{4-x^2}} \int_{x^2+y^2}^{8-x^2-y^2} dz\,dy\,dx = 4 \int_0^2 \int_0^{\sqrt{4-x^2}} \int_{x^2+y^2}^{8-x^2-y^2} dz\,dy\,dx$

$= 4 \int_0^2 \int_0^{\sqrt{4-x^2}} [8 - 2(x^2 + y^2)]\,dy\,dx$

$= 8 \int_0^2 \int_0^{\sqrt{4-x^2}} (4 - x^2 - y^2)\,dy\,dx$

$= 8 \int_0^{\pi/2} \int_0^2 (4 - r^2)\,r\,dr\,d\theta = 8 \int_0^{\pi/2} \left[2r^2 - \frac{r^4}{4}\right]_0^2 d\theta$

$= 32 \int_0^{\pi/2} d\theta = 32 \left(\frac{\pi}{2}\right) = 16\pi,$

$\int_{-2}^2 \int_{-\sqrt{4-y^2}}^{\sqrt{4-y^2}} \int_{x^2+y^2}^{8-x^2-y^2} dz\,dx\,dy,$

$\int_{-2}^2 \int_{y^2}^4 \int_{-\sqrt{z-y^2}}^{\sqrt{z-y^2}} dx\,dz\,dy + \int_{-2}^2 \int_4^{8-y^2} \int_{-\sqrt{8-z-y^2}}^{\sqrt{8-z-y^2}} dx\,dz\,dy,$

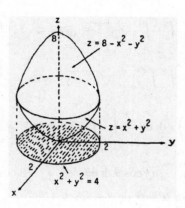

$\int_0^4 \int_{-\sqrt{z}}^{\sqrt{z}} \int_{-\sqrt{z-y^2}}^{\sqrt{z-y^2}} dx\,dy\,dz + \int_4^8 \int_{-\sqrt{8-z}}^{\sqrt{8-z}} \int_{-\sqrt{8-z-y^2}}^{\sqrt{8-z-y^2}} dx\,dy\,dz, \quad \int_{-2}^2 \int_{x^2}^4 \int_{-\sqrt{z-x^2}}^{\sqrt{z-x^2}} dy\,dz\,dx + \int_{-2}^2 \int_4^{8-x^2} \int_{-\sqrt{8-z-x^2}}^{\sqrt{8-z-x^2}} dy\,dz\,dx,$

$\int_0^4 \int_{-\sqrt{z}}^{\sqrt{z}} \int_{-\sqrt{z-x^2}}^{\sqrt{z-x^2}} dy\,dx\,dz + \int_4^8 \int_{-\sqrt{8-z}}^{\sqrt{8-z}} \int_{-\sqrt{8-z-x^2}}^{\sqrt{8-z-x^2}} dy\,dx\,dz$

7. $\int_0^1 \int_0^1 \int_0^1 (x^2 + y^2 + z^2)\,dz\,dy\,dx = \int_0^1 \int_0^1 \left(x^2 + y^2 + \frac{1}{3}\right) dy\,dx = \int_0^1 \left(x^2 + \frac{2}{3}\right) dx = 1$

9. $\int_1^e \int_1^e \int_1^e \frac{1}{xyz}\,dx\,dy\,dz = \int_1^e \int_1^e \left[\frac{\ln x}{yz}\right]_1^e dy\,dz = \int_1^e \int_1^e \frac{1}{yz}\,dy\,dz = \int_1^e \left[\frac{\ln y}{z}\right]_1^e dz = \int_1^e \frac{1}{z}\,dz = 1$

11. $\int_0^1 \int_0^\pi \int_0^\pi y \sin z\,dx\,dy\,dz = \int_0^1 \int_0^\pi \pi y \sin z\,dy\,dz = \frac{\pi^3}{2} \int_0^1 \sin z\,dz = \frac{\pi^3}{2}(1 - \cos 1)$

13. $\int_0^3 \int_0^{\sqrt{9-x^2}} \int_0^{\sqrt{9-x^2}} dz\,dy\,dx = \int_0^3 \int_0^{\sqrt{9-x^2}} \sqrt{9-x^2}\,dy\,dx = \int_0^3 (9 - x^2)\,dx = \left[9x - \frac{x^3}{3}\right]_0^3 = 18$

15. $\int_0^1 \int_0^{2-x} \int_0^{2-x-y} dz\,dy\,dx = \int_0^1 \int_0^{2-x} (2 - x - y)\,dy\,dx = \int_0^1 \left[(2-x)^2 - \frac{1}{2}(2-x)^2\right] dx = \frac{1}{2} \int_0^1 (2-x)^2\,dx$

$= \left[-\frac{1}{6}(2-x)^3\right]_0^1 = -\frac{1}{6} + \frac{8}{6} = \frac{7}{6}$

17. $\int_0^\pi \int_0^\pi \int_0^\pi \cos(u+v+w)\,du\,dv\,dw = \int_0^\pi \int_0^\pi [\sin(w+v+\pi) - \sin(w+v)]\,dv\,dw$

$= \int_0^\pi [(-\cos(w+2\pi) + \cos(w+\pi)) + (\cos(w+\pi) - \cos w)]\,dw$

$= [-\sin(w+2\pi) + \sin(w+\pi) - \sin w + \sin(w+\pi)]_0^\pi = 0$

19. $\int_0^{\pi/4} \int_0^{\ln \sec v} \int_{-\infty}^{2t} e^x\,dx\,dt\,dv = \int_0^{\pi/4} \int_0^{\ln \sec v} \lim_{b \to -\infty} (e^{2t} - e^b)\,dt\,dv = \int_0^{\pi/4} \int_0^{\ln \sec v} e^{2t}\,dt\,dv = \int_0^{\pi/4} \left(\frac{1}{2} e^{2\ln\sec v} - \frac{1}{2}\right) dv$

$= \int_0^{\pi/4} \left(\frac{\sec^2 v}{2} - \frac{1}{2}\right) dv = \left[\frac{\tan v}{2} - \frac{v}{2}\right]_0^{\pi/4} = \frac{1}{2} - \frac{\pi}{8}$

21. (a) $\int_{-1}^{1}\int_{0}^{1-x^2}\int_{x^2}^{1-z} dy\, dz\, dx$ (b) $\int_{0}^{1}\int_{-\sqrt{1-z}}^{\sqrt{1-z}}\int_{x^2}^{1-z} dy\, dx\, dz$ (c) $\int_{0}^{1}\int_{0}^{1-z}\int_{-\sqrt{y}}^{\sqrt{y}} dx\, dy\, dz$

(d) $\int_{0}^{1}\int_{0}^{1-y}\int_{-\sqrt{y}}^{\sqrt{y}} dx\, dz\, dy$ (e) $\int_{0}^{1}\int_{-\sqrt{y}}^{\sqrt{y}}\int_{0}^{1-y} dz\, dx\, dy$

23. $V = \int_{0}^{1}\int_{-1}^{1}\int_{0}^{y^2} dz\, dy\, dx = \int_{0}^{1}\int_{-1}^{1} y^2\, dy\, dx = \frac{2}{3}\int_{0}^{1} dx = \frac{2}{3}$

25. $V = \int_{0}^{4}\int_{0}^{\sqrt{4-x}}\int_{0}^{2-y} dz\, dy\, dx = \int_{0}^{4}\int_{0}^{\sqrt{4-x}} (2-y)\, dy\, dx = \int_{0}^{4}\left[2\sqrt{4-x} - \left(\frac{4-x}{2}\right)\right] dx$

$= \left[-\frac{4}{3}(4-x)^{3/2} + \frac{1}{4}(4-x)^2\right]_{0}^{4} = \frac{4}{3}(4)^{3/2} - \frac{1}{4}(16) = \frac{32}{3} - 4 = \frac{20}{3}$

27. $V = \int_{0}^{1}\int_{0}^{2-2x}\int_{0}^{3-3x-3y/2} dz\, dy\, dx = \int_{0}^{1}\int_{0}^{2-2x} \left(3 - 3x - \frac{3}{2}y\right) dy\, dx = \int_{0}^{1}\left[6(1-x)^2 - \frac{3}{4}\cdot 4(1-x)^2\right] dx$

$= \int_{0}^{1} 3(1-x)^2\, dx = \left[-(1-x)^3\right]_{0}^{1} = 1$

29. $V = 8\int_{0}^{1}\int_{0}^{\sqrt{1-x^2}}\int_{0}^{\sqrt{1-x^2}} dz\, dy\, dx = 8\int_{0}^{1}\int_{0}^{\sqrt{1-x^2}}\sqrt{1-x^2}\, dy\, dx = 8\int_{0}^{1}(1-x^2)\, dx = \frac{16}{3}$

31. $V = \int_{0}^{4}\int_{0}^{(\sqrt{16-y^2})/2}\int_{0}^{4-y} dx\, dz\, dy = \int_{0}^{4}\int_{0}^{(\sqrt{16-y^2})/2} (4-y)\, dz\, dy = \int_{0}^{4}\frac{\sqrt{16-y^2}}{2}(4-y)\, dy$

$= \int_{0}^{4} 2\sqrt{16-y^2}\, dy - \frac{1}{2}\int_{0}^{4} y\sqrt{16-y^2}\, dy = \left[y\sqrt{16-y^2} + 16\sin^{-1}\frac{y}{4}\right]_{0}^{4} + \left[\frac{1}{6}(16-y^2)^{3/2}\right]_{0}^{4}$

$= 16\left(\frac{\pi}{2}\right) - \frac{1}{6}(16)^{3/2} = 8\pi - \frac{32}{3}$

33. $\int_{0}^{2}\int_{0}^{2-x}\int_{(2-x-y)/2}^{4-2x-2y} dz\, dy\, dx = \int_{0}^{2}\int_{0}^{2-x}\left(3 - \frac{3x}{2} - \frac{3y}{2}\right) dy\, dx$

$= \int_{0}^{2}\left[3\left(1 - \frac{x}{2}\right)(2-x) - \frac{3}{4}(2-x)^2\right] dx$

$= \int_{0}^{2}\left[6 - 6x + \frac{3x^2}{2} - \frac{3(2-x)^2}{4}\right] dx$

$= \left[6x - 3x^2 + \frac{x^3}{2} + \frac{(2-x)^3}{4}\right]_{0}^{2} = (12 - 12 + 4 + 0) - \frac{2^3}{4} = 2$

35. $V = 2\int_{-2}^{2}\int_{0}^{\sqrt{4-x^2}/2}\int_{0}^{x+2} dz\, dy\, dx = 2\int_{-2}^{2}\int_{0}^{\sqrt{4-x^2}/2}(x+2)\, dy\, dx = \int_{-2}^{2}(x+2)\sqrt{4-x^2}\, dx$

$= \int_{-2}^{2} 2\sqrt{4-x^2}\, dx + \int_{-2}^{2} x\sqrt{4-x^2}\, dx = \left[x\sqrt{4-x^2} + 4\sin^{-1}\frac{x}{2}\right]_{-2}^{2} + \left[-\frac{1}{3}(4-x^2)^{3/2}\right]_{-2}^{2}$

$= 4\left(\frac{\pi}{2}\right) - 4\left(-\frac{\pi}{2}\right) = 4\pi$

37. average $= \frac{1}{8}\int_{0}^{2}\int_{0}^{2}\int_{0}^{2}(x^2+9)\, dz\, dy\, dx = \frac{1}{8}\int_{0}^{2}\int_{0}^{2}(2x^2+18)\, dy\, dx = \frac{1}{8}\int_{0}^{2}(4x^2+36)\, dx = \frac{31}{3}$

39. average $= \int_{0}^{1}\int_{0}^{1}\int_{0}^{1}(x^2+y^2+z^2)\, dz\, dy\, dx = \int_{0}^{1}\int_{0}^{1}\left(x^2+y^2+\frac{1}{3}\right) dy\, dx = \int_{0}^{1}\left(x^2+\frac{2}{3}\right) dx = 1$

41. $\int_{0}^{4}\int_{0}^{1}\int_{2y}^{2}\frac{4\cos(x^2)}{2\sqrt{z}}\, dx\, dy\, dz = \int_{0}^{4}\int_{0}^{2}\int_{0}^{x/2}\frac{4\cos(x^2)}{2\sqrt{z}}\, dy\, dx\, dz = \int_{0}^{4}\int_{0}^{2}\frac{x\cos(x^2)}{\sqrt{z}}\, dx\, dz = \int_{0}^{4}\left(\frac{\sin 4}{2}\right) z^{-1/2}\, dz$

$= \left[(\sin 4)z^{1/2}\right]_{0}^{4} = 2\sin 4$

43. $\int_0^1 \int_{\sqrt[3]{z}}^1 \int_0^{\ln 3} \frac{\pi e^{2x} \sin(\pi y^2)}{y^2} \, dx \, dy \, dz = \int_0^1 \int_{\sqrt[3]{z}}^1 \frac{4\pi \sin(\pi y^2)}{y^2} \, dy \, dz = \int_0^1 \int_0^{y^3} \frac{4\pi \sin(\pi y^2)}{y^2} \, dz \, dy$

$= \int_0^1 4\pi y \sin(\pi y^2) \, dy = [-2\cos(\pi y^2)]_0^1 = -2(-1) + 2(1) = 4$

45. $\int_0^1 \int_0^{4-a-x^2} \int_a^{4-x^2-y} dz \, dy \, dx = \frac{4}{15} \Rightarrow \int_0^1 \int_0^{4-a-x^2} (4 - x^2 - y - a) \, dy \, dx = \frac{4}{15}$

$\Rightarrow \int_0^1 \left[(4 - a - x^2)^2 - \frac{1}{2}(4 - a - x^2)^2 \right] dx = \frac{4}{15} \Rightarrow \frac{1}{2} \int_0^1 (4 - a - x^2)^2 \, dx = \frac{4}{15} \Rightarrow \int_0^1 [(4-a)^2 - 2x^2(4-a) + x^4] \, dx$

$= \frac{8}{15} \Rightarrow \left[(4-a)^2 x - \frac{2}{3} x^3 (4-a) + \frac{x^5}{5} \right]_0^1 = \frac{8}{15} \Rightarrow (4-a)^2 - \frac{2}{3}(4-a) + \frac{1}{5} = \frac{8}{15} \Rightarrow 15(4-a)^2 - 10(4-a) - 5 = 0$

$\Rightarrow 3(4-a)^2 - 2(4-a) - 1 = 0 \Rightarrow [3(4-a) + 1][(4-a) - 1] = 0 \Rightarrow 4 - a = -\frac{1}{3} \text{ or } 4 - a = 1 \Rightarrow a = \frac{13}{3} \text{ or } a = 3$

47. To minimize the integral, we want the domain to include all points where the integrand is negative and to exclude all points where it is positive. These criteria are met by the points (x, y, z) such that $4x^2 + 4y^2 + z^2 - 4 \le 0$ or $4x^2 + 4y^2 + z^2 \le 4$, which is a solid ellipsoid centered at the origin.

15.5 MASSES AND MOMENTS IN THREE DIMENSIONS

1. $I_x = \int_{-c/2}^{c/2} \int_{-b/2}^{b/2} \int_{-a/2}^{a/2} (y^2 + z^2) \, dx \, dy \, dz = a \int_{-c/2}^{c/2} \int_{-b/2}^{b/2} (y^2 + z^2) \, dy \, dz = a \int_{-c/2}^{c/2} \left[\frac{y^3}{3} + yz^2 \right]_{-b/2}^{b/2} dz$

$= a \int_{-c/2}^{c/2} \left(\frac{b^3}{12} + bz^2 \right) dz = ab \left[\frac{b^2}{12} z + \frac{z^3}{3} \right]_{-c/2}^{c/2} = ab \left(\frac{b^2 c}{12} + \frac{c^3}{12} \right) = \frac{abc}{12}(b^2 + c^2) = \frac{M}{12}(b^2 + c^2);$

$R_x = \sqrt{\frac{b^2+c^2}{12}}$; likewise $R_y = \sqrt{\frac{a^2+c^2}{12}}$ and $R_z = \sqrt{\frac{a^2+b^2}{12}}$, by symmetry

3. $I_x = \int_0^a \int_0^b \int_0^c (y^2 + z^2) \, dz \, dy \, dx = \int_0^a \int_0^b \left(cy^2 + \frac{c^3}{3} \right) dy \, dx = \int_0^a \left(\frac{cb^3}{3} + \frac{c^3 b}{3} \right) dx = \frac{abc(b^2+c^2)}{3}$

$= \frac{M}{3}(b^2 + c^2)$ where $M = abc$; $I_y = \frac{M}{3}(a^2 + c^2)$ and $I_z = \frac{M}{3}(a^2 + b^2)$, by symmetry

5. $M = 4 \int_0^1 \int_0^1 \int_{4y^2}^4 dz \, dy \, dx = 4 \int_0^1 \int_0^1 (4 - 4y^2) \, dy \, dx = 16 \int_0^1 \frac{2}{3} \, dx = \frac{32}{3}$; $M_{xy} = 4 \int_0^1 \int_0^1 \int_{4y^2}^4 z \, dz \, dy \, dx$

$= 2 \int_0^1 \int_0^1 (16 - 16y^4) \, dy \, dx = \frac{128}{5} \int_0^1 dx = \frac{128}{5} \Rightarrow \bar{z} = \frac{12}{5}$, and $\bar{x} = \bar{y} = 0$, by symmetry;

$I_x = 4 \int_0^1 \int_0^1 \int_{4y^2}^4 (y^2 + z^2) \, dz \, dy \, dx = 4 \int_0^1 \int_0^1 \left[\left(4y^2 + \frac{64}{3} \right) - \left(4y^4 + \frac{64y^6}{3} \right) \right] dy \, dx = 4 \int_0^1 \frac{1976}{105} \, dx = \frac{7904}{105};$

$I_y = 4 \int_0^1 \int_0^1 \int_{4y^2}^4 (x^2 + z^2) \, dz \, dy \, dx = 4 \int_0^1 \int_0^1 \left[\left(4x^2 + \frac{64}{3} \right) - \left(4x^2 y^2 + \frac{64y^6}{3} \right) \right] dy \, dx = 4 \int_0^1 \left(\frac{8}{3} x^2 + \frac{128}{7} \right) dx$

$= \frac{4832}{63}$; $I_z = 4 \int_0^1 \int_0^1 \int_{4y^2}^4 (x^2 + y^2) \, dz \, dy \, dx = 16 \int_0^1 \int_0^1 (x^2 - x^2 y^2 + y^2 - y^4) \, dy \, dx$

$= 16 \int_0^1 \left(\frac{2x^2}{3} + \frac{2}{15} \right) dx = \frac{256}{45}$

7. (a) $M = 4 \int_0^2 \int_0^{\sqrt{4-x^2}} \int_{x^2+y^2}^4 dz \, dy \, dx = 4 \int_0^{\pi/2} \int_0^2 \int_{r^2}^4 r \, dz \, dr \, d\theta = 4 \int_0^{\pi/2} \int_0^2 (4r - r^3) \, dr \, d\theta = 4 \int_0^{\pi/2} 4 \, d\theta = 8\pi;$

$M_{xy} = \int_0^{2\pi} \int_0^2 \int_{r^2}^4 zr \, dz \, dr \, d\theta = \int_0^{2\pi} \int_0^2 \frac{r}{2}(16 - r^4) \, dr \, d\theta = \frac{32}{3} \int_0^{2\pi} d\theta = \frac{64\pi}{3} \Rightarrow \bar{z} = \frac{8}{3}$, and $\bar{x} = \bar{y} = 0$, by symmetry

(b) $M = 8\pi \Rightarrow 4\pi = \int_0^{2\pi} \int_0^{\sqrt{c}} \int_{r^2}^c r \, dz \, dr \, d\theta = \int_0^{2\pi} \int_0^{\sqrt{c}} (cr - r^3) \, dr \, d\theta = \int_0^{2\pi} \frac{c^2}{4} \, d\theta = \frac{c^2 \pi}{2} \Rightarrow c^2 = 8 \Rightarrow c = 2\sqrt{2}$, since $c > 0$

9. The plane $y + 2z = 2$ is the top of the wedge $\Rightarrow I_L = \int_{-2}^2 \int_{-2}^4 \int_{-1}^{(2-y)/2} [(y-6)^2 + z^2] \, dz \, dy \, dx$

$= \int_{-2}^2 \int_{-2}^4 \left[\frac{(y-6)^2(4-y)}{2} + \frac{(2-y)^3}{24} + \frac{1}{3} \right] dy \, dx$; let $t = 2 - y \Rightarrow I_L = 4 \int_{-2}^4 \left(\frac{13t^3}{24} + 5t^2 + 16t + \frac{49}{3} \right) dt = 1386;$

$M = \frac{1}{2}(3)(6)(4) = 36 \Rightarrow R_L = \sqrt{\frac{I_L}{M}} = \sqrt{\frac{77}{2}}$

11. $M = 8$; $I_L = \int_0^4 \int_0^2 \int_0^1 [z^2 + (y-2)^2]\, dz\, dy\, dx = \int_0^4 \int_0^2 (y^2 - 4y + \frac{13}{3})\, dy\, dx = \frac{10}{3} \int_0^4 dx = \frac{40}{3}$

$\Rightarrow R_L = \sqrt{\frac{I_L}{M}} = \sqrt{\frac{5}{3}}$

13. (a) $M = \int_0^2 \int_0^{2-x} \int_0^{2-x-y} 2x\, dz\, dy\, dx = \int_0^2 \int_0^{2-x} (4x - 2x^2 - 2xy)\, dy\, dx = \int_0^2 (x^3 - 4x^2 + 4x)\, dx = \frac{4}{3}$

(b) $M_{xy} = \int_0^2 \int_0^{2-x} \int_0^{2-x-y} 2xz\, dz\, dy\, dx = \int_0^2 \int_0^{2-x} x(2-x-y)^2\, dy\, dx = \int_0^2 \frac{x(2-x)^3}{3}\, dx = \frac{8}{15}$; $M_{xz} = \frac{8}{15}$ by

symmetry; $M_{yz} = \int_0^2 \int_0^{2-x} \int_0^{2-x-y} 2x^2\, dz\, dy\, dx = \int_0^2 \int_0^{2-x} 2x^2(2-x-y)\, dy\, dx = \int_0^2 (2x - x^2)^2\, dx = \frac{16}{15}$

$\Rightarrow \bar{x} = \frac{4}{5}$, and $\bar{y} = \bar{z} = \frac{2}{5}$

15. (a) $M = \int_0^1 \int_0^1 \int_0^1 (x + y + z + 1)\, dz\, dy\, dx = \int_0^1 \int_0^1 (x + y + \frac{3}{2})\, dy\, dx = \int_0^1 (x + 2)\, dx = \frac{5}{2}$

(b) $M_{xy} = \int_0^1 \int_0^1 \int_0^1 z(x + y + z + 1)\, dz\, dy\, dx = \frac{1}{2} \int_0^1 \int_0^1 (x + y + \frac{5}{3})\, dy\, dx = \frac{1}{2} \int_0^1 (x + \frac{13}{6})\, dx = \frac{4}{3}$

$\Rightarrow M_{xy} = M_{yz} = M_{xz} = \frac{4}{3}$, by symmetry $\Rightarrow \bar{x} = \bar{y} = \bar{z} = \frac{8}{15}$

(c) $I_z = \int_0^1 \int_0^1 \int_0^1 (x^2 + y^2)(x + y + z + 1)\, dz\, dy\, dx = \int_0^1 \int_0^1 (x^2 + y^2)(x + y + \frac{3}{2})\, dy\, dx$

$= \int_0^1 (x^3 + 2x^2 + \frac{1}{3}x + \frac{3}{4})\, dx = \frac{11}{6} \Rightarrow I_x = I_y = I_z = \frac{11}{6}$, by symmetry

(d) $R_x = R_y = R_z = \sqrt{\frac{I_z}{M}} = \sqrt{\frac{11}{15}}$

17. $M = \int_0^1 \int_{z-1}^{1-z} \int_0^{\sqrt{z}} (2y + 5)\, dy\, dx\, dz = \int_0^1 \int_{z-1}^{1-z} (z + 5\sqrt{z})\, dx\, dz = \int_0^1 2(z + 5\sqrt{z})(1 - z)\, dz$

$= 2 \int_0^1 (5z^{1/2} + z - 5z^{3/2} - z^2)\, dz = 2 \left[\frac{10}{3} z^{3/2} + \frac{1}{2} z^2 - 2z^{5/2} - \frac{1}{3} z^3 \right]_0^1 = 2 \left(\frac{9}{3} - \frac{3}{2} \right) = 3$

19. (a) Let ΔV_i be the volume of the ith piece, and let (x_i, y_i, z_i) be a point in the ith piece. Then the work done by gravity in moving the ith piece to the xy-plane is approximately $W_i = m_i g z_i = (x_i + y_i + z_i + 1) g\, \Delta V_i z_i$

$\Rightarrow$ the total work done is the triple integral $W = \int_0^1 \int_0^1 \int_0^1 (x + y + z + 1) g z\, dz\, dy\, dx$

$= g \int_0^1 \int_0^1 \left[\frac{1}{2} x z^2 + \frac{1}{2} y z^2 + \frac{1}{3} z^3 + \frac{1}{2} z^2 \right]_0^1 dy\, dx = g \int_0^1 \int_0^1 \left(\frac{1}{2} x + \frac{1}{2} y + \frac{5}{6} \right) dy\, dx = g \int_0^1 \left[\frac{1}{2} xy + \frac{1}{4} y^2 + \frac{5}{6} y \right]_0^1 dx$

$= g \int_0^1 \left(\frac{1}{2} x + \frac{13}{12} \right) dx = g \left[\frac{x^2}{4} + \frac{13}{12} x \right]_0^1 = g \left(\frac{16}{12} \right) = \frac{4}{3} g$

(b) From Exercise 15 the center of mass is $\left(\frac{8}{15}, \frac{8}{15}, \frac{8}{15} \right)$ and the mass of the liquid is $\frac{5}{2} \Rightarrow$ the work done by gravity in moving the center of mass to the xy-plane is $W = mgd = \left(\frac{5}{2} \right) (g) \left(\frac{8}{15} \right) = \frac{4}{3} g$, which is the same as the work done in part (a).

21. (a) $\bar{x} = \frac{M_{yz}}{M} = 0 \Rightarrow \iiint\limits_R x \delta(x, y, z)\, dx\, dy\, dz = 0 \Rightarrow M_{yz} = 0$

(b) $I_L = \iiint\limits_D |\mathbf{v} - h\mathbf{i}|^2\, dm = \iiint\limits_D |(x - h)\mathbf{i} + y\mathbf{j}|^2\, dm = \iiint\limits_D (x^2 - 2xh + h^2 + y^2)\, dm$

$= \iiint\limits_D (x^2 + y^2)\, dm - 2h \iiint\limits_D x\, dm + h^2 \iiint\limits_D dm = I_x - 0 + h^2 m = I_{c.m.} + h^2 m$

23. (a) $(\bar{x}, \bar{y}, \bar{z}) = \left(\frac{a}{2}, \frac{b}{2}, \frac{c}{2} \right) \Rightarrow I_z = I_{c.m.} + abc \left(\sqrt{\frac{a^2}{4} + \frac{b^2}{4}} \right)^2 \Rightarrow I_{c.m.} = I_z - \frac{abc(a^2 + b^2)}{4}$

$= \frac{abc(a^2 + b^2)}{3} - \frac{abc(a^2 + b^2)}{4} = \frac{abc(a^2 + b^2)}{12}$; $R_{c.m.} = \sqrt{\frac{I_{c.m.}}{M}} = \sqrt{\frac{a^2 + b^2}{12}}$

(b) $I_L = I_{c.m.} + abc \left(\sqrt{\frac{a^2}{4} + \left(\frac{b}{2} - 2b \right)^2} \right)^2 = \frac{abc(a^2 + b^2)}{12} + \frac{abc(a^2 + 9b^2)}{4} = \frac{abc(4a^2 + 28b^2)}{12}$

$= \frac{abc(a^2 + 7b^2)}{3}$; $R_L = \sqrt{\frac{I_L}{M}} = \sqrt{\frac{a^2 + 7b^2}{3}}$

25. $M_{yz_{B_1 \cup B_2}} = \iiint\limits_{B_1} x \, dV_1 + \iiint\limits_{B_2} x \, dV_2 = M_{(yz)_1} + M_{(yz)_2} \Rightarrow \overline{x} = M_{(yz)_1} + M_{(yz)_2 m_1 + m_2}$; similarly,

$\overline{y} = M_{(xz)_1} + M_{(xz)_2 m_1 + m_2}$ and $\overline{z} = M_{(xy)_1} + M_{(xy)_2 m_1 + m_2} \Rightarrow \mathbf{c} = \overline{x}\mathbf{i} + \overline{y}\mathbf{j} + \overline{z}\mathbf{k}$

$= \frac{1}{m_1 + m_2} \left[\left(M_{(yz)_1} + M_{(yz)_2} \right) \mathbf{i} + \left(M_{(xz)_1} + M_{(xz)_2} \right) \mathbf{j} + \left(M_{(xy)_1} + M_{(xy)_2} \right) \mathbf{k} \right]$

$= \frac{1}{m_1 + m_2} \left[(m_1\overline{x}_1 + m_2\overline{x}_2)\mathbf{i} + (m_1\overline{y}_1 + m_2\overline{y}_2)\mathbf{j} + (m_1\overline{z}_1 + m_2\overline{z}_2)\mathbf{k} \right]$

$= \frac{1}{m_1 + m_2} \left[m_1 \left(\overline{x}_1\mathbf{i} + \overline{y}_1\mathbf{j} + \overline{z}_1\mathbf{k} \right) + m_2 \left(\overline{x}_2\mathbf{i} + \overline{y}_2\mathbf{j} + \overline{z}_2\mathbf{k} \right) \right] = \frac{m_1\mathbf{c}_1 + m_2\mathbf{c}_2}{m_1 + m_2}$

27. (a) $\mathbf{c} = \dfrac{\left(\frac{\pi a^2 h}{3} \right) \left(\frac{h}{4}\mathbf{k} \right) + \left(\frac{2\pi a^3}{3} \right) \left(-\frac{3a}{8}\mathbf{k} \right)}{m_1 + m_2} = \dfrac{\left(\frac{a^2 \pi}{3} \right) \left(\frac{h^2 - 3a^2}{4}\mathbf{k} \right)}{m_1 + m_2}$, where $m_1 = \frac{\pi a^2 h}{3}$ and $m_2 = \frac{2\pi a^3}{3}$; if

$\frac{h^2 - 3a^2}{4} = 0$, or $h = a\sqrt{3}$, then the centroid is on the common base

(b) See the solution to Exercise 55, Section 15.2, to see that $h = a\sqrt{2}$.

15.6 TRIPLE INTEGRALS IN CYLINDRICAL AND SPHERICAL COORDINATES

1. $\int_0^{2\pi} \int_0^1 \int_r^{\sqrt{2-r^2}} dz \, r \, dr \, d\theta = \int_0^{2\pi} \int_0^1 \left[r(2 - r^2)^{1/2} - r^2 \right] dr \, d\theta = \int_0^{2\pi} \left[-\frac{1}{3}(2 - r^2)^{3/2} - \frac{r^3}{3} \right]_0^1 d\theta$

$= \int_0^{2\pi} \left(\frac{2^{3/2}}{3} - \frac{2}{3} \right) d\theta = \frac{4\pi \left(\sqrt{2} - 1 \right)}{3}$

3. $\int_0^{2\pi} \int_0^{\theta/2\pi} \int_0^{3 + 24r^2} dz \, r \, dr \, d\theta = \int_0^{2\pi} \int_0^{\theta/2\pi} (3r + 24r^3) \, dr \, d\theta = \int_0^{2\pi} \left[\frac{3}{2}r^2 + 6r^4 \right]_0^{\theta/2\pi} d\theta = \frac{3}{2} \int_0^{2\pi} \left(\frac{\theta^2}{4\pi^2} + \frac{4\theta^4}{16\pi^4} \right) d\theta$

$= \frac{3}{2} \left[\frac{\theta^3}{12\pi^2} + \frac{\theta^5}{20\pi^4} \right]_0^{2\pi} = \frac{17\pi}{5}$

5. $\int_0^{2\pi} \int_0^1 \int_r^{(2-r^2)^{-1/2}} 3 \, dz \, r \, dr \, d\theta = 3 \int_0^{2\pi} \int_0^1 \left[r(2 - r^2)^{-1/2} - r^2 \right] dr \, d\theta = 3 \int_0^{2\pi} \left[-(2 - r^2)^{1/2} - \frac{r^3}{3} \right]_0^1 d\theta$

$= 3 \int_0^{2\pi} \left(\sqrt{2} - \frac{4}{3} \right) d\theta = \pi \left(6\sqrt{2} - 8 \right)$

7. $\int_0^{2\pi} \int_0^3 \int_0^{z/3} r^3 \, dr \, dz \, d\theta = \int_0^{2\pi} \int_0^3 \frac{z^4}{324} \, dz \, d\theta = \int_0^{2\pi} \frac{3}{20} \, d\theta = \frac{3\pi}{10}$

9. $\int_0^1 \int_0^{\sqrt{z}} \int_0^{2\pi} (r^2 \cos^2\theta + z^2) \, r \, d\theta \, dr \, dz = \int_0^1 \int_0^{\sqrt{z}} \left[\frac{r^2\theta}{2} + \frac{r^2\sin 2\theta}{4} + z^2\theta \right]_0^{2\pi} r \, dr \, dz = \int_0^1 \int_0^{\sqrt{z}} (\pi r^3 + 2\pi r z^2) \, dr \, dz$

$= \int_0^1 \left[\frac{\pi r^4}{4} + \pi r^2 z^2 \right]_0^{\sqrt{z}} dz = \int_0^1 \left(\frac{\pi z^2}{4} + \pi z^3 \right) dz = \left[\frac{\pi z^3}{12} + \frac{\pi z^4}{4} \right]_0^1 = \frac{\pi}{3}$

11. (a) $\int_0^{2\pi} \int_0^1 \int_0^{\sqrt{4-r^2}} dz \, r \, dr \, d\theta$

(b) $\int_0^{2\pi} \int_0^{\sqrt{3}} \int_0^1 r \, dr \, dz \, d\theta + \int_0^{2\pi} \int_{\sqrt{3}}^2 \int_0^{\sqrt{4-z^2}} r \, dr \, dz \, d\theta$

(c) $\int_0^1 \int_0^{\sqrt{4-r^2}} \int_0^{2\pi} r \, d\theta \, dz \, dr$

13. $\int_{-\pi/2}^{\pi/2} \int_0^{\cos\theta} \int_0^{3r^2} f(r, \theta, z) \, dz \, r \, dr \, d\theta$

15. $\int_0^{\pi} \int_0^{2\sin\theta} \int_0^{4 - r\sin\theta} f(r, \theta, z) \, dz \, r \, dr \, d\theta$

17. $\int_{-\pi/2}^{\pi/2}\int_{1}^{1+\cos\theta}\int_{0}^{4} f(r,\theta,z)\, dz\, r\, dr\, d\theta$

19. $\int_{0}^{\pi/4}\int_{0}^{\sec\theta}\int_{0}^{2-r\sin\theta} f(r,\theta,z)\, dz\, r\, dr\, d\theta$

21. $\int_{0}^{\pi}\int_{0}^{\pi}\int_{0}^{2\sin\phi}\rho^2\sin\phi\, d\rho\, d\phi\, d\theta = \frac{8}{3}\int_{0}^{\pi}\int_{0}^{\pi}\sin^4\phi\, d\phi\, d\theta = \frac{8}{3}\int_{0}^{\pi}\left(\left[-\frac{\sin^3\phi\cos\phi}{4}\right]_{0}^{\pi} + \frac{3}{4}\int_{0}^{\pi}\sin^2\phi\, d\phi\right)d\theta$

$= 2\int_{0}^{\pi}\int_{0}^{\pi}\sin^2\phi\, d\phi\, d\theta = \int_{0}^{\pi}\left[\theta - \frac{\sin 2\theta}{2}\right]_{0}^{\pi}d\theta = \int_{0}^{\pi}\pi\, d\theta = \pi^2$

23. $\int_{0}^{2\pi}\int_{0}^{\pi}\int_{0}^{(1-\cos\phi)/2}\rho^2\sin\phi\, d\rho\, d\phi\, d\theta = \frac{1}{24}\int_{0}^{2\pi}\int_{0}^{\pi}(1-\cos\phi)^3\sin\phi\, d\phi\, d\theta = \frac{1}{96}\int_{0}^{2\pi}[(1-\cos\phi)^4]_{0}^{\pi}\, d\theta$

$= \frac{1}{96}\int_{0}^{2\pi}(2^4 - 0)\, d\theta = \frac{16}{96}\int_{0}^{2\pi}d\theta = \frac{1}{6}(2\pi) = \frac{\pi}{3}$

25. $\int_{0}^{2\pi}\int_{0}^{\pi/3}\int_{\sec\phi}^{2}3\rho^2\sin\phi\, d\rho\, d\phi\, d\theta = \int_{0}^{2\pi}\int_{0}^{\pi/3}(8-\sec^3\phi)\sin\phi\, d\phi\, d\theta = \int_{0}^{2\pi}\left[-8\cos\phi - \frac{1}{2}\sec^2\phi\right]_{0}^{\pi/3}d\theta$

$= \int_{0}^{2\pi}\left[(-4-2) - (-8 - \frac{1}{2})\right]d\theta = \frac{5}{2}\int_{0}^{2\pi}d\theta = 5\pi$

27. $\int_{0}^{2}\int_{-\pi}^{0}\int_{\pi/4}^{\pi/2}\rho^3\sin 2\phi\, d\phi\, d\theta\, d\rho = \int_{0}^{2}\int_{-\pi}^{0}\rho^3\left[-\frac{\cos 2\phi}{2}\right]_{\pi/4}^{\pi/2}d\theta\, d\rho = \int_{0}^{2}\int_{-\pi}^{0}\frac{\rho^3}{2}\, d\theta\, d\rho = \int_{0}^{2}\frac{\rho^3\pi}{2}\, d\rho$

$= \left[\frac{\pi\rho^4}{8}\right]_{0}^{2} = 2\pi$

29. $\int_{0}^{1}\int_{0}^{\pi}\int_{0}^{\pi/4}12\rho\sin^3\phi\, d\phi\, d\theta\, d\rho = \int_{0}^{1}\int_{0}^{\pi}\left(12\rho\left[\frac{-\sin^2\phi\cos\phi}{3}\right]_{0}^{\pi/4} + 8\rho\int_{0}^{\pi/4}\sin\phi\, d\phi\right)d\theta\, d\rho$

$= \int_{0}^{1}\int_{0}^{\pi}\left(-\frac{2\rho}{\sqrt{2}} - 8\rho[\cos\phi]_{0}^{\pi/4}\right)d\theta\, d\rho = \int_{0}^{1}\int_{0}^{\pi}\left(8\rho - \frac{10\rho}{\sqrt{2}}\right)d\theta\, d\rho = \pi\int_{0}^{1}\left(8\rho - \frac{10\rho}{\sqrt{2}}\right)d\rho = \pi\left[4\rho^2 - \frac{5\rho^2}{\sqrt{2}}\right]_{0}^{1}$

$= \frac{(4\sqrt{2}-5)\pi}{\sqrt{2}}$

31. (a) $x^2 + y^2 = 1 \Rightarrow \rho^2\sin^2\phi = 1$, and $\rho\sin\phi = 1 \Rightarrow \rho = \csc\phi$; thus

$\int_{0}^{2\pi}\int_{0}^{\pi/6}\int_{0}^{2}\rho^2\sin\phi\, d\rho\, d\phi\, d\theta + \int_{0}^{2\pi}\int_{\pi/6}^{\pi/2}\int_{0}^{\csc\phi}\rho^2\sin\phi\, d\rho\, d\phi\, d\theta$

(b) $\int_{0}^{2\pi}\int_{1}^{2}\int_{\pi/6}^{\sin^{-1}(1/\rho)}\rho^2\sin\phi\, d\phi\, d\rho\, d\theta + \int_{0}^{2\pi}\int_{0}^{2}\int_{0}^{\pi/6}\rho^2\sin\phi\, d\phi\, d\rho\, d\theta$

33. $V = \int_{0}^{2\pi}\int_{0}^{\pi/2}\int_{\cos\phi}^{2}\rho^2\sin\phi\, d\rho\, d\phi\, d\theta = \frac{1}{3}\int_{0}^{2\pi}\int_{0}^{\pi/2}(8-\cos^3\phi)\sin\phi\, d\phi\, d\theta$

$= \frac{1}{3}\int_{0}^{2\pi}\left[-8\cos\phi + \frac{\cos^4\phi}{4}\right]_{0}^{\pi/2}d\theta = \frac{1}{3}\int_{0}^{2\pi}\left(8 - \frac{1}{4}\right)d\theta = \left(\frac{31}{12}\right)(2\pi) = \frac{31\pi}{6}$

35. $V = \int_{0}^{2\pi}\int_{0}^{\pi}\int_{0}^{1-\cos\phi}\rho^2\sin\phi\, d\rho\, d\phi\, d\theta = \frac{1}{3}\int_{0}^{2\pi}\int_{0}^{\pi}(1-\cos\phi)^3\sin\phi\, d\phi\, d\theta = \frac{1}{3}\int_{0}^{2\pi}\left[\frac{(1-\cos\phi)^4}{4}\right]_{0}^{\pi}d\theta$

$= \frac{1}{12}(2)^4\int_{0}^{2\pi}d\theta = \frac{4}{3}(2\pi) = \frac{8\pi}{3}$

37. $V = \int_{0}^{2\pi}\int_{\pi/4}^{\pi/2}\int_{0}^{2\cos\phi}\rho^2\sin\phi\, d\rho\, d\phi\, d\theta = \frac{8}{3}\int_{0}^{2\pi}\int_{\pi/4}^{\pi/2}\cos^3\phi\sin\phi\, d\phi\, d\theta = \frac{8}{3}\int_{0}^{2\pi}\left[-\frac{\cos^4\phi}{4}\right]_{\pi/4}^{\pi/2}d\theta$

$= \left(\frac{8}{3}\right)\left(\frac{1}{16}\right)\int_{0}^{2\pi}d\theta = \frac{1}{6}(2\pi) = \frac{\pi}{3}$

39. (a) $8\int_{0}^{\pi/2}\int_{0}^{\pi/2}\int_{0}^{2}\rho^2\sin\phi\, d\rho\, d\phi\, d\theta$ (b) $8\int_{0}^{\pi/2}\int_{0}^{2}\int_{0}^{\sqrt{4-r^2}} dz\, r\, dr\, d\theta$

(c) $8\int_{0}^{2}\int_{0}^{\sqrt{4-x^2}}\int_{0}^{\sqrt{4-x^2-y^2}} dz\, dy\, dx$

41. (a) $V = \int_0^{2\pi} \int_0^{\pi/3} \int_{\sec\phi}^2 \rho^2 \sin\phi \, d\rho \, d\phi \, d\theta$
    (b) $V = \int_0^{2\pi} \int_0^{\sqrt{3}} \int_1^{\sqrt{4-r^2}} dz \, r \, dr \, d\theta$

(c) $V = \int_{-\sqrt{3}}^{\sqrt{3}} \int_{-\sqrt{3-x^2}}^{\sqrt{3-x^2}} \int_1^{\sqrt{4-x^2-y^2}} dz \, dy \, dx$

(d) $V = \int_0^{2\pi} \int_0^{\sqrt{3}} \left[r(4-r^2)^{1/2} - r \right] dr \, d\theta = \int_0^{2\pi} \left[-\frac{(4-r^2)^{3/2}}{3} - \frac{r^2}{2} \right]_0^{\sqrt{3}} d\theta = \int_0^{2\pi} \left(-\frac{1}{3} - \frac{3}{2} + \frac{4^{3/2}}{3} \right) d\theta$

$\qquad = \frac{5}{6} \int_0^{2\pi} d\theta = \frac{5\pi}{3}$

43. $V = 4 \int_0^{\pi/2} \int_0^1 \int_{r^4-1}^{4-4r^2} dz \, r \, dr \, d\theta = 4 \int_0^{\pi/2} \int_0^1 (5r - 4r^3 - r^5) \, dr \, d\theta = 4 \int_0^{\pi/2} \left(\frac{5}{2} - 1 - \frac{1}{6} \right) d\theta$

$\qquad = 4 \int_0^{\pi/2} \frac{8}{6} \, d\theta = \frac{8\pi}{3}$

45. $V = \int_{3\pi/2}^{2\pi} \int_0^{3\cos\theta} \int_0^{-r\sin\theta} dz \, r \, dr \, d\theta = \int_{3\pi/2}^{2\pi} \int_0^{3\cos\theta} -r^2 \sin\theta \, dr \, d\theta = \int_{3\pi/2}^{2\pi} (-9\cos^3\theta)(\sin\theta) \, d\theta$

$\qquad = \left[\frac{9}{4} \cos^4\theta \right]_{3\pi/2}^{2\pi} = \frac{9}{4} - 0 = \frac{9}{4}$

47. $V = \int_0^{\pi/2} \int_0^{\sin\theta} \int_0^{\sqrt{1-r^2}} dz \, r \, dr \, d\theta = \int_0^{\pi/2} \int_0^{\sin\theta} r\sqrt{1-r^2} \, dr \, d\theta = \int_0^{\pi/2} \left[-\frac{1}{3}(1-r^2)^{3/2} \right]_0^{\sin\theta} d\theta$

$\qquad = -\frac{1}{3} \int_0^{\pi/2} \left[(1-\sin^2\theta)^{3/2} - 1 \right] d\theta = -\frac{1}{3} \int_0^{\pi/2} (\cos^3\theta - 1) \, d\theta = -\frac{1}{3} \left(\left[\frac{\cos^2\theta \sin\theta}{3} \right]_0^{\pi/2} + \frac{2}{3} \int_0^{\pi/2} \cos\theta \, d\theta \right) + \left[\frac{\theta}{3} \right]_0^{\pi/2}$

$\qquad = -\frac{2}{9} [\sin\theta]_0^{\pi/2} + \frac{\pi}{6} = \frac{-4+3\pi}{18}$

49. $V = \int_0^{2\pi} \int_{\pi/3}^{2\pi/3} \int_0^a \rho^2 \sin\phi \, d\rho \, d\phi \, d\theta = \int_0^{2\pi} \int_{\pi/3}^{2\pi/3} \frac{a^3}{3} \sin\phi \, d\phi \, d\theta = \frac{a^3}{3} \int_0^{2\pi} [-\cos\phi]_{\pi/3}^{2\pi/3} \, d\theta = \frac{a^3}{3} \int_0^{2\pi} \left(\frac{1}{2} + \frac{1}{2} \right) d\theta = \frac{2\pi a^3}{3}$

51. $V = \int_0^{2\pi} \int_0^{\pi/3} \int_{\sec\phi}^2 \rho^2 \sin\phi \, d\rho \, d\phi \, d\theta$

$\qquad = \frac{1}{3} \int_0^{2\pi} \int_0^{\pi/3} (8\sin\phi - \tan\phi\sec^2\phi) \, d\phi \, d\theta$

$\qquad = \frac{1}{3} \int_0^{2\pi} \left[-8\cos\phi - \frac{1}{2}\tan^2\phi \right]_0^{\pi/3} d\theta$

$\qquad = \frac{1}{3} \int_0^{2\pi} \left[-4 - \frac{1}{2}(3) + 8 \right] d\theta = \frac{1}{3} \int_0^{2\pi} \frac{5}{2} \, d\theta = \frac{5}{6}(2\pi) = \frac{5\pi}{3}$

53. $V = 4 \int_0^{\pi/2} \int_0^1 \int_0^{r^2} dz \, r \, dr \, d\theta = 4 \int_0^{\pi/2} \int_0^1 r^3 \, dr \, d\theta = \int_0^{\pi/2} d\theta = \frac{\pi}{2}$

55. $V = 8 \int_0^{\pi/2} \int_1^{\sqrt{2}} \int_0^r dz \, r \, dr \, d\theta = 8 \int_0^{\pi/2} \int_1^{\sqrt{2}} r^2 \, dr \, d\theta = 8 \left(\frac{2\sqrt{2}-1}{3} \right) \int_0^{\pi/2} d\theta = \frac{4\pi(2\sqrt{2}-1)}{3}$

57. $V = \int_0^{2\pi} \int_0^2 \int_0^{4-r\sin\theta} dz \, r \, dr \, d\theta = \int_0^{2\pi} \int_0^2 (4r - r^2\sin\theta) \, dr \, d\theta = 8 \int_0^{2\pi} \left(1 - \frac{\sin\theta}{3} \right) d\theta = 16\pi$

59. The paraboloids intersect when $4x^2 + 4y^2 = 5 - x^2 - y^2 \Rightarrow x^2 + y^2 = 1$ and $z = 4$

$\qquad \Rightarrow V = 4 \int_0^{\pi/2} \int_0^1 \int_{4r^2}^{5-r^2} dz \, r \, dr \, d\theta = 4 \int_0^{\pi/2} \int_0^1 (5r - 5r^3) \, dr \, d\theta = 20 \int_0^{\pi/2} \left[\frac{r^2}{2} - \frac{r^4}{4} \right]_0^1 d\theta = 5 \int_0^{\pi/2} d\theta = \frac{5\pi}{2}$

61. $V = 8 \int_0^{2\pi} \int_0^1 \int_0^{\sqrt{4-r^2}} dz \, r \, dr \, d\theta = 8 \int_0^{2\pi} \int_0^1 r(4-r^2)^{1/2} \, dr \, d\theta = 8 \int_0^{2\pi} \left[-\frac{1}{3}(4-r^2)^{3/2} \right]_0^1 d\theta$

$\qquad = -\frac{8}{3} \int_0^{2\pi} (3^{3/2} - 8) \, d\theta = \frac{4\pi(8-3\sqrt{3})}{3}$

63. average $= \frac{1}{2\pi} \int_0^{2\pi} \int_0^1 \int_{-1}^1 r^2 \, dz \, dr \, d\theta = \frac{1}{2\pi} \int_0^{2\pi} \int_0^1 2r^2 \, dr \, d\theta = \frac{1}{3\pi} \int_0^{2\pi} d\theta = \frac{2}{3}$

65. average $= \frac{1}{\left(\frac{4\pi}{3}\right)} \int_0^{2\pi} \int_0^{\pi} \int_0^1 \rho^3 \sin\phi \, d\rho \, d\phi \, d\theta = \frac{3}{16\pi} \int_0^{2\pi} \int_0^{\pi} \sin\phi \, d\phi \, d\theta = \frac{3}{8\pi} \int_0^{2\pi} d\theta = \frac{3}{4}$

67. $M = 4 \int_0^{\pi/2} \int_0^1 \int_0^r dz \, r \, dr \, d\theta = 4 \int_0^{\pi/2} \int_0^1 r^2 \, dr \, d\theta = \frac{4}{3} \int_0^{\pi/2} d\theta = \frac{2\pi}{3}$; $M_{xy} = \int_0^{2\pi} \int_0^1 \int_0^r z \, dz \, r \, dr \, d\theta$

$= \frac{1}{2} \int_0^{2\pi} \int_0^1 r^3 \, dr \, d\theta = \frac{1}{8} \int_0^{2\pi} d\theta = \frac{\pi}{4} \Rightarrow \bar{z} = \frac{M_{xy}}{M} = \left(\frac{\pi}{4}\right)\left(\frac{3}{2\pi}\right) = \frac{3}{8}$, and $\bar{x} = \bar{y} = 0$, by symmetry

69. $M = \frac{8\pi}{3}$; $M_{xy} = \int_0^{2\pi} \int_{\pi/3}^{\pi/2} \int_0^2 z\rho^2 \sin\phi \, d\rho \, d\phi \, d\theta = \int_0^{2\pi} \int_{\pi/3}^{\pi/2} \int_0^2 \rho^3 \cos\phi \sin\phi \, d\rho \, d\phi \, d\theta = 4 \int_0^{2\pi} \int_{\pi/3}^{\pi/2} \cos\phi \sin\phi \, d\phi \, d\theta$

$= 4 \int_0^{2\pi} \left[\frac{\sin^2\phi}{2}\right]_{\pi/3}^{\pi/2} d\theta = 4 \int_0^{2\pi} \left(\frac{1}{2} - \frac{3}{8}\right) d\theta = \frac{1}{2} \int_0^{2\pi} d\theta = \pi \Rightarrow \bar{z} = \frac{M_{xy}}{M} = (\pi)\left(\frac{3}{8\pi}\right) = \frac{3}{8}$, and $\bar{x} = \bar{y} = 0$,

by symmetry

71. $M = \int_0^{2\pi} \int_0^4 \int_0^{\sqrt{r}} dz \, r \, dr \, d\theta = \int_0^{2\pi} \int_0^4 r^{3/2} \, dr \, d\theta = \frac{64}{5} \int_0^{2\pi} d\theta = \frac{128\pi}{5}$; $M_{xy} = \int_0^{2\pi} \int_0^4 \int_0^{\sqrt{r}} z \, dz \, r \, dr \, d\theta$

$= \frac{1}{2} \int_0^{2\pi} \int_0^4 r^2 \, dr \, d\theta = \frac{32}{3} \int_0^{2\pi} d\theta = \frac{64\pi}{3} \Rightarrow \bar{z} = \frac{M_{xy}}{M} = \frac{5}{6}$, and $\bar{x} = \bar{y} = 0$, by symmetry

73. $I_z = \int_0^{2\pi} \int_1^2 \int_0^4 (x^2 + y^2) \, dz \, r \, dr \, d\theta = 4 \int_0^{2\pi} \int_1^2 r^3 \, dr \, d\theta = \int_0^{2\pi} 15 \, d\theta = 30\pi; \ M = \int_0^{2\pi} \int_1^2 \int_0^4 dz \, r \, dr \, d\theta$

$= \int_0^{2\pi} \int_1^2 4r \, dr \, d\theta = \int_0^{2\pi} 6 \, d\theta = 12\pi \Rightarrow R_z = \sqrt{\frac{I_z}{M}} = \sqrt{\frac{5}{2}}$

75. We orient the cone with its vertex at the origin and axis along the z-axis $\Rightarrow \phi = \frac{\pi}{4}$. We use the the x-axis

which is through the vertex and parallel to the base of the cone $\Rightarrow I_x = \int_0^{2\pi} \int_0^1 \int_r^1 (r^2 \sin^2\theta + z^2) \, dz \, r \, dr \, d\theta$

$= \int_0^{2\pi} \int_0^1 \left(r^3 \sin^2\theta - r^4 \sin^2\theta + \frac{r}{3} - \frac{r^4}{3}\right) dr \, d\theta = \int_0^{2\pi} \left(\frac{\sin^2\theta}{20} + \frac{1}{10}\right) d\theta = \left[\frac{\theta}{40} - \frac{\sin 2\theta}{80} + \frac{\theta}{10}\right]_0^{2\pi} = \frac{\pi}{20} + \frac{\pi}{5} = \frac{\pi}{4}$

77. $I_z = \int_0^{2\pi} \int_0^a \int_{\left(\frac{h}{a}\right)r}^h (x^2 + y^2) \, dz \, r \, dr \, d\theta = \int_0^{2\pi} \int_0^a \int_{\frac{hr}{a}}^h r^3 \, dz \, dr \, d\theta = \int_0^{2\pi} \int_0^a \left(hr^3 - \frac{hr^4}{a}\right) dr \, d\theta$

$= \int_0^{2\pi} h \left[\frac{r^4}{4} - \frac{r^5}{5a}\right]_0^a d\theta = \int_0^{2\pi} h \left(\frac{a^4}{4} - \frac{a^5}{5a}\right) d\theta = \frac{ha^4}{20} \int_0^{2\pi} d\theta = \frac{\pi ha^4}{10}$

79. (a) $M = \int_0^{2\pi} \int_0^1 \int_r^1 z \, dz \, r \, dr \, d\theta = \frac{1}{2} \int_0^{2\pi} \int_0^1 (r - r^3) \, dr \, d\theta = \frac{1}{8} \int_0^{2\pi} d\theta = \frac{\pi}{4}$; $M_{xy} = \int_0^{2\pi} \int_0^1 \int_r^1 z^2 \, dz \, r \, dr \, d\theta$

$= \frac{1}{3} \int_0^{2\pi} \int_0^1 (r - r^4) \, dr \, d\theta = \frac{1}{10} \int_0^{2\pi} d\theta = \frac{\pi}{5} \Rightarrow \bar{z} = \frac{4}{5}$, and $\bar{x} = \bar{y} = 0$, by symmetry; $I_z = \int_0^{2\pi} \int_0^1 \int_r^1 zr^3 \, dz \, dr \, d\theta$

$= \frac{1}{2} \int_0^{2\pi} \int_0^1 (r^3 - r^5) \, dr \, d\theta = \frac{1}{24} \int_0^{2\pi} d\theta = \frac{\pi}{12} \Rightarrow R_z = \sqrt{\frac{I_z}{M}} = \sqrt{\frac{1}{3}}$

(b) $M = \int_0^{2\pi} \int_0^1 \int_r^1 z^2 \, dz \, r \, dr \, d\theta = \frac{\pi}{5}$ from part (a); $M_{xy} = \int_0^{2\pi} \int_0^1 \int_r^1 z^3 \, dz \, r \, dr \, d\theta = \frac{1}{4} \int_0^{2\pi} \int_0^1 (r - r^5) \, dr \, d\theta$

$= \frac{1}{12} \int_0^{2\pi} d\theta = \frac{\pi}{6} \Rightarrow \bar{z} = \frac{5}{6}$, and $\bar{x} = \bar{y} = 0$, by symmetry; $I_z = \int_0^{2\pi} \int_0^1 \int_r^1 z^2 r^3 \, dz \, dr \, d\theta = \frac{1}{3} \int_0^{2\pi} \int_0^1 (r^3 - r^6) \, dr \, d\theta$

$= \frac{1}{28} \int_0^{2\pi} d\theta = \frac{\pi}{14} \Rightarrow R_z = \sqrt{\frac{I_z}{M}} = \sqrt{\frac{5}{14}}$

81. $M = \int_0^{2\pi} \int_0^a \int_0^{\frac{h}{a}\sqrt{a^2-r^2}} dz \, r \, dr \, d\theta = \int_0^{2\pi} \int_0^a \frac{h}{a} r\sqrt{a^2 - r^2} \, dr \, d\theta = \frac{h}{a} \int_0^{2\pi} \left[-\frac{1}{3}(a^2 - r^2)^{3/2}\right]_0^a d\theta$

$= \frac{h}{a} \int_0^{2\pi} \frac{a^3}{3} \, d\theta = \frac{2ha^2\pi}{3}$; $M_{xy} = \int_0^{2\pi} \int_0^a \int_0^{\frac{h}{a}\sqrt{a^2-r^2}} z \, dz \, r \, dr \, d\theta = \frac{h^2}{2a^2} \int_0^{2\pi} \int_0^a (a^2 r - r^3) \, dr \, d\theta$

$= \frac{h^2}{2a^2} \int_0^{2\pi} \left(\frac{a^4}{2} - \frac{a^4}{4}\right) d\theta = \frac{a^2 h^2 \pi}{4} \Rightarrow \bar{z} = \left(\frac{\pi a^2 h^2}{4}\right)\left(\frac{3}{2ha^2\pi}\right) = \frac{3}{8} h$, and $\bar{x} = \bar{y} = 0$, by symmetry

83. $M = \int_0^{2\pi} \int_0^a \int_0^h (z+1)\, dz\, r\, dr\, d\theta = \int_0^{2\pi} \int_0^a \left(\frac{h^2}{2}+h\right) r\, dr\, d\theta = \frac{a^2\,(h^2+2h)}{4} \int_0^{2\pi} d\theta = \frac{\pi a^2\,(h^2+2h)}{2}$;

$M_{xy} = \int_0^{2\pi} \int_0^a \int_0^h (z^2+z)\, dz\, r\, dr\, d\theta = \int_0^{2\pi} \int_0^a \left(\frac{h^3}{3}+\frac{h^2}{2}\right) r\, dr\, d\theta = \frac{a^2\,(2h^3+3h^2)}{12} \int_0^{2\pi} d\theta = \frac{\pi a^2\,(2h^3+3h^2)}{6}$

$\Rightarrow \bar{z} = \left[\frac{\pi a^2\,(2h^3+3h^2)}{6}\right]\left[\frac{2}{\pi a^2\,(h^2+2h)}\right] = \frac{2h^2+3h}{3h+6}$, and $\bar{x} = \bar{y} = 0$, by symmetry;

$I_z = \int_0^{2\pi} \int_0^a \int_0^h (z+1) r^3\, dz\, dr\, d\theta = \left(\frac{h^2+2h}{2}\right) \int_0^{2\pi} \int_0^a r^3\, dr\, d\theta = \left(\frac{h^2+2h}{2}\right)\left(\frac{a^4}{4}\right) \int_0^{2\pi} d\theta = \frac{\pi a^4\,(h^2+2h)}{4}$;

$R_z = \sqrt{\frac{I_z}{M}} = \sqrt{\frac{\pi a^4\,(h^2+2h)}{4} \cdot \frac{2}{\pi a^2\,(h^2+2h)}} = \frac{a}{\sqrt{2}}$

85. The density distribution function is linear so it has the form $\delta(\rho) = k\rho + C$, where ρ is the distance from the center of the planet. Now, $\delta(R) = 0 \Rightarrow kR + C = 0$, and $\delta(\rho) = k\rho - kR$. It remains to determine the constant

k: $M = \int_0^{2\pi} \int_0^\pi \int_0^R (k\rho - kR)\, \rho^2 \sin\phi\, d\rho\, d\phi\, d\theta = \int_0^{2\pi} \int_0^\pi \left[k\frac{\rho^4}{4} - kR\frac{\rho^3}{3}\right]_0^R \sin\phi\, d\phi\, d\theta$

$= \int_0^{2\pi} \int_0^\pi k\left(\frac{R^4}{4} - \frac{R^4}{3}\right) \sin\phi\, d\phi\, d\theta = \int_0^{2\pi} -\frac{k}{12} R^4 \left[-\cos\phi\right]_0^\pi d\theta = \int_0^{2\pi} -\frac{k}{6} R^4\, d\theta = -\frac{k\pi R^4}{3} \Rightarrow k = -\frac{3M}{\pi R^4}$

$\Rightarrow \delta(\rho) = -\frac{3M}{\pi R^4} \rho + \frac{3M}{\pi R^4} R$. At the center of the planet $\rho = 0 \Rightarrow \delta(0) = \left(\frac{3M}{\pi R^4}\right) R = \frac{3M}{\pi R^3}$.

87. (a) A plane perpendicular to the x-axis has the form $x = a$ in rectangular coordinates $\Rightarrow r\cos\theta = a \Rightarrow r = \frac{a}{\cos\theta}$
$\Rightarrow r = a\sec\theta$, in cylindrical coordinates.

(b) A plane perpendicular to the y-axis has the form $y = b$ in rectangular coordinates $\Rightarrow r\sin\theta = b \Rightarrow r = \frac{b}{\sin\theta}$
$\Rightarrow r = b\csc\theta$, in cylindrical coordinates.

89. The equation $r = f(z)$ implies that the point (r, θ, z)
$= (f(z), \theta, z)$ will lie on the surface for all θ. In particular
$(f(z), \theta + \pi, z)$ lies on the surface whenever $(f(z), \theta, z)$ does
$\Rightarrow$ the surface is symmetric with respect to the z-axis.

15.7 SUBSTITUTIONS IN MULTIPLE INTEGRALS

1. (a) $x - y = u$ and $2x + y = v \Rightarrow 3x = u + v$ and $y = x - u \Rightarrow x = \frac{1}{3}(u+v)$ and $y = \frac{1}{3}(-2u+v)$;

$\frac{\partial(x,y)}{\partial(u,v)} = \begin{vmatrix} \frac{1}{3} & \frac{1}{3} \\ -\frac{2}{3} & \frac{1}{3} \end{vmatrix} = \frac{1}{9} + \frac{2}{9} = \frac{1}{3}$

(b) The line segment $y = x$ from $(0,0)$ to $(1,1)$ is $x - y = 0$
$\Rightarrow u = 0$; the line segment $y = -2x$ from $(0,0)$ to
$(1,-2)$ is $2x + y = 0 \Rightarrow v = 0$; the line segment $x = 1$
from $(1,1)$ to $(1,-2)$ is $(x-y) + (2x+y) = 3$
$\Rightarrow u + v = 3$. The transformed region is sketched at the
right.

3. (a) $3x + 2y = u$ and $x + 4y = v \Rightarrow -5x = -2u + v$ and $y = \frac{1}{2}(u - 3x) \Rightarrow x = \frac{1}{5}(2u - v)$ and $y = \frac{1}{10}(3v - u)$;

$\frac{\partial(x,y)}{\partial(u,v)} = \begin{vmatrix} \frac{2}{5} & -\frac{1}{5} \\ -\frac{1}{10} & \frac{3}{10} \end{vmatrix} = \frac{6}{50} - \frac{1}{50} = \frac{1}{10}$

(b) The x-axis $y = 0 \Rightarrow u = 3v$; the y-axis $x = 0$
$\Rightarrow v = 2u$; the line $x + y = 1$
$\Rightarrow \frac{1}{5}(2u - v) + \frac{1}{10}(3v - u) = 1$
$\Rightarrow 2(2u - v) + (3v - u) = 10 \Rightarrow 3u + v = 10$. The
transformed region is sketched at the right.

5. $\int_0^4 \int_{y/2}^{(y/2)+1} \left(x - \frac{y}{2}\right) dx\, dy = \int_0^4 \left[\frac{x^2}{2} - \frac{xy}{2}\right]_{\frac{y}{2}}^{\frac{y}{2}+1} dy = \frac{1}{2}\int_0^4 \left[\left(\frac{y}{2}+1\right)^2 - \left(\frac{y}{2}\right)^2 - \left(\frac{y}{2}+1\right)y + \left(\frac{y}{2}\right)y\right] dy$

$= \frac{1}{2}\int_0^4 (y + 1 - y)\, dy = \frac{1}{2}\int_0^4 dy = \frac{1}{2}(4) = 2$

7. $\iint_R (3x^2 + 14xy + 8y^2)\, dx\, dy$

$= \iint_R (3x + 2y)(x + 4y)\, dx\, dy$

$= \iint_G uv \left|\frac{\partial(x,y)}{\partial(u,v)}\right| du\, dv = \frac{1}{10}\iint_G uv\, du\, dv;$

We find the boundaries of G from the boundaries of R,
shown in the accompanying figure:

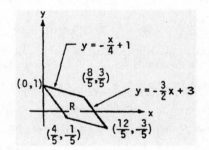

xy-equations for the boundary of R	Corresponding uv-equations for the boundary of G	Simplified uv-equations
$y = -\frac{3}{2}x + 1$	$\frac{1}{10}(3v - u) = -\frac{3}{10}(2u - v) + 1$	$u = 2$
$y = -\frac{3}{2}x + 3$	$\frac{1}{10}(3v - u) = -\frac{3}{10}(2u - v) + 3$	$u = 6$
$y = -\frac{1}{4}x$	$\frac{1}{10}(3v - u) = -\frac{1}{20}(2u - v)$	$v = 0$
$y = -\frac{1}{4}x + 1$	$\frac{1}{10}(3v - u) = -\frac{1}{20}(2u - v) + 1$	$v = 4$

$\Rightarrow \frac{1}{10}\iint_G uv\, du\, dv = \frac{1}{10}\int_2^6 \int_0^4 uv\, dv\, du = \frac{1}{10}\int_2^6 u\left[\frac{v^2}{2}\right]_0^4 du = \frac{4}{5}\int_2^6 u\, du = \left(\frac{4}{5}\right)\left[\frac{u^2}{2}\right]_2^6 = \left(\frac{4}{5}\right)(18 - 2) = \frac{64}{5}$

9. $x = \frac{u}{v}$ and $y = uv \Rightarrow \frac{y}{x} = v^2$ and $xy = u^2$; $\frac{\partial(x,y)}{\partial(u,v)} = J(u, v) = \begin{vmatrix} v^{-1} & -uv^{-2} \\ v & u \end{vmatrix} = v^{-1}u + v^{-1}u = \frac{2u}{v}$;

$y = x \Rightarrow uv = \frac{u}{v} \Rightarrow v = 1$, and $y = 4x \Rightarrow v = 2$; $xy = 1 \Rightarrow u = 1$, and $xy = 9 \Rightarrow u = 3$; thus

$\iint_R \left(\sqrt{\frac{y}{x}} + \sqrt{xy}\right) dx\, dy = \int_1^3 \int_1^2 (v + u)\left(\frac{2u}{v}\right) dv\, du = \int_1^3 \int_1^2 \left(2u + \frac{2u^2}{v}\right) dv\, du = \int_1^3 [2uv + 2u^2 \ln v]_1^2\, du$

$= \int_1^3 (2u + 2u^2 \ln 2)\, du = \left[u^2 + \frac{2}{3}u^2 \ln 2\right]_1^3 = 8 + \frac{2}{3}(26)(\ln 2) = 8 + \frac{52}{3}(\ln 2)$

11. $x = ar\cos\theta$ and $y = br\sin\theta \Rightarrow \frac{\partial(x,y)}{\partial(r,\theta)} = J(r, \theta) = \begin{vmatrix} a\cos\theta & -ar\sin\theta \\ b\sin\theta & br\cos\theta \end{vmatrix} = abr\cos^2\theta + abr\sin^2\theta = abr;$

$I_0 = \iint_R (x^2 + y^2)\, dA = \int_0^{2\pi}\int_0^1 r^2(a^2\cos^2\theta + b^2\sin^2\theta)\, |J(r, \theta)|\, dr\, d\theta = \int_0^{2\pi}\int_0^1 abr^3(a^2\cos^2\theta + b^2\sin^2\theta)\, dr\, d\theta$

$= \frac{ab}{4}\int_0^{2\pi}(a^2\cos^2\theta + b^2\sin^2\theta)\, d\theta = \frac{ab}{4}\left[\frac{a^2\theta}{2} + \frac{a^2\sin 2\theta}{4} + \frac{b^2\theta}{2} - \frac{b^2\sin 2\theta}{4}\right]_0^{2\pi} = \frac{ab\pi(a^2 + b^2)}{4}$

13. The region of integration R in the xy-plane is sketched in the figure at the right. The boundaries of the image G are obtained as follows, with G sketched at the right:

xy-equations for the boundary of R	Corresponding uv-equations for the boundary of G	Simplified uv-equations
$x = y$	$\frac{1}{3}(u + 2v) = \frac{1}{3}(u - v)$	$v = 0$
$x = 2 - 2y$	$\frac{1}{3}(u + 2v) = 2 - \frac{2}{3}(u - v)$	$u = 2$
$y = 0$	$0 = \frac{1}{3}(u - v)$	$v = u$

Also, from Exercise 2, $\frac{\partial(x,y)}{\partial(u,v)} = J(u, v) = -\frac{1}{3} \Rightarrow \int_0^{2/3} \int_y^{2-2y} (x + 2y) e^{(y-x)} \, dx \, dy = \int_0^2 \int_0^u u e^{-v} \left| -\frac{1}{3} \right| \, dv \, du$

$= \frac{1}{3} \int_0^2 u \left[-e^{-v} \right]_0^u \, du = \frac{1}{3} \int_0^2 u \left(1 - e^{-u} \right) \, du = \frac{1}{3} \left[u \left(u + e^{-u} \right) - \frac{u^2}{2} + e^{-u} \right]_0^2 = \frac{1}{3} \left[2 \left(2 + e^{-2} \right) - 2 + e^{-2} - 1 \right]$

$= \frac{1}{3} \left(3e^{-2} + 1 \right) \approx 0.4687$

15. (a) $x = u \cos v$ and $y = u \sin v \Rightarrow \frac{\partial(x,y)}{\partial(u,v)} = \begin{vmatrix} \cos v & -u \sin v \\ \sin v & u \cos v \end{vmatrix} = u \cos^2 v + u \sin^2 v = u$

(b) $x = u \sin v$ and $y = u \cos v \Rightarrow \frac{\partial(x,y)}{\partial(u,v)} = \begin{vmatrix} \sin v & u \cos v \\ \cos v & -u \sin v \end{vmatrix} = -u \sin^2 v - u \cos^2 v = -u$

17. $\begin{vmatrix} \sin \phi \cos \theta & \rho \cos \phi \cos \theta & -\rho \sin \phi \sin \theta \\ \sin \phi \sin \theta & \rho \cos \phi \sin \theta & \rho \sin \phi \cos \theta \\ \cos \phi & -\rho \sin \phi & 0 \end{vmatrix}$

$= (\cos \phi) \begin{vmatrix} \rho \cos \phi \cos \theta & -\rho \sin \phi \sin \theta \\ \rho \cos \phi \sin \theta & \rho \sin \phi \cos \theta \end{vmatrix} + (\rho \sin \phi) \begin{vmatrix} \sin \phi \cos \theta & -\rho \sin \phi \sin \theta \\ \sin \phi \sin \theta & \rho \sin \phi \cos \theta \end{vmatrix}$

$= (\rho^2 \cos \phi) (\sin \phi \cos \phi \cos^2 \theta + \sin \phi \cos \phi \sin^2 \theta) + (\rho^2 \sin \phi) (\sin^2 \phi \cos^2 \theta + \sin^2 \phi \sin^2 \theta)$

$= \rho^2 \sin \phi \cos^2 \phi + \rho^2 \sin^3 \phi = (\rho^2 \sin \phi) (\cos^2 \phi + \sin^2 \phi) = \rho^2 \sin \phi$

19. $\int_0^3 \int_0^4 \int_{y/2}^{1+(y/2)} \left(\frac{2x - y}{2} + \frac{z}{3} \right) dx \, dy \, dz = \int_0^3 \int_0^4 \left[\frac{x^2}{2} - \frac{xy}{2} + \frac{xz}{3} \right]_{y/2}^{1+(y/2)} dy \, dz = \int_0^3 \int_0^4 \left[\frac{1}{2}(y + 1) - \frac{y}{2} + \frac{z}{3} \right] dy \, dz$

$= \int_0^3 \left[\frac{(y+1)^2}{4} - \frac{y^2}{4} + \frac{yz}{3} \right]_0^4 dz = \int_0^3 \left(\frac{9}{4} + \frac{4z}{3} - \frac{1}{4} \right) dz = \int_0^3 \left(2 + \frac{4z}{3} \right) dz = \left[2z + \frac{2z^2}{3} \right]_0^3 = 12$

21. $J(u, v, w) = \begin{vmatrix} a & 0 & 0 \\ 0 & b & 0 \\ 0 & 0 & c \end{vmatrix} = abc$; for R and G as in Exercise 19, $\iiint_R |xyz| \, dx \, dy \, dz$

$= \iiint_G a^2 b^2 c^2 uvw \, dw \, dv \, du = 8a^2 b^2 c^2 \int_0^{\pi/2} \int_0^{\pi/2} \int_0^1 (\rho \sin \phi \cos \theta)(\rho \sin \phi \sin \theta)(\rho \cos \phi) (\rho^2 \sin \phi) \, d\rho \, d\phi \, d\theta$

$$= \frac{4a^2b^2c^2}{3} \int_0^{\pi/2} \int_0^{\pi/2} \sin\theta \cos\theta \sin^3\phi \cos\phi \, d\phi \, d\theta = \frac{a^2b^2c^2}{3} \int_0^{\pi/2} \sin\theta \cos\theta \, d\theta = \frac{a^2b^2c^2}{6}$$

23. The first moment about the xy-coordinate plane for the semi-ellipsoid, $\frac{x^2}{a^2} + \frac{y^2}{b^2} + \frac{z^2}{c^2} = 1$ using the transformation in Exercise 21 is, $M_{xy} = \iiint_D z \, dz \, dy \, dx = \iiint_G cw \, |J(u, v, w)| \, du \, dv \, dw$

$$= abc^2 \iiint_G w \, du \, dv \, dw = (abc^2) \cdot (M_{xy} \text{ of the hemisphere } x^2 + y^2 + z^2 = 1, z \geq 0) = \frac{abc^2\pi}{4};$$

the mass of the semi-ellipsoid is $\frac{2abc\pi}{3} \Rightarrow \bar{z} = \left(\frac{abc^2\pi}{4}\right)\left(\frac{3}{2abc\pi}\right) = \frac{3}{8}c$

CHAPTER 15 PRACTICE EXERCISES

1. $\int_1^{10} \int_0^{1/y} ye^{xy} \, dx \, dy = \int_1^{10} [e^{xy}]_0^{1/y} \, dy$

   $= \int_1^{10} (e - 1) \, dy = 9e - 9$

3. $\int_0^{3/2} \int_{-\sqrt{9-4t^2}}^{\sqrt{9-4t^2}} t \, ds \, dt = \int_0^{3/2} [ts]_{-\sqrt{9-4t^2}}^{\sqrt{9-4t^2}} \, dt$

   $= \int_0^{3/2} 2t\sqrt{9 - 4t^2} \, dt = \left[-\frac{1}{6}(9 - 4t^2)^{3/2}\right]_0^{3/2}$

   $= -\frac{1}{6}(0^{3/2} - 9^{3/2}) = \frac{27}{6} = \frac{9}{2}$

5. $\int_{-2}^0 \int_{2x+4}^{4-x^2} dy \, dx = \int_{-2}^0 (-x^2 - 2x) \, dx$

   $= \left[-\frac{x^3}{3} - x^2\right]_{-2}^0 = -\left(\frac{8}{3} - 4\right) = \frac{4}{3}$

   $\int_0^4 \int_{-\sqrt{4-y}}^{(y-4)/2} dx \, dy = \int_0^4 \left(\frac{y-4}{2} + \sqrt{4-y}\right) dy$

   $= \left[\frac{y^2}{2} - 2y - \frac{2}{3}(4-y)^{3/2}\right]_0^4 = 4 - 8 + \frac{2}{3} \cdot 4^{3/2}$

   $= -4 + \frac{16}{3} = \frac{4}{3}$

7. $\int_{-3}^3 \int_0^{(1/2)\sqrt{9-x^2}} y \, dy \, dx = \int_{-3}^3 \left[\frac{y^2}{2}\right]_0^{(1/2)\sqrt{9-x^2}} dx$

   $= \int_{-3}^3 \frac{1}{8}(9 - x^2) \, dx = \left[\frac{9x}{8} - \frac{x^3}{24}\right]_{-3}^3$

   $= \left(\frac{27}{8} - \frac{27}{24}\right) - \left(-\frac{27}{8} + \frac{27}{24}\right) = \frac{27}{6} = \frac{9}{2}$

   $\int_0^{3/2} \int_{-\sqrt{9-4y^2}}^{\sqrt{9-4y^2}} y \, dx \, dy = \int_0^{3/2} 2y\sqrt{9-4y^2} \, dy$

   $= -\frac{1}{4} \cdot \frac{2}{3}(9-4y^2)^{3/2}\Big|_0^{3/2} = \frac{1}{6} \cdot 9^{3/2} = \frac{27}{6} = \frac{9}{2}$

9. $\int_0^1 \int_{2y}^2 4\cos(x^2) \, dx \, dy = \int_0^2 \int_0^{x/2} 4\cos(x^2) \, dy \, dx = \int_0^2 2x\cos(x^2) \, dx = [\sin(x^2)]_0^2 = \sin 4$

11. $\int_0^8 \int_{\sqrt[3]{x}}^2 \frac{1}{y^4+1} \, dy \, dx = \int_0^2 \int_0^{y^3} \frac{1}{y^4+1} \, dx \, dy = \frac{1}{4} \int_0^2 \frac{4y^3}{y^4+1} \, dy = \frac{\ln 17}{4}$

13. $A = \int_{-2}^{0}\int_{2x+4}^{4-x^2} dy\, dx = \int_{-2}^{0}(-x^2 - 2x)\, dx = \frac{4}{3}$

15. $V = \int_{0}^{1}\int_{x}^{2-x}(x^2 + y^2)\, dy\, dx = \int_{0}^{1}\left[x^2 y + \frac{y^3}{3}\right]_{x}^{2-x} dx = \int_{0}^{1}\left[2x^2 + \frac{(2-x)^3}{3} - \frac{7x^3}{3}\right] dx = \left[\frac{2x^3}{3} - \frac{(2-x)^4}{12} - \frac{7x^4}{12}\right]_{0}^{1}$

  $= \left(\frac{2}{3} - \frac{1}{12} - \frac{7}{12}\right) + \frac{2^4}{12} = \frac{4}{3}$

17. average value $= \int_{0}^{1}\int_{0}^{1} xy\, dy\, dx = \int_{0}^{1}\left[\frac{xy^2}{2}\right]_{0}^{1} dx = \int_{0}^{1}\frac{x}{2}\, dx = \frac{1}{4}$

19. $M = \int_{1}^{2}\int_{2/x}^{2}dy\, dx = \int_{1}^{2}\left(2 - \frac{2}{x}\right) dx = 2 - \ln 4;\ M_y = \int_{1}^{2}\int_{2/x}^{2} x\, dy\, dx = \int_{1}^{2}x\left(2 - \frac{2}{x}\right) dx = 1;$

  $M_x = \int_{1}^{2}\int_{2/x}^{2} y\, dy\, dx = \int_{1}^{2}\left(2 - \frac{2}{x^2}\right) dx = 1 \Rightarrow \bar{x} = \bar{y} = \frac{1}{2 - \ln 4}$

21. $I_o = \int_{0}^{2}\int_{2x}^{4}(x^2 + y^2)\,(3)\, dy\, dx = 3\int_{0}^{2}\left(4x^2 + \frac{64}{3} - \frac{14x^3}{3}\right) dx = 104$

23. $M = \delta\int_{0}^{3}\int_{0}^{2x/3} dy\, dx = \delta\int_{0}^{3}\frac{2x}{3}\, dx = 3\delta;\ I_x = \delta\int_{0}^{3}\int_{0}^{2x/3} y^2\, dy\, dx = \frac{8\delta}{81}\int_{0}^{3}x^3\, dx = \left(\frac{8\delta}{81}\right)\left(\frac{3^4}{4}\right) = 2\delta \Rightarrow R_x = \sqrt{\frac{2}{3}}$

25. $M = \int_{-1}^{1}\int_{-1}^{1}\left(x^2 + y^2 + \frac{1}{3}\right) dy\, dx = \int_{-1}^{1}\left(2x^2 + \frac{4}{3}\right) dx = 4;\ M_x = \int_{-1}^{1}\int_{-1}^{1} y\left(x^2 + y^2 + \frac{1}{3}\right) dy\, dx = \int_{-1}^{1} 0\, dx = 0;$

  $M_y = \int_{-1}^{1}\int_{-1}^{1} x\left(x^2 + y^2 + \frac{1}{3}\right) dy\, dx = \int_{-1}^{1}\left(2x^3 + \frac{4}{3}x\right) dx = 0$

27. $\int_{-1}^{1}\int_{-\sqrt{1-x^2}}^{\sqrt{1-x^2}}\frac{2}{(1+x^2+y^2)}\, dy\, dx = \int_{0}^{2\pi}\int_{0}^{1}\frac{2r}{(1+r^2)^2}\, dr\, d\theta = \int_{0}^{2\pi}\left[-\frac{1}{1+r^2}\right]_{0}^{1} d\theta = \frac{1}{2}\int_{0}^{2\pi} d\theta = \pi$

29. $M = \int_{-\pi/3}^{\pi/3}\int_{0}^{3} r\, dr\, d\theta = \frac{9}{2}\int_{-\pi/3}^{\pi/3} d\theta = 3\pi;\ M_y = \int_{-\pi/3}^{\pi/3}\int_{0}^{3} r^2\cos\theta\, dr\, d\theta = 9\int_{-\pi/3}^{\pi/3}\cos\theta\, d\theta = 9\sqrt{3} \Rightarrow \bar{x} = \frac{3\sqrt{3}}{\pi},$

  and $\bar{y} = 0$ by symmetry

31. (a) $M = 2\int_{0}^{\pi/2}\int_{1}^{1+\cos\theta} r\, dr\, d\theta$ (b)

  $= \int_{0}^{\pi/2}\left(2\cos\theta + \frac{1+\cos 2\theta}{2}\right) d\theta = \frac{8+\pi}{4};$

  $M_y = \int_{-\pi/2}^{\pi/2}\int_{1}^{1+\cos\theta}(r\cos\theta)\, r\, dr\, d\theta$

  $= \int_{-\pi/2}^{\pi/2}\left(\cos^2\theta + \cos^3\theta + \frac{\cos^4\theta}{3}\right) d\theta$

  $= \frac{32+15\pi}{24} \Rightarrow \bar{x} = \frac{15\pi+32}{6\pi+48}$, and

  $\bar{y} = 0$ by symmetry

33. $(x^2 + y^2)^2 - (x^2 - y^2) = 0 \Rightarrow r^4 - r^2\cos 2\theta = 0 \Rightarrow r^2 = \cos 2\theta$ so the integral is $\int_{-\pi/4}^{\pi/4}\int_{0}^{\sqrt{\cos 2\theta}}\frac{r}{(1+r^2)^2}\, dr\, d\theta$

  $= \int_{-\pi/4}^{\pi/4}\left[-\frac{1}{2(1+r^2)}\right]_{0}^{\sqrt{\cos 2\theta}} d\theta = \frac{1}{2}\int_{-\pi/4}^{\pi/4}\left(1 - \frac{1}{1+\cos 2\theta}\right) d\theta = \frac{1}{2}\int_{-\pi/4}^{\pi/4}\left(1 - \frac{1}{2\cos^2\theta}\right) d\theta$

  $= \frac{1}{2}\int_{-\pi/4}^{\pi/4}\left(1 - \frac{\sec^2\theta}{2}\right) d\theta = \frac{1}{2}\left[\theta - \frac{\tan\theta}{2}\right]_{-\pi/4}^{\pi/4} = \frac{\pi-2}{4}$

35. $\int_{0}^{\pi}\int_{0}^{\pi}\int_{0}^{\pi}\cos(x+y+z)\, dx\, dy\, dz = \int_{0}^{\pi}\int_{0}^{\pi}[\sin(z+y+\pi) - \sin(z+y)]\, dy\, dz$

  $= \int_{0}^{\pi}[-\cos(z+2\pi) + \cos(z+\pi) - \cos z + \cos(z+\pi)]\, dz = 0$

37. $\int_0^1 \int_0^{x^2} \int_0^{x+y} (2x - y - z)\, dz\, dy\, dx = \int_0^1 \int_0^{x^2} \left(\frac{3x^2}{2} - \frac{3y^2}{2}\right) dy\, dx = \int_0^1 \left(\frac{3x^4}{2} - \frac{x^6}{2}\right) dx = \frac{8}{35}$

39. $V = 2\int_0^{\pi/2} \int_{-\cos y}^0 \int_0^{-2x} dz\, dx\, dy = 2\int_0^{\pi/2} \int_{-\cos y}^0 -2x\, dx\, dy = 2\int_0^{\pi/2} \cos^2 y\, dy = 2\left[\frac{y}{2} + \frac{\sin 2y}{4}\right]_0^{\pi/2} = \frac{\pi}{2}$

41. average $= \frac{1}{3}\int_0^1 \int_0^3 \int_0^1 30xz\sqrt{x^2 + y}\, dz\, dy\, dx = \frac{1}{3}\int_0^1 \int_0^3 15x\sqrt{x^2 + y}\, dy\, dx = \frac{1}{3}\int_0^3 \int_0^1 15x\sqrt{x^2 + y}\, dx\, dy$

$= \frac{1}{3}\int_0^3 \left[5(x^2 + y)^{3/2}\right]_0^1 dy = \frac{1}{3}\int_0^3 \left[5(1 + y)^{3/2} - 5y^{3/2}\right] dy = \frac{1}{3}\left[2(1 + y)^{5/2} - 2y^{5/2}\right]_0^3 = \frac{1}{3}\left[2(4)^{5/2} - 2(3)^{5/2} - 2\right]$

$= \frac{1}{3}\left[2\left(31 - 3^{5/2}\right)\right]$

43. (a) $\int_{-\sqrt{2}}^{\sqrt{2}} \int_{-\sqrt{2-y^2}}^{\sqrt{2-y^2}} \int_{\sqrt{x^2+y^2}}^{\sqrt{4-x^2-y^2}} 3\, dz\, dx\, dy$

(b) $\int_0^{2\pi} \int_0^{\pi/4} \int_0^2 3\rho^2 \sin\phi\, d\rho\, d\phi\, d\theta$

(c) $\int_0^{2\pi} \int_0^{\sqrt{2}} \int_r^{\sqrt{4-r^2}} 3\, dz\, r\, dr\, d\theta = 3\int_0^{2\pi} \int_0^{\sqrt{2}} \left[r(4 - r^2)^{1/2} - r^2\right] dr\, d\theta = 3\int_0^{2\pi} \left[-\frac{1}{3}(4 - r^2)^{3/2} - \frac{r^3}{3}\right]_0^{\sqrt{2}} d\theta$

$= \int_0^{2\pi} \left(-2^{3/2} - 2^{3/2} + 4^{3/2}\right) d\theta = \left(8 - 4\sqrt{2}\right)\int_0^{2\pi} d\theta = 2\pi\left(8 - 4\sqrt{2}\right)$

45. (a) $\int_0^{2\pi} \int_0^{\pi/4} \int_0^{\sec\phi} \rho^2 \sin\phi\, d\rho\, d\phi\, d\theta$

(b) $\int_0^{2\pi} \int_0^{\pi/4} \int_0^{\sec\phi} \rho^2 \sin\phi\, d\rho\, d\phi\, d\theta = \frac{1}{3}\int_0^{2\pi} \int_0^{\pi/4} (\sec\phi)(\sec\phi \tan\phi)\, d\phi\, d\theta = \frac{1}{3}\int_0^{2\pi} \left[\frac{1}{2}\tan^2\phi\right]_0^{\pi/4} d\theta = \frac{1}{6}\int_0^{2\pi} d\theta = \frac{\pi}{3}$

47. $\int_0^1 \int_{\sqrt{1-x^2}}^{\sqrt{3-x^2}} \int_1^{\sqrt{4-x^2-y^2}} z^2 yx\, dz\, dy\, dx + \int_1^{\sqrt{3}} \int_0^{\sqrt{3-x^2}} \int_1^{\sqrt{4-x^2-y^2}} z^2 yx\, dz\, dy\, dx$

49. (a) $V = \int_0^{2\pi} \int_0^2 \int_2^{\sqrt{8-r^2}} dz\, r\, dr\, d\theta = \int_0^{2\pi} \int_0^2 \left(r\sqrt{8 - r^2} - 2r\right) dr\, d\theta = \int_0^{2\pi} \left[-\frac{1}{3}(8 - r^2)^{3/2} - r^2\right]_0^2 d\theta$

$= \int_0^{2\pi} \left[-\frac{1}{3}(4)^{3/2} - 4 + \frac{1}{3}(8)^{3/2}\right] d\theta = \int_0^{2\pi} \frac{4}{3}\left(-2 - 3 + 2\sqrt{8}\right) d\theta = \frac{4}{3}\left(4\sqrt{2} - 5\right)\int_0^{2\pi} d\theta = \frac{8\pi\left(4\sqrt{2} - 5\right)}{3}$

(b) $V = \int_0^{2\pi} \int_0^{\pi/4} \int_{2\sec\phi}^{\sqrt{8}} \rho^2 \sin\phi\, d\rho\, d\phi\, d\theta = \frac{8}{3}\int_0^{2\pi} \int_0^{\pi/4} \left(2\sqrt{2}\sin\phi - \sec^3\phi \sin\phi\right) d\phi\, d\theta$

$= \frac{8}{3}\int_0^{2\pi} \int_0^{\pi/4} \left(2\sqrt{2}\sin\phi - \tan\phi \sec^2\phi\right) d\phi\, d\theta = \frac{8}{3}\int_0^{2\pi} \left[-2\sqrt{2}\cos\phi - \frac{1}{2}\tan^2\phi\right]_0^{\pi/4} d\theta$

$= \frac{8}{3}\int_0^{2\pi} \left(-2 - \frac{1}{2} + 2\sqrt{2}\right) d\theta = \frac{8}{3}\int_0^{2\pi} \left(\frac{-5 + 4\sqrt{2}}{2}\right) d\theta = \frac{8\pi\left(4\sqrt{2} - 5\right)}{3}$

51. With the centers of the spheres at the origin, $I_z = \int_0^{2\pi} \int_0^\pi \int_a^b \delta(\rho \sin\phi)^2 (\rho^2 \sin\phi)\, d\rho\, d\phi\, d\theta$

$= \frac{\delta(b^5 - a^5)}{5}\int_0^{2\pi} \int_0^\pi \sin^3\phi\, d\phi\, d\theta = \frac{\delta(b^5 - a^5)}{5}\int_0^{2\pi} \int_0^\pi (\sin\phi - \cos^2\phi \sin\phi)\, d\phi\, d\theta$

$= \frac{\delta(b^5 - a^5)}{5}\int_0^{2\pi} \left[-\cos\phi + \frac{\cos^3\phi}{3}\right]_0^\pi d\theta = \frac{4\delta(b^5 - a^5)}{15}\int_0^{2\pi} d\theta = \frac{8\pi\delta(b^5 - a^5)}{15}$

53. $x = u + y$ and $y = v \Rightarrow x = u + v$ and $y = v$

$\Rightarrow J(u, v) = \begin{vmatrix} 1 & 1 \\ 0 & 1 \end{vmatrix} = 1$; the boundary of the

image G is obtained from the boundary of R as follows:

xy-equations for the boundary of R	Corresponding uv-equations for the boundary of G	Simplified uv-equations
$y = x$	$v = u + v$	$u = 0$
$y = 0$	$v = 0$	$v = 0$

$\Rightarrow \int_0^\infty \int_0^x e^{-sx} f(x - y, y) \, dy \, dx = \int_0^\infty \int_0^\infty e^{-s(u+v)} f(u, v) \, du \, dv$

CHAPTER 15 ADDITIONAL AND ADVANCED EXERCISES

1. (a) $V = \int_{-3}^2 \int_x^{6-x^2} x^2 \, dy \, dx$ (b) $V = \int_{-3}^2 \int_x^{6-x^2} \int_0^{x^2} dz \, dy \, dx$

   (c) $V = \int_{-3}^2 \int_x^{6-x^2} x^2 \, dy \, dx = \int_{-3}^2 \int_x^{6-x^2} (6x^2 - x^4 - x^3) \, dx = \left[2x^3 - \frac{x^5}{5} - \frac{x^4}{4}\right]_{-3}^2 = \frac{125}{4}$

3. Using cylindrical coordinates, $V = \int_0^{2\pi} \int_0^1 \int_0^{2-r(\cos\theta + \sin\theta)} dz \, r \, dr \, d\theta = \int_0^{2\pi} \int_0^1 (2r - r^2 \cos\theta - r^2 \sin\theta) \, dr \, d\theta$

   $= \int_0^{2\pi} \left(1 - \frac{1}{3}\cos\theta - \frac{1}{3}\sin\theta\right) d\theta = \left[\theta - \frac{1}{3}\sin\theta + \frac{1}{3}\cos\theta\right]_0^{2\pi} = 2\pi$

5. The surfaces intersect when $3 - x^2 - y^2 = 2x^2 + 2y^2 \Rightarrow x^2 + y^2 = 1$. Thus the volume is

   $V = 4\int_0^1 \int_0^{\sqrt{1-x^2}} \int_{2x^2+2y^2}^{3-x^2-y^2} dz \, dy \, dx = 4\int_0^{\pi/2} \int_0^1 \int_{2r^2}^{3-r^2} dz \, r \, dr \, d\theta = 4\int_0^{\pi/2} \int_0^1 (3r - 3r^3) \, dr \, d\theta = 3\int_0^{\pi/2} d\theta = \frac{3\pi}{2}$

7. (a) The radius of the hole is 1, and the radius of the sphere is 2.

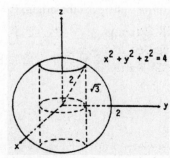

   (b) $V = 2\int_0^{2\pi} \int_0^{\sqrt{3}} \int_1^{\sqrt{4-z^2}} r \, dr \, dz \, d\theta = \int_0^{2\pi} \int_0^{\sqrt{3}} (3 - z^2) \, dz \, d\theta = 2\sqrt{3} \int_0^{2\pi} d\theta = 4\sqrt{3}\pi$

9. The surfaces intersect when $x^2 + y^2 = \frac{x^2+y^2+1}{2} \Rightarrow x^2 + y^2 = 1$. Thus the volume in cylindrical

   coordinates is $V = 4\int_0^{\pi/2} \int_0^1 \int_{r^2}^{(r^2+1)/2} dz \, r \, dr \, d\theta = 4\int_0^{\pi/2} \int_0^1 \left(\frac{r}{2} - \frac{r^3}{2}\right) dr \, d\theta = 4\int_0^{\pi/2} \left[\frac{r^2}{4} - \frac{r^4}{8}\right]_0^1 d\theta$

$= \frac{1}{2} \int_0^{\pi/2} d\theta = \frac{\pi}{4}$

11. $\int_0^\infty \frac{e^{-ax} - e^{-bx}}{x} dx = \int_0^\infty \int_a^b e^{-xy} dy\, dx = \int_a^b \int_0^\infty e^{-xy} dx\, dy = \int_a^b \left(\lim_{t \to \infty} \int_0^t e^{-xy} dx \right) dy$

$= \int_a^b \lim_{t \to \infty} \left[-\frac{e^{-xy}}{y} \right]_0^t dy = \int_a^b \lim_{t \to \infty} \left(\frac{1}{y} - \frac{e^{-yt}}{y} \right) dy = \int_a^b \frac{1}{y} dy = [\ln y]_a^b = \ln \left(\frac{b}{a} \right)$

13. $\int_0^x \int_0^u e^{m(x-t)} f(t)\, dt\, du = \int_0^x \int_t^x e^{m(x-t)} f(t)\, du\, dt = \int_0^x (x - t) e^{m(x-t)} f(t)\, dt;$ also

$\int_0^x \int_0^v \int_0^u e^{m(x-t)} f(t)\, dt\, du\, dv = \int_0^x \int_t^x \int_t^v e^{m(x-t)} f(t)\, du\, dv\, dt = \int_0^x \int_t^x (v - t) e^{m(x-t)} f(t)\, dv\, dt$

$= \int_0^x \left[\frac{1}{2} (v - t)^2 e^{m(x-t)} f(t) \right]_t^x dt = \int_0^x \frac{(x-t)^2}{2} e^{m(x-t)} f(t)\, dt$

15. $I_o(a) = \int_0^a \int_0^{x/a^2} (x^2 + y^2)\, dy\, dx = \int_0^a \left[x^2 y + \frac{y^3}{3} \right]_0^{x/a^2} dx = \int_0^a \left(\frac{x^3}{a^2} + \frac{x^3}{3a^6} \right) dx = \left[\frac{x^4}{4a^2} + \frac{x^4}{12a^6} \right]_0^a$

$= \frac{a^2}{4} + \frac{1}{12} a^{-2};\ I_o'(a) = \frac{1}{2} a - \frac{1}{6} a^{-3} = 0 \Rightarrow a^4 = \frac{1}{3} \Rightarrow a = \sqrt[4]{\frac{1}{3}} = \frac{1}{\sqrt[4]{3}}$. Since $I_o''(a) = \frac{1}{2} + \frac{1}{2} a^{-4} > 0$, the

value of a does provide a __minimum__ for the polar moment of inertia $I_o(a)$.

17. $M = \int_{-\theta}^\theta \int_{b \sec \theta}^a r\, dr\, d\theta = \int_{-\theta}^\theta \left(\frac{a^2}{2} - \frac{b^2}{2} \sec^2 \theta \right) d\theta$

$= a^2 \theta - b^2 \tan \theta = a^2 \cos^{-1} \left(\frac{b}{a} \right) - b^2 \left(\frac{\sqrt{a^2 - b^2}}{b} \right)$

$= a^2 \cos^{-1} \left(\frac{b}{a} \right) - b \sqrt{a^2 - b^2};\ I_o = \int_{-\theta}^\theta \int_{b \sec \theta}^a r^3\, dr\, d\theta$

$= \frac{1}{4} \int_{-\theta}^\theta (a^4 + b^4 \sec^4 \theta)\, d\theta$

$= \frac{1}{4} \int_{-\theta}^\theta [a^4 + b^4 (1 + \tan^2 \theta) (\sec^2 \theta)]\, d\theta$

$= \frac{1}{4} \left[a^4 \theta - b^4 \tan \theta - \frac{b^4 \tan^3 \theta}{3} \right]_{-\theta}^\theta$

$= \frac{a^4 \theta}{2} - \frac{b^4 \tan \theta}{2} - \frac{b^4 \tan^3 \theta}{6}$

$= \frac{1}{2} a^4 \cos^{-1} \left(\frac{b}{a} \right) - \frac{1}{2} b^3 \sqrt{a^2 - b^2} - \frac{1}{6} b^3 (a^2 - b^2)^{3/2}$

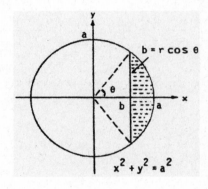

19. $\int_0^a \int_0^b e^{\max (b^2 x^2, a^2 y^2)}\, dy\, dx = \int_0^a \int_0^{bx/a} e^{b^2 x^2}\, dy\, dx + \int_0^b \int_0^{ay/b} e^{a^2 y^2}\, dx\, dy$

$= \int_0^a \left(\frac{b}{a} x \right) e^{b^2 x^2}\, dx + \int_0^b \left(\frac{a}{b} y \right) e^{a^2 y^2}\, dy = \left[\frac{1}{2ab} e^{b^2 x^2} \right]_0^a + \left[\frac{1}{2ba} e^{a^2 y^2} \right]_0^b = \frac{1}{2ab} \left(e^{b^2 a^2} - 1 \right) + \frac{1}{2ab} \left(e^{a^2 b^2} - 1 \right)$

$= \frac{1}{ab} \left(e^{a^2 b^2} - 1 \right)$

21. (a) (i)   Fubini's Theorem

(ii)   Treating G(y) as a constant

(iii)   Algebraic rearrangement

(iv)   The definite integral is a constant number

(b) $\int_0^{\ln 2} \int_0^{\pi/2} e^x \cos y\, dy\, dx = \left(\int_0^{\ln 2} e^x\, dx \right) \left(\int_0^{\pi/2} \cos y\, dy \right) = (e^{\ln 2} - e^0) \left(\sin \frac{\pi}{2} - \sin 0 \right) = (1)(1) = 1$

(c) $\int_1^2 \int_{-1}^1 \frac{x}{y^2}\, dx\, dy = \left(\int_1^2 \frac{1}{y^2}\, dy \right) \left(\int_{-1}^1 x\, dx \right) = \left[-\frac{1}{y} \right]_1^2 \left[\frac{x^2}{2} \right]_{-1}^1 = \left(-\frac{1}{2} + 1 \right) \left(\frac{1}{2} - \frac{1}{2} \right) = 0$

23. (a) $I^2 = \int_0^\infty \int_0^\infty e^{-(x^2 + y^2)}\, dx\, dy = \int_0^{\pi/2} \int_0^\infty (e^{-r^2})\, r\, dr\, d\theta = \int_0^{\pi/2} \left[\lim_{b \to \infty} \int_0^b r e^{-r^2}\, dr \right] d\theta$

$= -\frac{1}{2} \int_0^{\pi/2} \lim_{b \to \infty} (e^{-b^2} - 1)\, d\theta = \frac{1}{2} \int_0^{\pi/2} d\theta = \frac{\pi}{4} \Rightarrow I = \frac{\sqrt{\pi}}{2}$

(b) $\Gamma \left(\frac{1}{2} \right) = \int_0^\infty t^{-1/2} e^{-t}\, dt = \int_0^\infty (y^2)^{-1/2} e^{-y^2} (2y)\, dy = 2 \int_0^\infty e^{-y^2}\, dy = 2 \left(\frac{\sqrt{\pi}}{2} \right) = \sqrt{\pi}$, where $y = \sqrt{t}$

25. For a height h in the bowl the volume of water is $V = \int_{-\sqrt{h}}^{\sqrt{h}} \int_{-\sqrt{h-x^2}}^{\sqrt{h-x^2}} \int_{x^2+y^2}^{h} dz\,dy\,dx$

$= \int_{-\sqrt{h}}^{\sqrt{h}} \int_{-\sqrt{h-x^2}}^{\sqrt{h-x^2}} (h - x^2 - y^2)\,dy\,dx = \int_0^{2\pi} \int_0^{\sqrt{h}} (h - r^2)\,r\,dr\,d\theta = \int_0^{2\pi} \left[\frac{hr^2}{2} - \frac{r^4}{4}\right]_0^{\sqrt{h}} d\theta = \int_0^{2\pi} \frac{h^2}{4}\,d\theta = \frac{h^2\pi}{2}$.

Since the top of the bowl has area 10π, then we calibrate the bowl by comparing it to a right circular cylinder whose cross sectional area is 10π from $z = 0$ to $z = 10$. If such a cylinder contains $\frac{h^2\pi}{2}$ cubic inches of water to a depth w then we have $10\pi w = \frac{h^2\pi}{2} \Rightarrow w = \frac{h^2}{20}$. So for 1 inch of rain, $w = 1$ and $h = \sqrt{20}$; for 3 inches of rain, $w = 3$ and $h = \sqrt{60}$.

27. The cylinder is given by $x^2 + y^2 = 1$ from $z = 1$ to $\infty \Rightarrow \iiint_D z\,(r^2 + z^2)^{-5/2}\,dV$

$= \int_0^{2\pi} \int_0^1 \int_1^\infty \frac{z}{(r^2 + z^2)^{5/2}}\,dz\,r\,dr\,d\theta = \lim_{a \to \infty} \int_0^{2\pi} \int_0^1 \int_1^a \frac{rz}{(r^2 + z^2)^{5/2}}\,dz\,dr\,d\theta$

$= \lim_{a \to \infty} \int_0^{2\pi} \int_0^1 \left[\left(-\frac{1}{3}\right) \frac{r}{(r^2 + z^2)^{3/2}}\right]_1^a dr\,d\theta = \lim_{a \to \infty} \int_0^{2\pi} \int_0^1 \left[\left(-\frac{1}{3}\right) \frac{r}{(r^2 + a^2)^{3/2}} + \left(\frac{1}{3}\right) \frac{r}{(r^2 + 1)^{3/2}}\right] dr\,d\theta$

$= \lim_{a \to \infty} \int_0^{2\pi} \left[\frac{1}{3}(r^2 + a^2)^{-1/2} - \frac{1}{3}(r^2 + 1)^{-1/2}\right]_0^1 d\theta = \lim_{a \to \infty} \int_0^{2\pi} \left[\frac{1}{3}(1 + a^2)^{-1/2} - \frac{1}{3}(2^{-1/2}) - \frac{1}{3}(a^2)^{-1/2} + \frac{1}{3}\right] d\theta$

$= \lim_{a \to \infty} 2\pi \left[\frac{1}{3}(1 + a^2)^{-1/2} - \frac{1}{3}\left(\frac{\sqrt{2}}{2}\right) - \frac{1}{3}\left(\frac{1}{a}\right) + \frac{1}{3}\right] = 2\pi \left[\frac{1}{3} - \left(\frac{1}{3}\right) \frac{\sqrt{2}}{2}\right]$.

CHAPTER 16 INTEGRATION IN VECTOR FIELDS

16.1 LINE INTEGRALS

1. $\mathbf{r} = t\mathbf{i} + (1-t)\mathbf{j} \Rightarrow x = t$ and $y = 1-t \Rightarrow y = 1-x \Rightarrow$ (c)

3. $\mathbf{r} = (2\cos t)\mathbf{i} + (2\sin t)\mathbf{j} \Rightarrow x = 2\cos t$ and $y = 2\sin t \Rightarrow x^2 + y^2 = 4 \Rightarrow$ (g)

5. $\mathbf{r} = t\mathbf{i} + t\mathbf{j} + t\mathbf{k} \Rightarrow x = t,\, y = t,$ and $z = t \Rightarrow$ (d)

7. $\mathbf{r} = (t^2-1)\mathbf{j} + 2t\mathbf{k} \Rightarrow y = t^2 - 1$ and $z = 2t \Rightarrow y = \frac{z^2}{4} - 1 \Rightarrow$ (f)

9. $\mathbf{r}(t) = t\mathbf{i} + (1-t)\mathbf{j},\, 0 \le t \le 1 \Rightarrow \frac{d\mathbf{r}}{dt} = \mathbf{i} - \mathbf{j} \Rightarrow \left|\frac{d\mathbf{r}}{dt}\right| = \sqrt{2}\mathbf{j};\, x = t$ and $y = 1-t \Rightarrow x + y = t + (1-t) = 1$

$\Rightarrow \int_C f(x,y,z)\, ds = \int_0^1 f(t, 1-t, 0)\left|\frac{d\mathbf{r}}{dt}\right| dt = \int_0^1 (1)\left(\sqrt{2}\right) dt = \left[\sqrt{2}\,t\right]_0^1 = \sqrt{2}$

11. $\mathbf{r}(t) = 2t\mathbf{i} + t\mathbf{j} + (2-2t)\mathbf{k},\, 0 \le t \le 1 \Rightarrow \frac{d\mathbf{r}}{dt} = 2\mathbf{i} + \mathbf{j} - 2\mathbf{k} \Rightarrow \left|\frac{d\mathbf{r}}{dt}\right| = \sqrt{4+1+4} = 3;\, xy + y + z$

$= (2t)t + t + (2-2t) \Rightarrow \int_C f(x,y,z)\, ds = \int_0^1 (2t^2 - t + 2)\, 3\, dt = 3\left[\frac{2}{3}t^3 - \frac{1}{2}t^2 + 2t\right]_0^1 = 3\left(\frac{2}{3} - \frac{1}{2} + 2\right) = \frac{13}{2}$

13. $\mathbf{r}(t) = (\mathbf{i} + 2\mathbf{j} + 3\mathbf{k}) + t(-\mathbf{i} - 3\mathbf{j} - 2\mathbf{k}) = (1-t)\mathbf{i} + (2-3t)\mathbf{j} + (3-2t)\mathbf{k},\, 0 \le t \le 1 \Rightarrow \frac{d\mathbf{r}}{dt} = -\mathbf{i} - 3\mathbf{j} - 2\mathbf{k}$

$\Rightarrow \left|\frac{d\mathbf{r}}{dt}\right| = \sqrt{1+9+4} = \sqrt{14};\, x + y + z = (1-t) + (2-3t) + (3-2t) = 6 - 6t \Rightarrow \int_C f(x,y,z)\, ds$

$= \int_0^1 (6-6t)\sqrt{14}\, dt = 6\sqrt{14}\left[t - \frac{t^2}{2}\right]_0^1 = \left(6\sqrt{14}\right)\left(\frac{1}{2}\right) = 3\sqrt{14}$

15. C_1: $\mathbf{r}(t) = t\mathbf{i} + t^2\mathbf{j},\, 0 \le t \le 1 \Rightarrow \frac{d\mathbf{r}}{dt} = \mathbf{i} + 2t\mathbf{j} \Rightarrow \left|\frac{d\mathbf{r}}{dt}\right| = \sqrt{1+4t^2};\, x + \sqrt{y} - z^2 = t + \sqrt{t^2} - 0 = t + |t| = 2t$

since $t \ge 0 \Rightarrow \int_{C_1} f(x,y,z)\, ds = \int_0^1 2t\sqrt{1+4t^2}\, dt = \left[\frac{1}{6}(1+4t^2)^{3/2}\right]_0^1 = \frac{1}{6}(5)^{3/2} - \frac{1}{6} = \frac{1}{6}\left(5\sqrt{5} - 1\right);$

C_2: $\mathbf{r}(t) = \mathbf{i} + \mathbf{j} + t\mathbf{k},\, 0 \le t \le 1 \Rightarrow \frac{d\mathbf{r}}{dt} = \mathbf{k} \Rightarrow \left|\frac{d\mathbf{r}}{dt}\right| = 1;\, x + \sqrt{y} - z^2 = 1 + \sqrt{1} - t^2 = 2 - t^2$

$\Rightarrow \int_{C_2} f(x,y,z)\, ds = \int_0^1 (2-t^2)(1)\, dt = \left[2t - \frac{1}{3}t^3\right]_0^1 = 2 - \frac{1}{3} = \frac{5}{3};$ therefore $\int_C f(x,y,z)\, ds$

$= \int_{C_1} f(x,y,z)\, ds + \int_{C_2} f(x,y,z)\, ds = \frac{5}{6}\sqrt{5} + \frac{3}{2}$

17. $\mathbf{r}(t) = t\mathbf{i} + t\mathbf{j} + t\mathbf{k},\, 0 < a \le t \le b \Rightarrow \frac{d\mathbf{r}}{dt} = \mathbf{i} + \mathbf{j} + \mathbf{k} \Rightarrow \left|\frac{d\mathbf{r}}{dt}\right| = \sqrt{3};\, \frac{x+y+z}{x^2+y^2+z^2} = \frac{t+t+t}{t^2+t^2+t^2} = \frac{1}{t}$

$\Rightarrow \int_C f(x,y,z)\, ds = \int_a^b \left(\frac{1}{t}\right)\sqrt{3}\, dt = \left[\sqrt{3}\ln|t|\right]_a^b = \sqrt{3}\ln\left(\frac{b}{a}\right),$ since $0 < a \le b$

19. $\mathbf{r}(x) = x\mathbf{i} + y\mathbf{j} = x\mathbf{i} + \frac{x^2}{2}\mathbf{j},\, 0 \le x \le 2 \Rightarrow \frac{d\mathbf{r}}{dx} = \mathbf{i} + x\mathbf{j} \Rightarrow \left|\frac{d\mathbf{r}}{dx}\right| = \sqrt{1+x^2};\, f(x,y) = f\left(x, \frac{x^2}{2}\right) = \frac{x^3}{\left(\frac{x^2}{2}\right)} = 2x \Rightarrow \int_C f\, ds$

$= \int_0^2 (2x)\sqrt{1+x^2}\, dx = \left[\frac{2}{3}(1+x^2)^{3/2}\right]_0^2 = \frac{2}{3}\left(5^{3/2} - 1\right) = \frac{10\sqrt{5} - 2}{3}$

21. $\mathbf{r}(t) = (2\cos t)\mathbf{i} + (2\sin t)\mathbf{j},\, 0 \le t \le \frac{\pi}{2} \Rightarrow \frac{d\mathbf{r}}{dt} = (-2\sin t)\mathbf{i} + (2\cos t)\mathbf{j} \Rightarrow \left|\frac{d\mathbf{r}}{dt}\right| = 2;\, f(x,y) = f(2\cos t, 2\sin t)$

$= 2\cos t + 2\sin t \Rightarrow \int_C f\, ds = \int_0^{\pi/2} (2\cos t + 2\sin t)(2)\, dt = [4\sin t - 4\cos t]_0^{\pi/2} = 4 - (-4) = 8$

23. $\mathbf{r}(t) = (t^2 - 1)\mathbf{j} + 2t\mathbf{k}, 0 \le t \le 1 \;\Rightarrow\; \frac{d\mathbf{r}}{dt} = 2t\mathbf{j} + 2\mathbf{k} \Rightarrow \left|\frac{d\mathbf{r}}{dt}\right| = 2\sqrt{t^2 + 1}; M = \int_C \delta(x, y, z)\, ds = \int_0^1 \delta(t)\left(2\sqrt{t^2 + 1}\right) dt$

$= \int_0^1 \left(\tfrac{3}{2}t\right)\left(2\sqrt{t^2 + 1}\right) dt = \left[(t^2 + 1)^{3/2}\right]_0^1 = 2^{3/2} - 1 = 2\sqrt{2} - 1$

25. $\mathbf{r}(t) = \sqrt{2}t\mathbf{i} + \sqrt{2}t\mathbf{j} + (4 - t^2)\mathbf{k}, 0 \le t \le 1 \;\Rightarrow\; \frac{d\mathbf{r}}{dt} = \sqrt{2}\mathbf{i} + \sqrt{2}\mathbf{j} - 2t\mathbf{k} \;\Rightarrow\; \left|\frac{d\mathbf{r}}{dt}\right| = \sqrt{2 + 2 + 4t^2} = 2\sqrt{1 + t^2}\,;$

   (a) $M = \int_C \delta\, ds = \int_0^1 (3t)\left(2\sqrt{1 + t^2}\right) dt = \left[2\left(1 + t^2\right)^{3/2}\right]_0^1 = 2\left(2^{3/2} - 1\right) = 4\sqrt{2} - 2$

   (b) $M = \int_C \delta\, ds = \int_0^1 (1)\left(2\sqrt{1 + t^2}\right) dt = \left[t\sqrt{1 + t^2} + \ln\left(t + \sqrt{1 + t^2}\right)\right]_0^1 = \left[\sqrt{2} + \ln\left(1 + \sqrt{2}\right)\right] - (0 + \ln 1)$

   $= \sqrt{2} + \ln\left(1 + \sqrt{2}\right)$

27. Let $x = a\cos t$ and $y = a\sin t, 0 \le t \le 2\pi$. Then $\frac{dx}{dt} = -a\sin t, \frac{dy}{dt} = a\cos t, \frac{dz}{dt} = 0$

   $\Rightarrow \sqrt{\left(\frac{dx}{dt}\right)^2 + \left(\frac{dy}{dt}\right)^2 + \left(\frac{dz}{dt}\right)^2}\, dt = a\, dt; I_z = \int_C (x^2 + y^2)\delta\, ds = \int_0^{2\pi} (a^2 \sin^2 t + a^2 \cos^2 t)\, a\delta\, dt$

   $= \int_0^{2\pi} a^3 \delta\, dt = 2\pi\delta a^3; M = \int_C \delta(x, y, z)\, ds = \int_0^{2\pi} \delta a\, dt = 2\pi\delta a \;\Rightarrow\; R_z = \sqrt{\frac{I_z}{M}} = \sqrt{\frac{2\pi a^3 \delta}{2\pi a\delta}} = a.$

29. $\mathbf{r}(t) = (\cos t)\mathbf{i} + (\sin t)\mathbf{j} + t\mathbf{k}, 0 \le t \le 2\pi \;\Rightarrow\; \frac{d\mathbf{r}}{dt} = (-\sin t)\mathbf{i} + (\cos t)\mathbf{j} + \mathbf{k} \;\Rightarrow\; \left|\frac{d\mathbf{r}}{dt}\right| = \sqrt{\sin^2 t + \cos^2 t + 1} = \sqrt{2}\,;$

   (a) $M = \int_C \delta\, ds = \int_0^{2\pi} \delta\sqrt{2}\, dt = 2\pi\delta\sqrt{2}; I_z = \int_C (x^2 + y^2)\delta\, ds = \int_0^{2\pi} (\cos^2 t + \sin^2 t)\delta\sqrt{2}\, dt = 2\pi\delta\sqrt{2}$

   $\Rightarrow R_z = \sqrt{\frac{I_z}{M}} = 1$

   (b) $M = \int_C \delta(x, y, z)\, ds = \int_0^{4\pi} \delta\sqrt{2}\, dt = 4\pi\delta\sqrt{2}$ and $I_z = \int_C (x^2 + y^2)\delta\, ds = \int_0^{4\pi} \delta\sqrt{2}\, dt = 4\pi\delta\sqrt{2}$

   $\Rightarrow R_z = \sqrt{\frac{I_z}{M}} = 1$

31. $\delta(x, y, z) = 2 - z$ and $\mathbf{r}(t) = (\cos t)\mathbf{j} + (\sin t)\mathbf{k}, 0 \le t \le \pi \;\Rightarrow\; M = 2\pi - 2$ as found in Example 4 of the text;

   also $\left|\frac{d\mathbf{r}}{dt}\right| = 1; I_x = \int_C (y^2 + z^2)\delta\, ds = \int_0^{\pi} (\cos^2 t + \sin^2 t)(2 - \sin t)\, dt = \int_0^{\pi} (2 - \sin t)\, dt = 2\pi - 2 \;\Rightarrow\; R_x = \sqrt{\frac{I_x}{M}}$

   $= 1$

16.2 VECTOR FIELDS, WORK, CIRCULATION, AND FLUX

1. $f(x, y, z) = (x^2 + y^2 + z^2)^{-1/2} \;\Rightarrow\; \frac{\partial f}{\partial x} = -\frac{1}{2}(x^2 + y^2 + z^2)^{-3/2}(2x) = -x(x^2 + y^2 + z^2)^{-3/2};$ similarly,

   $\frac{\partial f}{\partial y} = -y(x^2 + y^2 + z^2)^{-3/2}$ and $\frac{\partial f}{\partial z} = -z(x^2 + y^2 + z^2)^{-3/2} \;\Rightarrow\; \nabla f = \frac{-x\mathbf{i} - y\mathbf{j} - z\mathbf{k}}{(x^2 + y^2 + z^2)^{3/2}}$

3. $g(x, y, z) = e^z - \ln(x^2 + y^2) \;\Rightarrow\; \frac{\partial g}{\partial x} = -\frac{2x}{x^2 + y^2}, \frac{\partial g}{\partial y} = -\frac{2y}{x^2 + y^2}$ and $\frac{\partial g}{\partial z} = e^z$

   $\Rightarrow \nabla g = \left(\frac{-2x}{x^2 + y^2}\right)\mathbf{i} - \left(\frac{2y}{x^2 + y^2}\right)\mathbf{j} + e^z\mathbf{k}$

5. $|\mathbf{F}|$ inversely proportional to the square of the distance from (x, y) to the origin $\;\Rightarrow\; \sqrt{(M(x, y))^2 + (N(x, y))^2}$

   $= \frac{k}{x^2 + y^2}, k > 0; \mathbf{F}$ points toward the origin $\;\Rightarrow\; \mathbf{F}$ is in the direction of $\mathbf{n} = \frac{-x}{\sqrt{x^2 + y^2}}\mathbf{i} - \frac{y}{\sqrt{x^2 + y^2}}\mathbf{j}$

   $\Rightarrow \mathbf{F} = a\mathbf{n}$, for some constant $a > 0$. Then $M(x, y) = \frac{-ax}{\sqrt{x^2 + y^2}}$ and $N(x, y) = \frac{-ay}{\sqrt{x^2 + y^2}}$

   $\Rightarrow \sqrt{(M(x, y))^2 + (N(x, y))^2} = a \;\Rightarrow\; a = \frac{k}{x^2 + y^2} \;\Rightarrow\; \mathbf{F} = \frac{-kx}{(x^2 + y^2)^{3/2}}\mathbf{i} - \frac{ky}{(x^2 + y^2)^{3/2}}\mathbf{j}$, for any constant $k > 0$

7. Substitute the parametric representations for $\mathbf{r}(t) = x(t)\mathbf{i} + y(t)\mathbf{j} + z(t)\mathbf{k}$ representing each path into the vector field $\mathbf{F}$, and calculate the work $W = \int_C \mathbf{F} \cdot \frac{d\mathbf{r}}{dt}$.

(a) $\mathbf{F} = 3t\mathbf{i} + 2t\mathbf{j} + 4t\mathbf{k}$ and $\frac{d\mathbf{r}}{dt} = \mathbf{i} + \mathbf{j} + \mathbf{k}$ $\Rightarrow$ $\mathbf{F} \cdot \frac{d\mathbf{r}}{dt} = 9t$ $\Rightarrow$ $W = \int_0^1 9t\, dt = \frac{9}{2}$

(b) $\mathbf{F} = 3t^2\mathbf{i} + 2t\mathbf{j} + 4t^4\mathbf{k}$ and $\frac{d\mathbf{r}}{dt} = \mathbf{i} + 2t\mathbf{j} + 4t^3\mathbf{k}$ $\Rightarrow$ $\mathbf{F} \cdot \frac{d\mathbf{r}}{dt} = 7t^2 + 16t^7$ $\Rightarrow$ $W = \int_0^1 (7t^2 + 16t^7)\, dt = \left[\frac{7}{3}t^3 + 2t^8\right]_0^1$
$= \frac{7}{3} + 2 = \frac{13}{3}$

(c) $\mathbf{r}_1 = t\mathbf{i} + t\mathbf{j}$ and $\mathbf{r}_2 = \mathbf{i} + \mathbf{j} + t\mathbf{k}$; $\mathbf{F}_1 = 3t\mathbf{i} + 2t\mathbf{j}$ and $\frac{d\mathbf{r}_1}{dt} = \mathbf{i} + \mathbf{j}$ $\Rightarrow$ $\mathbf{F}_1 \cdot \frac{d\mathbf{r}_1}{dt} = 5t$ $\Rightarrow$ $W_1 = \int_0^1 5t\, dt = \frac{5}{2}$;
$\mathbf{F}_2 = 3\mathbf{i} + 2\mathbf{j} + 4t\mathbf{k}$ and $\frac{d\mathbf{r}_2}{dt} = \mathbf{k}$ $\Rightarrow$ $\mathbf{F}_2 \cdot \frac{d\mathbf{r}_2}{dt} = 4t$ $\Rightarrow$ $W_2 = \int_0^1 4t\, dt = 2$ $\Rightarrow$ $W = W_1 + W_2 = \frac{9}{2}$

9. Substitute the parametric representation for $\mathbf{r}(t) = x(t)\mathbf{i} + y(t)\mathbf{j} + z(t)\mathbf{k}$ representing each path into the vector field $\mathbf{F}$, and calculate the work $W = \int_C \mathbf{F} \cdot \frac{d\mathbf{r}}{dt}$.

(a) $\mathbf{F} = \sqrt{t}\,\mathbf{i} - 2t\mathbf{j} + \sqrt{t}\,\mathbf{k}$ and $\frac{d\mathbf{r}}{dt} = \mathbf{i} + \mathbf{j} + \mathbf{k}$ $\Rightarrow$ $\mathbf{F} \cdot \frac{d\mathbf{r}}{dt} = 2\sqrt{t} - 2t$ $\Rightarrow$ $W = \int_0^1 (2\sqrt{t} - 2t)\, dt = \left[\frac{4}{3}t^{3/2} - t^2\right]_0^1 = \frac{1}{3}$

(b) $\mathbf{F} = t^2\mathbf{i} - 2t\mathbf{j} + t\mathbf{k}$ and $\frac{d\mathbf{r}}{dt} = \mathbf{i} + 2t\mathbf{j} + 4t^3\mathbf{k}$ $\Rightarrow$ $\mathbf{F} \cdot \frac{d\mathbf{r}}{dt} = 4t^4 - 3t^2$ $\Rightarrow$ $W = \int_0^1 (4t^4 - 3t^2)\, dt = \left[\frac{4}{5}t^5 - t^3\right]_0^1 = -\frac{1}{5}$

(c) $\mathbf{r}_1 = t\mathbf{i} + t\mathbf{j}$ and $\mathbf{r}_2 = \mathbf{i} + \mathbf{j} + t\mathbf{k}$; $\mathbf{F}_1 = -2t\mathbf{j} + \sqrt{t}\,\mathbf{k}$ and $\frac{d\mathbf{r}_1}{dt} = \mathbf{i} + \mathbf{j}$ $\Rightarrow$ $\mathbf{F}_1 \cdot \frac{d\mathbf{r}_1}{dt} = -2t$ $\Rightarrow$ $W_1 = \int_0^1 -2t\, dt$
$= -1$; $\mathbf{F}_2 = \sqrt{t}\,\mathbf{i} - 2\mathbf{j} + \mathbf{k}$ and $\frac{d\mathbf{r}_2}{dt} = \mathbf{k}$ $\Rightarrow$ $\mathbf{F}_2 \cdot \frac{d\mathbf{r}_2}{dt} = 1$ $\Rightarrow$ $W_2 = \int_0^1 dt = 1$ $\Rightarrow$ $W = W_1 + W_2 = 0$

11. Substitute the parametric representation for $\mathbf{r}(t) = x(t)\mathbf{i} + y(t)\mathbf{j} + z(t)\mathbf{k}$ representing each path into the vector field $\mathbf{F}$, and calculate the work $W = \int_C \mathbf{F} \cdot \frac{d\mathbf{r}}{dt}$.

(a) $\mathbf{F} = (3t^2 - 3t)\mathbf{i} + 3t\mathbf{j} + \mathbf{k}$ and $\frac{d\mathbf{r}}{dt} = \mathbf{i} + \mathbf{j} + \mathbf{k}$ $\Rightarrow$ $\mathbf{F} \cdot \frac{d\mathbf{r}}{dt} = 3t^2 + 1$ $\Rightarrow$ $W = \int_0^1 (3t^2 + 1)\, dt = [t^3 + t]_0^1 = 2$

(b) $\mathbf{F} = (3t^2 - 3t)\mathbf{i} + 3t^4\mathbf{j} + \mathbf{k}$ and $\frac{d\mathbf{r}}{dt} = \mathbf{i} + 2t\mathbf{j} + 4t^3\mathbf{k}$ $\Rightarrow$ $\mathbf{F} \cdot \frac{d\mathbf{r}}{dt} = 6t^5 + 4t^3 + 3t^2 - 3t$
$\Rightarrow$ $W = \int_0^1 (6t^5 + 4t^3 + 3t^2 - 3t)\, dt = \left[t^6 + t^4 + t^3 - \frac{3}{2}t^2\right]_0^1 = \frac{3}{2}$

(c) $\mathbf{r}_1 = t\mathbf{i} + t\mathbf{j}$ and $\mathbf{r}_2 = \mathbf{i} + \mathbf{j} + t\mathbf{k}$; $\mathbf{F}_1 = (3t^2 - 3t)\mathbf{i} + \mathbf{k}$ and $\frac{d\mathbf{r}_1}{dt} = \mathbf{i} + \mathbf{j}$ $\Rightarrow$ $\mathbf{F}_1 \cdot \frac{d\mathbf{r}_1}{dt} = 3t^2 - 3t$
$\Rightarrow$ $W_1 = \int_0^1 (3t^2 - 3t)\, dt = \left[t^3 - \frac{3}{2}t^2\right]_0^1 = -\frac{1}{2}$; $\mathbf{F}_2 = 3t\mathbf{j} + \mathbf{k}$ and $\frac{d\mathbf{r}_2}{dt} = \mathbf{k}$ $\Rightarrow$ $\mathbf{F}_2 \cdot \frac{d\mathbf{r}_2}{dt} = 1$ $\Rightarrow$ $W_2 = \int_0^1 dt = 1$
$\Rightarrow$ $W = W_1 + W_2 = \frac{1}{2}$

13. $\mathbf{r} = t\mathbf{i} + t^2\mathbf{j} + t\mathbf{k}, 0 \le t \le 1$, and $\mathbf{F} = xy\mathbf{i} + y\mathbf{j} - yz\mathbf{k}$ $\Rightarrow$ $\mathbf{F} = t^3\mathbf{i} + t^2\mathbf{j} - t^3\mathbf{k}$ and $\frac{d\mathbf{r}}{dt} = \mathbf{i} + 2t\mathbf{j} + \mathbf{k}$
$\Rightarrow$ $\mathbf{F} \cdot \frac{d\mathbf{r}}{dt} = 2t^3$ $\Rightarrow$ work $= \int_0^1 2t^3\, dt = \frac{1}{2}$

15. $\mathbf{r} = (\sin t)\mathbf{i} + (\cos t)\mathbf{j} + t\mathbf{k}, 0 \le t \le 2\pi$, and $\mathbf{F} = z\mathbf{i} + x\mathbf{j} + y\mathbf{k}$ $\Rightarrow$ $\mathbf{F} = t\mathbf{i} + (\sin t)\mathbf{j} + (\cos t)\mathbf{k}$ and
$\frac{d\mathbf{r}}{dt} = (\cos t)\mathbf{i} - (\sin t)\mathbf{j} + \mathbf{k}$ $\Rightarrow$ $\mathbf{F} \cdot \frac{d\mathbf{r}}{dt} = t\cos t - \sin^2 t + \cos t$ $\Rightarrow$ work $= \int_0^{2\pi} (t\cos t - \sin^2 t + \cos t)\, dt$
$= \left[\cos t + t\sin t - \frac{t}{2} + \frac{\sin 2t}{4} + \sin t\right]_0^{2\pi} = -\pi$

17. $x = t$ and $y = x^2 = t^2$ $\Rightarrow$ $\mathbf{r} = t\mathbf{i} + t^2\mathbf{j}, -1 \le t \le 2$, and $\mathbf{F} = xy\mathbf{i} + (x + y)\mathbf{j}$ $\Rightarrow$ $\mathbf{F} = t^3\mathbf{i} + (t + t^2)\mathbf{j}$ and
$\frac{d\mathbf{r}}{dt} = \mathbf{i} + 2t\mathbf{j}$ $\Rightarrow$ $\mathbf{F} \cdot \frac{d\mathbf{r}}{dt} = t^3 + (2t^2 + 2t^3) = 3t^3 + 2t^2$ $\Rightarrow$ $\int_C xy\, dx + (x + y)\, dy = \int_C \mathbf{F} \cdot \frac{d\mathbf{r}}{dt}\, dt = \int_{-1}^2 (3t^3 + 2t^2)\, dt$
$= \left[\frac{3}{4}t^4 + \frac{2}{3}t^3\right]_{-1}^2 = \left(12 + \frac{16}{3}\right) - \left(\frac{3}{4} - \frac{2}{3}\right) = \frac{45}{4} + \frac{18}{3} = \frac{69}{4}$

19. $\mathbf{r} = x\mathbf{i} + y\mathbf{j} = y^2\mathbf{i} + y\mathbf{j}, 2 \ge y \ge -1$, and $\mathbf{F} = x^2\mathbf{i} - y\mathbf{j} = y^4\mathbf{i} - y\mathbf{j}$ $\Rightarrow$ $\frac{d\mathbf{r}}{dy} = 2y\mathbf{i} + \mathbf{j}$ and $\mathbf{F} \cdot \frac{d\mathbf{r}}{dy} = 2y^5 - y$
$\Rightarrow$ $\int_C \mathbf{F} \cdot \mathbf{T}\, ds = \int_2^{-1} \mathbf{F} \cdot \frac{d\mathbf{r}}{dy}\, dy = \int_2^{-1} (2y^5 - y)\, dy = \left[\frac{1}{3}y^6 - \frac{1}{2}y^2\right]_2^{-1} = \left(\frac{1}{3} - \frac{1}{2}\right) - \left(\frac{64}{3} - \frac{4}{2}\right) = \frac{3}{2} - \frac{63}{3} = -\frac{39}{2}$

21. $\mathbf{r} = (\mathbf{i} + \mathbf{j}) + t(\mathbf{i} + 2\mathbf{j}) = (1 + t)\mathbf{i} + (1 + 2t)\mathbf{j}, 0 \le t \le 1$, and $\mathbf{F} = xy\mathbf{i} + (y - x)\mathbf{j}$ $\Rightarrow$ $\mathbf{F} = (1 + 3t + 2t^2)\mathbf{i} + t\mathbf{j}$ and
$\frac{d\mathbf{r}}{dt} = \mathbf{i} + 2\mathbf{j}$ $\Rightarrow$ $\mathbf{F} \cdot \frac{d\mathbf{r}}{dt} = 1 + 5t + 2t^2$ $\Rightarrow$ work $= \int_C \mathbf{F} \cdot \frac{d\mathbf{r}}{dt}\, dt = \int_0^1 (1 + 5t + 2t^2)\, dt = \left[t + \frac{5}{2}t^2 + \frac{2}{3}t^3\right]_0^1 = \frac{25}{6}$

23. (a) $\mathbf{r} = (\cos t)\mathbf{i} + (\sin t)\mathbf{j}$, $0 \leq t \leq 2\pi$, $\mathbf{F}_1 = x\mathbf{i} + y\mathbf{j}$, and $\mathbf{F}_2 = -y\mathbf{i} + x\mathbf{j}$ $\Rightarrow$ $\frac{d\mathbf{r}}{dt} = (-\sin t)\mathbf{i} + (\cos t)\mathbf{j}$,

   $\mathbf{F}_1 = (\cos t)\mathbf{i} + (\sin t)\mathbf{j}$, and $\mathbf{F}_2 = (-\sin t)\mathbf{i} + (\cos t)\mathbf{j}$ $\Rightarrow$ $\mathbf{F}_1 \cdot \frac{d\mathbf{r}}{dt} = 0$ and $\mathbf{F}_2 \cdot \frac{d\mathbf{r}}{dt} = \sin^2 t + \cos^2 t = 1$

   $\Rightarrow$ $\text{Circ}_1 = \int_0^{2\pi} 0\,dt = 0$ and $\text{Circ}_2 = \int_0^{2\pi} dt = 2\pi$; $\mathbf{n} = (\cos t)\mathbf{i} + (\sin t)\mathbf{j}$ $\Rightarrow$ $\mathbf{F}_1 \cdot \mathbf{n} = \cos^2 t + \sin^2 t = 1$ and

   $\mathbf{F}_2 \cdot \mathbf{n} = 0$ $\Rightarrow$ $\text{Flux}_1 = \int_0^{2\pi} dt = 2\pi$ and $\text{Flux}_2 = \int_0^{2\pi} 0\,dt = 0$

   (b) $\mathbf{r} = (\cos t)\mathbf{i} + (4\sin t)\mathbf{j}$, $0 \leq t \leq 2\pi$ $\Rightarrow$ $\frac{d\mathbf{r}}{dt} = (-\sin t)\mathbf{i} + (4\cos t)\mathbf{j}$, $\mathbf{F}_1 = (\cos t)\mathbf{i} + (4\sin t)\mathbf{j}$, and

   $\mathbf{F}_2 = (-4\sin t)\mathbf{i} + (\cos t)\mathbf{j}$ $\Rightarrow$ $\mathbf{F}_1 \cdot \frac{d\mathbf{r}}{dt} = 15\sin t\cos t$ and $\mathbf{F}_2 \cdot \frac{d\mathbf{r}}{dt} = 4$ $\Rightarrow$ $\text{Circ}_1 = \int_0^{2\pi} 15\sin t\cos t\,dt$

   $= \left[\frac{15}{2}\sin^2 t\right]_0^{2\pi} = 0$ and $\text{Circ}_2 = \int_0^{2\pi} 4\,dt = 8\pi$; $\mathbf{n} = \left(\frac{4}{\sqrt{17}}\cos t\right)\mathbf{i} + \left(\frac{1}{\sqrt{17}}\sin t\right)\mathbf{j}$ $\Rightarrow$ $\mathbf{F}_1 \cdot \mathbf{n}$

   $= \frac{4}{\sqrt{17}}\cos^2 t + \frac{4}{\sqrt{17}}\sin^2 t$ and $\mathbf{F}_2 \cdot \mathbf{n} = -\frac{15}{\sqrt{17}}\sin t\cos t$ $\Rightarrow$ $\text{Flux}_1 = \int_0^{2\pi}(\mathbf{F}_1 \cdot \mathbf{n})\,|\mathbf{v}|\,dt = \int_0^{2\pi}\left(\frac{4}{\sqrt{17}}\right)\sqrt{17}\,dt$

   $= 8\pi$ and $\text{Flux}_2 = \int_0^{2\pi}(\mathbf{F}_2 \cdot \mathbf{n})\,|\mathbf{v}|\,dt = \int_0^{2\pi}\left(-\frac{15}{\sqrt{17}}\sin t\cos t\right)\sqrt{17}\,dt = \left[-\frac{15}{2}\sin^2 t\right]_0^{2\pi} = 0$

25. $\mathbf{F}_1 = (a\cos t)\mathbf{i} + (a\sin t)\mathbf{j}$, $\frac{d\mathbf{r}_1}{dt} = (-a\sin t)\mathbf{i} + (a\cos t)\mathbf{j}$ $\Rightarrow$ $\mathbf{F}_1 \cdot \frac{d\mathbf{r}_1}{dt} = 0$ $\Rightarrow$ $\text{Circ}_1 = 0$; $M_1 = a\cos t$,

   $N_1 = a\sin t$, $dx = -a\sin t\,dt$, $dy = a\cos t\,dt$ $\Rightarrow$ $\text{Flux}_1 = \int_C M_1\,dy - N_1\,dx = \int_0^\pi(a^2\cos^2 t + a^2\sin^2 t)\,dt$

   $= \int_0^\pi a^2\,dt = a^2\pi$;

   $\mathbf{F}_2 = t\mathbf{i}$, $\frac{d\mathbf{r}_2}{dt} = \mathbf{i}$ $\Rightarrow$ $\mathbf{F}_2 \cdot \frac{d\mathbf{r}_2}{dt} = t$ $\Rightarrow$ $\text{Circ}_2 = \int_{-a}^a t\,dt = 0$; $M_2 = t$, $N_2 = 0$, $dx = dt$, $dy = 0$ $\Rightarrow$ Flux_2

   $= \int_C M_2\,dy - N_2\,dx = \int_{-a}^a 0\,dt = 0$; therefore, $\text{Circ} = \text{Circ}_1 + \text{Circ}_2 = 0$ and $\text{Flux} = \text{Flux}_1 + \text{Flux}_2 = a^2\pi$

27. $\mathbf{F}_1 = (-a\sin t)\mathbf{i} + (a\cos t)\mathbf{j}$, $\frac{d\mathbf{r}_1}{dt} = (-a\sin t)\mathbf{i} + (a\cos t)\mathbf{j}$ $\Rightarrow$ $\mathbf{F}_1 \cdot \frac{d\mathbf{r}_1}{dt} = a^2\sin^2 t + a^2\cos^2 t = a^2$

   $\Rightarrow$ $\text{Circ}_1 = \int_0^\pi a^2\,dt = a^2\pi$; $M_1 = -a\sin t$, $N_1 = a\cos t$, $dx = -a\sin t\,dt$, $dy = a\cos t\,dt$

   $\Rightarrow$ $\text{Flux}_1 = \int_C M_1\,dy - N_1\,dx = \int_0^\pi(-a^2\sin t\cos t + a^2\sin t\cos t)\,dt = 0$; $\mathbf{F}_2 = t\mathbf{j}$, $\frac{d\mathbf{r}_2}{dt} = \mathbf{i}$ $\Rightarrow$ $\mathbf{F}_2 \cdot \frac{d\mathbf{r}_2}{dt} = 0$

   $\Rightarrow$ $\text{Circ}_2 = 0$; $M_2 = 0$, $N_2 = t$, $dx = dt$, $dy = 0$ $\Rightarrow$ $\text{Flux}_2 = \int_C M_2\,dy - N_2\,dx = \int_{-a}^a -t\,dt = 0$; therefore,

   $\text{Circ} = \text{Circ}_1 + \text{Circ}_2 = a^2\pi$ and $\text{Flux} = \text{Flux}_1 + \text{Flux}_2 = 0$

29. (a) $\mathbf{r} = (\cos t)\mathbf{i} + (\sin t)\mathbf{j}$, $0 \leq t \leq \pi$, and $\mathbf{F} = (x+y)\mathbf{i} - (x^2+y^2)\mathbf{j}$ $\Rightarrow$ $\frac{d\mathbf{r}}{dt} = (-\sin t)\mathbf{i} + (\cos t)\mathbf{j}$ and

   $\mathbf{F} = (\cos t + \sin t)\mathbf{i} - (\cos^2 t + \sin^2 t)\mathbf{j}$ $\Rightarrow$ $\mathbf{F} \cdot \frac{d\mathbf{r}}{dt} = -\sin t\cos t - \sin^2 t - \cos t$ $\Rightarrow$ $\int_C \mathbf{F} \cdot \mathbf{T}\,ds$

   $= \int_0^\pi(-\sin t\cos t - \sin^2 t - \cos t)\,dt = \left[-\frac{1}{2}\sin^2 t - \frac{t}{2} + \frac{\sin 2t}{4} - \sin t\right]_0^\pi = -\frac{\pi}{2}$

   (b) $\mathbf{r} = (1-2t)\mathbf{i}$, $0 \leq t \leq 1$, and $\mathbf{F} = (x+y)\mathbf{i} - (x^2+y^2)\mathbf{j}$ $\Rightarrow$ $\frac{d\mathbf{r}}{dt} = -2\mathbf{i}$ and $\mathbf{F} = (1-2t)\mathbf{i} - (1-2t)^2\mathbf{j}$ $\Rightarrow$

   $\mathbf{F} \cdot \frac{d\mathbf{r}}{dt} = 4t - 2$ $\Rightarrow$ $\int_C \mathbf{F} \cdot \mathbf{T}\,ds = \int_0^1(4t-2)\,dt = [2t^2 - 2t]_0^1 = 0$

   (c) $\mathbf{r}_1 = (1-t)\mathbf{i} - t\mathbf{j}$, $0 \leq t \leq 1$, and $\mathbf{F} = (x+y)\mathbf{i} - (x^2+y^2)\mathbf{j}$ $\Rightarrow$ $\frac{d\mathbf{r}_1}{dt} = -\mathbf{i} - \mathbf{j}$ and $\mathbf{F} = (1-2t)\mathbf{i} - (1-2t+2t^2)\mathbf{j}$

   $\Rightarrow$ $\mathbf{F} \cdot \frac{d\mathbf{r}_1}{dt} = (2t-1) + (1-2t+2t^2) = 2t^2$ $\Rightarrow$ $\text{Flow}_1 = \int_{C_1}\mathbf{F} \cdot \frac{d\mathbf{r}_1}{dt} = \int_0^1 2t^2\,dt = \frac{2}{3}$; $\mathbf{r}_2 = -t\mathbf{i} + (t-1)\mathbf{j}$,

   $0 \leq t \leq 1$, and $\mathbf{F} = (x+y)\mathbf{i} - (x^2+y^2)\mathbf{j}$ $\Rightarrow$ $\frac{d\mathbf{r}_2}{dt} = -\mathbf{i} + \mathbf{j}$ and $\mathbf{F} = -\mathbf{i} - (t^2 + t^2 - 2t + 1)\mathbf{j}$

   $= -\mathbf{i} - (2t^2 - 2t + 1)\mathbf{j}$ $\Rightarrow$ $\mathbf{F} \cdot \frac{d\mathbf{r}_2}{dt} = 1 - (2t^2 - 2t + 1) = 2t - 2t^2$ $\Rightarrow$ $\text{Flow}_2 = \int_{C_2}\mathbf{F} \cdot \frac{d\mathbf{r}_2}{dt} = \int_0^1(2t - 2t^2)\,dt$

   $= \left[t^2 - \frac{2}{3}t^3\right]_0^1 = \frac{1}{3}$ $\Rightarrow$ $\text{Flow} = \text{Flow}_1 + \text{Flow}_2 = \frac{2}{3} + \frac{1}{3} = 1$

31. $\mathbf{F} = -\frac{y}{\sqrt{x^2+y^2}}\mathbf{i} + \frac{x}{\sqrt{x^2+y^2}}\mathbf{j}$ on $x^2 + y^2 = 4$;

at $(2,0)$, $\mathbf{F} = \mathbf{j}$; at $(0,2)$, $\mathbf{F} = -\mathbf{i}$; at $(-2,0)$,

$\mathbf{F} = -\mathbf{j}$; at $(0,-2)$, $\mathbf{F} = \mathbf{i}$; at $\left(\sqrt{2}, \sqrt{2}\right)$, $\mathbf{F} = -\frac{\sqrt{3}}{2}\mathbf{i} + \frac{1}{2}\mathbf{j}$;

at $\left(\sqrt{2}, -\sqrt{2}\right)$, $\mathbf{F} = \frac{\sqrt{3}}{2}\mathbf{i} + \frac{1}{2}\mathbf{j}$; at $\left(-\sqrt{2}, \sqrt{2}\right)$,

$\mathbf{F} = -\frac{\sqrt{3}}{2}\mathbf{i} - \frac{1}{2}\mathbf{j}$; at $\left(-\sqrt{2}, -\sqrt{2}\right)$, $\mathbf{F} = \frac{\sqrt{3}}{2}\mathbf{i} - \frac{1}{2}\mathbf{j}$

33. (a) $\mathbf{G} = P(x,y)\mathbf{i} + Q(x,y)\mathbf{j}$ is to have a magnitude $\sqrt{a^2+b^2}$ and to be tangent to $x^2 + y^2 = a^2 + b^2$ in a counterclockwise direction. Thus $x^2 + y^2 = a^2 + b^2 \Rightarrow 2x + 2yy' = 0 \Rightarrow y' = -\frac{x}{y}$ is the slope of the tangent line at any point on the circle $\Rightarrow y' = -\frac{a}{b}$ at (a,b). Let $\mathbf{v} = -b\mathbf{i} + a\mathbf{j} \Rightarrow |\mathbf{v}| = \sqrt{a^2+b^2}$, with $\mathbf{v}$ in a counterclockwise direction and tangent to the circle. Then let $P(x,y) = -y$ and $Q(x,y) = x$
$\Rightarrow \mathbf{G} = -y\mathbf{i} + x\mathbf{j} \Rightarrow$ for (a,b) on $x^2 + y^2 = a^2 + b^2$ we have $\mathbf{G} = -b\mathbf{i} + a\mathbf{j}$ and $|\mathbf{G}| = \sqrt{a^2+b^2}$.

(b) $\mathbf{G} = \left(\sqrt{x^2+y^2}\right)\mathbf{F} = \left(\sqrt{a^2+b^2}\right)\mathbf{F}$.

35. The slope of the line through (x,y) and the origin is $\frac{y}{x} \Rightarrow \mathbf{v} = x\mathbf{i} + y\mathbf{j}$ is a vector parallel to that line and pointing away from the origin $\Rightarrow \mathbf{F} = -\frac{x\mathbf{i}+y\mathbf{j}}{\sqrt{x^2+y^2}}$ is the unit vector pointing toward the origin.

37. $\mathbf{F} = -4t^3\mathbf{i} + 8t^2\mathbf{j} + 2\mathbf{k}$ and $\frac{d\mathbf{r}}{dt} = \mathbf{i} + 2t\mathbf{j} \Rightarrow \mathbf{F} \cdot \frac{d\mathbf{r}}{dt} = 12t^3 \Rightarrow \text{Flow} = \int_0^2 12t^3\, dt = [3t^4]_0^2 = 48$

39. $\mathbf{F} = (\cos t - \sin t)\mathbf{i} + (\cos t)\mathbf{k}$ and $\frac{d\mathbf{r}}{dt} = (-\sin t)\mathbf{i} + (\cos t)\mathbf{k} \Rightarrow \mathbf{F} \cdot \frac{d\mathbf{r}}{dt} = -\sin t \cos t + 1$
$\Rightarrow \text{Flow} = \int_0^\pi (-\sin t \cos t + 1)\, dt = \left[\frac{1}{2}\cos^2 t + t\right]_0^\pi = \left(\frac{1}{2} + \pi\right) - \left(\frac{1}{2} + 0\right) = \pi$

41. C_1: $\mathbf{r} = (\cos t)\mathbf{i} + (\sin t)\mathbf{j} + t\mathbf{k}$, $0 \le t \le \frac{\pi}{2} \Rightarrow \mathbf{F} = (2\cos t)\mathbf{i} + 2t\mathbf{j} + (2\sin t)\mathbf{k}$ and $\frac{d\mathbf{r}}{dt} = (-\sin t)\mathbf{i} + (\cos t)\mathbf{j} + \mathbf{k}$
$\Rightarrow \mathbf{F} \cdot \frac{d\mathbf{r}}{dt} = -2\cos t \sin t + 2t \cos t + 2\sin t = -\sin 2t + 2t \cos t + 2\sin t$
$\Rightarrow \text{Flow}_1 = \int_0^{\pi/2} (-\sin 2t + 2t \cos t + 2\sin t)\, dt = \left[\frac{1}{2}\cos 2t + 2t \sin t + 2\cos t - 2\cos t\right]_0^{\pi/2} = -1 + \pi$;
C_2: $\mathbf{r} = \mathbf{j} + \frac{\pi}{2}(1-t)\mathbf{k}$, $0 \le t \le 1 \Rightarrow \mathbf{F} = \pi(1-t)\mathbf{j} + 2\mathbf{k}$ and $\frac{d\mathbf{r}}{dt} = -\frac{\pi}{2}\mathbf{k} \Rightarrow \mathbf{F} \cdot \frac{d\mathbf{r}}{dt} = -\pi$
$\Rightarrow \text{Flow}_2 = \int_0^1 -\pi\, dt = [-\pi t]_0^1 = -\pi$;
C_3: $\mathbf{r} = t\mathbf{i} + (1-t)\mathbf{j}$, $0 \le t \le 1 \Rightarrow \mathbf{F} = 2t\mathbf{i} + 2(1-t)\mathbf{k}$ and $\frac{d\mathbf{r}}{dt} = \mathbf{i} - \mathbf{j} \Rightarrow \mathbf{F} \cdot \frac{d\mathbf{r}}{dt} = 2t$
$\Rightarrow \text{Flow}_3 = \int_0^1 2t\, dt = [t^2]_0^1 = 1 \Rightarrow \text{Circulation} = (-1 + \pi) - \pi + 1 = 0$

43. Let $x = t$ be the parameter $\Rightarrow y = x^2 = t^2$ and $z = x = t \Rightarrow \mathbf{r} = t\mathbf{i} + t^2\mathbf{j} + t\mathbf{k}$, $0 \le t \le 1$ from $(0,0,0)$ to $(1,1,1)$
$\Rightarrow \frac{d\mathbf{r}}{dt} = \mathbf{i} + 2t\mathbf{j} + \mathbf{k}$ and $\mathbf{F} = xy\mathbf{i} + y\mathbf{j} - yz\mathbf{k} = t^3\mathbf{i} + t^2\mathbf{j} - t^3\mathbf{k} \Rightarrow \mathbf{F} \cdot \frac{d\mathbf{r}}{dt} = t^3 + 2t^3 - t^3 = 2t^3 \Rightarrow \text{Flow} = \int_0^1 2t^3\, dt$
$= \frac{1}{2}$

45. Yes. The work and area have the same numerical value because work $= \int_C \mathbf{F} \cdot d\mathbf{r} = \int_C y\mathbf{i} \cdot d\mathbf{r}$
$= \int_b^a [f(t)\mathbf{i}] \cdot \left[\mathbf{i} + \frac{df}{dt}\mathbf{j}\right] dt$ [On the path, y equals $f(t)$]
$= \int_a^b f(t)\, dt = \text{Area under the curve}$ [because $f(t) > 0$]

16.3 PATH INDEPENDENCE, POTENTIAL FUNCTIONS, AND CONSERVATIVE FIELDS

1. $\frac{\partial P}{\partial y} = x = \frac{\partial N}{\partial z}$, $\frac{\partial M}{\partial z} = y = \frac{\partial P}{\partial x}$, $\frac{\partial N}{\partial x} = z = \frac{\partial M}{\partial y} \Rightarrow$ Conservative

3. $\frac{\partial P}{\partial y} = -1 \neq 1 = \frac{\partial N}{\partial z} \Rightarrow$ Not Conservative

5. $\frac{\partial N}{\partial x} = 0 \neq 1 = \frac{\partial M}{\partial y} \Rightarrow$ Not Conservative

7. $\frac{\partial f}{\partial x} = 2x \Rightarrow f(x, y, z) = x^2 + g(y, z) \Rightarrow \frac{\partial f}{\partial y} = \frac{\partial g}{\partial y} = 3y \Rightarrow g(y, z) = \frac{3y^2}{2} + h(z) \Rightarrow f(x, y, z) = x^2 + \frac{3y^2}{2} + h(z)$

   $\Rightarrow \frac{\partial f}{\partial z} = h'(z) = 4z \Rightarrow h(z) = 2z^2 + C \Rightarrow f(x, y, z) = x^2 + \frac{3y^2}{2} + 2z^2 + C$

9. $\frac{\partial f}{\partial x} = e^{y+2z} \Rightarrow f(x, y, z) = xe^{y+2z} + g(y, z) \Rightarrow \frac{\partial f}{\partial y} = xe^{y+2z} + \frac{\partial g}{\partial y} = xe^{y+2z} \Rightarrow \frac{\partial g}{\partial y} = 0 \Rightarrow f(x, y, z)$

   $= xe^{y+2z} + h(z) \Rightarrow \frac{\partial f}{\partial z} = 2xe^{y+2z} + h'(z) = 2xe^{y+2z} \Rightarrow h'(z) = 0 \Rightarrow h(z) = C \Rightarrow f(x, y, z) = xe^{y+2z} + C$

11. $\frac{\partial f}{\partial z} = \frac{z}{y^2+z^2} \Rightarrow f(x, y, z) = \frac{1}{2} \ln(y^2 + z^2) + g(x, y) \Rightarrow \frac{\partial f}{\partial x} = \frac{\partial g}{\partial x} = \ln x + \sec^2(x + y) \Rightarrow g(x, y)$

    $= (x \ln x - x) + \tan(x + y) + h(y) \Rightarrow f(x, y, z) = \frac{1}{2} \ln(y^2 + z^2) + (x \ln x - x) + \tan(x + y) + h(y)$

    $\Rightarrow \frac{\partial f}{\partial y} = \frac{y}{y^2+z^2} + \sec^2(x + y) + h'(y) = \sec^2(x + y) + \frac{y}{y^2+z^2} \Rightarrow h'(y) = 0 \Rightarrow h(y) = C \Rightarrow f(x, y, z)$

    $= \frac{1}{2} \ln(y^2 + z^2) + (x \ln x - x) + \tan(x + y) + C$

13. Let $\mathbf{F}(x, y, z) = 2x\mathbf{i} + 2y\mathbf{j} + 2z\mathbf{k} \Rightarrow \frac{\partial P}{\partial y} = 0 = \frac{\partial N}{\partial z}, \frac{\partial M}{\partial z} = 0 = \frac{\partial P}{\partial x}, \frac{\partial N}{\partial x} = 0 = \frac{\partial M}{\partial y} \Rightarrow M\,dx + N\,dy + P\,dz$ is

    exact; $\frac{\partial f}{\partial x} = 2x \Rightarrow f(x, y, z) = x^2 + g(y, z) \Rightarrow \frac{\partial f}{\partial y} = \frac{\partial g}{\partial y} = 2y \Rightarrow g(y, z) = y^2 + h(z) \Rightarrow f(x, y, z) = x^2 + y^2 = h(z)$

    $\Rightarrow \frac{\partial f}{\partial z} = h'(z) = 2z \Rightarrow h(z) = z^2 + C \Rightarrow f(x, y, z) = x^2 + y^2 + z^2 + C \Rightarrow \int_{(0,0,0)}^{(2,3,-6)} 2x\,dx + 2y\,dy + 2z\,dz$

    $= f(2, 3, -6) - f(0, 0, 0) = 2^2 + 3^2 + (-6)^2 = 49$

15. Let $\mathbf{F}(x, y, z) = 2xy\mathbf{i} + (x^2 - z^2)\mathbf{j} - 2yz\mathbf{k} \Rightarrow \frac{\partial P}{\partial y} = -2z = \frac{\partial N}{\partial z}, \frac{\partial M}{\partial z} = 0 = \frac{\partial P}{\partial x}, \frac{\partial N}{\partial x} = 2x = \frac{\partial M}{\partial y}$

    $\Rightarrow M\,dx + N\,dy + P\,dz$ is exact; $\frac{\partial f}{\partial x} = 2xy \Rightarrow f(x, y, z) = x^2y + g(y, z) \Rightarrow \frac{\partial f}{\partial y} = x^2 + \frac{\partial g}{\partial y} = x^2 - z^2 \Rightarrow \frac{\partial g}{\partial y} = -z^2$

    $\Rightarrow g(y, z) = -yz^2 + h(z) \Rightarrow f(x, y, z) = x^2y - yz^2 + h(z) \Rightarrow \frac{\partial f}{\partial z} = -2yz + h'(z) = -2yz \Rightarrow h'(z) = 0 \Rightarrow h(z) = C$

    $\Rightarrow f(x, y, z) = x^2y - yz^2 + C \Rightarrow \int_{(0,0,0)}^{(1,2,3)} 2xy\,dx + (x^2 - z^2)\,dy - 2yz\,dz = f(1, 2, 3) - f(0, 0, 0) = 2 - 2(3)^2 = -16$

17. Let $\mathbf{F}(x, y, z) = (\sin y \cos x)\mathbf{i} + (\cos y \sin x)\mathbf{j} + \mathbf{k} \Rightarrow \frac{\partial P}{\partial y} = 0 = \frac{\partial N}{\partial z}, \frac{\partial M}{\partial z} = 0 = \frac{\partial P}{\partial x}, \frac{\partial N}{\partial x} = \cos y \cos x = \frac{\partial M}{\partial y}$

    $\Rightarrow M\,dx + N\,dy + P\,dz$ is exact; $\frac{\partial f}{\partial x} = \sin y \cos x \Rightarrow f(x, y, z) = \sin y \sin x + g(y, z) \Rightarrow \frac{\partial f}{\partial y} = \cos y \sin x + \frac{\partial g}{\partial y}$

    $= \cos y \sin x \Rightarrow \frac{\partial g}{\partial y} = 0 \Rightarrow g(y, z) = h(z) \Rightarrow f(x, y, z) = \sin y \sin x + h(z) \Rightarrow \frac{\partial f}{\partial z} = h'(z) = 1 \Rightarrow h(z) = z + C$

    $\Rightarrow f(x, y, z) = \sin y \sin x + z + C \Rightarrow \int_{(1,0,0)}^{(0,1,1)} \sin y \cos x\,dx + \cos y \sin x\,dy + dz = f(0, 1, 1) - f(1, 0, 0)$

    $= (0 + 1) - (0 + 0) = 1$

19. Let $\mathbf{F}(x, y, z) = 3x^2\mathbf{i} + \left(\frac{z^2}{y}\right)\mathbf{j} + (2z \ln y)\mathbf{k} \Rightarrow \frac{\partial P}{\partial y} = \frac{2z}{y} = \frac{\partial N}{\partial z}, \frac{\partial M}{\partial z} = 0 = \frac{\partial P}{\partial x}, \frac{\partial N}{\partial x} = 0 = \frac{\partial M}{\partial y}$

    $\Rightarrow M\,dx + N\,dy + P\,dz$ is exact; $\frac{\partial f}{\partial x} = 3x^2 \Rightarrow f(x, y, z) = x^3 + g(y, z) \Rightarrow \frac{\partial f}{\partial y} = \frac{\partial g}{\partial y} = \frac{z^2}{y} \Rightarrow g(y, z) = z^2 \ln y + h(z)$

    $\Rightarrow f(x, y, z) = x^3 + z^2 \ln y + h(z) \Rightarrow \frac{\partial f}{\partial z} = 2z \ln y + h'(z) = 2z \ln y \Rightarrow h'(z) = 0 \Rightarrow h(z) = C \Rightarrow f(x, y, z)$

    $= x^3 + z^2 \ln y + C \Rightarrow \int_{(1,1,1)}^{(1,2,3)} 3x^2\,dx + \frac{z^2}{y}\,dy + 2z \ln y\,dz = f(1, 2, 3) - f(1, 1, 1)$

    $= (1 + 9 \ln 2 + C) - (1 + 0 + C) = 9 \ln 2$

21. Let $\mathbf{F}(x, y, z) = \left(\frac{1}{y}\right)\mathbf{i} + \left(\frac{1}{z} - \frac{x}{y^2}\right)\mathbf{j} - \left(\frac{y}{z^2}\right)\mathbf{k} \Rightarrow \frac{\partial P}{\partial y} = -\frac{1}{z^2} = \frac{\partial N}{\partial z}, \frac{\partial M}{\partial z} = 0 = \frac{\partial P}{\partial x}, \frac{\partial N}{\partial x} = -\frac{1}{y^2} = \frac{\partial M}{\partial y}$

    $\Rightarrow M\,dx + N\,dy + P\,dz$ is exact; $\frac{\partial f}{\partial x} = \frac{1}{y} \Rightarrow f(x, y, z) = \frac{x}{y} + g(y, z) \Rightarrow \frac{\partial f}{\partial y} = -\frac{x}{y^2} + \frac{\partial g}{\partial y} = \frac{1}{z} - \frac{x}{y^2}$

    $\Rightarrow \frac{\partial g}{\partial y} = \frac{1}{z} \Rightarrow g(y, z) = \frac{y}{z} + h(z) \Rightarrow f(x, y, z) = \frac{x}{y} + \frac{y}{z} + h(z) \Rightarrow \frac{\partial f}{\partial z} = -\frac{y}{z^2} + h'(z) = -\frac{y}{z^2} \Rightarrow h'(z) = 0 \Rightarrow h(z) = C$

$\Rightarrow f(x, y, z) = \frac{x}{y} + \frac{y}{z} + C \Rightarrow \int_{(1,1,1)}^{(2,2,2)} \frac{1}{y} dx + \left(\frac{1}{z} - \frac{x}{y^2}\right) dy - \frac{y}{z^2} dz = f(2, 2, 2) - f(1, 1, 1) = \left(\frac{2}{2} + \frac{2}{2} + C\right) - \left(\frac{1}{1} + \frac{1}{1} + C\right)$

$= 0$

23. $\mathbf{r} = (\mathbf{i} + \mathbf{j} + \mathbf{k}) + t(\mathbf{i} + 2\mathbf{j} - 2\mathbf{k}) = (1 + t)\mathbf{i} + (1 + 2t)\mathbf{j} + (1 - 2t)\mathbf{k}, 0 \le t \le 1 \Rightarrow dx = dt, dy = 2 dt, dz = -2 dt$

$\Rightarrow \int_{(1,1,1)}^{(2,3,-1)} y \, dx + x \, dy + 4 \, dz = \int_0^1 (2t + 1) \, dt + (t + 1)(2 \, dt) + 4(-2) \, dt = \int_0^1 (4t - 5) \, dt = [2t^2 - 5t]_0^1 = -3$

25. $\frac{\partial P}{\partial y} = 0 = \frac{\partial N}{\partial z}, \frac{\partial M}{\partial z} = 2z = \frac{\partial P}{\partial x}, \frac{\partial N}{\partial x} = 0 = \frac{\partial M}{\partial y} \Rightarrow M \, dx + N \, dy + P \, dz$ is exact $\Rightarrow \mathbf{F}$ is conservative

$\Rightarrow$ path independence

27. $\frac{\partial P}{\partial y} = 0 = \frac{\partial N}{\partial z}, \frac{\partial M}{\partial z} = 0 = \frac{\partial P}{\partial x}, \frac{\partial N}{\partial x} = -\frac{2x}{y^2} = \frac{\partial M}{\partial y} \Rightarrow \mathbf{F}$ is conservative $\Rightarrow$ there exists an f so that $\mathbf{F} = \nabla f$;

$\frac{\partial f}{\partial x} = \frac{2x}{y} \Rightarrow f(x, y) = \frac{x^2}{y} + g(y) \Rightarrow \frac{\partial f}{\partial y} = -\frac{x^2}{y^2} + g'(y) = \frac{1-x^2}{y^2} \Rightarrow g'(y) = \frac{1}{y^2} \Rightarrow g(y) = -\frac{1}{y} + C$

$\Rightarrow f(x, y) = \frac{x^2}{y} - \frac{1}{y} + C \Rightarrow \mathbf{F} = \nabla\left(\frac{x^2-1}{y}\right)$

29. $\frac{\partial P}{\partial y} = 0 = \frac{\partial N}{\partial z}, \frac{\partial M}{\partial z} = 0 = \frac{\partial P}{\partial x}, \frac{\partial N}{\partial x} = 1 = \frac{\partial M}{\partial y} \Rightarrow \mathbf{F}$ is conservative $\Rightarrow$ there exists an f so that $\mathbf{F} = \nabla f$;

$\frac{\partial f}{\partial x} = x^2 + y \Rightarrow f(x, y, z) = \frac{1}{3}x^3 + xy + g(y, z) \Rightarrow \frac{\partial f}{\partial y} = x + \frac{\partial g}{\partial y} = y^2 + x \Rightarrow \frac{\partial g}{\partial y} = y^2 \Rightarrow g(y, z) = \frac{1}{3}y^3 + h(z)$

$\Rightarrow f(x, y, z) = \frac{1}{3}x^3 + xy + \frac{1}{3}y^3 + h(z) \Rightarrow \frac{\partial f}{\partial z} = h'(z) = ze^z \Rightarrow h(z) = ze^z - e^z + C \Rightarrow f(x, y, z)$

$= \frac{1}{3}x^3 + xy + \frac{1}{3}y^3 + ze^z - e^z + C \Rightarrow \mathbf{F} = \nabla\left(\frac{1}{3}x^3 + xy + \frac{1}{3}y^3 + ze^z - e^z\right)$

(a) work $= \int_A^B \mathbf{F} \cdot \frac{d\mathbf{r}}{dt} dt = \int_A^B \mathbf{F} \cdot d\mathbf{r} = \left[\frac{1}{3}x^3 + xy + \frac{1}{3}y^3 + ze^z - e^z\right]_{(1,0,0)}^{(1,0,1)} = \left(\frac{1}{3} + 0 + 0 + e - e\right) - \left(\frac{1}{3} + 0 + 0 - 1\right)$

$= 1$

(b) work $= \int_A^B \mathbf{F} \cdot d\mathbf{r} = \left[\frac{1}{3}x^3 + xy + \frac{1}{3}y^3 + ze^z - e^z\right]_{(1,0,0)}^{(1,0,1)} = 1$

(c) work $= \int_A^B \mathbf{F} \cdot d\mathbf{r} = \left[\frac{1}{3}x^3 + xy + \frac{1}{3}y^3 + ze^z - e^z\right]_{(1,0,0)}^{(1,0,1)} = 1$

Note: Since $\mathbf{F}$ is conservative, $\int_A^B \mathbf{F} \cdot d\mathbf{r}$ is independent of the path from $(1, 0, 0)$ to $(1, 0, 1)$.

31. (a) $\mathbf{F} = \nabla(x^3y^2) \Rightarrow \mathbf{F} = 3x^2y^2\mathbf{i} + 2x^3y\mathbf{j}$; let C_1 be the path from $(-1, 1)$ to $(0, 0) \Rightarrow x = t - 1$ and

$y = -t + 1, 0 \le t \le 1 \Rightarrow \mathbf{F} = 3(t-1)^2(-t+1)^2\mathbf{i} + 2(t-1)^3(-t+1)\mathbf{j} = 3(t-1)^4\mathbf{i} - 2(t-1)^4\mathbf{j}$

and $\mathbf{r}_1 = (t-1)\mathbf{i} + (-t+1)\mathbf{j} \Rightarrow d\mathbf{r}_1 = dt\,\mathbf{i} - dt\,\mathbf{j} \Rightarrow \int_{C_1} \mathbf{F} \cdot d\mathbf{r}_1 = \int_0^1 [3(t-1)^4 + 2(t-1)^4] \, dt$

$= \int_0^1 5(t-1)^4 \, dt = [(t-1)^5]_0^1 = 1$; let C_2 be the path from $(0, 0)$ to $(1, 1) \Rightarrow x = t$ and $y = t$,

$0 \le t \le 1 \Rightarrow \mathbf{F} = 3t^4\mathbf{i} + 2t^4\mathbf{j}$ and $\mathbf{r}_2 = t\mathbf{i} + t\mathbf{j} \Rightarrow d\mathbf{r}_2 = dt\,\mathbf{i} + dt\,\mathbf{j} \Rightarrow \int_{C_2} \mathbf{F} \cdot d\mathbf{r}_2 = \int_0^1 (3t^4 + 2t^4) \, dt$

$= \int_0^1 5t^4 \, dt = 1 \Rightarrow \int_C \mathbf{F} \cdot d\mathbf{r} = \int_{C_1} \mathbf{F} \cdot d\mathbf{r}_1 + \int_{C_2} \mathbf{F} \cdot d\mathbf{r}_2 = 2$

(b) Since $f(x, y) = x^3y^2$ is a potential function for $\mathbf{F}$, $\int_{(-1,1)}^{(1,1)} \mathbf{F} \cdot d\mathbf{r} = f(1, 1) - f(-1, 1) = 2$

33. (a) If the differential form is exact, then $\frac{\partial P}{\partial y} = \frac{\partial N}{\partial z} \Rightarrow 2ay = cy$ for all $y \Rightarrow 2a = c, \frac{\partial M}{\partial z} = \frac{\partial P}{\partial x} \Rightarrow 2cx = 2cx$ for

all x, and $\frac{\partial N}{\partial x} = \frac{\partial M}{\partial y} \Rightarrow by = 2ay$ for all $y \Rightarrow b = 2a$ and $c = 2a$

(b) $\mathbf{F} = \nabla f \Rightarrow$ the differential form with $a = 1$ in part (a) is exact $\Rightarrow b = 2$ and $c = 2$

35. The path will not matter; the work along any path will be the same because the field is conservative.

37. Let the coordinates of points A and B be (x_A, y_A, z_A) and (x_B, y_B, z_B), respectively. The force $\mathbf{F} = a\mathbf{i} + b\mathbf{j} + c\mathbf{k}$ is conservative because all the partial derivatives of M, N, and P are zero. Therefore, the potential function is $f(x, y, z) = ax + by + cz + C$, and the work done by the force in moving a particle along any path from A to B is $f(B) - f(A) = f(x_B, y_B, z_B) - f(x_A, y_A, z_A) = (ax_B + by_B + cz_B + C) - (ax_A + by_A + cz_A + C)$

$$= a(x_B - x_A) + b(y_B - y_A) + c(z_B - z_A) = \mathbf{F} \cdot \vec{BA}$$

16.4 GREEN'S THEOREM IN THE PLANE

1. $M = -y = -a \sin t$, $N = x = a \cos t$, $dx = -a \sin t\, dt$, $dy = a \cos t\, dt \Rightarrow \frac{\partial M}{\partial x} = 0$, $\frac{\partial M}{\partial y} = -1$, $\frac{\partial N}{\partial x} = 1$, and

$\frac{\partial N}{\partial y} = 0$;

Equation (11): $\oint_C M\, dy - N\, dx = \int_0^{2\pi} [(-a \sin t)(a \cos t) - (a \cos t)(-a \sin t)]\, dt = \int_0^{2\pi} 0\, dt = 0$;

$\iint_R \left(\frac{\partial M}{\partial x} + \frac{\partial N}{\partial y} \right) dx\, dy = \iint_R 0\, dx\, dy = 0$, Flux

Equation (12): $\oint_C M\, dx + N\, dy = \int_0^{2\pi} [(-a \sin t)(-a \sin t) - (a \cos t)(a \cos t)]\, dt = \int_0^{2\pi} a^2\, dt = 2\pi a^2$;

$\iint_R \left(\frac{\partial N}{\partial x} - \frac{\partial M}{\partial y} \right) dx\, dy = \int_{-a}^{a} \int_{-c}^{\sqrt{a^2-x^2}} 2\, dy\, dx = \int_{-a}^{a} 4\sqrt{a^2-x^2}\, dx = 4 \left[\frac{x}{2}\sqrt{a^2-x^2} + \frac{a^2}{2} \sin^{-1} \frac{x}{a} \right]_{-a}^{a}$

$= 2a^2 \left(\frac{\pi}{2} + \frac{\pi}{2} \right) = 2a^2 \pi$, Circulation

3. $M = 2x = 2a \cos t$, $N = -3y = -3a \sin t$, $dx = -a \sin t\, dt$, $dy = a \cos t\, dt \Rightarrow \frac{\partial M}{\partial x} = 2$, $\frac{\partial M}{\partial y} = 0$, $\frac{\partial N}{\partial x} = 0$, and

$\frac{\partial N}{\partial y} = -3$;

Equation (11): $\oint_C M\, dy - N\, dx = \int_0^{2\pi} [(2a \cos t)(a \cos t) + (3a \sin t)(-a \sin t)]\, dt$

$= \int_0^{2\pi} (2a^2 \cos^2 t - 3a^2 \sin^2 t)\, dt = 2a^2 \left[\frac{t}{2} + \frac{\sin 2t}{4} \right]_0^{2\pi} - 3a^2 \left[\frac{t}{2} - \frac{\sin 2t}{4} \right]_0^{2\pi} = 2\pi a^2 - 3\pi a^2 = -\pi a^2$;

$\iint_R \left(\frac{\partial M}{\partial x} + \frac{\partial N}{\partial y} \right) = \iint_R -1\, dx\, dy = \int_0^{2\pi} \int_0^{a} -r\, dr\, d\theta = \int_0^{2\pi} -\frac{a^2}{2}\, d\theta = -\pi a^2$, Flux

Equation (12): $\oint_C M\, dx + N\, dy = \int_0^{2\pi} [(2a \cos t)(-a \sin t) + (-3a \sin t)(a \cos t)]\, dt$

$= \int_0^{2\pi} (-2a^2 \sin t \cos t - 3a^2 \sin t \cos t)\, dt = -5a^2 \left[\frac{1}{2} \sin^2 t \right]_0^{2\pi} = 0$; $\iint_R 0\, dx\, dy = 0$, Circulation

5. $M = x - y$, $N = y - x \Rightarrow \frac{\partial M}{\partial x} = 1$, $\frac{\partial M}{\partial y} = -1$, $\frac{\partial N}{\partial x} = -1$, $\frac{\partial N}{\partial y} = 1 \Rightarrow$ Flux $= \iint_R 2\, dx\, dy = \int_0^1 \int_0^1 2\, dx\, dy = 2$;

Circ $= \iint_R [-1 - (-1)]\, dx\, dy = 0$

7. $M = y^2 - x^2$, $N = x^2 + y^2 \Rightarrow \frac{\partial M}{\partial x} = -2x$, $\frac{\partial M}{\partial y} = 2y$, $\frac{\partial N}{\partial x} = 2x$, $\frac{\partial N}{\partial y} = 2y \Rightarrow$ Flux $= \iint_R (-2x + 2y)\, dx\, dy$

$= \int_0^3 \int_0^x (-2x + 2y)\, dy\, dx = \int_0^3 (-2x^2 + x^2)\, dx = \left[-\frac{1}{3} x^3 \right]_0^3 = -9$; Circ $= \iint_R (2x - 2y)\, dx\, dy$

$= \int_0^3 \int_0^x (2x - 2y)\, dy\, dx = \int_0^3 x^2\, dx = 9$

9. $M = x + e^x \sin y$, $N = x + e^x \cos y \Rightarrow \frac{\partial M}{\partial x} = 1 + e^x \sin y$, $\frac{\partial M}{\partial y} = e^x \cos y$, $\frac{\partial N}{\partial x} = 1 + e^x \cos y$, $\frac{\partial N}{\partial y} = -e^x \sin y$

$\Rightarrow$ Flux $= \iint_R dx\, dy = \int_{-\pi/4}^{\pi/4} \int_0^{\sqrt{\cos 2\theta}} r\, dr\, d\theta = \int_{-\pi/4}^{\pi/4} \left(\frac{1}{2} \cos 2\theta \right) d\theta = \left[\frac{1}{4} \sin 2\theta \right]_{-\pi/4}^{\pi/4} = \frac{1}{2}$;

Circ $= \iint_R (1 + e^x \cos y - e^x \cos y)\, dx\, dy = \iint_R dx\, dy = \int_{-\pi/4}^{\pi/4} \int_0^{\sqrt{\cos 2\theta}} r\, dr\, d\theta = \int_{-\pi/4}^{\pi/4} \left(\frac{1}{2} \cos 2\theta \right) d\theta = \frac{1}{2}$

11. $M = xy$, $N = y^2 \Rightarrow \frac{\partial M}{\partial x} = y$, $\frac{\partial M}{\partial y} = x$, $\frac{\partial N}{\partial x} = 0$, $\frac{\partial N}{\partial y} = 2y \Rightarrow$ Flux $= \iint_R (y + 2y)\, dy\, dx = \int_0^1 \int_{x^2}^x 3y\, dy\, dx$

$= \int_0^1 \left(\frac{3x^2}{2} - \frac{3x^4}{2} \right) dx = \frac{1}{5}$; Circ $= \iint_R -x\, dy\, dx = \int_0^1 \int_{x^2}^x -x\, dy\, dx = \int_0^1 (-x^2 + x^3)\, dx = -\frac{1}{12}$

13. $M = 3xy - \frac{x}{1+y^2}$, $N = e^x + \tan^{-1} y \Rightarrow \frac{\partial M}{\partial x} = 3y - \frac{1}{1+y^2}$, $\frac{\partial N}{\partial y} = \frac{1}{1+y^2}$

$\Rightarrow$ Flux $= \iint_R \left(3y - \frac{1}{1+y^2} + \frac{1}{1+y^2}\right) dx\, dy = \iint_R 3y\, dx\, dy = \int_0^{2\pi} \int_0^{a(1+\cos\theta)} (3r\sin\theta)\, r\, dr\, d\theta$

$= \int_0^{2\pi} a^3(1+\cos\theta)^3(\sin\theta)\, d\theta = \left[-\frac{a^3}{4}(1+\cos\theta)^4\right]_0^{2\pi} = -4a^3 - (-4a^3) = 0$

15. $M = 2xy^3$, $N = 4x^2y^2 \Rightarrow \frac{\partial M}{\partial y} = 6xy^2$, $\frac{\partial N}{\partial x} = 8xy^2 \Rightarrow$ work $= \oint_C 2xy^3\, dx + 4x^2y^2\, dy = \iint_R (8xy^2 - 6xy^2)\, dx\, dy$

$= \int_0^1 \int_0^{x^3} 2xy^2\, dy\, dx = \int_0^1 \frac{2}{3} x^{10}\, dx = \frac{2}{33}$

17. $M = y^2$, $N = x^2 \Rightarrow \frac{\partial M}{\partial y} = 2y$, $\frac{\partial N}{\partial x} = 2x \Rightarrow \oint_C y^2\, dx + x^2\, dy = \iint_R (2x - 2y)\, dy\, dx$

$= \int_0^1 \int_0^{1-x} (2x - 2y)\, dy\, dx = \int_0^1 (-3x^2 + 4x - 1)\, dx = [-x^3 + 2x^2 - x]_0^1 = -1 + 2 - 1 = 0$

19. $M = 6y + x$, $N = y + 2x \Rightarrow \frac{\partial M}{\partial y} = 6$, $\frac{\partial N}{\partial x} = 2 \Rightarrow \oint_C (6y + x)\, dx + (y + 2x)\, dy = \iint_R (2 - 6)\, dy\, dx$

$= -4(\text{Area of the circle}) = -16\pi$

21. $M = x = a\cos t$, $N = y = a\sin t \Rightarrow dx = -a\sin t\, dt$, $dy = a\cos t\, dt \Rightarrow$ Area $= \frac{1}{2} \oint_C x\, dy - y\, dx$

$= \frac{1}{2} \int_0^{2\pi} (a^2\cos^2 t + a^2\sin^2 t)\, dt = \frac{1}{2} \int_0^{2\pi} a^2\, dt = \pi a^2$

23. $M = x = a\cos^3 t$, $N = y = \sin^3 t \Rightarrow dx = -3\cos^2 t \sin t\, dt$, $dy = 3\sin^2 t \cos t\, dt \Rightarrow$ Area $= \frac{1}{2} \oint_C x\, dy - y\, dx$

$= \frac{1}{2} \int_0^{2\pi} (3\sin^2 t \cos^2 t)(\cos^2 t + \sin^2 t)\, dt = \frac{1}{2} \int_0^{2\pi} (3\sin^2 t \cos^2 t)\, dt = \frac{3}{8} \int_0^{2\pi} \sin^2 2t\, dt = \frac{3}{16} \int_0^{4\pi} \sin^2 u\, du$

$= \frac{3}{16} \left[\frac{u}{2} - \frac{\sin 2u}{4}\right]_0^{4\pi} = \frac{3}{8}\pi$

25. (a) $M = f(x)$, $N = g(y) \Rightarrow \frac{\partial M}{\partial y} = 0$, $\frac{\partial N}{\partial x} = 0 \Rightarrow \oint_C f(x)\, dx + g(y)\, dy = \iint_R \left(\frac{\partial N}{\partial x} - \frac{\partial M}{\partial y}\right) dx\, dy$

$= \iint_R 0\, dx\, dy = 0$

(b) $M = ky$, $N = hx \Rightarrow \frac{\partial M}{\partial y} = k$, $\frac{\partial N}{\partial x} = h \Rightarrow \oint_C ky\, dx + hx\, dy = \iint_R \left(\frac{\partial N}{\partial x} - \frac{\partial M}{\partial y}\right) dx\, dy$

$= \iint_R (h - k)\, dx\, dy = (h - k)(\text{Area of the region})$

27. The integral is 0 for any simple closed plane curve C. The reasoning: By the tangential form of Green's

Theorem, with $M = 4x^3y$ and $N = x^4$, $\oint_C 4x^3y\, dx + x^4\, dy = \iint_R \left[\frac{\partial}{\partial x}(x^4) - \frac{\partial}{\partial y}(4x^3y)\right] dx\, dy$

$= \iint_R \underbrace{(4x^3 - 4x^3)}_{0}\, dx\, dy = 0$.

29. Let $M = x$ and $N = 0 \Rightarrow \frac{\partial M}{\partial x} = 1$ and $\frac{\partial N}{\partial y} = 0 \Rightarrow \oint_C M\, dy - N\, dx = \iint_R \left(\frac{\partial M}{\partial x} + \frac{\partial N}{\partial y}\right) dx\, dy \Rightarrow \oint_C x\, dy$

$= \iint_R (1 + 0)\, dx\, dy \Rightarrow$ Area of $R = \iint_R dx\, dy = \oint_C x\, dy$; similarly, $M = y$ and $N = 0 \Rightarrow \frac{\partial M}{\partial y} = 1$ and

$\frac{\partial N}{\partial x} = 0 \Rightarrow \oint_C M\, dx + N\, dy = \iint_R \left(\frac{\partial N}{\partial x} + \frac{\partial M}{\partial y}\right) dy\, dx \Rightarrow \oint_C y\, dx = \iint_R (0 - 1)\, dy\, dx \Rightarrow -\oint_C y\, dx$

$= \iint_R dx\, dy = \text{Area of } R$

31. Let $\delta(x, y) = 1 \Rightarrow \bar{x} = \frac{M_y}{M} = \frac{\iint_R x \delta(x,y)\,dA}{\iint_R \delta(x,y)\,dA} = \frac{\iint_R x\,dA}{\iint_R dA} = \frac{\iint_R x\,dA}{A} \Rightarrow A\bar{x} = \iint_R x\,dA = \iint_R (x + 0)\,dx\,dy$

$= \oint_C \frac{x^2}{2}\,dy$, $A\bar{x} = \iint_R x\,dA = \iint_R (0 + x)\,dx\,dy = -\oint_C xy\,dx$, and $A\bar{x} = \iint_R x\,dA = \iint_R \left(\frac{2}{3}x + \frac{1}{3}x\right)dx\,dy$

$= \oint_C \frac{1}{3}x^2\,dy - \frac{1}{3}xy\,dx \Rightarrow \frac{1}{2}\oint_C x^2\,dy = -\oint_C xy\,dx = \frac{1}{3}\oint_C x^2\,dy - xy\,dx = A\bar{x}$

33. $M = \frac{\partial f}{\partial y}$, $N = -\frac{\partial f}{\partial x} \Rightarrow \frac{\partial M}{\partial y} = \frac{\partial^2 f}{\partial y^2}$, $\frac{\partial N}{\partial x} = -\frac{\partial^2 f}{\partial x^2} \Rightarrow \oint_C \frac{\partial f}{\partial y}\,dx - \frac{\partial f}{\partial x}\,dy = \iint_R \left(-\frac{\partial^2 f}{\partial x^2} - \frac{\partial^2 f}{\partial y^2}\right)dx\,dy = 0$ for such

curves C

35. (a) $\nabla f = \left(\frac{2x}{x^2+y^2}\right)\mathbf{i} + \left(\frac{2y}{x^2+y^2}\right)\mathbf{j} \Rightarrow M = \frac{2x}{x^2+y^2}$, $N = \frac{2y}{x^2+y^2}$; since M, N are discontinuous at $(0,0)$, we

compute $\int_C \nabla f \cdot \mathbf{n}\,ds$ directly since Green's Theorem does not apply. Let $x = a\cos t$, $y = a\sin t \Rightarrow dx = -a\sin t\,dt$,

$dy = a\cos t\,dt$, $M = \frac{2}{a}\cos t$, $N = \frac{2}{a}\sin t$, $0 \le t \le 2\pi$, so $\int_C \nabla f \cdot \mathbf{n}\,ds = \int_C M\,dy - N\,dx$

$= \int_0^{2\pi} \left[\left(\frac{2}{a}\cos t\right)(a\cos t) - \left(\frac{2}{a}\sin t\right)(-a\sin t)\right]dt = \int_0^{2\pi} 2(\cos^2 t + \sin^2 t)dt = 4\pi$. Note that this holds for any

$a > 0$, so $\int_C \nabla f \cdot \mathbf{n}\,ds = 4\pi$ for any circle C centered at $(0, 0)$ traversed counterclockwise and $\int_C \nabla f \cdot \mathbf{n}\,ds = -4\pi$

if C is traversed clockwise.

(b) If K does not enclose the point $(0, 0)$ we may apply Green's Theorem: $\int_C \nabla f \cdot \mathbf{n}\,ds = \int_C M\,dy - N\,dx$

$= \iint_R \left(\frac{\partial M}{\partial x} + \frac{\partial N}{\partial y}\right)dx\,dy = \iint_R \left(\frac{2(y^2 - x^2)}{(x^2+y^2)^2} + \frac{2(x^2 - y^2)}{(x^2+y^2)^2}\right)dx\,dy = \iint_R 0\,dx\,dy = 0$. If K does enclose the point

$(0, 0)$ we proceed as in Example 6:

Choose a small enough so that the circle C centered at $(0, 0)$ of radius a lies entirely within K. Green's Theorem

applies to the region R that lies between K and C. Thus, as before, $0 = \iint_R \left(\frac{\partial M}{\partial x} + \frac{\partial N}{\partial y}\right)dx\,dy$

$= \int_K M\,dy - N\,dx + \int_C M\,dy - N\,dx$ where K is traversed counterclockwise and C is traversed clockwise.

Hence by part (a) $0 = \left[\int_K M\,dy - N\,dx\right] - 4\pi \Rightarrow 4\pi = \int_K M\,dy - N\,dx = \int_K \nabla f \cdot \mathbf{n}\,ds$. We have shown:

$\int_K \nabla f \cdot \mathbf{n}\,ds = \begin{cases} 0 & \text{if } (0,0) \text{ lies inside K} \\ 4\pi & \text{if } (0,0) \text{ lies outside K} \end{cases}$

37. $\int_{g_1(y)}^{g_2(y)} \frac{\partial N}{\partial x}\,dx\,dy = N(g_2(y), y) - N(g_1(y), y) \Rightarrow \int_c^d \int_{g_1(y)}^{g_2(y)} \left(\frac{\partial N}{\partial x}\,dx\right)dy = \int_c^d [N(g_2(y), y) - N(g_1(y), y)]\,dy$

$= \int_c^d N(g_2(y), y)\,dy - \int_c^d N(g_1(y), y)\,dy = \int_c^d N(g_2(y), y)\,dy + \int_d^c N(g_1(y), y)\,dy = \int_{C_2} N\,dy + \int_{C_1} N\,dy$

$= \oint_C dy \Rightarrow \oint_C N\,dy = \iint_R \frac{\partial N}{\partial x}\,dx\,dy$

39. The curl of a conservative two-dimensional field is zero. The reasoning: A two-dimensional field $\mathbf{F} = M\mathbf{i} + N\mathbf{j}$
    can be considered to be the restriction to the xy-plane of a three-dimensional field whose k component is zero,
    and whose **i** and **j** components are independent of z. For such a field to be conservative, we must have
    $\frac{\partial N}{\partial x} = \frac{\partial M}{\partial y}$ by the component test in Section 16.3 $\Rightarrow$ curl $\mathbf{F} = \frac{\partial N}{\partial x} - \frac{\partial M}{\partial y} = 0$.

16.5 SURFACE AREA AND SURFACE INTEGRALS

1. $\mathbf{p} = \mathbf{k}$, $\nabla f = 2x\mathbf{i} + 2y\mathbf{j} - \mathbf{k} \Rightarrow |\nabla f| = \sqrt{(2x)^2 + (2y)^2 + (-1)^2} = \sqrt{4x^2 + 4y^2 + 1}$ and $|\nabla f \cdot \mathbf{p}| = 1$;
    $z = 2 \Rightarrow x^2 + y^2 = 2$; thus $S = \iint_R \frac{|\nabla f|}{|\nabla f \cdot \mathbf{p}|}\,dA = \iint_R \sqrt{4x^2 + 4y^2 + 1}\,dx\,dy$

    $= \iint_R \sqrt{4r^2\cos^2\theta + 4r^2\sin^2\theta + 1}\,r\,dr\,d\theta = \int_0^{2\pi}\int_0^{\sqrt{2}} \sqrt{4r^2 + 1}\,r\,dr\,d\theta = \int_0^{2\pi} \left[\frac{1}{12}(4r^2 + 1)^{3/2}\right]_0^{\sqrt{2}} d\theta$

$$= \int_0^{2\pi} \frac{13}{6} \, d\theta = \frac{13}{3} \pi$$

3. $\mathbf{p} = \mathbf{k}$, $\nabla f = \mathbf{i} + 2\mathbf{j} + 2\mathbf{k} \Rightarrow |\nabla f| = 3$ and $|\nabla f \cdot \mathbf{p}| = 2$; $x = y^2$ and $x = 2 - y^2$ intersect at $(1, 1)$ and $(1, -1)$

$\Rightarrow S = \iint_R \frac{|\nabla f|}{|\nabla f \cdot \mathbf{p}|} \, dA = \iint_R \frac{3}{2} \, dx \, dy = \int_{-1}^1 \int_{y^2}^{2-y^2} \frac{3}{2} \, dx \, dy = \int_{-1}^1 (3 - 3y^2) \, dy = 4$

5. $\mathbf{p} = \mathbf{k}$, $\nabla f = 2x\mathbf{i} - 2\mathbf{j} - 2\mathbf{k} \Rightarrow |\nabla f| = \sqrt{(2x)^2 + (-2)^2 + (-2)^2} = \sqrt{4x^2 + 8} = 2\sqrt{x^2 + 2}$ and $|\nabla f \cdot \mathbf{p}| = 2$

$\Rightarrow S = \iint_R \frac{|\nabla f|}{|\nabla f \cdot \mathbf{p}|} \, dA = \iint_R \frac{2\sqrt{x^2+2}}{2} \, dx \, dy = \int_0^2 \int_0^{3x} \sqrt{x^2 + 2} \, dy \, dx = \int_0^2 3x \sqrt{x^2 + 2} \, dx = \left[(x^2 + 2)^{3/2} \right]_0^2$

$= 6\sqrt{6} - 2\sqrt{2}$

7. $\mathbf{p} = \mathbf{k}$, $\nabla f = c\mathbf{i} - \mathbf{k} \Rightarrow |\nabla f| = \sqrt{c^2 + 1}$ and $|\nabla f \cdot \mathbf{p}| = 1 \Rightarrow S = \iint_R \frac{|\nabla f|}{|\nabla f \cdot \mathbf{p}|} \, dA = \iint_R \sqrt{c^2 + 1} \, dx \, dy$

$= \int_0^{2\pi} \int_0^1 \sqrt{c^2 + 1} \, r \, dr \, d\theta = \int_0^{2\pi} \frac{\sqrt{c^2+1}}{2} \, d\theta = \pi\sqrt{c^2 + 1}$

9. $\mathbf{p} = \mathbf{i}$, $\nabla f = \mathbf{i} + 2y\mathbf{j} + 2z\mathbf{k} \Rightarrow |\nabla f| = \sqrt{1^2 + (2y)^2 + (2z)^2} = \sqrt{1 + 4y^2 + 4z^2}$ and $|\nabla f \cdot \mathbf{p}| = 1$; $1 \le y^2 + z^2 \le 4$

$\Rightarrow S = \iint_R \frac{|\nabla f|}{|\nabla f \cdot \mathbf{p}|} \, dA = \iint_R \sqrt{1 + 4y^2 + 4z^2} \, dy \, dz = \int_0^{2\pi} \int_1^2 \sqrt{1 + 4r^2 \cos^2 \theta + 4r^2 \sin^2 \theta} \, r \, dr \, d\theta$

$= \int_0^{2\pi} \int_1^2 \sqrt{1 + 4r^2} \, r \, dr \, d\theta = \int_0^{2\pi} \left[\frac{1}{12} (1 + 4r^2)^{3/2} \right]_1^2 \, d\theta = \int_0^{2\pi} \frac{1}{12} \left(17\sqrt{17} - 5\sqrt{5} \right) \, d\theta = \frac{\pi}{6} \left(17\sqrt{17} - 5\sqrt{5} \right)$

11. $\mathbf{p} = \mathbf{k}$, $\nabla f = \left(2x - \frac{2}{x} \right) \mathbf{i} + \sqrt{15}\mathbf{j} - \mathbf{k} \Rightarrow |\nabla f| = \sqrt{\left(2x - \frac{2}{x} \right)^2 + \left(\sqrt{15} \right)^2 + (-1)^2} = \sqrt{4x^2 + 8 + \frac{4}{x^2}} = \sqrt{\left(2x + \frac{2}{x} \right)^2}$

$= 2x + \frac{2}{x}$, on $1 \le x \le 2$ and $|\nabla f \cdot \mathbf{p}| = 1 \Rightarrow S = \iint_R \frac{|\nabla f|}{|\nabla f \cdot \mathbf{p}|} \, dA = \iint_R (2x + 2x^{-1}) \, dx \, dy$

$= \int_0^1 \int_1^2 (2x + 2x^{-1}) \, dx \, dy = \int_0^1 [x^2 + 2 \ln x]_1^2 \, dy = \int_0^1 (3 + 2 \ln 2) \, dy = 3 + 2 \ln 2$

13. The bottom face S of the cube is in the xy-plane $\Rightarrow z = 0 \Rightarrow g(x, y, 0) = x + y$ and $f(x, y, z) = z = 0 \Rightarrow \mathbf{p} = \mathbf{k}$

and $\nabla f = \mathbf{k} \Rightarrow |\nabla f| = 1$ and $|\nabla f \cdot \mathbf{p}| = 1 \Rightarrow d\sigma = dx \, dy \Rightarrow \iint_S g \, d\sigma = \iint_R (x + y) \, dx \, dy$

$= \int_0^a \int_0^a (x + y) \, dx \, dy = \int_0^a \left(\frac{a^2}{2} + ay \right) \, dy = a^3$. Because of symmetry, we also get a^3 over the face of the cube

in the xz-plane and a^3 over the face of the cube in the yz-plane. Next, on the top of the cube, $g(x, y, z)$

$= g(x, y, a) = x + y + a$ and $f(x, y, z) = z = a \Rightarrow \mathbf{p} = \mathbf{k}$ and $\nabla f = \mathbf{k} \Rightarrow |\nabla f| = 1$ and $|\nabla f \cdot \mathbf{p}| = 1 \Rightarrow d\sigma = dx \, dy$

$\iint_S g \, d\sigma = \iint_R (x + y + a) \, dx \, dy = \int_0^a \int_0^a (x + y + a) \, dx \, dy = \int_0^a \int_0^a (x + y) \, dx \, dy + \int_0^a \int_0^a a \, dx \, dy = 2a^3$.

Because of symmetry, the integral is also $2a^3$ over each of the other two faces. Therefore,

$\iint_{\text{cube}} (x + y + z) \, d\sigma = 3 (a^3 + 2a^3) = 9a^3$.

15. On the faces in the coordinate planes, $g(x, y, z) = 0 \Rightarrow$ the integral over these faces is 0.

On the face $x = a$, we have $f(x, y, z) = x = a$ and $g(x, y, z) = g(a, y, z) = ayz \Rightarrow \mathbf{p} = \mathbf{i}$ and $\nabla f = \mathbf{i} \Rightarrow |\nabla f| = 1$

and $|\nabla f \cdot \mathbf{p}| = 1 \Rightarrow d\sigma = dy \, dz \Rightarrow \iint_S g \, d\sigma = \iint_S ayz \, d\sigma = \int_0^c \int_0^b ayz \, dy \, dz = \frac{ab^2c^2}{4}$.

On the face $y = b$, we have $f(x, y, z) = y = b$ and $g(x, y, z) = g(x, b, z) = bxz \Rightarrow \mathbf{p} = \mathbf{j}$ and $\nabla f = \mathbf{j} \Rightarrow |\nabla f| = 1$

and $|\nabla f \cdot \mathbf{p}| = 1 \Rightarrow d\sigma = dx \, dz \Rightarrow \iint_S g \, d\sigma = \iint_S bxz \, d\sigma = \int_0^c \int_0^a bxz \, dx \, dz = \frac{a^2bc^2}{4}$.

On the face $z = c$, we have $f(x, y, z) = z = c$ and $g(x, y, z) = g(x, y, c) = cxy \Rightarrow \mathbf{p} = \mathbf{k}$ and $\nabla f = \mathbf{k} \Rightarrow |\nabla f| = 1$

and $|\nabla f \cdot \mathbf{p}| = 1 \Rightarrow d\sigma = dy \, dx \Rightarrow \iint_S g \, d\sigma = \iint_S cxy \, d\sigma = \int_0^b \int_0^a cxy \, dx \, dy = \frac{a^2b^2c}{4}$. Therefore,

$$\iint_S g(x, y, z) \, d\sigma = \frac{abc(ab + ac + bc)}{4}.$$

17. $f(x, y, z) = 2x + 2y + z = 2 \Rightarrow \nabla f = 2\mathbf{i} + 2\mathbf{j} + \mathbf{k}$ and $g(x, y, z) = x + y + (2 - 2x - 2y) = 2 - x - y \Rightarrow \mathbf{p} = \mathbf{k}$,

$|\nabla f| = 3$ and $|\nabla f \cdot \mathbf{p}| = 1 \Rightarrow d\sigma = 3 \, dy \, dx; z = 0 \Rightarrow 2x + 2y = 2 \Rightarrow y = 1 - x \Rightarrow \iint_S g \, d\sigma = \iint_S (2 - x - y) \, d\sigma$

$= 3 \int_0^1 \int_0^{1-x} (2 - x - y) \, dy \, dx = 3 \int_0^1 \left[(2 - x)(1 - x) - \frac{1}{2}(1 - x)^2 \right] dx = 3 \int_0^1 \left(\frac{3}{2} - 2x + \frac{x^2}{2} \right) dx = 2$

19. $g(x, y, z) = z, \mathbf{p} = \mathbf{k} \Rightarrow \nabla g = \mathbf{k} \Rightarrow |\nabla g| = 1$ and $|\nabla g \cdot \mathbf{p}| = 1 \Rightarrow \text{Flux} = \iint_S \mathbf{F} \cdot \mathbf{n} \, d\sigma = \iint_R (\mathbf{F} \cdot \mathbf{k}) \, dA$

$= \int_0^2 \int_0^3 3 \, dy \, dx = 18$

21. $\nabla g = 2x\mathbf{i} + 2y\mathbf{j} + 2z\mathbf{k} \Rightarrow |\nabla g| = \sqrt{4x^2 + 4y^2 + 4z^2} = 2a; \mathbf{n} = \frac{2x\mathbf{i} + 2y\mathbf{j} + 2z\mathbf{k}}{2\sqrt{x^2 + y^2 + z^2}} = \frac{x\mathbf{i} + y\mathbf{j} + z\mathbf{k}}{a} \Rightarrow \mathbf{F} \cdot \mathbf{n} = \frac{z^2}{a};$

$|\nabla g \cdot \mathbf{k}| = 2z \Rightarrow d\sigma = \frac{2a}{2z} \, dA \Rightarrow \text{Flux} = \iint_R \left(\frac{z^2}{a} \right) \left(\frac{a}{z} \right) dA = \iint_R z \, dA = \iint_R \sqrt{a^2 - (x^2 + y^2)} \, dx \, dy$

$= \int_0^{\pi/2} \int_0^a \sqrt{a^2 - r^2} \, r \, dr \, d\theta = \frac{\pi a^3}{6}$

23. From Exercise 21, $\mathbf{n} = \frac{x\mathbf{i} + y\mathbf{j} + z\mathbf{k}}{a}$ and $d\sigma = \frac{a}{z} \, dA \Rightarrow \mathbf{F} \cdot \mathbf{n} = \frac{xy}{a} - \frac{xy}{a} + \frac{z}{a} = \frac{z}{a} \Rightarrow \text{Flux} = \iint_R \left(\frac{z}{a} \right) \left(\frac{a}{z} \right) dA$

$= \iint_R 1 \, dA = \frac{\pi a^2}{4}$

25. From Exercise 21, $\mathbf{n} = \frac{x\mathbf{i} + y\mathbf{j} + z\mathbf{k}}{a}$ and $d\sigma = \frac{a}{z} \, dA \Rightarrow \mathbf{F} \cdot \mathbf{n} = \frac{x^2}{a} + \frac{y^2}{a} + \frac{z^2}{a} = a \Rightarrow \text{Flux}$

$= \iint_R a \left(\frac{a}{z} \right) dA = \iint_R \frac{a^2}{z} \, dA = \iint_R \frac{a^2}{\sqrt{a^2 - (x^2 + y^2)}} \, dA = \int_0^{\pi/2} \int_0^a \frac{a^2}{\sqrt{a^2 - r^2}} \, r \, dr \, d\theta$

$= \int_0^{\pi/2} a^2 \left[-\sqrt{a^2 - r^2} \right]_0^a d\theta = \frac{\pi a^3}{2}$

27. $g(x, y, z) = y^2 + z = 4 \Rightarrow \nabla g = 2y\mathbf{j} + \mathbf{k} \Rightarrow |\nabla g| = \sqrt{4y^2 + 1} \Rightarrow \mathbf{n} = \frac{2y\mathbf{j} + \mathbf{k}}{\sqrt{4y^2 + 1}}$

$\Rightarrow \mathbf{F} \cdot \mathbf{n} = \frac{2xy - 3z}{\sqrt{4y^2 + 1}}; \mathbf{p} = \mathbf{k} \Rightarrow |\nabla g \cdot \mathbf{p}| = 1 \Rightarrow d\sigma = \sqrt{4y^2 + 1} \, dA \Rightarrow \text{Flux}$

$= \iint_R \left(\frac{2xy - 3z}{\sqrt{4y^2 + 1}} \right) \sqrt{4y^2 + 1} \, dA = \iint_R (2xy - 3z) \, dA; z = 0 \text{ and } z = 4 - y^2 \Rightarrow y^2 = 4$

$\Rightarrow \text{Flux} = \iint_R [2xy - 3(4 - y^2)] \, dA = \int_0^1 \int_{-2}^2 (2xy - 12 + 3y^2) \, dy \, dx = \int_0^1 [xy^2 - 12y + y^3]_{-2}^2 \, dx$

$= \int_0^1 -32 \, dx = -32$

29. $g(x, y, z) = y - e^x = 0 \Rightarrow \nabla g = -e^x\mathbf{i} + \mathbf{j} \Rightarrow |\nabla g| = \sqrt{e^{2x} + 1} \Rightarrow \mathbf{n} = \frac{e^x\mathbf{i} - \mathbf{j}}{\sqrt{e^{2x} + 1}} \Rightarrow \mathbf{F} \cdot \mathbf{n} = \frac{-2e^x - 2y}{\sqrt{e^{2x} + 1}}; \mathbf{p} = \mathbf{i}$

$\Rightarrow |\nabla g \cdot \mathbf{p}| = e^x \Rightarrow d\sigma = \frac{\sqrt{e^{2x} + 1}}{e^x} \, dA \Rightarrow \text{Flux} = \iint_R \left(\frac{-2e^x - 2y}{\sqrt{e^{2x} + 1}} \right) \left(\frac{\sqrt{e^{2x} + 1}}{e^x} \right) dA = \iint_R \frac{-2e^x - 2e^x}{e^x} \, dA$

$= \iint_R -4 \, dA = \int_0^1 \int_1^2 -4 \, dy \, dz = -4$

31. On the face $z = a$: $g(x, y, z) = z \Rightarrow \nabla g = \mathbf{k} \Rightarrow |\nabla g| = 1; \mathbf{n} = \mathbf{k} \Rightarrow \mathbf{F} \cdot \mathbf{n} = 2xz = 2ax$ since $z = a$;

$d\sigma = dx \, dy \Rightarrow \text{Flux} = \iint_R 2ax \, dx \, dy = \int_0^a \int_0^a 2ax \, dx \, dy = a^4.$

On the face $z = 0$: $g(x, y, z) = z \Rightarrow \nabla g = \mathbf{k} \Rightarrow |\nabla g| = 1; \mathbf{n} = -\mathbf{k} \Rightarrow \mathbf{F} \cdot \mathbf{n} = -2xz = 0$ since $z = 0$;

$d\sigma = dx \, dy \Rightarrow \text{Flux} = \iint_R 0 \, dx \, dy = 0.$

On the face $x = a$: $g(x, y, z) = x \Rightarrow \nabla g = \mathbf{i} \Rightarrow |\nabla g| = 1; \mathbf{n} = \mathbf{i} \Rightarrow \mathbf{F} \cdot \mathbf{n} = 2xy = 2ay$ since $x = a$;

$d\sigma = dy\,dz \Rightarrow$ Flux $= \int_0^a \int_0^a 2ay\,dy\,dz = a^4.$

On the face $x = 0$: $g(x,y,z) = x \Rightarrow \nabla g = \mathbf{i} \Rightarrow |\nabla g| = 1; \mathbf{n} = -\mathbf{i} \Rightarrow \mathbf{F} \cdot \mathbf{n} = -2xy = 0$ since $x = 0$
$\Rightarrow$ Flux $= 0$.

On the face $y = a$: $g(x,y,z) = y \Rightarrow \nabla g = \mathbf{j} \Rightarrow |\nabla g| = 1; \mathbf{n} = \mathbf{j} \Rightarrow \mathbf{F} \cdot \mathbf{n} = 2yz = 2az$ since $y = a$;

$d\sigma = dz\,dx \Rightarrow$ Flux $= \int_0^a \int_0^a 2az\,dz\,dx = a^4.$

On the face $y = 0$: $g(x,y,z) = y \Rightarrow \nabla g = \mathbf{j} \Rightarrow |\nabla g| = 1; \mathbf{n} = -\mathbf{j} \Rightarrow \mathbf{F} \cdot \mathbf{n} = -2yz = 0$ since $y = 0$
$\Rightarrow$ Flux $= 0$. Therefore, Total Flux $= 3a^4$.

33. $\nabla f = 2x\mathbf{i} + 2y\mathbf{j} + 2z\mathbf{k} \Rightarrow |\nabla f| = \sqrt{4x^2 + 4y^2 + 4z^2} = 2a; \mathbf{p} = \mathbf{k} \Rightarrow |\nabla f \cdot \mathbf{p}| = 2z$ since $z \geq 0 \Rightarrow d\sigma = \frac{2a}{2z}dA$

$= \frac{a}{z}dA; M = \iint_S \delta\,d\sigma = \frac{\delta}{8}$ (surface area of sphere) $= \frac{\delta\pi a^2}{2}; M_{xy} = \iint_S z\delta\,d\sigma = \delta \iint_R z\left(\frac{a}{z}\right)dA$

$= a\delta \iint_R dA = a\delta \int_0^{\pi/2}\int_0^a r\,dr\,d\theta = \frac{\delta\pi a^3}{4} \Rightarrow \bar{z} = \frac{M_{xy}}{M} = \left(\frac{\delta\pi a^3}{4}\right)\left(\frac{2}{\delta\pi a^2}\right) = \frac{a}{2}$. Because of symmetry, $\bar{x} = \bar{y}$

$= \frac{a}{2} \Rightarrow$ the centroid is $\left(\frac{a}{2}, \frac{a}{2}, \frac{a}{2}\right)$.

35. Because of symmetry, $\bar{x} = \bar{y} = 0; M = \iint_S \delta\,d\sigma = \delta \iint_S d\sigma = $ (Area of S)$\delta = 3\pi\sqrt{2}\,\delta; \nabla f = 2x\mathbf{i} + 2y\mathbf{j} - 2z\mathbf{k}$

$\Rightarrow |\nabla f| = \sqrt{4x^2 + 4y^2 + 4z^2} = 2\sqrt{x^2 + y^2 + z^2}; \mathbf{p} = \mathbf{k} \Rightarrow |\nabla f \cdot \mathbf{p}| = 2z \Rightarrow d\sigma = \frac{2\sqrt{x^2+y^2+z^2}}{2z}dA$

$= \frac{\sqrt{x^2+y^2+(x^2+y^2)}}{z}dA = \frac{\sqrt{2}\sqrt{x^2+y^2}}{z}dA \Rightarrow M_{xy} = \delta \iint_R z\left(\frac{\sqrt{2}\sqrt{x^2+y^2}}{z}\right)dA$

$= \delta \iint_R \sqrt{2}\sqrt{x^2+y^2}\,dA = \delta \int_0^{2\pi}\int_1^2 \sqrt{2}\,r^2\,dr\,d\theta = \frac{14\pi\sqrt{2}}{3}\delta \Rightarrow \bar{z} = \frac{\left(\frac{14\pi\sqrt{2}}{3}\delta\right)}{3\pi\sqrt{2}\,\delta} = \frac{14}{9}$

$\Rightarrow (\bar{x}, \bar{y}, \bar{z}) = \left(0, 0, \frac{14}{9}\right)$. Next, $I_z = \iint_S (x^2 + y^2)\delta\,d\sigma = \iint_R (x^2+y^2)\left(\frac{\sqrt{2}\sqrt{x^2+y^2}}{z}\right)\delta\,dA$

$= \delta\sqrt{2} \iint_R (x^2+y^2)\,dA = \delta\sqrt{2}\int_0^{2\pi}\int_1^2 r^3\,dr\,d\theta = \frac{15\pi\sqrt{2}}{2}\delta \Rightarrow R_z = \sqrt{\frac{I_z}{M}} = \frac{\sqrt{10}}{2}$

37. (a) Let the diameter lie on the z-axis and let $f(x,y,z) = x^2 + y^2 + z^2 = a^2, z \geq 0$ be the upper hemisphere
$\Rightarrow \nabla f = 2x\mathbf{i} + 2y\mathbf{j} + 2z\mathbf{k} \Rightarrow |\nabla f| = \sqrt{4x^2 + 4y^2 + 4z^2} = 2a, a > 0; \mathbf{p} = \mathbf{k} \Rightarrow |\nabla f \cdot \mathbf{p}| = 2z$ since $z \geq 0$

$\Rightarrow d\sigma = \frac{a}{z}dA \Rightarrow I_z = \iint_S \delta(x^2+y^2)\left(\frac{a}{z}\right)d\sigma = a\delta \iint_R \frac{x^2+y^2}{\sqrt{a^2-(x^2+y^2)}}dA = a\delta \int_0^{2\pi}\int_0^a \frac{r^2}{\sqrt{a^2-r^2}}r\,dr\,d\theta$

$= a\delta \int_0^{2\pi}\left[-r^2\sqrt{a^2-r^2} - \frac{2}{3}(a^2-r^2)^{3/2}\right]_0^a d\theta = a\delta \int_0^{2\pi}\frac{2}{3}a^3\,d\theta = \frac{4\pi}{3}a^4\delta \Rightarrow$ the moment of inertia is $\frac{8\pi}{3}a^4\delta$ for
the whole sphere

(b) $I_L = I_{c.m.} + mh^2$, where m is the mass of the body and h is the distance between the parallel lines; now,
$I_{c.m.} = \frac{8\pi}{3}a^4\delta$ (from part a) and $\frac{m}{2} = \iint_S \delta\,d\sigma = \delta\iint_R \left(\frac{a}{z}\right)dA = a\delta \iint_R \frac{1}{\sqrt{a^2-(x^2+y^2)}}dy\,dx$

$= a\delta \int_0^{2\pi}\int_0^a \frac{1}{\sqrt{a^2-r^2}}r\,dr\,d\theta = a\delta \int_0^{2\pi}\left[-\sqrt{a^2-r^2}\right]_0^a d\theta = a\delta\int_0^{2\pi} a\,d\theta = 2\pi a^2\delta$ and $h = a$

$\Rightarrow I_L = \frac{8\pi}{3}a^4\delta + 4\pi a^2\delta a^2 = \frac{20\pi}{3}a^4\delta$

39. $f_x(x,y) = 2x, f_y(x,y) = 2y \Rightarrow \sqrt{f_x^2 + f_y^2 + 1} = \sqrt{4x^2 + 4y^2 + 1} \Rightarrow$ Area $= \iint_R \sqrt{4x^2 + 4y^2 + 1}\,dx\,dy$

$= \int_0^{2\pi}\int_0^{\sqrt{3}} \sqrt{4r^2 + 1}\,r\,dr\,d\theta = \frac{\pi}{6}\left(13\sqrt{13} - 1\right)$

41. $f_x(x,y) = \frac{x}{\sqrt{x^2+y^2}}, f_y(x,y) = \frac{y}{\sqrt{x^2+y^2}} \Rightarrow \sqrt{f_x^2 + f_y^2 + 1} = \sqrt{\frac{x^2}{x^2+y^2} + \frac{y^2}{x^2+y^2} + 1} = \sqrt{2}$

$\Rightarrow$ Area $= \iint_{R_{xy}} \sqrt{2}\,dx\,dy = \sqrt{2}$(Area between the ellipse and the circle) $= \sqrt{2}(6\pi - \pi) = 5\pi\sqrt{2}$

43. $y = \frac{2}{3} z^{3/2} \Rightarrow f_x(x, z) = 0, f_z(x, z) = z^{1/2} \Rightarrow \sqrt{f_x^2 + f_z^2 + 1} = \sqrt{z + 1}; y = \frac{16}{3} \Rightarrow \frac{16}{3} = \frac{2}{3} z^{3/2} \Rightarrow z = 4$

$\Rightarrow$ Area $= \int_0^4 \int_0^1 \sqrt{z + 1} \, dx \, dz = \int_0^4 \sqrt{z + 1} \, dz = \frac{2}{3} \left(5\sqrt{5} - 1 \right)$

16.6 PARAMETRIZED SURFACES

1. In cylindrical coordinates, let $x = r \cos \theta, y = r \sin \theta, z = \left(\sqrt{x^2 + y^2} \right)^2 = r^2$. Then
   $\mathbf{r}(r, \theta) = (r \cos \theta)\mathbf{i} + (r \sin \theta)\mathbf{j} + r^2 \mathbf{k}, 0 \leq r \leq 2, 0 \leq \theta \leq 2\pi$.

3. In cylindrical coordinates, let $x = r \cos \theta, y = r \sin \theta, z = \frac{\sqrt{x^2 + y^2}}{2} \Rightarrow z = \frac{r}{2}$. Then
   $\mathbf{r}(r, \theta) = (r \cos \theta)\mathbf{i} + (r \sin \theta)\mathbf{j} + \left(\frac{r}{2} \right) \mathbf{k}$. For $0 \leq z \leq 3, 0 \leq \frac{r}{2} \leq 3 \Rightarrow 0 \leq r \leq 6$; to get only the first octant, let
   $0 \leq \theta \leq \frac{\pi}{2}$.

5. In cylindrical coordinates, let $x = r \cos \theta, y = r \sin \theta$ since $x^2 + y^2 = r^2 \Rightarrow z^2 = 9 - (x^2 + y^2) = 9 - r^2$
   $\Rightarrow z = \sqrt{9 - r^2}, z \geq 0$. Then $\mathbf{r}(r, \theta) = (r \cos \theta)\mathbf{i} + (r \sin \theta)\mathbf{j} + \sqrt{9 - r^2}\mathbf{k}$. Let $0 \leq \theta \leq 2\pi$. For the domain
   of r: $z = \sqrt{x^2 + y^2}$ and $x^2 + y^2 + z^2 = 9 \Rightarrow x^2 + y^2 + \left(\sqrt{x^2 + y^2} \right)^2 = 9 \Rightarrow 2(x^2 + y^2) = 9 \Rightarrow 2r^2 = 9$
   $\Rightarrow r = \frac{3}{\sqrt{2}} \Rightarrow 0 \leq r \leq \frac{3}{\sqrt{2}}$.

7. In spherical coordinates, $x = \rho \sin \phi \cos \theta, y = \rho \sin \phi \sin \theta, \rho = \sqrt{x^2 + y^2 + z^2} \Rightarrow \rho^2 = 3 \Rightarrow \rho = \sqrt{3}$
   $\Rightarrow z = \sqrt{3} \cos \phi$ for the sphere; $z = \frac{\sqrt{3}}{2} = \sqrt{3} \cos \phi \Rightarrow \cos \phi = \frac{1}{2} \Rightarrow \phi = \frac{\pi}{3}; z = -\frac{\sqrt{3}}{2} \Rightarrow -\frac{\sqrt{3}}{2} = \sqrt{3} \cos \phi$
   $\Rightarrow \cos \phi = -\frac{1}{2} \Rightarrow \phi = \frac{2\pi}{3}$. Then $\mathbf{r}(\phi, \theta) = \left(\sqrt{3} \sin \phi \cos \theta \right)\mathbf{i} + \left(\sqrt{3} \sin \phi \sin \theta \right)\mathbf{j} + \left(\sqrt{3} \cos \phi \right)\mathbf{k}$,
   $\frac{\pi}{3} \leq \phi \leq \frac{2\pi}{3}$ and $0 \leq \theta \leq 2\pi$.

9. Since $z = 4 - y^2$, we can let $\mathbf{r}$ be a function of x and $y \Rightarrow \mathbf{r}(x, y) = x\mathbf{i} + y\mathbf{j} + (4 - y^2)\mathbf{k}$. Then $z = 0$
   $\Rightarrow 0 = 4 - y^2 \Rightarrow y = \pm 2$. Thus, let $-2 \leq y \leq 2$ and $0 \leq x \leq 2$.

11. When $x = 0$, let $y^2 + z^2 = 9$ be the circular section in the yz-plane. Use polar coordinates in the yz-plane
    $\Rightarrow y = 3 \cos \theta$ and $z = 3 \sin \theta$. Thus let $x = u$ and $\theta = v \Rightarrow \mathbf{r}(u, v) = u\mathbf{i} + (3 \cos v)\mathbf{j} + (3 \sin v)\mathbf{k}$ where
    $0 \leq u \leq 3$, and $0 \leq v \leq 2\pi$.

13. (a) $x + y + z = 1 \Rightarrow z = 1 - x - y$. In cylindrical coordinates, let $x = r \cos \theta$ and $y = r \sin \theta$
    $\Rightarrow z = 1 - r \cos \theta - r \sin \theta \Rightarrow \mathbf{r}(r, \theta) = (r \cos \theta)\mathbf{i} + (r \sin \theta)\mathbf{j} + (1 - r \cos \theta - r \sin \theta)\mathbf{k}, 0 \leq \theta \leq 2\pi$ and
    $0 \leq r \leq 3$.

    (b) In a fashion similar to cylindrical coordinates, but working in the yz-plane instead of the xy-plane, let
    $y = u \cos v, z = u \sin v$ where $u = \sqrt{y^2 + z^2}$ and v is the angle formed by $(x, y, z), (x, 0, 0)$, and $(x, y, 0)$
    with $(x, 0, 0)$ as vertex. Since $x + y + z = 1 \Rightarrow x = 1 - y - z \Rightarrow x = 1 - u \cos v - u \sin v$, then $\mathbf{r}$ is a
    function of u and $v \Rightarrow \mathbf{r}(u, v) = (1 - u \cos v - u \sin v)\mathbf{i} + (u \cos v)\mathbf{j} + (u \sin v)\mathbf{k}, 0 \leq u \leq 3$ and $0 \leq v \leq 2\pi$.

15. Let $x = w \cos v$ and $z = w \sin v$. Then $(x - 2)^2 + z^2 = 4 \Rightarrow x^2 - 4x + z^2 = 0 \Rightarrow w^2 \cos^2 v - 4w \cos v + w^2 \sin^2 v$
    $= 0 \Rightarrow w^2 - 4w \cos v = 0 \Rightarrow w = 0$ or $w - 4 \cos v = 0 \Rightarrow w = 0$ or $w = 4 \cos v$. Now $w = 0 \Rightarrow x = 0$ and $y = 0$,
    which is a line not a cylinder. Therefore, let $w = 4 \cos v \Rightarrow x = (4 \cos v)(\cos v) = 4 \cos^2 v$ and $z = 4 \cos v \sin v$.
    Finally, let $y = u$. Then $\mathbf{r}(u, v) = (4 \cos^2 v)\mathbf{i} + u\mathbf{j} + (4 \cos v \sin v)\mathbf{k}, -\frac{\pi}{2} \leq v \leq \frac{\pi}{2}$ and $0 \leq u \leq 3$.

17. Let $x = r \cos \theta$ and $y = r \sin \theta$. Then $\mathbf{r}(r, \theta) = (r \cos \theta)\mathbf{i} + (r \sin \theta)\mathbf{j} + \left(\frac{2 - r \sin \theta}{2} \right)\mathbf{k}, 0 \leq r \leq 1$ and $0 \leq \theta \leq 2\pi$
    $\Rightarrow \mathbf{r}_r = (\cos \theta)\mathbf{i} + (\sin \theta)\mathbf{j} - \left(\frac{\sin \theta}{2} \right)\mathbf{k}$ and $\mathbf{r}_\theta = (-r \sin \theta)\mathbf{i} + (r \cos \theta)\mathbf{j} - \left(\frac{r \cos \theta}{2} \right)\mathbf{k}$

$$\Rightarrow \mathbf{r_r} \times \mathbf{r_\theta} = \begin{vmatrix} \mathbf{i} & \mathbf{j} & \mathbf{k} \\ \cos\theta & \sin\theta & -\frac{\sin\theta}{2} \\ -r\sin\theta & r\cos\theta & -\frac{r\cos\theta}{2} \end{vmatrix}$$

$$= \left(\frac{-r\sin\theta\cos\theta}{2} + \frac{(\sin\theta)(r\cos\theta)}{2}\right)\mathbf{i} + \left(\frac{r\sin^2\theta}{2} + \frac{r\cos^2\theta}{2}\right)\mathbf{j} + (r\cos^2\theta + r\sin^2\theta)\mathbf{k} = \frac{r}{2}\mathbf{j} + r\mathbf{k}$$

$$\Rightarrow |\mathbf{r_r} \times \mathbf{r_\theta}| = \sqrt{\frac{r^2}{4} + r^2} = \frac{\sqrt{5}r}{2} \Rightarrow A = \int_0^{2\pi}\int_0^1 \frac{\sqrt{5}r}{2}\,dr\,d\theta = \int_0^{2\pi}\left[\frac{\sqrt{5}r^2}{4}\right]_0^1 d\theta = \int_0^{2\pi} d\theta = \frac{\pi\sqrt{5}}{2}$$

19. Let $x = r\cos\theta$ and $y = r\sin\theta \Rightarrow z = 2\sqrt{x^2 + y^2} = 2r$, $1 \le r \le 3$ and $0 \le \theta \le 2\pi$. Then
$\mathbf{r}(r,\theta) = (r\cos\theta)\mathbf{i} + (r\sin\theta)\mathbf{j} + 2r\mathbf{k} \Rightarrow \mathbf{r_r} = (\cos\theta)\mathbf{i} + (\sin\theta)\mathbf{j} + 2\mathbf{k}$ and $\mathbf{r_\theta} = (-r\sin\theta)\mathbf{i} + (r\cos\theta)\mathbf{j}$

$$\Rightarrow \mathbf{r_r} \times \mathbf{r_\theta} = \begin{vmatrix} \mathbf{i} & \mathbf{j} & \mathbf{k} \\ \cos\theta & \sin\theta & 2 \\ -r\sin\theta & r\cos\theta & 0 \end{vmatrix} = (-2r\cos\theta)\mathbf{i} - (2r\sin\theta)\mathbf{j} + (r\cos^2\theta + r\sin^2\theta)\mathbf{k}$$

$= (-2r\cos\theta)\mathbf{i} - (2r\sin\theta)\mathbf{j} + r\mathbf{k} \Rightarrow |\mathbf{r_r} \times \mathbf{r_\theta}| = \sqrt{4r^2\cos^2\theta + 4r^2\sin^2\theta + r^2} = \sqrt{5r^2} = r\sqrt{5}$

$$\Rightarrow A = \int_0^{2\pi}\int_1^3 r\sqrt{5}\,dr\,d\theta = \int_0^{2\pi}\left[\frac{r^2\sqrt{5}}{2}\right]_1^3 d\theta = \int_0^{2\pi} 4\sqrt{5}\,d\theta = 8\pi\sqrt{5}$$

21. Let $x = r\cos\theta$ and $y = r\sin\theta \Rightarrow r^2 = x^2 + y^2 = 1$, $1 \le z \le 4$ and $0 \le \theta \le 2\pi$. Then
$\mathbf{r}(z,\theta) = (\cos\theta)\mathbf{i} + (\sin\theta)\mathbf{j} + z\mathbf{k} \Rightarrow \mathbf{r_z} = \mathbf{k}$ and $\mathbf{r_\theta} = (-\sin\theta)\mathbf{i} + (\cos\theta)\mathbf{j}$

$$\Rightarrow \mathbf{r_\theta} \times \mathbf{r_z} = \begin{vmatrix} \mathbf{i} & \mathbf{j} & \mathbf{k} \\ -\sin\theta & \cos\theta & 0 \\ 0 & 0 & 1 \end{vmatrix} = (\cos\theta)\mathbf{i} + (\sin\theta)\,\mathbf{j} \Rightarrow |\mathbf{r_\theta} \times \mathbf{r_z}| = \sqrt{\cos^2\theta + \sin^2\theta} = 1$$

$$\Rightarrow A = \int_0^{2\pi}\int_1^4 1\,dr\,d\theta = \int_0^{2\pi} 3\,d\theta = 6\pi$$

23. $z = 2 - x^2 - y^2$ and $z = \sqrt{x^2 + y^2} \Rightarrow z = 2 - z^2 \Rightarrow z^2 + z - 2 = 0 \Rightarrow z = -2$ or $z = 1$. Since $z = \sqrt{x^2 + y^2} \ge 0$,
we get $z = 1$ where the cone intersects the paraboloid. When $x = 0$ and $y = 0$, $z = 2 \Rightarrow$ the vertex of the
paraboloid is $(0, 0, 2)$. Therefore, z ranges from 1 to 2 on the "cap" $\Rightarrow$ r ranges from 1 (when $x^2 + y^2 = 1$) to 0
(when $x = 0$ and $y = 0$ at the vertex). Let $x = r\cos\theta$, $y = r\sin\theta$, and $z = 2 - r^2$. Then
$\mathbf{r}(r,\theta) = (r\cos\theta)\mathbf{i} + (r\sin\theta)\mathbf{j} + (2 - r^2)\,\mathbf{k}$, $0 \le r \le 1$, $0 \le \theta \le 2\pi \Rightarrow \mathbf{r_r} = (\cos\theta)\mathbf{i} + (\sin\theta)\mathbf{j} - 2r\mathbf{k}$ and

$$\mathbf{r_\theta} = (-r\sin\theta)\mathbf{i} + (r\cos\theta)\mathbf{j} \Rightarrow \mathbf{r_r} \times \mathbf{r_\theta} = \begin{vmatrix} \mathbf{i} & \mathbf{j} & \mathbf{k} \\ \cos\theta & \sin\theta & -2r \\ -r\sin\theta & r\cos\theta & 0 \end{vmatrix}$$

$= (2r^2\cos\theta)\,\mathbf{i} + (2r^2\sin\theta)\,\mathbf{j} + r\mathbf{k} \Rightarrow |\mathbf{r_r} \times \mathbf{r_\theta}| = \sqrt{4r^4\cos^2\theta + 4r^4\sin^2\theta + r^2} = r\sqrt{4r^2 + 1}$

$$\Rightarrow A = \int_0^{2\pi}\int_0^1 r\sqrt{4r^2 + 1}\,dr\,d\theta = \int_0^{2\pi}\left[\frac{1}{12}\left(4r^2 + 1\right)^{3/2}\right]_0^1 d\theta = \int_0^{2\pi}\left(\frac{5\sqrt{5}-1}{12}\right)d\theta = \frac{\pi}{6}\left(5\sqrt{5} - 1\right)$$

25. Let $x = \rho\sin\phi\cos\theta$, $y = \rho\sin\phi\sin\theta$, and $z = \rho\cos\phi \Rightarrow \rho = \sqrt{x^2 + y^2 + z^2} = \sqrt{2}$ on the sphere. Next,
$x^2 + y^2 + z^2 = 2$ and $z = \sqrt{x^2 + y^2} \Rightarrow z^2 + z^2 = 2 \Rightarrow z^2 = 1 \Rightarrow z = 1$ since $z \ge 0 \Rightarrow \phi = \frac{\pi}{4}$. For the lower
portion of the sphere cut by the cone, we get $\phi = \pi$. Then
$\mathbf{r}(\phi,\theta) = \left(\sqrt{2}\sin\phi\cos\theta\right)\mathbf{i} + \left(\sqrt{2}\sin\phi\sin\theta\right)\mathbf{j} + \left(\sqrt{2}\cos\phi\right)\mathbf{k}$, $\frac{\pi}{4} \le \phi \le \pi$, $0 \le \theta \le 2\pi$

$\Rightarrow \mathbf{r_\phi} = \left(\sqrt{2}\cos\phi\cos\theta\right)\mathbf{i} + \left(\sqrt{2}\cos\phi\sin\theta\right)\mathbf{j} - \left(\sqrt{2}\sin\phi\right)\mathbf{k}$ and $\mathbf{r_\theta} = \left(-\sqrt{2}\sin\phi\sin\theta\right)\mathbf{i} + \left(\sqrt{2}\sin\phi\cos\theta\right)\mathbf{j}$

$$\Rightarrow \mathbf{r_\phi} \times \mathbf{r_\theta} = \begin{vmatrix} \mathbf{i} & \mathbf{j} & \mathbf{k} \\ \sqrt{2}\cos\phi\cos\theta & \sqrt{2}\cos\phi\sin\theta & -\sqrt{2}\sin\phi \\ -\sqrt{2}\sin\phi\sin\theta & \sqrt{2}\sin\phi\cos\theta & 0 \end{vmatrix}$$

$= (2\sin^2\phi\cos\theta)\,\mathbf{i} + (2\sin^2\phi\sin\theta)\,\mathbf{j} + (2\sin\phi\cos\phi)\mathbf{k}$

$\Rightarrow |\mathbf{r_\phi} \times \mathbf{r_\theta}| = \sqrt{4\sin^4\phi\cos^2\theta + 4\sin^4\phi\sin^2\theta + 4\sin^2\phi\cos^2\phi} = \sqrt{4\sin^2\phi} = 2\,|\sin\phi| = 2\sin\phi$

$$\Rightarrow A = \int_0^{2\pi}\int_{\pi/4}^{\pi} 2\sin\phi\,d\phi\,d\theta = \int_0^{2\pi}\left(2 + \sqrt{2}\right)d\theta = \left(4 + 2\sqrt{2}\right)\pi$$

27. Let the parametrization be $r(x, z) = xi + x^2j + zk \Rightarrow r_x = i + 2xj$ and $r_z = k \Rightarrow r_x \times r_z = \begin{vmatrix} i & j & k \\ 1 & 2x & 0 \\ 0 & 0 & 1 \end{vmatrix}$

$= 2xi + j \Rightarrow |r_x \times r_z| = \sqrt{4x^2 + 1} \Rightarrow \iint_S G(x, y, z)\, d\sigma = \int_0^3 \int_0^2 x\sqrt{4x^2 + 1}\, dx\, dz = \int_0^3 \left[\frac{1}{12}(4x^2 + 1)^{3/2} \right]_0^2 dz$

$= \int_0^3 \frac{1}{12}\left(17\sqrt{17} - 1 \right) dz = \frac{17\sqrt{17} - 1}{4}$

29. Let the parametrization be $r(\phi, \theta) = (\sin\phi\cos\theta)i + (\sin\phi\sin\theta)j + (\cos\phi)k$ (spherical coordinates with $\rho = 1$ on the sphere), $0 \le \phi \le \pi$, $0 \le \theta \le 2\pi \Rightarrow r_\phi = (\cos\phi\cos\theta)i + (\cos\phi\sin\theta)j - (\sin\phi)k$ and

$r_\theta = (-\sin\phi\sin\theta)i + (\sin\phi\cos\theta)j \Rightarrow r_\phi \times r_\theta = \begin{vmatrix} i & j & k \\ \cos\phi\cos\theta & \cos\phi\sin\theta & -\sin\phi \\ -\sin\phi\sin\theta & \sin\phi\cos\theta & 0 \end{vmatrix}$

$= (\sin^2\phi\cos\theta)\,i + (\sin^2\phi\sin\theta)\,j + (\sin\phi\cos\phi)k \Rightarrow |r_\phi \times r_\theta| = \sqrt{\sin^4\phi\cos^2\theta + \sin^4\phi\sin^2\theta + \sin^2\phi\cos^2\phi}$

$= \sin\phi; \ x = \sin\phi\cos\theta \Rightarrow G(x, y, z) = \cos^2\theta\sin^2\phi \Rightarrow \iint_S G(x, y, z)\, d\sigma = \int_0^{2\pi}\int_0^\pi (\cos^2\theta\sin^2\phi)(\sin\phi)\, d\phi\, d\theta$

$= \int_0^{2\pi}\int_0^\pi (\cos^2\theta)(1 - \cos^2\phi)(\sin\phi)\, d\phi\, d\theta; \ \begin{bmatrix} u = \cos\phi \\ du = -\sin\phi\, d\phi \end{bmatrix} \to \int_0^{2\pi}\int_1^{-1} (\cos^2\theta)(u^2 - 1)\, du\, d\theta$

$= \int_0^{2\pi} (\cos^2\theta)\left[\frac{u^3}{3} - u \right]_1^{-1} d\theta = \frac{4}{3}\int_0^{2\pi}\cos^2\theta\, d\theta = \frac{4}{3}\left[\frac{\theta}{2} + \frac{\sin 2\theta}{4} \right]_0^{2\pi} = \frac{4\pi}{3}$

31. Let the parametrization be $r(x, y) = xi + yj + (4 - x - y)k \Rightarrow r_x = i - k$ and $r_y = j - k$

$\Rightarrow r_x \times r_y = \begin{vmatrix} i & j & k \\ 1 & 0 & -1 \\ 0 & 1 & -1 \end{vmatrix} = i + j + k \Rightarrow |r_x \times r_y| = \sqrt{3} \Rightarrow \iint_S F(x, y, z)\, d\sigma = \int_0^1 \int_0^1 (4 - x - y)\sqrt{3}\, dy\, dx$

$= \int_0^1 \sqrt{3}\left[4y - xy - \frac{y^2}{2} \right]_0^1 dx = \int_0^1 \sqrt{3}\left(\frac{7}{2} - x \right) dx = \sqrt{3}\left[\frac{7}{2}x - \frac{x^2}{2} \right]_0^1 = 3\sqrt{3}$

33. Let the parametrization be $r(r, \theta) = (r\cos\theta)i + (r\sin\theta)j + (1 - r^2)k$, $0 \le r \le 1$ (since $0 \le z \le 1$) and $0 \le \theta \le 2\pi$

$\Rightarrow r_r = (\cos\theta)i + (\sin\theta)j - 2rk$ and $r_\theta = (-r\sin\theta)i + (r\cos\theta)j \Rightarrow r_r \times r_\theta = \begin{vmatrix} i & j & k \\ \cos\theta & \sin\theta & -2r \\ -r\sin\theta & r\cos\theta & 0 \end{vmatrix}$

$= (2r^2\cos\theta)i + (2r^2\sin\theta)j + rk \Rightarrow |r_r \times r_\theta| = \sqrt{(2r^2\cos\theta)^2 + (2r^2\sin\theta) + r^2} = r\sqrt{1 + 4r^2}; \ z = 1 - r^2$ and

$x = r\cos\theta \Rightarrow H(x, y, z) = (r^2\cos^2\theta)\sqrt{1 + 4r^2} \Rightarrow \iint_S H(x, y, z)\, d\sigma$

$= \int_0^{2\pi}\int_0^1 (r^2\cos^2\theta)\left(\sqrt{1 + 4r^2} \right)\left(r\sqrt{1 + 4r^2} \right) dr\, d\theta = \int_0^{2\pi}\int_0^1 r^3(1 + 4r^2)\cos^2\theta\, dr\, d\theta = \frac{11\pi}{12}$

35. Let the parametrization be $r(x, y) = xi + yj + (4 - y^2)k$, $0 \le x \le 1$, $-2 \le y \le 2$; $z = 0 \Rightarrow 0 = 4 - y^2$

$\Rightarrow y = \pm 2; \ r_x = i$ and $r_y = j - 2yk \Rightarrow r_x \times r_y = \begin{vmatrix} i & j & k \\ 1 & 0 & 0 \\ 0 & 1 & -2y \end{vmatrix} = 2yj + k \Rightarrow F \cdot n\, d\sigma$

$= F \cdot \frac{r_x \times r_y}{|r_x \times r_y|} |r_x \times r_y|\, dy\, dx = (2xy - 3z)\, dy\, dx = [2xy - 3(4 - y^2)]\, dy\, dx \Rightarrow \iint_S F \cdot n\, d\sigma$

$= \int_0^1 \int_{-2}^2 (2xy + 3y^2 - 12)\, dy\, dx = \int_0^1 [xy^2 + y^3 - 12y]_{-2}^2\, dx = \int_0^1 -32\, dx = -32$

37. Let the parametrization be $r(\phi, \theta) = (a\sin\phi\cos\theta)i + (a\sin\phi\sin\theta)j + (a\cos\phi)k$ (spherical coordinates with $\rho = a$, $a \ge 0$, on the sphere), $0 \le \phi \le \frac{\pi}{2}$ (for the first octant), $0 \le \theta \le \frac{\pi}{2}$ (for the first octant)

$\Rightarrow r_\phi = (a\cos\phi\cos\theta)i + (a\cos\phi\sin\theta)j - (a\sin\phi)k$ and $r_\theta = (-a\sin\phi\sin\theta)i + (a\sin\phi\cos\theta)j$

$$\Rightarrow \mathbf{r}_\phi \times \mathbf{r}_\theta = \begin{vmatrix} \mathbf{i} & \mathbf{j} & \mathbf{k} \\ a\cos\phi\cos\theta & a\cos\phi\sin\theta & -a\sin\phi \\ -a\sin\phi\sin\theta & a\sin\phi\cos\theta & 0 \end{vmatrix}$$

$$= (a^2\sin^2\phi\cos\theta)\mathbf{i} + (a^2\sin^2\phi\sin\theta)\mathbf{j} + (a^2\sin\phi\cos\phi)\mathbf{k} \Rightarrow \mathbf{F}\cdot\mathbf{n}\,d\sigma = \mathbf{F}\cdot\frac{\mathbf{r}_\phi\times\mathbf{r}_\theta}{|\mathbf{r}_\phi\times\mathbf{r}_\theta|}\,|\mathbf{r}_\phi\times\mathbf{r}_\theta|\,d\theta\,d\phi$$

$$= a^3\cos^2\phi\sin\phi\,d\theta\,d\phi \text{ since } \mathbf{F} = z\mathbf{k} = (a\cos\phi)\mathbf{k} \Rightarrow \iint_S \mathbf{F}\cdot\mathbf{n}\,d\sigma = \int_0^{\pi/2}\int_0^{\pi/2} a^3\cos^2\phi\sin\phi\,d\phi\,d\theta = \frac{\pi a^3}{6}$$

39. Let the parametrization be $\mathbf{r}(x,y) = x\mathbf{i} + y\mathbf{j} + (2a - x - y)\mathbf{k}, 0 \le x \le a, 0 \le y \le a \Rightarrow \mathbf{r}_x = \mathbf{i} - \mathbf{k}$ and $\mathbf{r}_y = \mathbf{j} - \mathbf{k}$

$$\Rightarrow \mathbf{r}_x \times \mathbf{r}_y = \begin{vmatrix} \mathbf{i} & \mathbf{j} & \mathbf{k} \\ 1 & 0 & -1 \\ 0 & 1 & -1 \end{vmatrix} = \mathbf{i} + \mathbf{j} + \mathbf{k} \Rightarrow \mathbf{F}\cdot\mathbf{n}\,d\sigma = \mathbf{F}\cdot\frac{\mathbf{r}_x\times\mathbf{r}_y}{|\mathbf{r}_x\times\mathbf{r}_y|}\,|\mathbf{r}_x\times\mathbf{r}_y|\,dy\,dx$$

$$= [2xy + 2y(2a - x - y) + 2x(2a - x - y)]\,dy\,dx \text{ since } \mathbf{F} = 2xy\mathbf{i} + 2yz\mathbf{j} + 2xz\mathbf{k}$$

$$= 2xy\mathbf{i} + 2y(2a - x - y)\mathbf{j} + 2x(2a - x - y)\mathbf{k} \Rightarrow \iint_S \mathbf{F}\cdot\mathbf{n}\,d\sigma$$

$$= \int_0^a\int_0^a [2xy + 2y(2a - x - y) + 2x(2a - x - y)]\,dy\,dx = \int_0^a\int_0^a (4ay - 2y^2 + 4ax - 2x^2 - 2xy)\,dy\,dx$$

$$= \int_0^a \left(\tfrac{4}{3}a^3 + 3a^2x - 2ax^2\right)dx = \left(\tfrac{4}{3} + \tfrac{3}{2} - \tfrac{2}{3}\right)a^4 = \frac{13a^4}{6}$$

41. Let the parametrization be $\mathbf{r}(r,\theta) = (r\cos\theta)\mathbf{i} + (r\sin\theta)\mathbf{j} + r\mathbf{k}, 0 \le r \le 1$ (since $0 \le z \le 1$) and $0 \le \theta \le 2\pi$

$$\Rightarrow \mathbf{r}_r = (\cos\theta)\mathbf{i} + (\sin\theta)\mathbf{j} + \mathbf{k} \text{ and } \mathbf{r}_\theta = (-r\sin\theta)\mathbf{i} + (r\cos\theta)\mathbf{j} \Rightarrow \mathbf{r}_\theta \times \mathbf{r}_r = \begin{vmatrix} \mathbf{i} & \mathbf{j} & \mathbf{k} \\ -r\sin\theta & r\cos\theta & 0 \\ \cos\theta & \sin\theta & 1 \end{vmatrix}$$

$$= (r\cos\theta)\mathbf{i} + (r\sin\theta)\mathbf{j} - r\mathbf{k} \Rightarrow \mathbf{F}\cdot\mathbf{n}\,d\sigma = \mathbf{F}\cdot\frac{\mathbf{r}_\theta\times\mathbf{r}_r}{|\mathbf{r}_\theta\times\mathbf{r}_r|}\,|\mathbf{r}_\theta\times\mathbf{r}_r|\,d\theta\,dr = (r^3\sin\theta\cos^2\theta + r^2)\,d\theta\,dr \text{ since}$$

$$\mathbf{F} = (r^2\sin\theta\cos\theta)\mathbf{i} - r\mathbf{k} \Rightarrow \iint_S \mathbf{F}\cdot\mathbf{n}\,d\sigma = \int_0^{2\pi}\int_0^1 (r^3\sin\theta\cos^2\theta + r^2)\,dr\,d\theta = \int_0^{2\pi} \left(\tfrac{1}{4}\sin\theta\cos^2\theta + \tfrac{1}{3}\right)d\theta$$

$$= \left[-\tfrac{1}{12}\cos^3\theta + \tfrac{\theta}{3}\right]_0^{2\pi} = \frac{2\pi}{3}$$

43. Let the parametrization be $\mathbf{r}(r,\theta) = (r\cos\theta)\mathbf{i} + (r\sin\theta)\mathbf{j} + r\mathbf{k}, 1 \le r \le 2$ (since $1 \le z \le 2$) and $0 \le \theta \le 2\pi$

$$\Rightarrow \mathbf{r}_r = (\cos\theta)\mathbf{i} + (\sin\theta)\mathbf{j} + \mathbf{k} \text{ and } \mathbf{r}_\theta = (-r\sin\theta)\mathbf{i} + (r\cos\theta)\mathbf{j} \Rightarrow \mathbf{r}_\theta \times \mathbf{r}_r = \begin{vmatrix} \mathbf{i} & \mathbf{j} & \mathbf{k} \\ -r\sin\theta & r\cos\theta & 0 \\ \cos\theta & \sin\theta & 1 \end{vmatrix}$$

$$= (r\cos\theta)\mathbf{i} + (r\sin\theta)\mathbf{j} - r\mathbf{k} \Rightarrow \mathbf{F}\cdot\mathbf{n}\,d\sigma = \mathbf{F}\cdot\frac{\mathbf{r}_\theta\times\mathbf{r}_r}{|\mathbf{r}_\theta\times\mathbf{r}_r|}\,|\mathbf{r}_\theta\times\mathbf{r}_r|\,d\theta\,dr = (-r^2\cos^2\theta - r^2\sin^2\theta - r^3)\,d\theta\,dr$$

$$= (-r^2 - r^3)\,d\theta\,dr \text{ since } \mathbf{F} = (-r\cos\theta)\mathbf{i} - (r\sin\theta)\mathbf{j} + r^2\mathbf{k} \Rightarrow \iint_S \mathbf{F}\cdot\mathbf{n}\,d\sigma = \int_0^{2\pi}\int_1^2 (-r^2 - r^3)\,dr\,d\theta = -\frac{73\pi}{6}$$

45. Let the parametrization be $\mathbf{r}(\phi,\theta) = (a\sin\phi\cos\theta)\mathbf{i} + (a\sin\phi\sin\theta)\mathbf{j} + (a\cos\phi)\mathbf{k}, 0 \le \phi \le \tfrac{\pi}{2}, 0 \le \theta \le \tfrac{\pi}{2}$

$$\Rightarrow \mathbf{r}_\phi = (a\cos\phi\cos\theta)\mathbf{i} + (a\cos\phi\sin\theta)\mathbf{j} - (a\sin\phi)\mathbf{k} \text{ and } \mathbf{r}_\theta = (-a\sin\phi\sin\theta)\mathbf{i} + (a\sin\phi\cos\theta)\mathbf{j}$$

$$\Rightarrow \mathbf{r}_\phi \times \mathbf{r}_\theta = \begin{vmatrix} \mathbf{i} & \mathbf{j} & \mathbf{k} \\ a\cos\phi\cos\theta & a\cos\phi\sin\theta & -a\sin\phi \\ -a\sin\phi\sin\theta & a\sin\phi\cos\theta & 0 \end{vmatrix}$$

$$= (a^2\sin^2\phi\cos\theta)\,\mathbf{i} + (a^2\sin^2\phi\sin\theta)\,\mathbf{j} + (a^2\sin\phi\cos\phi)\,\mathbf{k}$$

$$\Rightarrow |\mathbf{r}_\phi \times \mathbf{r}_\theta| = \sqrt{a^4\sin^4\phi\cos^2\theta + a^4\sin^4\phi\sin^2\theta + a^4\sin^2\phi\cos^2\phi} = \sqrt{a^4\sin^2\phi} = a^2\sin\phi. \text{ The mass is}$$

$$M = \iint_S d\sigma = \int_0^{\pi/2}\int_0^{\pi/2} (a^2\sin\phi)\,d\phi\,d\theta = \frac{a^2\pi}{2}; \text{ the first moment is } M_{yz} = \iint_S x\,d\sigma$$

$$= \int_0^{\pi/2}\int_0^{\pi/2} (a\sin\phi\cos\theta)(a^2\sin\phi)\,d\phi\,d\theta = \frac{a^3\pi}{4} \Rightarrow \bar{x} = \frac{\left(\frac{a^3\pi}{4}\right)}{\left(\frac{a^2\pi}{2}\right)} = \frac{a}{2} \Rightarrow \text{ the centroid is located at } \left(\tfrac{a}{2}, \tfrac{a}{2}, \tfrac{a}{2}\right) \text{ by}$$

symmetry

47. Let the parametrization be $\mathbf{r}(\phi, \theta) = (a \sin \phi \cos \theta)\mathbf{i} + (a \sin \phi \sin \theta)\mathbf{j} + (a \cos \phi)\mathbf{k}$, $0 \le \phi \le \pi, 0 \le \theta \le 2\pi$

$\Rightarrow \mathbf{r}_\phi = (a \cos \phi \cos \theta)\mathbf{i} + (a \cos \phi \sin \theta)\mathbf{j} - (a \sin \phi)\mathbf{k}$ and $\mathbf{r}_\theta = (-a \sin \phi \sin \theta)\mathbf{i} + (a \sin \phi \cos \theta)\mathbf{j}$

$\Rightarrow \mathbf{r}_\phi \times \mathbf{r}_\theta = \begin{vmatrix} \mathbf{i} & \mathbf{j} & \mathbf{k} \\ a \cos \phi \cos \theta & a \cos \phi \sin \theta & -a \sin \phi \\ -a \sin \phi \sin \theta & a \sin \phi \cos \theta & 0 \end{vmatrix}$

$= (a^2 \sin^2 \phi \cos \theta)\,\mathbf{i} + (a^2 \sin^2 \phi \sin \theta)\,\mathbf{j} + (a^2 \sin \phi \cos \phi)\,\mathbf{k}$

$\Rightarrow |\mathbf{r}_\phi \times \mathbf{r}_\theta| = \sqrt{a^4 \sin^4 \phi \cos^2 \theta + a^4 \sin^4 \phi \sin^2 \theta + a^4 \sin^2 \phi \cos^2 \phi} = \sqrt{a^4 \sin^2 \phi} = a^2 \sin \phi$. The moment of

inertia is $I_z = \iint_S \delta (x^2 + y^2)\, d\sigma = \int_0^{2\pi} \int_0^\pi \delta\,[(a \sin \phi \cos \theta)^2 + (a \sin \phi \sin \theta)^2]\,(a^2 \sin \phi)\, d\phi\, d\theta$

$= \int_0^{2\pi} \int_0^\pi \delta\,(a^2 \sin^2 \phi)\,(a^2 \sin \phi)\, d\phi\, d\theta = \int_0^{2\pi} \int_0^\pi \delta a^4 \sin^3 \phi\, d\phi\, d\theta = \int_0^{2\pi} \delta a^4 \left[\left(-\tfrac{1}{3} \cos \phi\right)(\sin^2 \phi + 2)\right]_0^\pi d\theta = \frac{8\delta\pi a^4}{3}$

49. The parametrization $\mathbf{r}(r, \theta) = (r \cos \theta)\mathbf{i} + (r \sin \theta)\mathbf{j} + r\mathbf{k}$

at $P_0 = \left(\sqrt{2}, \sqrt{2}, 2\right) \Rightarrow \theta = \frac{\pi}{4}, r = 2$,

$\mathbf{r}_r = (\cos \theta)\mathbf{i} + (\sin \theta)\mathbf{j} + \mathbf{k} = \frac{\sqrt{2}}{2}\mathbf{i} + \frac{\sqrt{2}}{2}\mathbf{j} + \mathbf{k}$ and

$\mathbf{r}_\theta = (-r \sin \theta)\mathbf{i} + (r \cos \theta)\mathbf{j} = -\sqrt{2}\mathbf{i} + \sqrt{2}\mathbf{j}$

$\Rightarrow \mathbf{r}_r \times \mathbf{r}_\theta = \begin{vmatrix} \mathbf{i} & \mathbf{j} & \mathbf{k} \\ \sqrt{2}/2 & \sqrt{2}/2 & 1 \\ -\sqrt{2} & \sqrt{2} & 0 \end{vmatrix}$

$= -\sqrt{2}\mathbf{i} - \sqrt{2}\mathbf{j} + 2\mathbf{k} \Rightarrow$ the tangent plane is

$0 = \left(-\sqrt{2}\mathbf{i} - \sqrt{2}\mathbf{j} + 2\mathbf{k}\right) \cdot \left[\left(x - \sqrt{2}\right)\mathbf{i} + \left(y - \sqrt{2}\right)\mathbf{j} + (z - 2)\mathbf{k}\right] \Rightarrow \sqrt{2}x + \sqrt{2}y - 2z = 0$, or $x + y - \sqrt{2}z = 0$.

The parametrization $\mathbf{r}(r, \theta) \Rightarrow x = r \cos \theta, y = r \sin \theta$ and $z = r \Rightarrow x^2 + y^2 = r^2 = z^2 \Rightarrow$ the surface is $z = \sqrt{x^2 + y^2}$.

51. The parametrization $\mathbf{r}(\theta, z) = (3 \sin 2\theta)\mathbf{i} + (6 \sin^2 \theta)\,\mathbf{j} + z\mathbf{k}$

at $P_0 = \left(\frac{3\sqrt{3}}{2}, \frac{9}{2}, 0\right) \Rightarrow \theta = \frac{\pi}{3}$ and $z = 0$. Then

$\mathbf{r}_\theta = (6 \cos 2\theta)\mathbf{i} + (12 \sin \theta \cos \theta)\mathbf{j}$

$= -3\mathbf{i} + 3\sqrt{3}\mathbf{j}$ and $\mathbf{r}_z = \mathbf{k}$ at P_0

$\Rightarrow \mathbf{r}_\theta \times \mathbf{r}_z = \begin{vmatrix} \mathbf{i} & \mathbf{j} & \mathbf{k} \\ -3 & 3\sqrt{3} & 0 \\ 0 & 0 & 1 \end{vmatrix} = 3\sqrt{3}\mathbf{i} + 3\mathbf{j}$

$\Rightarrow$ the tangent plane is

$\left(3\sqrt{3}\mathbf{i} + 3\mathbf{j}\right) \cdot \left[\left(x - \frac{3\sqrt{3}}{2}\right)\mathbf{i} + \left(y - \frac{9}{2}\right)\mathbf{j} + (z - 0)\mathbf{k}\right] = 0$

$\Rightarrow \sqrt{3}x + y = 9$. The parametrization $\Rightarrow x = 3 \sin 2\theta$

and $y = 6 \sin^2 \theta \Rightarrow x^2 + y^2 = 9 \sin^2 2\theta + \left(6 \sin^2 \theta\right)^2$

$= 9\left(4 \sin^2 \theta \cos^2 \theta\right) + 36 \sin^4 \theta = 6\left(6 \sin^2 \theta\right) = 6y \Rightarrow x^2 + y^2 - 6y + 9 = 9 \Rightarrow x^2 + (y - 3)^2 = 9$

53. (a) An arbitrary point on the circle C is $(x, z) = (R + r \cos u, r \sin u) \Rightarrow (x, y, z)$ is on the torus with

$x = (R + r \cos u) \cos v, y = (R + r \cos u) \sin v$, and $z = r \sin u, 0 \le u \le 2\pi, 0 \le v \le 2\pi$

(b) $\mathbf{r}_u = (-r \sin u \cos v)\mathbf{i} - (r \sin u \sin v)\mathbf{j} + (r \cos u)\mathbf{k}$ and $\mathbf{r}_v = (-(R + r \cos u) \sin v)\mathbf{i} + ((R + r \cos u) \cos v)\mathbf{j}$

$\Rightarrow \mathbf{r}_u \times \mathbf{r}_v = \begin{vmatrix} \mathbf{i} & \mathbf{j} & \mathbf{k} \\ -r \sin u \cos v & -r \sin u \sin v & r \cos u \\ -(R + r \cos u) \sin v & (R + r \cos u) \cos v & 0 \end{vmatrix}$

$= -(R + r \cos u)(r \cos v \cos u)\mathbf{i} - (R + r \cos u)(r \sin v \cos u)\mathbf{j} + (-r \sin u)(R + r \cos u)\mathbf{k}$

$\Rightarrow |\mathbf{r}_u \times \mathbf{r}_v|^2 = (R + r \cos u)^2 \left(r^2 \cos^2 v \cos^2 u + r^2 \sin^2 v \cos^2 u + r^2 \sin^2 u\right) \Rightarrow |\mathbf{r}_u \times \mathbf{r}_v| = r(R + r \cos u)$

$\Rightarrow A = \int_0^{2\pi} \int_0^{2\pi} \left(rR + r^2 \cos u\right) du\, dv = \int_0^{2\pi} 2\pi rR\, dv = 4\pi^2 rR$

55. (a) Let $w^2 + \frac{z^2}{c^2} = 1$ where $w = \cos\phi$ and $\frac{z}{c} = \sin\phi \Rightarrow \frac{x^2}{a^2} + \frac{y^2}{b^2} = \cos^2\phi \Rightarrow \frac{x}{a} = \cos\phi\cos\theta$ and $\frac{y}{b} = \cos\phi\sin\theta$

$\Rightarrow x = a\cos\theta\cos\phi$, $y = b\sin\theta\cos\phi$, and $z = c\sin\phi$

$\Rightarrow \mathbf{r}(\theta, \phi) = (a\cos\theta\cos\phi)\mathbf{i} + (b\sin\theta\cos\phi)\mathbf{j} + (c\sin\phi)\mathbf{k}$

(b) $\mathbf{r}_\theta = (-a\sin\theta\cos\phi)\mathbf{i} + (b\cos\theta\cos\phi)\mathbf{j}$ and $\mathbf{r}_\phi = (-a\cos\theta\sin\phi)\mathbf{i} - (b\sin\theta\sin\phi)\mathbf{j} + (c\cos\phi)\mathbf{k}$

$\Rightarrow \mathbf{r}_\theta \times \mathbf{r}_\phi = \begin{vmatrix} \mathbf{i} & \mathbf{j} & \mathbf{k} \\ -a\sin\theta\cos\phi & b\cos\theta\cos\phi & 0 \\ -a\cos\theta\sin\phi & -b\sin\theta\sin\phi & c\cos\phi \end{vmatrix}$

$= (bc\cos\theta\cos^2\phi)\mathbf{i} + (ac\sin\theta\cos^2\phi)\mathbf{j} + (ab\sin\phi\cos\phi)\mathbf{k}$

$\Rightarrow |\mathbf{r}_\theta \times \mathbf{r}_\phi|^2 = b^2c^2\cos^2\theta\cos^4\phi + a^2c^2\sin^2\theta\cos^4\phi + a^2b^2\sin^2\phi\cos^2\phi$, and the result follows.

$A \Rightarrow \int_0^{2\pi} \int_0^\pi |\mathbf{r}_\theta \times \mathbf{r}_\phi| \, d\phi \, d\theta = \int_0^{2\pi} \int_0^\pi [a^2b^2\sin^2\phi\cos^2\phi + b^2c^2\cos^2\theta\cos^4\phi + a^2c^2\sin^2\theta\cos^4\phi]^{1/2} \, d\phi \, d\theta$

57. $\mathbf{r}(\theta, u) = (5\cosh u \cos\theta)\mathbf{i} + (5\cosh u \sin\theta)\mathbf{j} + (5\sinh u)\mathbf{k} \Rightarrow \mathbf{r}_\theta = (-5\cosh u \sin\theta)\mathbf{i} + (5\cosh u \cos\theta)\mathbf{j}$ and
$\mathbf{r}_u = (5\sinh u \cos\theta)\mathbf{i} + (5\sinh u \sin\theta)\mathbf{j} + (5\cosh u)\mathbf{k}$

$\Rightarrow \mathbf{r}_\theta \times \mathbf{r}_u = \begin{vmatrix} \mathbf{i} & \mathbf{j} & \mathbf{k} \\ -5\cosh u \sin\theta & 5\cosh u \cos\theta & 0 \\ 5\sinh u \cos\theta & 5\sinh u \sin\theta & 5\cosh u \end{vmatrix}$

$= (25\cosh^2 u \cos\theta)\mathbf{i} + (25\cosh^2 u \sin\theta)\mathbf{j} - (25\cosh u \sinh u)\mathbf{k}$. At the point $(x_0, y_0, 0)$, where $x_0^2 + y_0^2 = 25$
we have $5\sinh u = 0 \Rightarrow u = 0$ and $x_0 = 25\cos\theta$, $y_0 = 25\sin\theta \Rightarrow$ the tangent plane is
$5(x_0\mathbf{i} + y_0\mathbf{j}) \cdot [(x - x_0)\mathbf{i} + (y - y_0)\mathbf{j} + z\mathbf{k}] = 0 \Rightarrow x_0x - x_0^2 + y_0y - y_0^2 = 0 \Rightarrow x_0x + y_0y = 25$

16.7 STOKES' THEOREM

1. $\text{curl } \mathbf{F} = \nabla \times \mathbf{F} = \begin{vmatrix} \mathbf{i} & \mathbf{j} & \mathbf{k} \\ \frac{\partial}{\partial x} & \frac{\partial}{\partial y} & \frac{\partial}{\partial z} \\ x^2 & 2x & z^2 \end{vmatrix} = 0\mathbf{i} + 0\mathbf{j} + (2-0)\mathbf{k} = 2\mathbf{k}$ and $\mathbf{n} = \mathbf{k} \Rightarrow \text{curl } \mathbf{F} \cdot \mathbf{n} = 2 \Rightarrow d\sigma = dx\,dy$

$\Rightarrow \oint_C \mathbf{F} \cdot d\mathbf{r} = \iint_R 2\,dA = 2(\text{Area of the ellipse}) = 4\pi$

3. $\text{curl } \mathbf{F} = \nabla \times \mathbf{F} = \begin{vmatrix} \mathbf{i} & \mathbf{j} & \mathbf{k} \\ \frac{\partial}{\partial x} & \frac{\partial}{\partial y} & \frac{\partial}{\partial z} \\ y & xz & x^2 \end{vmatrix} = -x\mathbf{i} - 2x\mathbf{j} + (z-1)\mathbf{k}$ and $\mathbf{n} = \frac{\mathbf{i}+\mathbf{j}+\mathbf{k}}{\sqrt{3}} \Rightarrow \text{curl } \mathbf{F} \cdot \mathbf{n}$

$= \frac{1}{\sqrt{3}}(-x - 2x + z - 1) \Rightarrow d\sigma = \frac{\sqrt{3}}{1}\,dA \Rightarrow \oint_C \mathbf{F} \cdot d\mathbf{r} = \iint_R \frac{1}{\sqrt{3}}(-3x + z - 1)\sqrt{3}\,dA$

$= \int_0^1 \int_0^{1-x} [-3x + (1 - x - y) - 1]\,dy\,dx = \int_0^1 \int_0^{1-x} (-4x - y)\,dy\,dx = \int_0^1 -\left[4x(1-x) + \frac{1}{2}(1-x)^2\right]\,dx$

$= -\int_0^1 \left(\frac{1}{2} + 3x - \frac{7}{2}x^2\right)\,dx = -\frac{5}{6}$

5. $\text{curl } \mathbf{F} = \nabla \times \mathbf{F} = \begin{vmatrix} \mathbf{i} & \mathbf{j} & \mathbf{k} \\ \frac{\partial}{\partial x} & \frac{\partial}{\partial y} & \frac{\partial}{\partial z} \\ y^2 + z^2 & x^2 + y^2 & x^2 + y^2 \end{vmatrix} = 2y\mathbf{i} + (2z - 2x)\mathbf{j} + (2x - 2y)\mathbf{k}$ and $\mathbf{n} = \mathbf{k}$

$\Rightarrow \text{curl } \mathbf{F} \cdot \mathbf{n} = 2x - 2y \Rightarrow d\sigma = dx\,dy \Rightarrow \oint_C \mathbf{F} \cdot d\mathbf{r} = \int_{-1}^1 \int_{-1}^1 (2x - 2y)\,dx\,dy = \int_{-1}^1 [x^2 - 2xy]_{-1}^1 \,dy$

$= \int_{-1}^1 -4y\,dy = 0$

7. $x = 3\cos t$ and $y = 2\sin t \Rightarrow \mathbf{F} = (2\sin t)\mathbf{i} + (9\cos^2 t)\mathbf{j} + (9\cos^2 t + 16\sin^4 t)\sin e^{\sqrt{(6\sin t \cos t)(0)}}\mathbf{k}$ at the
base of the shell; $\mathbf{r} = (3\cos t)\mathbf{i} + (2\sin t)\mathbf{j} \Rightarrow d\mathbf{r} = (-3\sin t)\mathbf{i} + (2\cos t)\mathbf{j} \Rightarrow \mathbf{F} \cdot \frac{d\mathbf{r}}{dt} = -6\sin^2 t + 18\cos^3 t$

$\Rightarrow \iint_S \nabla \times \mathbf{F} \cdot \mathbf{n}\,d\sigma = \int_0^{2\pi} (-6\sin^2 t + 18\cos^3 t)\,dt = \left[-3t + \frac{3}{2}\sin 2t + 6(\sin t)(\cos^2 t + 2)\right]_0^{2\pi} = -6\pi$

9. Flux of $\nabla \times \mathbf{F} = \iint\limits_{S} \nabla \times \mathbf{F} \cdot \mathbf{n}\, d\sigma = \oint_{C} \mathbf{F} \cdot d\mathbf{r}$, so let C be parametrized by $\mathbf{r} = (a \cos t)\mathbf{i} + (a \sin t)\mathbf{j}$,

$0 \le t \le 2\pi \Rightarrow \frac{d\mathbf{r}}{dt} = (-a \sin t)\mathbf{i} + (a \cos t)\mathbf{j} \Rightarrow \mathbf{F} \cdot \frac{d\mathbf{r}}{dt} = ay \sin t + ax \cos t = a^2 \sin^2 t + a^2 \cos^2 t = a^2$

$\Rightarrow$ Flux of $\nabla \times \mathbf{F} = \oint_{C} \mathbf{F} \cdot d\mathbf{r} = \int_{0}^{2\pi} a^2\, dt = 2\pi a^2$

11. Let S_1 and S_2 be oriented surfaces that span C and that induce the same positive direction on C. Then

$\iint\limits_{S_1} \nabla \times \mathbf{F} \cdot \mathbf{n}_1\, d\sigma_1 = \oint_{C} \mathbf{F} \cdot d\mathbf{r} = \iint\limits_{S_2} \nabla \times \mathbf{F} \cdot \mathbf{n}_2\, d\sigma_2$

13. $\nabla \times \mathbf{F} = \begin{vmatrix} \mathbf{i} & \mathbf{j} & \mathbf{k} \\ \frac{\partial}{\partial x} & \frac{\partial}{\partial y} & \frac{\partial}{\partial z} \\ 2z & 3x & 5y \end{vmatrix} = 5\mathbf{i} + 2\mathbf{j} + 3\mathbf{k}$; $\mathbf{r}_r = (\cos\theta)\mathbf{i} + (\sin\theta)\mathbf{j} - 2r\mathbf{k}$ and $\mathbf{r}_\theta = (-r \sin\theta)\mathbf{i} + (r \cos\theta)\mathbf{j}$

$\Rightarrow \mathbf{r}_r \times \mathbf{r}_\theta = \begin{vmatrix} \mathbf{i} & \mathbf{j} & \mathbf{k} \\ \cos\theta & \sin\theta & -2r \\ -r \sin\theta & r \cos\theta & 0 \end{vmatrix} = (2r^2 \cos\theta)\mathbf{i} + (2r^2 \sin\theta)\mathbf{j} + r\mathbf{k}$; $\mathbf{n} = \frac{\mathbf{r}_r \times \mathbf{r}_\theta}{|\mathbf{r}_r \times \mathbf{r}_\theta|}$ and $d\sigma = |\mathbf{r}_r \times \mathbf{r}_\theta|\, dr\, d\theta$

$\Rightarrow \nabla \times \mathbf{F} \cdot \mathbf{n}\, d\sigma = (\nabla \times \mathbf{F}) \cdot (\mathbf{r}_r \times \mathbf{r}_\theta)\, dr\, d\theta = (10r^2 \cos\theta + 4r^2 \sin\theta + 3r)\, dr\, d\theta \Rightarrow \iint\limits_{S} \nabla \times \mathbf{F} \cdot \mathbf{n}\, d\sigma$

$= \int_{0}^{2\pi} \int_{0}^{2} (10r^2 \cos\theta + 4r^2 \sin\theta + 3r)\, dr\, d\theta = \int_{0}^{2\pi} \left[\frac{10}{3} r^3 \cos\theta + \frac{4}{3} r^3 \sin\theta + \frac{3}{2} r^2\right]_{0}^{2}\, d\theta$

$= \int_{0}^{2\pi} \left(\frac{80}{3} \cos\theta + \frac{32}{3} \sin\theta + 6\right)\, d\theta = 6(2\pi) = 12\pi$

15. $\nabla \times \mathbf{F} = \begin{vmatrix} \mathbf{i} & \mathbf{j} & \mathbf{k} \\ \frac{\partial}{\partial x} & \frac{\partial}{\partial y} & \frac{\partial}{\partial z} \\ x^2 y & 2y^3 z & 3z \end{vmatrix} = -2y^3\mathbf{i} + 0\mathbf{j} - x^2\mathbf{k}$; $\mathbf{r}_r \times \mathbf{r}_\theta = \begin{vmatrix} \mathbf{i} & \mathbf{j} & \mathbf{k} \\ \cos\theta & \sin\theta & 1 \\ -r \sin\theta & r \cos\theta & 0 \end{vmatrix}$

$= (-r \cos\theta)\mathbf{i} - (r \sin\theta)\mathbf{j} + r\mathbf{k}$ and $\nabla \times \mathbf{F} \cdot \mathbf{n}\, d\sigma = (\nabla \times \mathbf{F}) \cdot (\mathbf{r}_r \times \mathbf{r}_\theta)\, dr\, d\theta$ (see Exercise 13 above)

$\Rightarrow \iint\limits_{S} \nabla \times \mathbf{F} \cdot \mathbf{n}\, d\sigma = \iint\limits_{R} (2ry^3 \cos\theta - rx^2)\, dr\, d\theta = \int_{0}^{2\pi} \int_{0}^{1} (2r^4 \sin^3\theta \cos\theta - r^3 \cos^2\theta)\, dr\, d\theta$

$= \int_{0}^{2\pi} \left(\frac{2}{5} \sin^3\theta \cos\theta - \frac{1}{4} \cos^2\theta\right)\, d\theta = \left[\frac{1}{10} \sin^4\theta - \frac{1}{4}\left(\frac{\theta}{2} + \frac{\sin 2\theta}{4}\right)\right]_{0}^{2\pi} = -\frac{\pi}{4}$

17. $\nabla \times \mathbf{F} = \begin{vmatrix} \mathbf{i} & \mathbf{j} & \mathbf{k} \\ \frac{\partial}{\partial x} & \frac{\partial}{\partial y} & \frac{\partial}{\partial z} \\ 3y & 5 - 2x & z^2 - 2 \end{vmatrix} = 0\mathbf{i} + 0\mathbf{j} - 5\mathbf{k}$;

$\mathbf{r}_\phi \times \mathbf{r}_\theta = \begin{vmatrix} \mathbf{i} & \mathbf{j} & \mathbf{k} \\ \sqrt{3} \cos\phi \cos\theta & \sqrt{3} \cos\phi \sin\theta & -\sqrt{3} \sin\phi \\ -\sqrt{3} \sin\phi \sin\theta & \sqrt{3} \sin\phi \cos\theta & 0 \end{vmatrix}$

$= (3 \sin^2\phi \cos\theta)\mathbf{i} + (3 \sin^2\phi \sin\theta)\mathbf{j} + (3 \sin\phi \cos\phi)\mathbf{k}$; $\nabla \times \mathbf{F} \cdot \mathbf{n}\, d\sigma = (\nabla \times \mathbf{F}) \cdot (\mathbf{r}_\phi \times \mathbf{r}_\theta)\, d\phi\, d\theta$ (see Exercise

13 above) $\Rightarrow \iint\limits_{S} \nabla \times \mathbf{F} \cdot \mathbf{n}\, d\sigma = \int_{0}^{2\pi} \int_{0}^{\pi/2} -15 \cos\phi \sin\phi\, d\phi\, d\theta = \int_{0}^{2\pi} \left[\frac{15}{2} \cos^2\phi\right]_{0}^{\pi/2}\, d\theta = \int_{0}^{2\pi} -\frac{15}{2}\, d\theta = -15\pi$

19. (a) $\mathbf{F} = 2x\mathbf{i} + 2y\mathbf{j} + 2z\mathbf{k} \Rightarrow \text{curl } \mathbf{F} = \mathbf{0} \Rightarrow \oint_{C} \mathbf{F} \cdot d\mathbf{r} = \iint\limits_{S} \nabla \times \mathbf{F} \cdot \mathbf{n}\, d\sigma = \iint\limits_{S} 0\, d\sigma = 0$

(b) Let $f(x, y, z) = x^2 y^2 z^3 \Rightarrow \nabla \times \mathbf{F} = \nabla \times \nabla f = \mathbf{0} \Rightarrow \text{curl } \mathbf{F} = \mathbf{0} \Rightarrow \oint_{C} \mathbf{F} \cdot d\mathbf{r} = \iint\limits_{S} \nabla \times \mathbf{F} \cdot \mathbf{n}\, d\sigma$

$= \iint\limits_{S} 0\, d\sigma = 0$

(c) $\mathbf{F} = \nabla \times (x\mathbf{i} + y\mathbf{j} + z\mathbf{k}) = \mathbf{0} \Rightarrow \nabla \times \mathbf{F} = \mathbf{0} \Rightarrow \oint_{C} \mathbf{F} \cdot d\mathbf{r} = \iint\limits_{S} \nabla \times \mathbf{F} \cdot \mathbf{n}\, d\sigma = \iint\limits_{S} 0\, d\sigma = 0$

(d) $\mathbf{F} = \nabla f \Rightarrow \nabla \times \mathbf{F} = \nabla \times \nabla f = \mathbf{0} \Rightarrow \oint_{C} \mathbf{F} \cdot d\mathbf{r} = \iint\limits_{S} \nabla \times \mathbf{F} \cdot \mathbf{n}\, d\sigma = \iint\limits_{S} 0\, d\sigma = 0$

21. Let $\mathbf{F} = 2y\mathbf{i} + 3z\mathbf{j} - x\mathbf{k} \Rightarrow \nabla \times \mathbf{F} = \begin{vmatrix} \mathbf{i} & \mathbf{j} & \mathbf{k} \\ \frac{\partial}{\partial x} & \frac{\partial}{\partial y} & \frac{\partial}{\partial z} \\ 2y & 3z & -x \end{vmatrix} = -3\mathbf{i} + \mathbf{j} - 2\mathbf{k}; \mathbf{n} = \frac{2\mathbf{i} + 2\mathbf{j} + \mathbf{k}}{3}$

$\Rightarrow \nabla \times \mathbf{F} \cdot \mathbf{n} = -2 \Rightarrow \oint_C 2y\,dx + 3z\,dy - x\,dz = \oint_C \mathbf{F} \cdot d\mathbf{r} = \iint_S \nabla \times \mathbf{F} \cdot \mathbf{n}\,d\sigma = \iint_S -2\,d\sigma$

$= -2 \iint_S d\sigma$, where $\iint_S d\sigma$ is the area of the region enclosed by C on the plane S: $2x + 2y + z = 2$

23. Suppose $\mathbf{F} = M\mathbf{i} + N\mathbf{j} + P\mathbf{k}$ exists such that $\nabla \times \mathbf{F} = \left(\frac{\partial P}{\partial y} - \frac{\partial N}{\partial z} \right) \mathbf{i} + \left(\frac{\partial M}{\partial z} - \frac{\partial P}{\partial x} \right) \mathbf{j} + \left(\frac{\partial N}{\partial x} - \frac{\partial M}{\partial y} \right) \mathbf{k}$

$= x\mathbf{i} + y\mathbf{j} + z\mathbf{k}$. Then $\frac{\partial}{\partial x} \left(\frac{\partial P}{\partial y} - \frac{\partial N}{\partial z} \right) = \frac{\partial}{\partial x}(x) \Rightarrow \frac{\partial^2 P}{\partial x \partial y} - \frac{\partial^2 N}{\partial x \partial z} = 1$. Likewise, $\frac{\partial}{\partial y} \left(\frac{\partial M}{\partial z} - \frac{\partial P}{\partial x} \right) = \frac{\partial}{\partial y}(y)$

$\Rightarrow \frac{\partial^2 M}{\partial y \partial z} - \frac{\partial^2 P}{\partial y \partial x} = 1$ and $\frac{\partial}{\partial z} \left(\frac{\partial N}{\partial x} - \frac{\partial M}{\partial y} \right) = \frac{\partial}{\partial z}(z) \Rightarrow \frac{\partial^2 N}{\partial z \partial x} - \frac{\partial^2 M}{\partial z \partial y} = 1$. Summing the calculated equations

$\Rightarrow \left(\frac{\partial^2 P}{\partial x \partial y} - \frac{\partial^2 P}{\partial y \partial x} \right) + \left(\frac{\partial^2 N}{\partial z \partial x} - \frac{\partial^2 N}{\partial x \partial z} \right) + \left(\frac{\partial^2 M}{\partial y \partial z} - \frac{\partial^2 M}{\partial z \partial y} \right) = 3$ or $0 = 3$ (assuming the second mixed partials are

equal). This result is a contradiction, so there is no field $\mathbf{F}$ such that curl $\mathbf{F} = x\mathbf{i} + y\mathbf{j} + z\mathbf{k}$.

25. $r = \sqrt{x^2 + y^2} \Rightarrow r^4 = (x^2 + y^2)^2 \Rightarrow \mathbf{F} = \nabla (r^4) = 4x(x^2 + y^2)\mathbf{i} + 4y(x^2 + y^2)\mathbf{j} = M\mathbf{i} + N\mathbf{j}$

$\Rightarrow \oint_C \nabla (r^4) \cdot \mathbf{n}\,ds = \oint_C \mathbf{F} \cdot \mathbf{n}\,ds = \oint_C M\,dy - N\,dx = \iint_R \left(\frac{\partial M}{\partial x} + \frac{\partial N}{\partial y} \right) dx\,dy$

$= \iint_R [4(x^2 + y^2) + 8x^2 + 4(x^2 + y^2) + 8y^2]\,dA = \iint_R 16(x^2 + y^2)\,dA = 16 \iint_R x^2\,dA + 16 \iint_R y^2\,dA$

$= 16 I_y + 16 I_x$.

16.8 THE DIVERGENCE THEOREM AND A UNIFIED THEORY

1. $\mathbf{F} = \frac{-y\mathbf{i} + x\mathbf{j}}{\sqrt{x^2 + y^2}} \Rightarrow$ div $\mathbf{F} = \frac{xy - xy}{(x^2 + y^2)^{3/2}} = 0$

3. $\mathbf{F} = -\frac{GM(x\mathbf{i} + y\mathbf{j} + z\mathbf{k})}{(x^2 + y^2 + z^2)^{3/2}} \Rightarrow$ div $\mathbf{F} = -GM \left[\frac{(x^2 + y^2 + z^2)^{3/2} - 3x^2(x^2 + y^2 + z^2)^{1/2}}{(x^2 + y^2 + z^2)^3} \right]$

$- GM \left[\frac{(x^2 + y^2 + z^2)^{3/2} - 3y^2(x^2 + y^2 + z^2)^{1/2}}{(x^2 + y^2 + z^2)^3} \right] - GM \left[\frac{(x^2 + y^2 + z^2)^{3/2} - 3z^2(x^2 + y^2 + z^2)^{1/2}}{(x^2 + y^2 + z^2)^3} \right]$

$= -GM \left[\frac{3(x^2 + y^2 + z^2)^2 - 3(x^2 + y^2 + z^2)(x^2 + y^2 + z^2)}{(x^2 + y^2 + z^2)^{7/2}} \right] = 0$

5. $\frac{\partial}{\partial x}(y - x) = -1, \frac{\partial}{\partial y}(z - y) = -1, \frac{\partial}{\partial z}(y - x) = 0 \Rightarrow \nabla \cdot \mathbf{F} = -2 \Rightarrow$ Flux $= \int_{-1}^1 \int_{-1}^1 \int_{-1}^1 -2\,dx\,dy\,dz = -2(2^3)$

$= -16$

7. $\frac{\partial}{\partial x}(y) = 0, \frac{\partial}{\partial y}(xy) = x, \frac{\partial}{\partial z}(-z) = -1 \Rightarrow \nabla \cdot \mathbf{F} = x - 1; z = x^2 + y^2 \Rightarrow z = r^2$ in cylindrical coordinates

$\Rightarrow$ Flux $= \iiint_D (x - 1)\,dz\,dy\,dx = \int_0^{2\pi} \int_0^2 \int_0^{r^2} (r\cos\theta - 1)\,dz\,r\,dr\,d\theta = \int_0^{2\pi} \int_0^2 (r^3 \cos\theta - r^2)\,r\,dr\,d\theta$

$= \int_0^{2\pi} \left[\frac{r^5}{5} \cos\theta - \frac{r^4}{4} \right]_0^2 d\theta = \int_0^{2\pi} \left(\frac{32}{5} \cos\theta - 4 \right) d\theta = \left[\frac{32}{5} \sin\theta - 4\theta \right]_0^{2\pi} = -8\pi$

9. $\frac{\partial}{\partial x}(x^2) = 2x, \frac{\partial}{\partial y}(-2xy) = -2x, \frac{\partial}{\partial z}(3xz) = 3x \Rightarrow$ Flux $= \iiint_D 3x\,dx\,dy\,dz$

$= \int_0^{\pi/2} \int_0^{\pi/2} \int_0^2 (3\rho \sin\phi \cos\theta)(\rho^2 \sin\phi)\,d\rho\,d\phi\,d\theta = \int_0^{\pi/2} \int_0^{\pi/2} 12 \sin^2\phi \cos\theta\,d\phi\,d\theta = \int_0^{\pi/2} 3\pi \cos\theta\,d\theta = 3\pi$

11. $\frac{\partial}{\partial x}(2xz) = 2z$, $\frac{\partial}{\partial y}(-xy) = -x$, $\frac{\partial}{\partial z}(-z^2) = -2z$ $\Rightarrow$ $\nabla \cdot \mathbf{F} = -x$ $\Rightarrow$ Flux $= \iiint_D -x\,dV$

$= \int_0^2 \int_0^{\sqrt{16-4x^2}} \int_0^{4-y} -x\,dz\,dy\,dx = \int_0^2 \int_0^{\sqrt{16-4x^2}} (xy - 4x)\,dy\,dx = \int_0^2 \left[\frac{1}{2}x(16 - 4x^2) - 4x\sqrt{16 - 4x^2}\right]dx$

$= \left[4x^2 - \frac{1}{2}x^4 + \frac{1}{3}(16 - 4x^2)^{3/2}\right]_0^2 = -\frac{40}{3}$

13. Let $\rho = \sqrt{x^2 + y^2 + z^2}$. Then $\frac{\partial \rho}{\partial x} = \frac{x}{\rho}$, $\frac{\partial \rho}{\partial y} = \frac{y}{\rho}$, $\frac{\partial \rho}{\partial z} = \frac{z}{\rho}$ $\Rightarrow$ $\frac{\partial}{\partial x}(\rho x) = \left(\frac{\partial \rho}{\partial x}\right)x + \rho = \frac{x^2}{\rho} + \rho$, $\frac{\partial}{\partial y}(\rho y) = \left(\frac{\partial \rho}{\partial y}\right)y + \rho$

$= \frac{y^2}{\rho} + \rho$, $\frac{\partial}{\partial z}(\rho z) = \left(\frac{\partial \rho}{\partial z}\right)z + \rho = \frac{z^2}{\rho} + \rho$ $\Rightarrow$ $\nabla \cdot \mathbf{F} = \frac{x^2+y^2+z^2}{\rho} + 3\rho = 4\rho$, since $\rho = \sqrt{x^2 + y^2 + z^2}$

$\Rightarrow$ Flux $= \iiint_D 4\rho\,dV = \int_0^{2\pi}\int_0^{\pi}\int_1^{\sqrt{2}}(4\rho)(\rho^2 \sin\phi)\,d\rho\,d\phi\,d\theta = \int_0^{2\pi}\int_0^{\pi} 3\sin\phi\,d\phi\,d\theta = \int_0^{2\pi} 6\,d\theta = 12\pi$

15. $\frac{\partial}{\partial x}(5x^3 + 12xy^2) = 15x^2 + 12y^2$, $\frac{\partial}{\partial y}(y^3 + e^y \sin z) = 3y^2 + e^y \sin z$, $\frac{\partial}{\partial z}(5z^3 + e^y \cos z) = 15z^2 - e^y \sin z$

$\Rightarrow$ $\nabla \cdot \mathbf{F} = 15x^2 + 15y^2 + 15z^2 = 15\rho^2$ $\Rightarrow$ Flux $= \iiint_D 15\rho^2\,dV = \int_0^{2\pi}\int_0^{\pi}\int_1^{\sqrt{2}}(15\rho^2)(\rho^2 \sin\phi)\,d\rho\,d\phi\,d\theta$

$= \int_0^{2\pi}\int_0^{\pi}\left(12\sqrt{2} - 3\right)\sin\phi\,d\phi\,d\theta = \int_0^{2\pi}\left(24\sqrt{2} - 6\right)d\theta = \left(48\sqrt{2} - 12\right)\pi$

17. (a) $\mathbf{G} = M\mathbf{i} + N\mathbf{j} + P\mathbf{k}$ $\Rightarrow$ $\nabla \times \mathbf{G} = \text{curl } \mathbf{G} = \left(\frac{\partial P}{\partial y} - \frac{\partial N}{\partial z}\right)\mathbf{i} + \left(\frac{\partial M}{\partial z} - \frac{\partial P}{\partial x}\right)\mathbf{k} + \left(\frac{\partial N}{\partial x} - \frac{\partial M}{\partial y}\right)\mathbf{k}$ $\Rightarrow$ $\nabla \cdot \nabla \times \mathbf{G}$

$= \text{div}(\text{curl } \mathbf{G}) = \frac{\partial}{\partial x}\left(\frac{\partial P}{\partial y} - \frac{\partial N}{\partial z}\right) + \frac{\partial}{\partial y}\left(\frac{\partial M}{\partial z} - \frac{\partial P}{\partial x}\right) + \frac{\partial}{\partial z}\left(\frac{\partial N}{\partial x} - \frac{\partial M}{\partial y}\right)$

$= \frac{\partial^2 P}{\partial x \partial y} - \frac{\partial^2 N}{\partial x \partial z} + \frac{\partial^2 M}{\partial y \partial z} - \frac{\partial^2 P}{\partial y \partial x} + \frac{\partial^2 N}{\partial z \partial x} - \frac{\partial^2 M}{\partial z \partial y} = 0$ if all first and second partial derivatives are continuous

(b) By the Divergence Theorem, the outward flux of $\nabla \times \mathbf{G}$ across a closed surface is zero because

outward flux of $\nabla \times \mathbf{G} = \iint_S (\nabla \times \mathbf{G}) \cdot \mathbf{n}\,d\sigma$

$= \iiint_D \nabla \cdot \nabla \times \mathbf{G}\,dV$ [Divergence Theorem with $\mathbf{F} = \nabla \times \mathbf{G}$]

$= \iiint_D (0)\,dV = 0$ [by part (a)]

19. (a) $\text{div}(g\mathbf{F}) = \nabla \cdot g\mathbf{F} = \frac{\partial}{\partial x}(gM) + \frac{\partial}{\partial y}(gN) + \frac{\partial}{\partial z}(gP) = \left(g\frac{\partial M}{\partial x} + M\frac{\partial g}{\partial x}\right) + \left(g\frac{\partial N}{\partial y} + N\frac{\partial g}{\partial y}\right) + \left(g\frac{\partial P}{\partial z} + P\frac{\partial g}{\partial z}\right)$

$= \left(M\frac{\partial g}{\partial x} + N\frac{\partial g}{\partial y} + P\frac{\partial g}{\partial z}\right) + g\left(\frac{\partial M}{\partial x} + \frac{\partial N}{\partial y} + \frac{\partial P}{\partial z}\right) = g\nabla \cdot \mathbf{F} + \nabla g \cdot \mathbf{F}$

(b) $\nabla \times (g\mathbf{F}) = \left[\frac{\partial}{\partial y}(gP) - \frac{\partial}{\partial z}(gN)\right]\mathbf{i} + \left[\frac{\partial}{\partial z}(gM) - \frac{\partial}{\partial x}(gP)\right]\mathbf{j} + \left[\frac{\partial}{\partial x}(gN) - \frac{\partial}{\partial y}(gM)\right]\mathbf{k}$

$= \left(P\frac{\partial g}{\partial y} + g\frac{\partial P}{\partial y} - N\frac{\partial g}{\partial z} - g\frac{\partial N}{\partial z}\right)\mathbf{i} + \left(M\frac{\partial g}{\partial z} + g\frac{\partial M}{\partial z} - P\frac{\partial g}{\partial x} - g\frac{\partial P}{\partial x}\right)\mathbf{j} + \left(N\frac{\partial g}{\partial x} + g\frac{\partial N}{\partial x} - M\frac{\partial g}{\partial y} - g\frac{\partial M}{\partial y}\right)\mathbf{k}$

$= \left(P\frac{\partial g}{\partial y} - N\frac{\partial g}{\partial z}\right)\mathbf{i} + \left(g\frac{\partial P}{\partial y} - g\frac{\partial N}{\partial z}\right)\mathbf{i} + \left(M\frac{\partial g}{\partial z} - P\frac{\partial g}{\partial x}\right)\mathbf{j} + \left(g\frac{\partial M}{\partial z} - g\frac{\partial P}{\partial x}\right)\mathbf{j} + \left(N\frac{\partial g}{\partial x} - M\frac{\partial g}{\partial y}\right)\mathbf{k}$

$+ \left(g\frac{\partial N}{\partial x} - g\frac{\partial M}{\partial y}\right)\mathbf{k} = g\nabla \times \mathbf{F} + \nabla g \times \mathbf{F}$

21. The integral's value never exceeds the surface area of S. Since $|\mathbf{F}| \leq 1$, we have $|\mathbf{F} \cdot \mathbf{n}| = |\mathbf{F}|\,|\mathbf{n}| \leq (1)(1) = 1$ and

$\iiint_D \nabla \cdot \mathbf{F}\,d\sigma = \iint_S \mathbf{F} \cdot \mathbf{n}\,d\sigma$ [Divergence Theorem]

$\leq \iint_S |\mathbf{F} \cdot \mathbf{n}|\,d\sigma$ [A property of integrals]

$\leq \iint_S (1)\,d\sigma$ [$|\mathbf{F} \cdot \mathbf{n}| \leq 1$]

$= \text{Area of S}.$

23. (a) $\frac{\partial}{\partial x}(x)=1,\ \frac{\partial}{\partial y}(y)=1,\ \frac{\partial}{\partial z}(z)=1 \Rightarrow \nabla\cdot F=3 \Rightarrow$ Flux $=\iiint_D 3\,dV = 3\iiint_D dV$

$= 3(\text{Volume of the solid})$

(b) If F is orthogonal to n at every point of S, then $F\cdot n=0$ everywhere $\Rightarrow$ Flux $=\iint_S F\cdot n\,d\sigma = 0.$

But the flux is 3(Volume of the solid) $\neq 0$, so F is not orthogonal to n at every point.

25. $\iint_S F\cdot n\,d\sigma = \iiint_D \nabla\cdot F\,dV = \iiint_D 3\,dV \Rightarrow \frac{1}{3}\iint_S F\cdot n\,d\sigma = \iiint_D dV = \text{Volume of D}$

27. (a) From the Divergence Theorem, $\iint_S \nabla f\cdot n\,d\sigma = \iiint_D \nabla\cdot\nabla f\,dV = \iiint_D \nabla^2 f\,dV = \iiint_D 0\,dV = 0$

(b) From the Divergence Theorem, $\iint_S f\nabla f\cdot n\,d\sigma = \iiint_D \nabla\cdot f\nabla f\,dV.$ Now,

$f\nabla f = \left(f\frac{\partial f}{\partial x}\right)i + \left(f\frac{\partial f}{\partial y}\right)j + \left(f\frac{\partial f}{\partial z}\right)k \Rightarrow \nabla\cdot f\nabla f = \left[f\frac{\partial^2 f}{\partial x^2}+\left(\frac{\partial f}{\partial x}\right)^2\right]+\left[f\frac{\partial^2 f}{\partial y^2}+\left(\frac{\partial f}{\partial y}\right)^2\right]+\left[f\frac{\partial^2 f}{\partial z^2}+\left(\frac{\partial f}{\partial z}\right)^2\right]$

$= f\nabla^2 f+|\nabla f|^2 = 0+|\nabla f|^2$ since f is harmonic $\Rightarrow \iint_S f\nabla f\cdot n\,d\sigma = \iiint_D |\nabla f|^2\,dV$, as claimed.

29. $\iint_S f\nabla g\cdot n\,d\sigma = \iiint_D \nabla\cdot f\nabla g\,dV = \iiint_D \nabla\cdot\left(f\frac{\partial g}{\partial x}i+f\frac{\partial g}{\partial y}j+f\frac{\partial g}{\partial z}k\right)dV$

$= \iiint_D \left(f\frac{\partial^2 g}{\partial x^2}+\frac{\partial f}{\partial x}\frac{\partial g}{\partial x}+f\frac{\partial^2 g}{\partial y^2}+\frac{\partial f}{\partial y}\frac{\partial g}{\partial y}+f\frac{\partial^2 g}{\partial z^2}+\frac{\partial f}{\partial z}\frac{\partial g}{\partial z}\right)dV$

$= \iiint_D \left[f\left(\frac{\partial^2 g}{\partial x^2}+\frac{\partial^2 g}{\partial y^2}+\frac{\partial^2 g}{\partial z^2}\right)+\left(\frac{\partial f}{\partial x}\frac{\partial g}{\partial x}+\frac{\partial f}{\partial y}\frac{\partial g}{\partial y}+\frac{\partial f}{\partial z}\frac{\partial g}{\partial z}\right)\right]dV = \iiint_D (f\nabla^2 g+\nabla f\cdot\nabla g)\,dV$

31. (a) The integral $\iiint_D p(t,x,y,z)\,dV$ represents the mass of the fluid at any time t. The equation says that

the instantaneous rate of change of mass is flux of the fluid through the surface S enclosing the region D: the mass decreases if the flux is outward (so the fluid flows out of D), and increases if the flow is inward (interpreting n as the outward pointing unit normal to the surface).

(b) $\iiint_D \frac{\partial p}{\partial t}\,dV = \frac{d}{dt}\iiint_D p\,dV = -\iint_S pv\cdot n\,d\sigma = -\iiint_D \nabla\cdot pv\,dV \Rightarrow \frac{\partial p}{\partial t}=-\nabla\cdot pv$

Since the law is to hold for all regions D, $\nabla\cdot pv+\frac{\partial p}{\partial t}=0$, as claimed

CHAPTER 16 PRACTICE EXERCISES

1. Path 1: $r=ti+tj+tk \Rightarrow x=t,\,y=t,\,z=t,\,0\le t\le 1 \Rightarrow f(g(t),h(t),k(t))=3-3t^2$ and $\frac{dx}{dt}=1,\frac{dy}{dt}=1,$

$\frac{dz}{dt}=1 \Rightarrow \sqrt{\left(\frac{dx}{dt}\right)^2+\left(\frac{dy}{dt}\right)^2+\left(\frac{dz}{dt}\right)^2}\,dt = \sqrt{3}\,dt \Rightarrow \int_C f(x,y,z)\,ds = \int_0^1 \sqrt{3}\,(3-3t^2)\,dt = 2\sqrt{3}$

Path 2: $r_1=ti+tj,\,0\le t\le 1 \Rightarrow x=t,\,y=t,\,z=0 \Rightarrow f(g(t),h(t),k(t))=2t-3t^2+3$ and $\frac{dx}{dt}=1,\frac{dy}{dt}=1,$

$\frac{dz}{dt}=0 \Rightarrow \sqrt{\left(\frac{dx}{dt}\right)^2+\left(\frac{dy}{dt}\right)^2+\left(\frac{dz}{dt}\right)^2}\,dt = \sqrt{2}\,dt \Rightarrow \int_{C_1} f(x,y,z)\,ds = \int_0^1 \sqrt{2}\,(2t-3t^2+3)\,dt = 3\sqrt{2};$

$r_2=i+j+tk \Rightarrow x=1,\,y=1,\,z=t \Rightarrow f(g(t),h(t),k(t))=2-2t$ and $\frac{dx}{dt}=0,\frac{dy}{dt}=0,\frac{dz}{dt}=1$

$\Rightarrow \sqrt{\left(\frac{dx}{dt}\right)^2+\left(\frac{dy}{dt}\right)^2+\left(\frac{dz}{dt}\right)^2}\,dt = dt \Rightarrow \int_{C_2} f(x,y,z)\,ds = \int_0^1 (2-2t)\,dt = 1$

$\Rightarrow \int_C f(x,y,z)\,ds = \int_{C_1} f(x,y,z)\,ds + \int_{C_2} f(x,y,z)\,ds = 3\sqrt{2}+1$

3. $r=(a\cos t)j+(a\sin t)k \Rightarrow x=0,\,y=a\cos t,\,z=a\sin t \Rightarrow f(g(t),h(t),k(t))=\sqrt{a^2\sin^2 t}=a\,|\sin t|$ and

$\frac{dx}{dt}=0,\frac{dy}{dt}=-a\sin t,\frac{dz}{dt}=a\cos t \Rightarrow \sqrt{\left(\frac{dx}{dt}\right)^2+\left(\frac{dy}{dt}\right)^2+\left(\frac{dz}{dt}\right)^2}\,dt = a\,dt$

$$\Rightarrow \int_C f(x, y, z)\, ds = \int_0^{2\pi} a^2 \,|\sin t|\, dt = \int_0^{\pi} a^2 \sin t\, dt + \int_{\pi}^{2\pi} -a^2 \sin t\, dt = 4a^2$$

5. $\frac{\partial P}{\partial y} = -\frac{1}{2}(x+y+z)^{-3/2} = \frac{\partial N}{\partial z}$, $\frac{\partial M}{\partial z} = -\frac{1}{2}(x+y+z)^{-3/2} = \frac{\partial P}{\partial x}$, $\frac{\partial N}{\partial x} = -\frac{1}{2}(x+y+z)^{-3/2} = \frac{\partial M}{\partial y}$

$\Rightarrow$ M dx + N dy + P dz is exact; $\frac{\partial f}{\partial x} = \frac{1}{\sqrt{x+y+z}} \Rightarrow f(x, y, z) = 2\sqrt{x+y+z} + g(y, z) \Rightarrow \frac{\partial f}{\partial y} = \frac{1}{\sqrt{x+y+z}} + \frac{\partial g}{\partial y}$

$= \frac{1}{\sqrt{x+y+z}} \Rightarrow \frac{\partial g}{\partial y} = 0 \Rightarrow g(y, z) = h(z) \Rightarrow f(x, y, z) = 2\sqrt{x+y+z} + h(z) \Rightarrow \frac{\partial f}{\partial z} = \frac{1}{\sqrt{x+y+z}} + h'(z)$

$= \frac{1}{\sqrt{x+y+z}} \Rightarrow h'(x) = 0 \Rightarrow h(z) = C \Rightarrow f(x, y, z) = 2\sqrt{x+y+z} + C \Rightarrow \int_{(-1,1,1)}^{(4,-3,0)} \frac{dx+dy+dz}{\sqrt{x+y+z}}$

$= f(4, -3, 0) - f(-1, 1, 1) = 2\sqrt{1} - 2\sqrt{1} = 0$

7. $\frac{\partial M}{\partial z} = -y \cos z \neq y \cos z = \frac{\partial P}{\partial x} \Rightarrow \mathbf{F}$ is not conservative; $\mathbf{r} = (2\cos t)\mathbf{i} + (2\sin t)\mathbf{j} - \mathbf{k}, 0 \leq t \leq 2\pi$

$\Rightarrow d\mathbf{r} = (-2\sin t)\mathbf{i} - (2\cos t)\mathbf{j} \Rightarrow \int_C \mathbf{F} \cdot d\mathbf{r} = \int_0^{2\pi}[-(-2\sin t)(\sin(-1))(-2\sin t) + (2\cos t)(\sin(-1))(-2\cos t)]\, dt$

$= 4\sin(1) \int_0^{2\pi}(\sin^2 t + \cos^2 t)dt = 8\pi \sin(1)$

9. Let M = 8x sin y and N = $-8y\cos x \Rightarrow \frac{\partial M}{\partial y} = 8x\cos y$ and $\frac{\partial N}{\partial x} = 8y\sin x \Rightarrow \int_C 8x \sin y\, dx - 8y\cos x\, dy$

$= \iint_R (8y\sin x - 8x\cos y)\, dy\, dx = \int_0^{\pi/2}\int_0^{\pi/2}(8y\sin x - 8x\cos y)\, dy\, dx = \int_0^{\pi/2}(\pi^2 \sin x - 8x)\, dx$

$= -\pi^2 + \pi^2 = 0$

11. Let $z = 1 - x - y \Rightarrow f_x(x, y) = -1$ and $f_y(x, y) = -1 \Rightarrow \sqrt{f_x^2 + f_y^2 + 1} = \sqrt{3} \Rightarrow$ Surface Area $= \iint_R \sqrt{3}\, dx\, dy$

$= \sqrt{3}$(Area of the circular region in the xy-plane) $= \pi\sqrt{3}$

13. $\nabla f = 2x\mathbf{i} + 2y\mathbf{j} + 2z\mathbf{k}, \mathbf{p} = \mathbf{k} \Rightarrow |\nabla f| = \sqrt{4x^2 + 4y^2 + 4z^2} = 2\sqrt{x^2 + y^2 + z^2} = 2$ and $|\nabla f \cdot \mathbf{p}| = |2z| = 2z$ since

$z \geq 0 \Rightarrow$ Surface Area $= \iint_R \frac{2}{2z}\, dA = \iint_R \frac{1}{z}\, dA = \iint_R \frac{1}{\sqrt{1-x^2-y^2}}\, dx\, dy = \int_0^{2\pi}\int_0^{1/\sqrt{2}} \frac{1}{\sqrt{1-r^2}}\, r\, dr\, d\theta$

$\int_0^{2\pi}\left[-\sqrt{1-r^2}\right]_0^{1/\sqrt{2}} d\theta = \int_0^{2\pi}\left(1 - \frac{1}{\sqrt{2}}\right)d\theta = 2\pi\left(1 - \frac{1}{\sqrt{2}}\right)$

15. $f(x, y, z) = \frac{x}{a} + \frac{y}{b} + \frac{z}{c} = 1 \Rightarrow \nabla f = \left(\frac{1}{a}\right)\mathbf{i} + \left(\frac{1}{b}\right)\mathbf{j} + \left(\frac{1}{c}\right)\mathbf{k} \Rightarrow |\nabla f| = \sqrt{\frac{1}{a^2} + \frac{1}{b^2} + \frac{1}{c^2}}$ and $\mathbf{p} = \mathbf{k} \Rightarrow |\nabla f \cdot \mathbf{p}| = \frac{1}{c}$

since $c > 0 \Rightarrow$ Surface Area $= \iint_R \frac{\sqrt{\frac{1}{a^2} + \frac{1}{b^2} + \frac{1}{c^2}}}{\left(\frac{1}{c}\right)}\, dA = c\sqrt{\frac{1}{a^2} + \frac{1}{b^2} + \frac{1}{c^2}} \iint_R dA = \frac{1}{2}\, abc\sqrt{\frac{1}{a^2} + \frac{1}{b^2} + \frac{1}{c^2}}$,

since the area of the triangular region R is $\frac{1}{2}$ ab. To check this result, let $\mathbf{v} = a\mathbf{i} + c\mathbf{k}$ and $\mathbf{w} = -a\mathbf{i} + b\mathbf{j}$; the area can be found by computing $\frac{1}{2}|\mathbf{v} \times \mathbf{w}|$.

17. $\nabla f = 2y\mathbf{j} + 2z\mathbf{k}, \mathbf{p} = \mathbf{k} \Rightarrow |\nabla f| = \sqrt{4y^2 + 4z^2} = 2\sqrt{y^2 + z^2} = 10$ and $|\nabla f \cdot \mathbf{p}| = 2z$ since $z \geq 0$

$\Rightarrow d\sigma = \frac{10}{2z}\, dx\, dy = \frac{5}{z}\, dx\, dy \Rightarrow \iint_S g(x, y, z)\, d\sigma = \iint_R (x^4 y)(y^2 + z^2)\left(\frac{5}{z}\right)dx\, dy$

$= \iint_R (x^4 y)(25)\left(\frac{5}{\sqrt{25-y^2}}\right)dx\, dy = \int_0^4 \int_0^1 \frac{125y}{\sqrt{25-y^2}} x^4\, dx\, dy = \int_0^4 \frac{25y}{\sqrt{25-y^2}}\, dy = 50$

19. A possible parametrization is $\mathbf{r}(\phi, \theta) = (6\sin\phi\cos\theta)\mathbf{i} + (6\sin\phi\sin\theta)\mathbf{j} + (6\cos\phi)\mathbf{k}$ (spherical coordinates);

now $\rho = 6$ and $z = -3 \Rightarrow -3 = 6\cos\phi \Rightarrow \cos\phi = -\frac{1}{2} \Rightarrow \phi = \frac{2\pi}{3}$ and $z = 3\sqrt{3} \Rightarrow 3\sqrt{3} = 6\cos\phi$

$\Rightarrow \cos\phi = \frac{\sqrt{3}}{2} \Rightarrow \phi = \frac{\pi}{6} \Rightarrow \frac{\pi}{6} \leq \phi \leq \frac{2\pi}{3}$; also $0 \leq \theta \leq 2\pi$

21. A possible parametrization is $\mathbf{r}(r, \theta) = (r\cos\theta)\mathbf{i} + (r\sin\theta)\mathbf{j} + (1+r)\mathbf{k}$ (cylindrical coordinates);

now $r = \sqrt{x^2 + y^2} \Rightarrow z = 1 + r$ and $1 \leq z \leq 3 \Rightarrow 1 \leq 1 + r \leq 3 \Rightarrow 0 \leq r \leq 2$; also $0 \leq \theta \leq 2\pi$

23. Let $x = u \cos v$ and $z = u \sin v$, where $u = \sqrt{x^2 + z^2}$ and v is the angle in the xz-plane with the x-axis

$\Rightarrow \mathbf{r}(u, v) = (u \cos v)\mathbf{i} + 2u^2\mathbf{j} + (u \sin v)\mathbf{k}$ is a possible parametrization; $0 \le y \le 2 \Rightarrow 2u^2 \le 2 \Rightarrow u^2 \le 1$

$\Rightarrow 0 \le u \le 1$ since $u \ge 0$; also, for just the upper half of the paraboloid, $0 \le v \le \pi$

25. $\mathbf{r}_u = \mathbf{i} + \mathbf{j}, \mathbf{r}_v = \mathbf{i} - \mathbf{j} + \mathbf{k} \Rightarrow \mathbf{r}_u \times \mathbf{r}_v = \begin{vmatrix} \mathbf{i} & \mathbf{j} & \mathbf{k} \\ 1 & 1 & 0 \\ 1 & -1 & 1 \end{vmatrix} = \mathbf{i} - \mathbf{j} - 2\mathbf{k} \Rightarrow |\mathbf{r}_u \times \mathbf{r}_v| = \sqrt{6}$

$\Rightarrow$ Surface Area $= \iint\limits_{R_{uv}} |\mathbf{r}_u \times \mathbf{r}_v| \, du \, dv = \int_0^1 \int_0^1 \sqrt{6} \, du \, dv = \sqrt{6}$

27. $\mathbf{r}_r = (\cos \theta)\mathbf{i} + (\sin \theta)\mathbf{j}, \mathbf{r}_\theta = (-r \sin \theta)\mathbf{i} + (r \cos \theta)\mathbf{j} + \mathbf{k} \Rightarrow \mathbf{r}_r \times \mathbf{r}_\theta = \begin{vmatrix} \mathbf{i} & \mathbf{j} & \mathbf{k} \\ \cos \theta & \sin \theta & 0 \\ -r \sin \theta & r \cos \theta & 1 \end{vmatrix}$

$= (\sin \theta)\mathbf{i} - (\cos \theta)\mathbf{j} + r\mathbf{k} \Rightarrow |\mathbf{r}_r \times \mathbf{r}_\theta| = \sqrt{\sin^2 \theta + \cos^2 \theta + r^2} = \sqrt{1 + r^2} \Rightarrow$ Surface Area $= \iint\limits_{R_{r\theta}} |\mathbf{r}_r \times \mathbf{r}_\theta| \, dr \, d\theta$

$= \int_0^{2\pi} \int_0^1 \sqrt{1 + r^2} \, dr \, d\theta = \int_0^{2\pi} \left[\frac{r}{2} \sqrt{1 + r^2} + \frac{1}{2} \ln \left(r + \sqrt{1 + r^2} \right) \right]_0^1 d\theta = \int_0^{2\pi} \left[\frac{1}{2} \sqrt{2} + \frac{1}{2} \ln \left(1 + \sqrt{2} \right) \right] d\theta$

$= \pi \left[\sqrt{2} + \ln \left(1 + \sqrt{2} \right) \right]$

29. $\frac{\partial P}{\partial y} = 0 = \frac{\partial N}{\partial z}, \frac{\partial M}{\partial z} = 0 = \frac{\partial P}{\partial x}, \frac{\partial N}{\partial x} = 0 = \frac{\partial M}{\partial y} \Rightarrow$ Conservative

31. $\frac{\partial P}{\partial y} = 0 \ne ye^z = \frac{\partial N}{\partial z} \Rightarrow$ Not Conservative

33. $\frac{\partial f}{\partial x} = 2 \Rightarrow f(x, y, z) = 2x + g(y, z) \Rightarrow \frac{\partial f}{\partial y} = \frac{\partial g}{\partial y} = 2y + z \Rightarrow g(y, z) = y^2 + zy + h(z)$

$\Rightarrow f(x, y, z) = 2x + y^2 + zy + h(z) \Rightarrow \frac{\partial f}{\partial z} = y + h'(z) = y + 1 \Rightarrow h'(z) = 1 \Rightarrow h(z) = z + C$

$\Rightarrow f(x, y, z) = 2x + y^2 + zy + z + C$

35. Over Path 1: $\mathbf{r} = t\mathbf{i} + t\mathbf{j} + t\mathbf{k}, 0 \le t \le 1 \Rightarrow x = t, y = t, z = t$ and $d\mathbf{r} = (\mathbf{i} + \mathbf{j} + \mathbf{k}) \, dt \Rightarrow \mathbf{F} = 2t^2\mathbf{i} + \mathbf{j} + t^2\mathbf{k}$

$\Rightarrow \mathbf{F} \cdot d\mathbf{r} = (3t^2 + 1) \, dt \Rightarrow$ Work $= \int_0^1 (3t^2 + 1) \, dt = 2;$

Over Path 2: $\mathbf{r}_1 = t\mathbf{i} + t\mathbf{j}, 0 \le t \le 1 \Rightarrow x = t, y = t, z = 0$ and $d\mathbf{r}_1 = (\mathbf{i} + \mathbf{j}) \, dt \Rightarrow \mathbf{F}_1 = 2t^2\mathbf{i} + \mathbf{j} + t^2\mathbf{k}$

$\Rightarrow \mathbf{F}_1 \cdot d\mathbf{r}_1 = (2t^2 + 1) \, dt \Rightarrow$ Work$_1 = \int_0^1 (2t^2 + 1) \, dt = \frac{5}{3}; \mathbf{r}_2 = \mathbf{i} + \mathbf{j} + t\mathbf{k}, 0 \le t \le 1 \Rightarrow x = 1, y = 1, z = t$ and

$d\mathbf{r}_2 = \mathbf{k} \, dt \Rightarrow \mathbf{F}_2 = 2\mathbf{i} + \mathbf{j} + \mathbf{k} \Rightarrow \mathbf{F}_2 \cdot d\mathbf{r}_2 = dt \Rightarrow$ Work$_2 = \int_0^1 dt = 1 \Rightarrow$ Work $=$ Work$_1 +$ Work$_2 = \frac{5}{3} + 1 = \frac{8}{3}$

37. (a) $\mathbf{r} = (e^t \cos t)\mathbf{i} + (e^t \sin t)\mathbf{j} \Rightarrow x = e^t \cos t, y = e^t \sin t$ from $(1, 0)$ to $(e^{2\pi}, 0) \Rightarrow 0 \le t \le 2\pi$

$\Rightarrow \frac{d\mathbf{r}}{dt} = (e^t \cos t - e^t \sin t)\mathbf{i} + (e^t \sin t + e^t \cos t)\mathbf{j}$ and $\mathbf{F} = \frac{x\mathbf{i} + y\mathbf{j}}{(x^2 + y^2)^{3/2}} = \frac{(e^t \cos t)\mathbf{i} + (e^t \sin t)\mathbf{j}}{(e^{2t} \cos^2 t + e^{2t} \sin^2 t)^{3/2}}$

$= \left(\frac{\cos t}{e^{2t}} \right)\mathbf{i} + \left(\frac{\sin t}{e^{2t}} \right)\mathbf{j} \Rightarrow \mathbf{F} \cdot \frac{d\mathbf{r}}{dt} = \left(\frac{\cos^2 t}{e^t} - \frac{\sin t \cos t}{e^t} + \frac{\sin^2 t}{e^t} + \frac{\sin t \cos t}{e^t} \right) = e^{-t}$

$\Rightarrow$ Work $= \int_0^{2\pi} e^{-t} \, dt = 1 - e^{-2\pi}$

(b) $\mathbf{F} = \frac{x\mathbf{i} + y\mathbf{j}}{(x^2 + y^2)^{3/2}} \Rightarrow \frac{\partial f}{\partial x} = \frac{x}{(x^2 + y^2)^{3/2}} \Rightarrow f(x, y, z) = -(x^2 + y^2)^{-1/2} + g(y, z) \Rightarrow \frac{\partial f}{\partial y} = \frac{y}{(x^2 + y^2)^{3/2}} + \frac{\partial g}{\partial y}$

$= \frac{y}{(x^2 + y^2)^{3/2}} \Rightarrow g(y, z) = C \Rightarrow f(x, y, z) = -(x^2 + y^2)^{-1/2}$ is a potential function for $\mathbf{F} \Rightarrow \int_C \mathbf{F} \cdot d\mathbf{r}$

$= f(e^{2\pi}, 0) - f(1, 0) = 1 - e^{-2\pi}$

39. $\nabla \times \mathbf{F} = \begin{vmatrix} \mathbf{i} & \mathbf{j} & \mathbf{k} \\ \frac{\partial}{\partial x} & \frac{\partial}{\partial y} & \frac{\partial}{\partial z} \\ y^2 & -y & 3z^2 \end{vmatrix} = -2y\mathbf{k}$; unit normal to the plane is $\mathbf{n} = \frac{2\mathbf{i} + 6\mathbf{j} - 3\mathbf{k}}{\sqrt{4 + 36 + 9}} = \frac{2}{7}\mathbf{i} + \frac{6}{7}\mathbf{j} - \frac{3}{7}\mathbf{k}$

$\Rightarrow \quad \nabla \times \mathbf{F} \cdot \mathbf{n} = \frac{6}{7} y; \; \mathbf{p} = \mathbf{k}$ and $f(x, y, z) = 2x + 6y - 3z \Rightarrow |\nabla f \cdot \mathbf{p}| = 3 \Rightarrow d\sigma = \frac{|\nabla f|}{|\nabla f \cdot \mathbf{p}|} dA = \frac{7}{3} dA$

$\Rightarrow \oint_C \mathbf{F} \cdot d\mathbf{r} = \iint_R \frac{6}{7} y \, d\sigma = \iint_R \left(\frac{6}{7} y\right)\left(\frac{7}{3} dA\right) = \iint_R 2y \, dA = \int_0^{2\pi} \int_0^1 2r \sin\theta \, r \, dr \, d\theta = \int_0^{2\pi} \frac{2}{3} \sin\theta \, d\theta = 0$

41. (a) $\mathbf{r} = \sqrt{2} t \mathbf{i} + \sqrt{2} t \mathbf{j} + (4 - t^2) \mathbf{k}, \, 0 \le t \le 1 \Rightarrow x = \sqrt{2} t, \, y = \sqrt{2} t, \, z = 4 - t^2 \Rightarrow \frac{dx}{dt} = \sqrt{2}, \frac{dy}{dt} = \sqrt{2}, \frac{dz}{dt} = -2t$

$\Rightarrow \sqrt{\left(\frac{dx}{dt}\right)^2 + \left(\frac{dy}{dt}\right)^2 + \left(\frac{dz}{dt}\right)^2} \, dt = \sqrt{4 + 4t^2} \, dt \Rightarrow M = \int_C \delta(x, y, z) \, ds = \int_0^1 3t\sqrt{4 + 4t^2} \, dt = \left[\frac{1}{4}(4 + 4t)^{3/2}\right]_0^1$

$= 4\sqrt{2} - 2$

(b) $M = \int_C \delta(x, y, z) \, ds = \int_0^1 \sqrt{4 + 4t^2} \, dt = \left[t\sqrt{1 + t^2} + \ln\left(t + \sqrt{1 + t^2}\right)\right]_0^1 = \sqrt{2} + \ln\left(1 + \sqrt{2}\right)$

43. $\mathbf{r} = t\mathbf{i} + \left(\frac{2\sqrt{2}}{3} t^{3/2}\right)\mathbf{j} + \left(\frac{t^2}{2}\right)\mathbf{k}, \, 0 \le t \le 2 \Rightarrow x = t, \, y = \frac{2\sqrt{2}}{3} t^{3/2}, \, z = \frac{t^2}{2} \Rightarrow \frac{dx}{dt} = 1, \frac{dy}{dt} = \sqrt{2} t^{1/2}, \frac{dz}{dt} = t$

$\Rightarrow \sqrt{\left(\frac{dx}{dt}\right)^2 + \left(\frac{dy}{dt}\right)^2 + \left(\frac{dz}{dt}\right)^2} \, dt = \sqrt{1 + 2t + t^2} \, dt = \sqrt{(t + 1)^2} \, dt = |t + 1| \, dt = (t + 1) \, dt$ on the domain given.

Then $M = \int_C \delta \, ds = \int_0^2 \left(\frac{1}{t+1}\right)(t + 1) \, dt = \int_0^2 dt = 2; \; M_{yz} = \int_C x\delta \, ds = \int_0^2 t\left(\frac{1}{t+1}\right)(t + 1) \, dt = \int_0^2 t \, dt = 2;$

$M_{xz} = \int_C y\delta \, ds = \int_0^2 \left(\frac{2\sqrt{2}}{3} t^{3/2}\right)\left(\frac{1}{t+1}\right)(t + 1) \, dt = \int_0^2 \frac{2\sqrt{2}}{3} t^{3/2} \, dt = \frac{32}{15}; \; M_{xy} = \int_C z\delta \, ds$

$= \int_0^2 \left(\frac{t^2}{2}\right)\left(\frac{1}{t+1}\right)(t + 1) \, dt = \int_0^2 \frac{t^2}{2} \, dt = \frac{4}{3} \Rightarrow \bar{x} = \frac{M_{yz}}{M} = \frac{2}{2} = 1; \; \bar{y} = \frac{M_{xz}}{M} = \frac{\left(\frac{32}{15}\right)}{2} = \frac{16}{15}; \; \bar{z} = \frac{M_{xy}}{M}$

$= \frac{\left(\frac{4}{3}\right)}{2} = \frac{2}{3}; \; I_x = \int_C (y^2 + z^2)\delta \, ds = \int_0^2 \left(\frac{8}{9} t^3 + \frac{t^4}{4}\right) dt = \frac{232}{45}; \; I_y = \int_C (x^2 + z^2)\delta \, ds = \int_0^2 \left(t^2 + \frac{t^4}{4}\right) dt = \frac{64}{15};$

$I_z = \int_C (y^2 + x^2)\delta \, ds = \int_0^2 \left(t^2 + \frac{8}{9} t^3\right) dt = \frac{56}{9}; \; R_x = \sqrt{\frac{I_x}{M}} = \sqrt{\frac{\left(\frac{232}{45}\right)}{2}} = \frac{2\sqrt{29}}{3\sqrt{5}}; \; R_y = \sqrt{\frac{I_y}{M}} = \sqrt{\frac{\left(\frac{64}{15}\right)}{2}} = \frac{4\sqrt{2}}{\sqrt{15}};$

$R_z = \sqrt{\frac{I_z}{M}} = \sqrt{\frac{\left(\frac{56}{9}\right)}{2}} = \frac{2\sqrt{7}}{3}$

45. $\mathbf{r}(t) = (e^t \cos t)\mathbf{i} + (e^t \sin t)\mathbf{j} + e^t \mathbf{k}, \, 0 \le t \le \ln 2 \Rightarrow x = e^t \cos t, \, y = e^t \sin t, \, z = e^t \Rightarrow \frac{dx}{dt} = (e^t \cos t - e^t \sin t),$

$\frac{dy}{dt} = (e^t \sin t + e^t \cos t), \frac{dz}{dt} = e^t \Rightarrow \sqrt{\left(\frac{dx}{dt}\right)^2 + \left(\frac{dy}{dt}\right)^2 + \left(\frac{dz}{dt}\right)^2} \, dt$

$= \sqrt{(e^t \cos t - e^t \sin t)^2 + (e^t \sin t + e^t \cos t)^2 + (e^t)^2} \, dt = \sqrt{3e^{2t}} \, dt = \sqrt{3} e^t \, dt; \; M = \int_C \delta \, ds = \int_0^{\ln 2} \sqrt{3} e^t \, dt$

$= \sqrt{3}; \; M_{xy} = \int_C z\delta \, ds = \int_0^{\ln 2} \left(\sqrt{3} e^t\right)(e^t) \, dt = \int_0^{\ln 2} \sqrt{3} e^{2t} \, dt = \frac{3\sqrt{3}}{2} \Rightarrow \bar{z} = \frac{M_{xy}}{M} = \frac{\left(\frac{3\sqrt{3}}{2}\right)}{\sqrt{3}} = \frac{3}{2};$

$I_z = \int_C (x^2 + y^2)\delta \, ds = \int_0^{\ln 2} (e^{2t} \cos^2 t + e^{2t} \sin^2 t)\left(\sqrt{3} e^t\right) dt = \int_0^{\ln 2} \sqrt{3} e^{3t} \, dt = \frac{7\sqrt{3}}{3} \Rightarrow R_z = \sqrt{\frac{I_z}{M}}$

$= \sqrt{\frac{7\sqrt{3}}{3\sqrt{3}}} = \sqrt{\frac{7}{3}}$

47. Because of symmetry $\bar{x} = \bar{y} = 0$. Let $f(x, y, z) = x^2 + y^2 + z^2 = 25 \Rightarrow \nabla f = 2x\mathbf{i} + 2y\mathbf{j} + 2z\mathbf{k}$

$\Rightarrow |\nabla f| = \sqrt{4x^2 + 4y^2 + 4z^2} = 10$ and $\mathbf{p} = \mathbf{k} \Rightarrow |\nabla f \cdot \mathbf{p}| = 2z$, since $z \ge 0 \Rightarrow M = \iint_R \delta(x, y, z) \, d\sigma$

$= \iint_R z\left(\frac{10}{2z}\right) dA = \iint_R 5 \, dA = 5(\text{Area of the circular region}) = 80\pi; \; M_{xy} = \iint_R z\delta \, d\sigma = \iint_R 5z \, dA$

$= \iint_R 5\sqrt{25 - x^2 - y^2} \, dx \, dy = \int_0^{2\pi} \int_0^4 \left(5\sqrt{25 - r^2}\right) r \, dr \, d\theta = \int_0^{2\pi} \frac{490}{3} \, d\theta = \frac{980}{3} \pi \Rightarrow \bar{z} = \frac{\left(\frac{980}{3}\pi\right)}{80\pi} = \frac{49}{12}$

$\Rightarrow (\bar{x}, \bar{y}, \bar{z}) = \left(0, 0, \frac{49}{12}\right); \; I_z = \iint_R (x^2 + y^2)\delta \, d\sigma = \iint_R 5(x^2 + y^2) \, dx \, dy = \int_0^{2\pi} \int_0^4 5r^3 \, dr \, d\theta = \int_0^{2\pi} 320 \, d\theta = 640\pi;$

$R_z = \sqrt{\frac{I_z}{M}} = \sqrt{\frac{640\pi}{80\pi}} = 2\sqrt{2}$

49. $M = 2xy + x$ and $N = xy - y \Rightarrow \frac{\partial M}{\partial x} = 2y + 1$, $\frac{\partial M}{\partial y} = 2x$, $\frac{\partial N}{\partial x} = y$, $\frac{\partial N}{\partial y} = x - 1 \Rightarrow \text{Flux} = \iint\limits_R \left(\frac{\partial M}{\partial x} + \frac{\partial N}{\partial y} \right) dx\, dy$

$= \iint\limits_R (2y + 1 + x - 1)\, dy\, dx = \int_0^1 \int_0^1 (2y + x)\, dy\, dx = \frac{3}{2}$; $\text{Circ} = \iint\limits_R \left(\frac{\partial N}{\partial x} - \frac{\partial M}{\partial y} \right) dx\, dy$

$= \iint\limits_R (y - 2x)\, dy\, dx = \int_0^1 \int_0^1 (y - 2x)\, dy\, dx = -\frac{1}{2}$

51. $M = -\frac{\cos y}{x}$ and $N = \ln x \sin y \Rightarrow \frac{\partial M}{\partial y} = \frac{\sin y}{x}$ and $\frac{\partial N}{\partial x} = \frac{\sin y}{x} \Rightarrow \oint_C \ln x \sin y\, dy - \frac{\cos y}{x}\, dx$

$= \iint\limits_R \left(\frac{\partial N}{\partial x} - \frac{\partial M}{\partial y} \right) dx\, dy = \iint\limits_R \left(\frac{\sin y}{x} - \frac{\sin y}{x} \right) dx\, dy = 0$

53. $\frac{\partial}{\partial x} (2xy) = 2y$, $\frac{\partial}{\partial y} (2yz) = 2z$, $\frac{\partial}{\partial z} (2xz) = 2x \Rightarrow \nabla \cdot \mathbf{F} = 2y + 2z + 2x \Rightarrow \text{Flux} = \iiint\limits_D (2x + 2y + 2z)\, dV$

$= \int_0^1 \int_0^1 \int_0^1 (2x + 2y + 2z)\, dx\, dy\, dz = \int_0^1 \int_0^1 (1 + 2y + 2z)\, dy\, dz = \int_0^1 (2 + 2z)\, dz = 3$

55. $\frac{\partial}{\partial x} (-2x) = -2$, $\frac{\partial}{\partial y} (-3y) = -3$, $\frac{\partial}{\partial z} (z) = 1 \Rightarrow \nabla \cdot \mathbf{F} = -4$; $x^2 + y^2 + z^2 = 2$ and $x^2 + y^2 = z \Rightarrow z = 1$

$\Rightarrow x^2 + y^2 = 1 \Rightarrow \text{Flux} = \iiint\limits_D -4\, dV = -4 \int_0^{2\pi} \int_0^1 \int_{r^2}^{\sqrt{2-r^2}} dz\, r\, dr\, d\theta = -4 \int_0^{2\pi} \int_0^1 \left(r\sqrt{2 - r^2} - r^3 \right) dr\, d\theta$

$= -4 \int_0^{2\pi} \left(-\frac{7}{12} + \frac{2}{3}\sqrt{2} \right) d\theta = \frac{2}{3} \pi \left(7 - 8\sqrt{2} \right)$

57. $\mathbf{F} = y\mathbf{i} + z\mathbf{j} + x\mathbf{k} \Rightarrow \nabla \cdot \mathbf{F} = 0 \Rightarrow \text{Flux} = \iint\limits_S \mathbf{F} \cdot \mathbf{n}\, d\sigma = \iiint\limits_D \nabla \cdot \mathbf{F}\, dV = 0$

59. $\mathbf{F} = xy^2\mathbf{i} + x^2y\mathbf{j} + y\mathbf{k} \Rightarrow \nabla \cdot \mathbf{F} = y^2 + x^2 + 0 \Rightarrow \text{Flux} = \iint\limits_S \mathbf{F} \cdot \mathbf{n}\, d\sigma = \iiint\limits_D \nabla \cdot \mathbf{F}\, dV$

$= \iiint\limits_D (x^2 + y^2)\, dV = \int_0^{2\pi} \int_0^1 \int_{-1}^1 r^2\, dz\, r\, dr\, d\theta = \int_0^{2\pi} \int_0^1 2r^3\, dr\, d\theta = \int_0^{2\pi} \frac{1}{2}\, d\theta = \pi$

CHAPTER 16 ADDITIONAL AND ADVANCED EXERCISES

1. $dx = (-2 \sin t + 2 \sin 2t)\, dt$ and $dy = (2 \cos t - 2 \cos 2t)\, dt$; $\text{Area} = \frac{1}{2} \oint_C x\, dy - y\, dx$

$= \frac{1}{2} \int_0^{2\pi} [(2 \cos t - \cos 2t)(2 \cos t - 2 \cos 2t) - (2 \sin t - \sin 2t)(-2 \sin t + 2 \sin 2t)]\, dt$

$= \frac{1}{2} \int_0^{2\pi} [6 - (6 \cos t \cos 2t + 6 \sin t \sin 2t)]\, dt = \frac{1}{2} \int_0^{2\pi} (6 - 6 \cos t)\, dt = 6\pi$

3. $dx = \cos 2t\, dt$ and $dy = \cos t\, dt$; $\text{Area} = \frac{1}{2} \oint_C x\, dy - y\, dx = \frac{1}{2} \int_0^\pi \left(\frac{1}{2} \sin 2t \cos t - \sin t \cos 2t \right) dt$

$= \frac{1}{2} \int_0^\pi [\sin t \cos^2 t - (\sin t)(2 \cos^2 t - 1)]\, dt = \frac{1}{2} \int_0^\pi (-\sin t \cos^2 t + \sin t)\, dt = \frac{1}{2} \left[\frac{1}{3} \cos^3 t - \cos t \right]_0^\pi = -\frac{1}{3} + 1 = \frac{2}{3}$

5. (a) $\mathbf{F}(x, y, z) = z\mathbf{i} + x\mathbf{j} + y\mathbf{k}$ is $\mathbf{0}$ only at the point $(0, 0, 0)$, and curl $\mathbf{F}(x, y, z) = \mathbf{i} + \mathbf{j} + \mathbf{k}$ is never $\mathbf{0}$.

(b) $\mathbf{F}(x, y, z) = z\mathbf{i} + y\mathbf{k}$ is $\mathbf{0}$ only on the line $x = t$, $y = 0$, $z = 0$ and curl $\mathbf{F}(x, y, z) = \mathbf{i} + \mathbf{j}$ is never $\mathbf{0}$.

(c) $\mathbf{F}(x, y, z) = z\mathbf{i}$ is $\mathbf{0}$ only when $z = 0$ (the xy-plane) and curl $\mathbf{F}(x, y, z) = \mathbf{j}$ is never $\mathbf{0}$.

7. Set up the coordinate system so that $(a, b, c) = (0, R, 0) \Rightarrow \delta(x, y, z) = \sqrt{x^2 + (y - R)^2 + z^2}$

$= \sqrt{x^2 + y^2 + z^2 - 2Ry + R^2} = \sqrt{2R^2 - 2Ry}$; let $f(x, y, z) = x^2 + y^2 + z^2 - R^2$ and $\mathbf{p} = \mathbf{i}$

$\Rightarrow \nabla f = 2x\mathbf{i} + 2y\mathbf{j} + 2z\mathbf{k} \Rightarrow |\nabla f| = 2\sqrt{x^2 + y^2 + z^2} = 2R \Rightarrow d\sigma = \frac{|\nabla f|}{|\nabla f \cdot \mathbf{i}|}\, dz\, dy = \frac{2R}{2x}\, dz\, dy$

$\Rightarrow \text{Mass} = \iint\limits_S \delta(x, y, z)\, d\sigma = \iint\limits_{R_{yz}} \sqrt{2R^2 - 2Ry} \left(\frac{R}{x} \right) dz\, dy = R \iint\limits_{R_{yz}} \frac{\sqrt{2R^2 - 2Ry}}{\sqrt{R^2 - y^2 - z^2}}\, dz\, dy$

$$= 4R \int_{-R}^{R} \int_{0}^{\sqrt{R^2-y^2}} \frac{\sqrt{2R^2-2Ry}}{\sqrt{R^2-y^2-z^2}} \, dz \, dy = 4R \int_{-R}^{R} \sqrt{2R^2-2Ry} \, \sin^{-1}\left(\frac{z}{\sqrt{R^2-y^2}}\right)\Big|_{0}^{\sqrt{R^2-y^2}} \, dy$$

$$= 2\pi R \int_{-R}^{R} \sqrt{2R^2-2Ry} \, dy = 2\pi R \left(\tfrac{-1}{3R}\right)(2R^2-2Ry)^{3/2}\Big|_{-R}^{R} = \tfrac{16\pi R^3}{3}$$

9. $M = x^2 + 4xy$ and $N = -6y \Rightarrow \frac{\partial M}{\partial x} = 2x + 4y$ and $\frac{\partial N}{\partial x} = -6 \Rightarrow$ Flux $= \int_{0}^{b}\int_{0}^{a}(2x+4y-6)\,dx\,dy$

$= \int_{0}^{b}(a^2 + 4ay - 6a)\,dy = a^2b + 2ab^2 - 6ab$. We want to minimize $f(a, b) = a^2b + 2ab^2 - 6ab = ab(a + 2b - 6)$.

Thus, $f_a(a, b) = 2ab + 2b^2 - 6b = 0$ and $f_b(a, b) = a^2 + 4ab - 6a = 0 \Rightarrow b(2a + 2b - 6) = 0 \Rightarrow b = 0$ or

$b = -a + 3$. Now $b = 0 \Rightarrow a^2 - 6a = 0 \Rightarrow a = 0$ or $a = 6 \Rightarrow (0, 0)$ and $(6, 0)$ are critical points. On the other

hand, $b = -a + 3 \Rightarrow a^2 + 4a(-a + 3) - 6a = 0 \Rightarrow -3a^2 + 6a = 0 \Rightarrow a = 0$ or $a = 2 \Rightarrow (0, 3)$ and $(2, 1)$ are also

critical points. The flux at $(0, 0) = 0$, the flux at $(6, 0) = 0$, the flux at $(0, 3) = 0$ and the flux at $(2, 1) = -4$.

Therefore, the flux is minimized at $(2, 1)$ with value -4.

11. (a) Partition the string into small pieces. Let $\Delta_i s$ be the length of the ith piece. Let (x_i, y_i) be a point in the
    i^{th} piece. The work done by gravity in moving the ith piece to the x-axis is approximately
    $W_i = (gx_i y_i \Delta_i s)y_i$ where $x_i y_i \Delta_i s$ is approximately the mass of the ith piece. The total work done by
    gravity in moving the string to the x-axis is $\sum_i W_i = \sum_i g x_i y_i^2 \Delta_i s \Rightarrow$ Work $= \int_C g x y^2 \, ds$

    (b) Work $= \int_C g x y^2 \, ds = \int_{0}^{\pi/2} g(2 \cos t)(4 \sin^2 t)\sqrt{4 \sin^2 t + 4 \cos^2 t} \, dt = 16g \int_{0}^{\pi/2} \cos t \sin^2 t \, dt$

    $= \left[16g\left(\frac{\sin^3 t}{3}\right)\right]_{0}^{\pi/2} = \frac{16}{3} g$

    (c) $\bar{x} = \frac{\int_C x(xy) \, ds}{\int_C xy \, ds}$ and $\bar{y} = \frac{\int_C y(xy) \, ds}{\int_C xy \, ds}$; the mass of the string is $\int_C xy \, ds$ and the weight of the string is

    $g \int_C xy \, ds$. Therefore, the work done in moving the point mass at $(\bar{x}, \bar{y})$ to the x-axis is

    $W = \left(g \int_C xy \, ds\right)\bar{y} = g \int_C xy^2 \, ds = \frac{16}{3} g$.

13. (a) Partition the sphere $x^2 + y^2 + (z - 2)^2 = 1$ into small pieces. Let $\Delta_i \sigma$ be the surface area of the ith piece
    and let (x_i, y_i, z_i) be a point on the ith piece. The force due to pressure on the ith piece is approximately
    $w(4 - z_i)\Delta_i \sigma$. The total force on S is approximately $\sum_i w(4 - z_i)\Delta_i \sigma$. This gives the actual force to be
    $\iint_S w(4 - z) \, d\sigma$.

    (b) The upward buoyant force is a result of the **k**-component of the force on the ball due to liquid pressure.
    The force on the ball at (x, y, z) is $w(4 - z)(-\mathbf{n}) = w(z - 4)\mathbf{n}$, where **n** is the outer unit normal at (x, y, z).
    Hence the **k**-component of this force is $w(z - 4)\mathbf{n} \cdot \mathbf{k} = w(z - 4)\mathbf{k} \cdot \mathbf{n}$. The (magnitude of the) buoyant force
    on the ball is obtained by adding up all these **k**-components to obtain $\iint_S w(z - 4)\mathbf{k} \cdot \mathbf{n} \, d\sigma$.

    (c) The Divergence Theorem says $\iint_S w(z - 4)\mathbf{k} \cdot \mathbf{n} \, d\sigma = \iiint_D \text{div}(w(z - 4)\mathbf{k}) \, dV = \iiint_D w \, dV$, where D
    is $x^2 + y^2 + (z - 2)^2 \leq 1 \Rightarrow \iint_S w(z - 4)\mathbf{k} \cdot \mathbf{n} \, d\sigma = w \iiint_D 1 \, dV = \frac{4}{3}\pi w$, the weight of the fluid if it
    were to occupy the region D.

15. Assume that S is a surface to which Stokes's Theorem applies. Then $\oint_C \mathbf{E} \cdot d\mathbf{r} = \iint_S (\nabla \times \mathbf{E}) \cdot \mathbf{n} \, d\sigma$

    $= \iint_S -\frac{\partial \mathbf{B}}{\partial t} \cdot \mathbf{n} \, d\sigma = -\frac{\partial}{\partial t} \iint_S \mathbf{B} \cdot \mathbf{n} \, d\sigma$. Thus the voltage around a loop equals the negative of the rate

    of change of magnetic flux through the loop.

17. $\oint_C f \nabla g \cdot d\mathbf{r} = \iint\limits_S \nabla \times (f \nabla g) \cdot \mathbf{n} \, d\sigma$ (Stokes's Theorem)

$\quad = \iint\limits_S (f \nabla \times \nabla g + \nabla f \times \nabla g) \cdot \mathbf{n} \, d\sigma$ (Section 16.8, Exercise 19b)

$\quad = \iint\limits_S [(f)(\mathbf{0}) + \nabla f \times \nabla g] \cdot \mathbf{n} \, d\sigma$ (Section 16.7, Equation 8)

$\quad = \iint\limits_S (\nabla f \times \nabla g) \cdot \mathbf{n} \, d\sigma$

19. False; let $\mathbf{F} = y\mathbf{i} + x\mathbf{j} \neq \mathbf{0} \Rightarrow \nabla \cdot \mathbf{F} = \frac{\partial}{\partial x}(y) + \frac{\partial}{\partial y}(x) = 0$ and $\nabla \times \mathbf{F} = \begin{vmatrix} \mathbf{i} & \mathbf{j} & \mathbf{k} \\ \frac{\partial}{\partial x} & \frac{\partial}{\partial y} & \frac{\partial}{\partial z} \\ x & y & 0 \end{vmatrix} = 0\mathbf{i} + 0\mathbf{j} + 0\mathbf{k} = \mathbf{0}$

21. $\mathbf{r} = x\mathbf{i} + y\mathbf{j} + z\mathbf{k} \Rightarrow \nabla \cdot \mathbf{r} = 1 + 1 + 1 = 3 \Rightarrow \iiint\limits_D \nabla \cdot \mathbf{r} \, dV = 3 \iiint\limits_D dV = 3V \Rightarrow V = \frac{1}{3} \iiint\limits_D \nabla \cdot \mathbf{r} \, dV$

$\quad = \frac{1}{3} \iint\limits_S \mathbf{r} \cdot \mathbf{n} \, d\sigma$, by the Divergence Theorem

NOTES: